M000205577

Transcriptional Regulation

Transcriptional Regulation: Molecules, Involved Mechanisms and Misregulation

Special Issue Editors

Amelia Casamassimi
Alfredo Ciccodicola

MDPI • Basel • Beijing • Wuhan • Barcelona • Belgrade

Special Issue Editors
Amelia Casamassimi
University of Campania "Luigi Vanvitelli"
Italy

Alfredo Ciccodicola
IGB-CNR and University of Naples "Parthenope"
Italy

Editorial Office
MDPI
St. Alban-Anlage 66
4052 Basel, Switzerland

This is a reprint of articles from the Special Issue published online in the open access journal *International Journal of Molecular Sciences* (ISSN 1422-0067) from 2018 to 2019 (available at: https://www.mdpi.com/journal/ijms/special_issues/transcriptional_regulation_biophys)

For citation purposes, cite each article independently as indicated on the article page online and as indicated below:

LastName, A.A.; LastName, B.B.; LastName, C.C. Article Title. *Journal Name* **Year**, *Article Number, Page Range*.

ISBN 978-3-03921-265-1 (Pbk)
ISBN 978-3-03921-266-8 (PDF)

Cover image courtesy of Amelia Casamassimi and Alfredo Ciccodicola.

© 2019 by the authors. Articles in this book are Open Access and distributed under the Creative Commons Attribution (CC BY) license, which allows users to download, copy and build upon published articles, as long as the author and publisher are properly credited, which ensures maximum dissemination and a wider impact of our publications.

The book as a whole is distributed by MDPI under the terms and conditions of the Creative Commons license CC BY-NC-ND.

Contents

About the Special Issue Editors

Amelia Casamassimi obtained her Biological Sciences degree in 1989 at the University of Naples, Federico II (Italy). She has worked at IGB-CNR Institute and Pascale Foundation (IRCSS) in Naples and is currently working at the Department of Precision Medicine of University of Campania "Luigi Vanvitelli". Casamassimi is interested in the application of genomics and post-genomics approaches, particularly transcriptome analysis, to study human diseases. She is co-author of several scientific papers in this research field.

Alfredo Ciccodicola is a graduate in Biological Sciences at the Federico II University of Naples and Professor of Molecular Biology at the Department of Science and Technology of the Parthenope University of Naples. Ciccodicola is Research Director at the "A. Buzzati-Traverso" Institute of Genetics and Biophysics of the National Research Council of Naples. His current scientific interests include human genetic diseases, molecular mechanism pathogenesis, whole-transcriptome analysis, and non-coding RNAs.

Preface to "Transcriptional Regulation: Molecules, Involved Mechanisms and Misregulation"

Transcriptional regulation is a critical biological process involved in the response of a cell, tissue, or organism to a variety of intra- and extracellular signals. Moreover, it controls the establishment and maintenance of cell identity throughout developmental and differentiation programs. This highly complex and dynamic process is orchestrated by a vast number of molecules and protein networks and occurs through multiple temporal and functional steps. Of note, many human disorders are characterized by misregulation of global transcription, since most of the signaling pathways ultimately target components of the transcriptional machinery. This book includes a selection of papers that illustrate recent advances in our understanding of transcriptional regulation and focuses on many important topics, from cis-regulatory elements to transcription factors, chromatin regulators, and non-coding RNAs, in addition to multiple transcriptome studies and computational analyses.

Amelia Casamassimi, Alfredo Ciccodicola
Special Issue Editors

International Journal of
Molecular Sciences

Editorial

Transcriptional Regulation: Molecules, Involved Mechanisms, and Misregulation

Amelia Casamassimi [1],* and Alfredo Ciccodicola [2,3],*

1   Department of Precision Medicine, University of Campania "Luigi Vanvitelli", Via L. De Crecchio, 80138 Naples, Italy
2   Institute of Genetics and Biophysics "Adriano Buzzati Traverso", CNR, 80131 Naples, Italy
3   Department of Science and Technology, University of Naples "Parthenope", 80143 Naples, Italy
*   Correspondence: amelia.casamassimi@unicampania.it (A.C.); alfredo.ciccodicola@igb.cnr.it (A.C.)

Received: 26 February 2019; Accepted: 11 March 2019; Published: 14 March 2019

Transcriptional regulation is a critical biological process that allows the cell or an organism to respond to a variety of intra- and extra-cellular signals, to define cell identity during development, to maintain it throughout its lifetime, and to coordinate cellular activity. This highly dynamic mechanism includes a series of biophysical events orchestrated by a huge number of molecules establishing larger networks and occurring through multiple temporal and functional steps that range from specific DNA-protein interactions to the recruitment and assembly of nucleoprotein complexes. Essentially, the key transcription levels include the recruitment and assembly of the entire transcription machinery, the initiation step, pause release and elongation phases, as well as termination of transcription. Additionally, these steps are interconnected with governing chromatin accessibility (such as the unwrapping process, which is controlled by histone modification and chromatin remodeling proteins), and other epigenetic mechanisms (such as enhancer-promoter looping, which is necessary for a successful gene transcription). Finally, various RNA maturation events, such as the splicing that occurs with transcription, constitute an additional level of complexity. Numerous molecules and molecular factors, including transcription factors, cofactors (both coactivators and corepressors), and chromatin regulators, are known to participate to this process [1]. Essential components of the basal transcription machinery comprise the RNA polymerase II holoenzyme, the general initiation transcription factors (TFIIA, -IIB, -IID, -IIE, -IIF, and -IIH) and the Mediator complex, a multi-subunit compound that joins transcription factors bound at the upstream regulatory elements—such as nuclear receptors—and all the remaining apparatus at the promoter region. It is noteworthy that it also works in close interplay between the basal machinery and factors responsible for the epigenetic modifications; for instance, together with cohesin, it facilitates DNA looping [2]. More recently, a novel multi-subunit complex named Integrator was added as one of the components of the RNA Polymerase II-mediated transcription apparatus. It is also involved in many stages of eukaryotic transcription for most regulated genes [3].

Additionally, the high complexity of transcriptional regulation is also derived from the involvement of non-coding RNAs (ncRNAs). Indeed, research over the last two decades has revealed new classes of ncRNAs, including microRNAs (miRNAs), small nucleolar RNAs (snoRNAs), long ncRNAs (lncRNAs), circular RNAs (circRNAs), and enhancer RNAs (eRNAs), each with different regulatory functions and altogether belonging to a larger RNA communication network ultimately controlling the production of the final protein [4].

Recent advances in "omics" and computational biology have provided novel tools that allow one to integrate different layers of information from biophysical, biochemical, and molecular cell biology studies. In turn, these novel strategies provided a fuller understanding of how DNA sequence information, epigenetic modifications, and transcription machinery cooperate to regulate gene expression. Of note, most of the new molecular biomarkers and therapeutic targets for several human pathologies derive from transcriptome profiling studies, and their number is continuously increasing. Next Generation Sequencing (NGS), mainly RNA-Sequencing (RNA-Seq), has completely revolutionized transcriptome analysis, allowing the quantification of gene expression levels and allele-specific expression in a single experiment, as well as the identification of novel genes, splice isoforms, fusion transcripts, and the entire world of ncRNAs at an unprecedented level [4].

It is well known that many human disorders are characterized by global transcriptional dysregulation because most of the signaling pathways ultimately target transcription machinery. Indeed, many syndromes and genetic and complex diseases—cancer, autoimmunity, neurological and developmental disorders, metabolic and cardiovascular diseases—can be caused by mutations/alterations in regulatory sequences, transcription factors, cofactors, chromatin regulators, ncRNAs, and other components of transcription apparatus [1–4]. Thus, advances in our understanding of molecules and mechanisms involved in the transcriptional circuitry and apparatus lead to new insights into the pathogenetic mechanisms of various human diseases and disorders.

In this special issue, a total of 19 excellent and interesting papers consisting of 11 original research studies, seven reviews, and one communication are published [5–23]. They cover all subjects of transcriptional regulation, from cis-regulatory elements to transcription factors, chromatin regulators, and ncRNAs. Additionally, several transcriptome studies and computational analyses are also included in this issue.

Huang et al. analyzed the transcriptional regulation of the gene coding for the Chloride intracellular channel 4 (*CLIC4*). This is a multifunctional protein with diverse physiological functions. Differential expression of *CLIC4* between cancer cells and the surrounding stroma has been reported in various tumor types [11]. Here, the authors found an alternative G-quadruplex (G4) structure, PG4-3, in its promoter region. Through the use of the CRISPR/Cas9 system, they provided evidence that this element could play an important role in regulating the *CLIC4* transcription levels [11].

Regarding transcription factors, a comprehensive review summarized the structures and functions of these regulators in both model and non-model insects, including Drosophila, and appraises the importance of transcription factors in orchestrating diverse insect physiological and biochemical processes [17]. An original article examined the paired-box 3 (*Pax3*) transcription factor in the winged pearl oyster *Pteria penguin*. More precisely, this study investigated the role of *PpPax3* in melanin synthesis and used RNA interference to provide evidence that this function is exerted in this important marine species through the tyrosinase pathway [18]. A bioinformatics approach was used to identify the significant genes responsible for the human Patau syndrome (PS), a rare congenital anomaly due to chromosome 13 trisomy. This molecular network analysis and protein-protein interaction study indicated *FOXO1* (Forkhead Box O1) as a strong transcription factor interacting with other key genes associated with lethal heart disorders in PS. [15].

As expected in the NGS era, transcriptome analysis by RNA-Seq has been widely used in many studies to elucidate the most varied mechanisms of pathophysiology as well as other relevant biological processes in diverse organisms [5,9,20,21]. Actually, a small number of studies still utilize microarray as a useful approach. Indeed, this platform allows one to identify the common pathway(s) of Major Depressive Disorder and glioblastoma [5]. Otherwise, most of the studies employ RNA-Seq to, for example, understand the regulatory system of stringent response in sphingomonads [9] or to unravel molecular insights of phase-specific pollen-pistil interaction during self-incompatibility and fertilization in tea [21]. Additionally, in silico analyses of available transcriptome databases are often very useful when the biological material is scarce or difficult to isolate, as in the case of a study aimed to identify genes that could have a potential role in the oyster larval adhesion at the pediveliger stage [20]. Additionally, the availability of multi-omics datasets from patient tissues represents a unique source to study human diseases. Particularly, The Cancer Genome Atlas (TCGA) collects data from thousands of subjects with human malignancies, thus enabling the in silico analysis of genes or families of genes of interest. For example, in an effort to obtain a pan-cancer overview of the genomic and transcriptomic alterations of the PR/SET domain gene family (PRDM) members in cancer, our group reanalyzed the

Exome- and RNA-Seq datasets from the TCGA portal [12]. Likewise, to date, a lot of similar studies have led to a better comprehension of the pathogenetic mechanisms as well as the discovery of novel biomarkers and/or therapeutic targets for these human disorders, as cited in a review dissecting the role of Adiponectin as a link factor between adipose tissue and cancer [23].

In the field of cancer research, an interesting pathogenetic mechanism involving dysregulation of transcription is represented by the destabilization of the messenger RNAs of critical genes implicated in both tumor onset and tumor progression exerted by tristetraprolin (TTP). Indeed, as reviewed in a paper of this special issue, the tumor suppressor TTP can negatively regulate tumorigenesis. In turn, TTP expression is frequently downregulated in several tumors by various mechanisms [13].

Several papers have described novelties in the field of ncRNAs. For instance, a study investigated the possible role in cell metabolism of miR-25-3p. This miRNA is highly conserved in mammals and was previously found to be involved in many biological processes and in some cancer and cardiovascular related diseases. Specifically, in the C2C12 cell line derived from mouse muscle myoblasts, it is positively regulated by the transcription factor AP-2α and enhances cell metabolism by directly targeting the 3' untranslated region of AKT serine/threonine kinase 1 (*Akt1*), a gene related to metabolism [6].

LncRNAs play an important role as epigenetic and transcriptional regulators. Evidence of their importance in the pathophysiology of many malignancies has drastically increased in the last decade. In their excellent contribution, Cruz-Miranda et al. describe the functional classification, biogenesis, and role of lncRNAs in leukemogenesis, highlighting the evidence that lncRNAs could be useful as biomarkers in the diagnosis, prognosis, and therapeutic response of leukemia patients, as well as showing that they could represent potential therapeutic targets in these tumors [22]. In a preliminary study, RNA-Seq data were used to profile, quantify, and classify (for the first time) lncRNAs in human term placenta [8]. Although the obtained lncRNAs still need to be functionally characterized, they could expand the current knowledge of the essential mechanisms in pregnancy maintenance and fetal development.

Lei et al. proposed a new computational path weighted method for predicting circRNA-disease associations, the PWCDA method. Despite some limitations, it showed a much better performance than other computational models [14].

A remarkable study explored the utility of eRNA expression as a causal anchor in predicting transcription regulatory networks based on the observation that eRNAs mark the activity of regulatory regions [16]. In their work, the authors developed a novel statistical framework to infer causal gene networks (named Findr-A) by extending the Findr software for causal inference through the use of cap analysis of gene expression (CAGE) data from the FANTOM5 consortium [16].

Numerous epigenetic mechanisms other than regulation by ncRNAs take place during RNA polymerase II-transcription and may be involved in human pathophysiology. An outstanding review on the Cyclin Dependent Kinase Inhibitor 1C (*CDKN1C*) gene summarizes all the possible (epi)-genetic alterations leading to diseases. This gene encodes the p57Kip2 protein, the third member of the CIP/Kip family, and its alterations are known to cause three human hereditary syndromes characterized by altered growth rate. Interestingly, *CDKN1C* is positioned in a genomic region characterized by a remarkable regional imprinting that results in the transcription of only the maternal allele. Moreover, this gene is also down-regulated in human cancers. Of note, its transcriptional regulation is linked to several mechanisms, including DNA methylation and specific histone modifications. Finally, ncRNAs also play important roles in controlling p57Kip2 levels [7].

Selenium-related transcriptional regulation is the topic of a comprehensive review [10]. Selenium is a trace element controlling the expression levels of numerous genes; it is essential to human health, and its deficiency is related to several diseases. It is incorporated as seleno-cysteine to the so-called seleno-proteins via an uncommon mechanism. Indeed, the codon for seleno-cysteine is a regular in-frame stop codon, which can be passed by a specific complex translation machinery in the presence of a signal sequence in the $3'$-untranslated part of the seleno-protein mRNAs. Nonsense-mediated decay and other mechanisms are able to regulate seleno-protein mRNA levels [10].

It is well-known that DNA methylation contributes to the gene expression regulation without changing the DNA sequence. Abnormal DNA methylation has been associated with improper gene expression and may lead to several disorders. Both genetic factors and modifiable factors, including nutrition, are able to alter methylation pathways. An interesting review of this special issue carefully describes molecular mechanisms underlying the link between diet and DNA methylation [19].

Finally, we hope the readers enjoy this Special Issue of IJMS and the effort to present the current advances and promising results in the field of transcriptional regulation and its involvement in all of the relevant biological processes and in pathophysiology.

Acknowledgments: We would like to thank all the participating assistant editors and reviewers for their important contribution to this Special Issue.

Conflicts of Interest: The authors declare no conflict of interest.

References

1. Lee, T.I.; Young, R.A. Transcriptional regulation and its misregulation in disease. *Cell* **2013**, *152*, 1237–1251. [CrossRef]
2. Schiano, C.; Casamassimi, A.; Vietri, M.T.; Rienzo, M.; Napoli, C. The roles of mediator complex in cardiovascular diseases. *Biochim. Biophys. Acta* **2014**, *1839*, 444–451. [CrossRef] [PubMed]
3. Rienzo, M.; Casamassimi, A. Integrator complex and transcription regulation: Recent findings and pathophysiology. *Biochim. Biophys. Acta* **2016**, *1859*, 1269–1280. [CrossRef] [PubMed]
4. Casamassimi, A.; Federico, A.; Rienzo, M.; Esposito, S.; Ciccodicola, A. Transcriptome Profiling in Human Diseases: New Advances and Perspectives. *Int. J. Mol. Sci.* **2017**, *18*, 1652. [CrossRef] [PubMed]
5. Xie, Y.; Wang, L.; Xie, Z.; Zeng, C.; Shu, K. Transcriptomics Evidence for Common Pathways in Human Major Depressive Disorder and Glioblastoma. *Int. J. Mol. Sci.* **2018**, *19*, 234. [CrossRef] [PubMed]
6. Zhang, F.; Chen, K.; Tao, H.; Kang, T.; Xiong, Q.; Zeng, Q.; Liu, Y.; Jiang, S.; Chen, M. miR-25-3p, Positively Regulated by Transcription Factor AP-2α, Regulates the Metabolism of C2C12 Cells by Targeting Akt1. *Int. J. Mol. Sci.* **2018**, *19*, 773. [CrossRef] [PubMed]
7. Stampone, E.; Caldarelli, I.; Zullo, A.; Bencivenga, D.; Mancini, F.; Della Ragione, F.; Borriello, A. Genetic and Epigenetic Control of CDKN1C Expression: Importance in Cell Commitment and Differentiation, Tissue Homeostasis and Human Diseases. *Int. J. Mol. Sci.* **2018**, *19*, 1055. [CrossRef]
8. Majewska, M.; Lipka, A.; Paukszto, L.; Jastrzebski, J.; Gowkielewicz, M.; Jozwik, M.; Majewski, M. Preliminary RNA-Seq Analysis of Long Non-Coding RNAs Expressed in Human Term Placenta. *Int. J. Mol. Sci.* **2018**, *19*, 1894. [CrossRef] [PubMed]
9. Lu, H.; Huang, Y. Transcriptome Analysis of Novosphingobium pentaromativorans US6-1 Reveals the Rsh Regulon and Potential Molecular Mechanisms of N-acyl-l-homoserine Lactone Accumulation. *Int. J. Mol. Sci.* **2018**, *19*, 2631. [CrossRef]
10. Lammi, M.; Qu, C. Selenium-Related Transcriptional Regulation of Gene Expression. *Int. J. Mol. Sci.* **2018**, *19*, 2665. [CrossRef]
11. Huang, M.; Chu, I.; Wang, Z.; Lin, S.; Chang, T.; Chen, C. A G-Quadruplex Structure in the Promoter Region of CLIC4 Functions as a Regulatory Element for Gene Expression. *Int. J. Mol. Sci.* **2018**, *19*, 2678. [CrossRef] [PubMed]
12. Sorrentino, A.; Federico, A.; Rienzo, M.; Gazzerro, P.; Bifulco, M.; Ciccodicola, A.; Casamassimi, A.; Abbondanza, C. PR/SET Domain Family and Cancer: Novel Insights from The Cancer Genome Atlas. *Int. J. Mol. Sci.* **2018**, *19*, 3250. [CrossRef] [PubMed]
13. Park, J.; Lee, T.; Kang, T. Roles of Tristetraprolin in Tumorigenesis. *Int. J. Mol. Sci.* **2018**, *19*, 3384. [CrossRef]
14. Lei, X.; Fang, Z.; Chen, L.; Wu, F. PWCDA: Path Weighted Method for Predicting circRNA-Disease Associations. *Int. J. Mol. Sci.* **2018**, *19*, 3410. [CrossRef] [PubMed]
15. Abuzenadah, A.; Alsaedi, S.; Karim, S.; Al-Qahtani, M. Role of Overexpressed Transcription Factor FOXO1 in Fatal Cardiovascular Septal Defects in Patau Syndrome: Molecular and Therapeutic Strategies. *Int. J. Mol. Sci.* **2018**, *19*, 3547. [CrossRef]

16. Vipin, D.; Wang, L.; Devailly, G.; Michoel, T.; Joshi, A. Causal Transcription Regulatory Network Inference Using Enhancer Activity as a Causal Anchor. *Int. J. Mol. Sci.* **2018**, *19*, 3609. [CrossRef] [PubMed]

17. Guo, Z.; Qin, J.; Zhou, X.; Zhang, Y. Insect Transcription Factors: A Landscape of Their Structures and Biological Functions in Drosophila and beyond. *Int. J. Mol. Sci.* **2018**, *19*, 3691. [CrossRef] [PubMed]

18. Yu, F.; Qu, B.; Lin, D.; Deng, Y.; Huang, R.; Zhong, Z. Pax3 Gene Regulated Melanin Synthesis by Tyrosinase Pathway in Pteria penguin. *Int. J. Mol. Sci.* **2018**, *19*, 3700. [CrossRef]

19. Kadayifci, F.; Zheng, S.; Pan, Y. Molecular Mechanisms Underlying the Link between Diet and DNA Methylation. *Int. J. Mol. Sci.* **2018**, *19*, 4055. [CrossRef]

20. Foulon, V.; Boudry, P.; Artigaud, S.; Guérard, F.; Hellio, C. In Silico Analysis of Pacific Oyster (*Crassostrea gigas*) Transcriptome over Developmental Stages Reveals Candidate Genes for Larval Settlement. *Int. J. Mol. Sci.* **2019**, *20*, 197. [CrossRef]

21. Seth, R.; Bhandawat, A.; Parmar, R.; Singh, P.; Kumar, S.; Sharma, R. Global Transcriptional Insights of Pollen-Pistil Interactions Commencing Self-Incompatibility and Fertilization in Tea [*Camellia sinensis* (L.) O. Kuntze]. *Int. J. Mol. Sci.* **2019**, *20*, 539. [CrossRef] [PubMed]

22. Cruz-Miranda, G.; Hidalgo-Miranda, A.; Bárcenas-López, D.; Núñez-Enríquez, J.; Ramírez-Bello, J.; Mejía-Aranguré, J.; Jiménez-Morales, S. Long Non-Coding RNA and Acute Leukemia. *Int. J. Mol. Sci.* **2019**, *20*, 735. [CrossRef] [PubMed]

23. Di Zazzo, E.; Polito, R.; Bartollino, S.; Nigro, E.; Porcile, C.; Bianco, A.; Daniele, A.; Moncharmont, B. Adiponectin as Link Factor between Adipose Tissue and Cancer. *Int. J. Mol. Sci.* **2019**, *20*, 839. [CrossRef] [PubMed]

 © 2019 by the authors. Licensee MDPI, Basel, Switzerland. This article is an open access article distributed under the terms and conditions of the Creative Commons Attribution (CC BY) license (http://creativecommons.org/licenses/by/4.0/).

International Journal of
Molecular Sciences

Article

Transcriptomics Evidence for Common Pathways in Human Major Depressive Disorder and Glioblastoma

Yongfang Xie, Ling Wang *, Zengyan Xie, Chuisheng Zeng and Kunxian Shu *

Institute of Bioinformatics, Chongqing University of Posts and Telecommunications, Chongqing 400065, China; xieyf@cqupt.edu.cn (Y.X.); xiezy@cqupt.edu.cn (Z.X.); zengcs@cqupt.edu.cn (C.Z.)
* Correspondence: wangling685@gmail.com (L.W.); shukx@cqupt.edu.cn (K.S.); Tel.:+86-23-6246-0025 (K.S.)

Received: 14 December 2017; Accepted: 10 January 2018; Published: 12 January 2018

Abstract: Depression as a common complication of brain tumors. Is there a possible common pathogenesis for depression and glioma? The most serious major depressive disorder (MDD) and glioblastoma (GBM) in both diseases are studied, to explore the common pathogenesis between the two diseases. In this article, we first rely on transcriptome data to obtain reliable and useful differentially expressed genes (DEGs) by differential expression analysis. Then, we used the transcriptomics of DEGs to find out and analyze the common pathway of MDD and GBM from three directions. Finally, we determine the important biological pathways that are common to MDD and GBM by statistical knowledge. Our findings provide the first direct transcriptomic evidence that common pathway in two diseases for the common pathogenesis of the human MDD and GBM. Our results provide a new reference methods and values for the study of the pathogenesis of depression and glioblastoma.

Keywords: major depressive disorder; glioblastoma; differentially expressed genes; transcriptomics; common pathway

1. Introduction

Glioma is the most common tumor in the central nervous system, mostly occurring in the brain, and the diagnosis and treatment of glioma are incomplete, inaccurate, and easily reappeared. The current study [1,2] shows that most patients with glioma can get better diagnosis and treatment, but the diagnosis and treatment results are still unsatisfactory, even with depression. Moreover, the pathogenesis of depression is still unknown, which seriously hinders the prevention, diagnosis, and treatment of depression. Therefore, depression is one of the major causes of global disability and has considerable risks in patients with gliomas. Depression has become a common complication of brain tumors [3], and has become the first clinical manifestation of gliomas in clinical diagnosis. Seddighi et al.'s studies have shown that depressive symptoms are shown to be common signs in patients with brain tumors [4]. They suggest that statistical analysis of the deterioration of psychiatric symptoms mentioned in the later stages of tumorigenesis is not feasible due to the high variability of tumor staging. Glioblastoma (GBM) is a rare malignant tumor that arises from astrocytes—the star-shaped cells that make up the "glue-like" or supportive tissue of the brain and is the most malignant glioma in astrocytic tumors. Despite all therapeutic efforts, GBM remains largely incurable.

Aiming at this problem, this study uses GBM and major depressive disorder (MDD) as the research object to study the overlapping genes, miRNA, biological pathways, and so on. Is the statistical analysis of the correlation between MDD and GBM feasible? With the implementation of the human genome project (HGP), the Human Proteome Project (HPP), and the Human Connectome Project (HCP), more and more ion channels, cytokines, growth factors, neurotransmitters and neurotransmitter receptors, enzymes, other proteins, and miRNA associated with the development of depression and glioblastoma diseases, have been identified and validated [5]. Therefore, it is feasible to analyze the correlation between MDD and GBM by the method of omics. But, few new and effective treatments appear. At present, RNA interference has enormous therapeutic potential for two diseases. Therefore, it is the best way to explore the pathogenesis of the disease through transcriptome data. This study designs a set of transcriptomics in three directions to study the common pathways of disease programs, the flowchart can be found in Figure 1. The process is mainly to analyze the function of RNA in coding region and non-coding region. It mainly divided into three parts. (1) The differentially expressed genes (DEGs) were screened from the gene expression profile data by R software and its corresponding expansion kit [6–9], and the gene ontology (GO), Kyoto Encyclopedia of Genes and Genomes(KEGG) results were significantly correlated with functional enrichment analysis; (2) Using the STRING [10] and Cytoscape [11] tools to construct the protein—protein interaction (PPI) network, the core gene module was excavated by MCODE [12] algorithm, and the GO and KEGG results of MDD and GBM overlap were obtained by functional enrichment analysis; and (3) Targetscan [13] tool was used to predict the miRNA of differentially expressed genes in two diseases, and to enrich, analyze, and annotate the overlapped miRNA in two diseases by miEAA [14]. This study finds from another direction the pathogenesis of the disease. It is hoped that these findings will provide new ideas for the diagnosis and treatment of MDD and GBM.

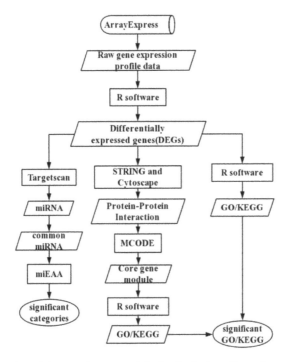

Figure 1. The flowchart of the research program. Cylinder: the database; Rectangle: method or software; Parallelogram: data or result; Ellipse indicates the finally result.

2. Results

2.1. The Common Co-Occurrence Gene by Text-Mining

Through COREMINE platform text mining tools, MDD and 1826 genes have co-occurrence relationship, Glioma and 1826 genes have co-occurrence relationship, GBM and 4510 genes have co-occurrence relationship. Among them, 57% of MDD co-occurrence genes and 23.1% of GBM co-occurrence genes were identified as common genes, with a total of 1041 genes (Table 1). Besides, it is shared with 78 co-occurrence GO cellular component (CC), 317 co-occurrence GO biology process (BP), and 52 co-occurrence GO molecular function(MF) betweenthe two diseases. Our finds speculated that may have common biological pathways or the occurrence of the same mutation between MDD and GBM.

Table 1. The results of text-mining in COREMINE platform. MDD: Major Depressive Disorder; GBM: glioblastoma; Related articles: Pubmed search with a concept or expert name to generate a list of articles; BP: Biology Process; CC: Cellular Component; MF: Molecular Function.

Disease	Related Articles	Gene/Protein	Chemical	CC	BP	MF
MDD	34377	1826	3511	110	498	104
GBM	30193	4510	7779	229	834	244
GBM ∩ MDD	4	1041	2248	78	317	52

2.2. Differentially Expressed Genes

After the DEGs was screened out, the DEGs of different platforms of the same disease were combined as the final DEGs of the disease. There are 463 DEGs (p-value < 0.01) significantly associated with MDD, and 823 DEGs (p-value < 0.05 and fold change $\geq$ 4) were significantly associated with GBM. A simple statistical analysis of DEGs revealed that a total of 27 genes were not only significantly associated with MDD but also closely related to GBM. It was found that five genes (*GRK3, SHANK3, EGR4, CRH, GNB5*) in these 27 genes are down-regulated genes, and six genes (*IGF2BP3, MGP, LOX, KCNE4, DLGAP5, MS4A7*) are up-regulated genes.

Statistics were found through literature mining, in 463 MDD DEGs, 80 genes have been reported related to MDD, there are 201 genes associated with depression; in 823 GBM DEGs, 452 genes are reported with GBM; 27 DEGs overlap in MDD and GBM, eight genes has been reported related to MDD, 14 genes have been reported related to GBM. Moreover, four genes in the reported gene are associated with both MDD and GBM. The four genes are *LOX, NPY1R, SHANK3, VEGFA*. The study finds that LOX expression and activity increased positively correlated with GBM [15]. MDD treatment of electroconvulsive shock (ECS) can be induced by activity-dependent induction of genes (FOX) that are associated with plasticity of the brain, such as neuronal signaling-induced neurogenesis and tissue remodeling [16]. Berent et al. found that higher VEGFA concentrations may have antidepressant effects [17]. Therefore, VEGFA may play a potentially important role in the pathogenesis of MDD. However, Stefano et al. suggest that VEGFA triggers an angiogenic response and promotes GBM vascular growth [18]. There are indications that have been screened for differentially expressed genes that are reliable. We can carry out the next step of the functional analysis.

2.3. Functional Enrichment of DEGs

The R tool is used to analyze and enrich the DEGs. DEGs in MDD were significantly enriched in 804 terms (count $\geq$ 2 and p-value < 0.05), including 704 GO biology process terms, 35 GO cellular component terms, 47 GO molecular function terms, and 18 KEGG pathway terms. DEGs in GBM are significantly enriched in 1681 terms, involving 1207 GO biology process terms, 201 GO cellular component terms, 224 GO molecular function terms, and 48 KEGG pathway terms. These results show that MDD and GBM have 264 BP, 18 CC, 16 MF functional annotations overlap in GO, and seven biological pathways overlap in KEGG. Figure 2 shows the same functional enrichment results for the Wein diagram and its proportion in both diseases. It can be found that the enrichment of the two diseases has some common ground. The same GO or KEGG of the two diseases is approximately 1/3 of the MDD functional enrichment results, approximately 1/10 of the GBM functional enrichment.

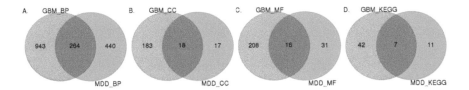

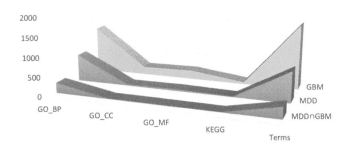

Figure 2. Differentially Expressed Genes Enrichment Venn Diagram and Its 3D Area Map. Figure (**A–D**) indicate similarities and differences in the functional enrichment results of two diseases. They are GO_BP, GO_CC, GO_MF, KEGG. In Figure (**E**), the Yellow: GBM enrichment results; blue: MDD functional enrichment results; green: MDD and GBM common enrichment results. MDD: Major Depressive Disorder; GBM: glioblastoma; BP: Biology Process; CC: Cellular Component; MF: Molecular Function.

The 1680 GBM function enrichment results and 804 MDD function enrichment results are summarized, and 305 common data of MDD and GBM are extracted. Pearson's method was used to calculate the correlation coefficient of the respective differentially expressed genes of MDD and GBM in the common data. Finally, calculated the correlation coefficient between the two is 0.9525328, close to 1, the relevance and high. That is, even though only 27 of the two diseases overlap, almost completely different differentially expressed genes. There is also an extremely high correlation in this common functional enrichment data, suggesting that MDD may also have some relevance to the underlying pathogenesis of GBM. Of course, we also functionally enrich 27 co-differentially expressed genes and obtain three significant KEGG pathways.

2.4. Protein-Protein Interaction Network of DEGs

In this study, we use the STRING online tool to construct the PPI of 402 nodes and 512 sides for MDD, as well as PPI of 794 nodes and 4443 sides for GBM (Figure 3). The Cytoscape tools are used to build the interaction of MDD and GBM PPI. Based on PPI (the elimination of independent protein), 74 and 64 HUB genes (Table 2) (Betweenness Centrality (BC) $\geq$ 0.01, degree $\geq$ 2) were closely related to MDD and GBM. Among the key genes in MDD and GBM, four genes exist together. Namely, *CXCR4, VEGFA, MGP, GNB5,* and *MGP* genes are down-regulated in two diseases, while *GNB5* gene is co-up-regulated.

A. PPI of MDD

B. PPI of GBM

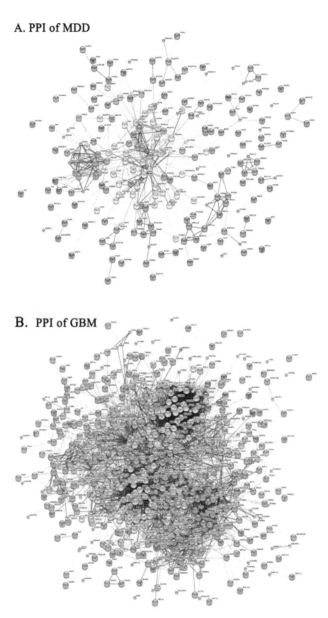

Figure 3. The protein-protein interaction network of MDD and GBM. (**A**) The protein—protein interaction (PPI) of MDD. (**B**) The PPI of GBM. Nodes of the same color represent proteins that are aggregated into the same class; large nodes indicate that the three-dimensional structure of the protein is known and that the small nodes are unknown; the line represents the interaction between proteins; there are seven kinds of relationship. Red, fusion gene; Green, adjacent interaction; Blue, coexistence relationship; Purple, experimental study of validation interactions; Yellow, literature digging to the interaction; and, Light blue, the database included interaction; Black, shared expression. MDD, Major Depressive Disorder. GBM, glioblastoma.

Table 2. MDD and GBM Hub genes.

GBM	MDD
HS6ST3; ZNF385B; VSTM2L; EGFR; VEGFA; TOP2A; CDC42; MYC; IL8; FN1; PRKACB; CD44; CDK1; VIM; GFAP; SYP; PPP3CA; SNCA; STX1A; DNM1; GNAO1; CACNA1B; LPL; PCSK2; PRKCB; SYT1; SNAP25; TUBB4A;DLG2; PVALB; CAMK2A; CALB1; CDK2; CAV1; CXCR4; GNB5; VAMP2; NPY; VCAM1; PRKCE; C3; EZH2; CDC20; SST; GAD2; ITPR1; ADCY2; LUM; TAC1; AURKA; CD163; SYN1; SPARC; BIRC5; GABBR2; ANXA1; MGP; GAD1; TYMS; GNG3; SCN2A; MCL1; CNKSR2; NDE1	JUN; VEGFA; EGF; CXCR4; GNAI3; EGR1; FLT1; CDKN1A; ATF3; CXCL1; GNG11; GNB5; ACTA2; MET; FGFR2; SH3GL1; CXCL5; MGP; THBS1; FAS; IRF5; JUP; RAP2A; TCF7L2; MRPL23; TNFAIP3; CCND3; SLC25A1; MAPK4; BATF3; CD55; CDC25C; MPP3; PPM1D; ILF3; HIST1H4D; CDK13; SSU72; PTPN6; CREM; OCLN; ADORA2B; HIBCH; DYNC1I2; CTSB; MAFF; RYR2; DLGAP5; DCLRE1C; SSR4; ADRBK2; COPS2; COX6A1; LOX; SNAP29; BRD4; DDX11; KRR1; AKAP9; SMC5; ZFP36L1; AIMP1; CFDP1; GAS7; MYO10; GP5; SYT7; ESCO2; MSI2; CLEC1B; FECH; B4GALT1; TKT

MCODE algorithm is used to cluster MDD and GBM PPI, respectively. The PPI of MDD can be clustered into 11 categories, and the GBM of PPI is clustered into 20 categories (Table S1). In the MDD's 11 core gene module, the functional enrichment of the most significant class (Figure 4 and Table 3) found that GABAergic synapse, Serotonergic synapse, Cholinergic synapse, Glutamatergic synapse, Dopaminergic synapse, and Morphine addiction affect the development of depression. In the GBM's 20 core gene module, the functional enrichment of the second significant group (Figure 4 and Table 3) also found that GABAergic synapse, Serotonergic synapse, Cholinergic synapse, Glutamatergic synapse, and Morphine addiction, were associated with the development of GBM. Therefore, the two core gene modules with high significance and overlapping biological pathways are regarded as the significant core gene modules of disease. Moreover, there are two common key genes in the core gene module, that is CXCR4 and GNB5. The accumulation of two significant core gene modules revealed that 10 biological pathways overlap, accounting for 71.4% of MDD enrichment, and 50% of GBM enrichment. The two core gene modules with higher saliency and overlapping biological pathways as the significant core gene modules of the disease. In addition, there are two overlapping key genes in significant core gene modules—CXCR4 and GNB5, accounting for half of the overlapping key genes of both diseases. These two significant core gene modules may play an important role in the underlying pathogenesis of MDD or GBM. Ten common KEGG biological pathways of significant core gene modules are the likely common pathways of MDD and GBM.

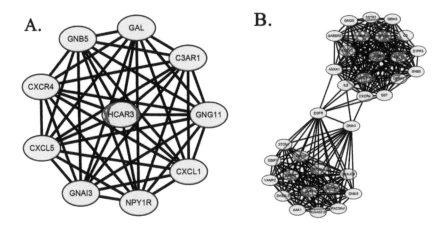

Figure 4. PPI of the core gene module. (**A**) The PPI of the most significant core module about MDD. (**B**) The PPI of the secondary core gene module about GBM. Node: protein; connection: the interaction between proteins. MDD, Major Depressive Disorder. GBM, glioblastoma.

Table 3. Functional enrichment of significant core gene module—KEGG. MDD, Major Depressive Disorder. GBM, glioblastoma. MDD KEGG: the unique KEGG pathways of significant core gene module of Major Depressive Disorder; GBM KEGG: the unique KEGG pathways of significant core gene module of glioblastoma; Common KEGG: the common KEGG pathways of significant core gene module in Major Depressive Disorder and glioblastoma.

MDD KEGG	GBM KEGG	Common KEGG
Alcoholism	Endocytosis	GABAergic synapse
Pertussis	Gap junction	Cholinergic synapse
Serotonergic synapse	Insulin secretion	Pathways in cancer
	Salivary secretion	Morphine addiction
	Synaptic vesicle cycle	Circadian entrainment
	Gastric acid secretion	Glutamatergic synapse
Dopaminergic synapse	GnRH signaling pathway	cAMP signaling pathway
	Estrogen signaling pathway	Chemokine signaling pathway
	Oxytocin signaling pathway	Retrograde endocannabinoid signaling
	Neuroactive ligand-receptor interaction	Regulation of lipolysis in adipocytes

2.5. miRNA

This study uses the TargetScanHuman tool to predict the related miRNAs of DEGs (mRNA). After a series of screening criteria, 18,656 pairs for MDD mRNA—miRNA, 52,413 pairs for GBM mRNA—miRNA are obtained. Data analysis of MDD reveals that there are 370 different mRNA corresponding to 2455 different miRNA. Data analysis of GBM reveals that there are 754 different mRNA, corresponding to 2586 different miRNA. Figure 5 intuitively shows the miRNA of two diseases, overlapping up to 2453. In the figure, 99.9% of the miRNAs predicted by MDD-related DEGs are the same as the miRNA predicted by GBM. Moreover, the predicted miRNAs of the 27 common DEGs to the two diseases completely overlap with the common miRNAs predicted by the two diseases.

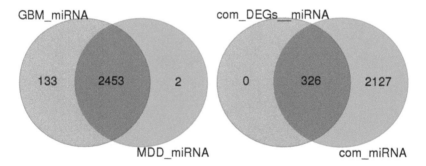

Figure 5. The predicted miRNA Venn Diagram. GBM_miRNA: the miRNAs predicted by GBM-related DEGs; MDD_miRNA: the miRNAs predicted by MDD-related DEGs; com_DEGs_miRNA: the miRNAs predicted by common 27 DEGs; com_miRNA: common miRNA between GBM_miRNA and MDD_miRNA.

2.6. Functional Enrichment of Common DEGs and Common miRNA

Through the enrichment of 27 common DEGs in MDD and GBM, three KEGG biological pathways are enriched, which are Cytokine—cytokine receptor interaction, Chemokine signaling pathway, Bladder cancer. There are 128 GO biology process (BP) terms, two GO cellular component (CC) terms, three GO molecular function (MF) terms are significantly enriched (Table S2). The results showed that the cellular components of common DEGs in the two diseases are related to the activity of transcription factors. Its molecular function is related to the extracellular matrix, and its biological process is mainly involved in the regulation of multicellular biological processes, the regulation of ion transport, the regulation of growth and development, and the response to some stimuli, and so on.

The common miRNAs are analyzed by miEAA, and their enrichment and annotation results are shown in Table 2, where the enrichment results of GO are not fully shown. These miRNAs are closely related to the expression of CD3, CD14, CD19, and CD56 in four immune cells, indicating that both MDD and GBM can cause immune system disorders. 2453 common miRNAs have 102 miRNAs located on chromosome 7, indicating that chromosome 7 is not only associated with mental illness, such as depression and schizophrenia, but are also closely related to the pathogenesis of glioblastoma [19,20]. The results show that there are seven biological pathways (Table 4), 356 gene ontologies are enriched. In the 7 pathways, two are related to amino acid metabolism, two are related to carbohydrate metabolism, two are related to mRNA processing, and one is the Notch signaling pathway that affects multiple processes that occur in cells.

Table 4. Results of the common miRNA enrichment and annotation.

Category	Subcategory	p-Value	Observed
Pathways	WP411 mRNA processing	0.060337	145
Pathways	hsa00260 Glycine serine and threonine metabolism	0.060337	30
Pathways	hsa00562 Inositol phosphate metabolism	0.060337	99
Pathways	hsa03040 Spliceosome	0.060337	138
Pathways	hsa00330 Arginine and proline metabolism	0.081737	67
Pathways	hsa04330 Notch signaling pathway	0.081737	94
Pathways	P02756 N acetylglucosamine metabolism	0.095531	14
Immune cells	CD3 expressed	0.008542	205
Immune cells	CD19 expressed	0.029347	182
Immune cells	CD14 expressed	0.040935	235
Immune cells	CD56 expressed	0.055024	252
Chromosomal location	Chromosome 7	0.036298	102
Gene Ontology	GO0042832 defense response to protozoan	0.0626691	11
Gene Ontology	GO0045859 regulation of protein kinase activity	0.0626691	41
Gene Ontology	GO0048304 positive regulation of isotype switching to igg isotypes	0.0626691	11
Gene Ontology	GO0016290 palmitoyl coa hydrolase activity	0.0665877	24
Gene Ontology	GO0044130 negative regulation of growth of symbiont in host	0.0681013	19
Gene Ontology	GO0004439 phosphatidylinositol 4 5 bisphosphate 5 phosphatase activity	0.0753516	16
Gene Ontology	GO0004523 ribonuclease h activity	0.0753516	19
Gene Ontology	GO0031848 protection from non homologous end joining at telomere	0.0753516	16

3. Discussion

To comprehensively and accurately identify the pathogenesis of MDD and GBM, all available transcript data for both of the diseases are downloaded. The purpose is to horizontally merge large amounts of transcriptome data to expand the sample size and obtain a larger sample size dataset. Functional analysis of the differential expression genes of the two diseases is carried out from three aspects.

From the perspective of coding genes, MDD and GBM differentially expressed genes are enriched in seven common biological pathways, namely Melanoma, Pathways in cancer, mitogen-activated protein kinase (MAPK) signaling pathway, Endocytosis, p53 signaling, Focal adhesion, Bladder cancer. As a result, it has been found that five common pathways are associated with the development of cancer, suggesting that the two diseases may also be closely related to other diseases, particularly cancer. Due to the complexity of cancer, the five pathways are temporarily serve as the common pathway for the two diseases. At the same time, it can see that the other two common pathways, MAPK signaling pathway and Endocytosis, are reported to be associated with both diseases. The MAPK pathway may be initiated at the cell surface and continue during endosomal sorting, while more recent studies suggest that MAPK signaling is a required element of endocytosis [21]. Li Kai et al. found the disturbance mechanism of MAPK and cell cycle signaling pathway in GBM by bioinformatics analysis [22]. The study has found that the MAPK signaling pathway is impaired in MDD and plays a key role in neuronal plasticity and neurogenesis, and is shown to be stimulated by an antidepressant treatment [23]. It is suggested by the results that MAPK signaling pathways may be one of the common pathways for MDD and GBM. Cytokine—cytokine receptor interaction, Chemokine signaling pathway and Bladder cancer are enriched by MDD and GBM common DEGs. These three biological pathways also belong to the common biological pathway of MDD and GBM.

From the perspective of miRNA, the corresponding miRNAs are predicted by the mRNA of the two diseases, and the number of common miRNAs has been found to be 2453. In addition, various miRNAs have been demonstrated to be either upregulated or downregulated in glioma tumors, and played critical roles in regulating glioblastoma proliferation, migration, and chemosensitivity [24]. Several recent studies have suggested the possible role of miRNAs in synaptic plasticity, neurogenesis, and stress response, all implicated in MDD [25]. Most of these miRNAs are contained in common miRNAs. For example, Hsa-miR-21 is not only involved in the alterations of white matter in depression and alcoholism [26], but it also plays a key role in the pathogenesis of GBM and can be used as a biomarker for the diagnosis and treatment of GBM patients [27]. The miRNAs were found to be closely related to the abnormal expression of CD3, CD14, CD19, and CD56 in immune cells category. CD3 and CD4 are protein mixtures present on the surface of T cells, CD19 is a protein present on the surface of B cells, and CD56 is an affinity binding glycoprotein expressed on the surface of neurons, glial, and skeletal muscle. CD3, CD4, CD19 are related to the immune process, CD56 role in the p59Fyn signaling pathway. Are the common pathogenesis of the two diseases related to the immune system? There is no confirmation here. It can only be said that the pathogenesis of MDD and GBM may be associated with the immune system. In the seven biological pathways enriched by common miRNA, four pathways are metabolically related, two pathways are associated with mRNA processes, and one is Notch signaling pathway. On the one hand, the possible reason is that MDD and GBM patients biochemical environment affects the brain tissue, metabolic changes occur. On the other hand, Irshad et al. have identified the key molecular cluster characteristics of the Notch pathway response in hypoxic GBM and glial cell spheres [28]. Moreover, Ning et al. also determined that differential expression of Notch-associated miRNAs in peripheral blood may be involved in the development of depression [29]. Thus, glycine—serine and threonine metabolism, inositol phosphate metabolism, arginine and proline metabolism, N-acetylglucosamine metabolism, and Notch signaling pathways are also common biological pathways for MDD and GBM.

From the perspective of protein interaction, the significant core gene modules in the MDD and GBM protein interaction networks were enriched to 10 common biological pathways. The discovery of significant core gene modules in protein—protein interaction networks allows for a more accurate and comprehensive understanding of the function of DEGs in disease. The γ-aminobutyric acid (GABA), glutamic acid, and acetylcholine (Ach) are three common amino acid neurotransmitters, which are specific chemicals that act as "messengers" in synaptic transmission. Salvadore et al. confirm that amino-acid neurotransmitter system dysfunction plays a major role in the pathophysiology of major depressive disorder [30]. Panosyan et al. have found that these three neurotransmitters are involved in the metabolic pathways underlying the potential targets of GBM therapy, but the hypothesis that they have a significant antitumor effect on GBM has not been demonstrated [31]. Hence, GABAergic synapse, Cholinergic synapse, and Glutamatergic synapse may be common pathways for MDD and GBM. Retrograde endocannabinoid signaling has been shown to be related to the pathophysiological mechanisms of MDD and GBM [32,33]. Pathways in cancer, Chemokine signaling pathway is also a common pathway when using genomics enrichment. Therefore, Pathways in cancer, Retrograde endocannabinoid signaling, Chemokine signaling pathway may also be a common pathway for MDD and GBM.

4. Materials and Methods

4.1. Text Mining

In the biomedical field, text mining has been widely used to identify biological terms, such as genes, disease names in the literature, and even reveal the relationship between biological terms. In this study, COREMINE [34] (Available online: http://www.coremine.com/medical/), a medical ontology information retrieval platform, was used to search for key words, such as major depressive disorder and glioblastoma.

4.2. Data and Data Preprocessing

The original gene expression profiling data was based on the GPL570 and GPL17027 platform developed by Affymetrix, derived from EBI's common library database ArrayExpress [35] (Available online: http://www.ebi.uk/ArrayExpress). Including transcriptome data sets of major depressive disorder (excluding bipolar disorder) and glioblastoma. A total of 47 series, 2093 samples of raw data, 11 series, 367 samples were associated with MDD, 36 series, and 1726 samples were associated with GBM (Table S3 for data sources). In this study, we used the RMA (Robust Multichip Average) method in the Affy package of the R tool to normalized the raw data and then obtained the corresponding gene expression matrix.

4.3. Differential Expression Analysis

Studies had found that disease is associated with genes, even if only a small change in a subunit in the genome. For example, the duplication or absence of a dose-sensitive gene [36] is associated with disease, including heart disease, cancer, and neuropsychiatric disorders. Therefore, the use of differential expression analysis method to identify the disease-related genes is essential. In this study, the linear regression and classical Bayesian method in the limma package of R language were used to analyzed and screened differentially expressed genes of the two diseases. Since the two diseases do not belong to the same type of disease, the screening criteria for differentially expressed genes use different thresholds. MDD differential expression gene screening criteria were p-value threshold of 0.01, GBM differential expression gene screening criteria were p-value threshold is 0.05, and the difference in expression was greater than 4.

4.4. Functional Enrichment Analysis

Functional enrichment analysis is a method of cross-integration between biology and mathematics, which is the best choice to solve the massive data of gene chip. In this study, we used the GOstats and KEGG.db toolkit in the R language to perform functional enrichment analysis on the significantly differentially expressed genes and select the GO entry with a Count value greater than or equal to 2 and a p-value of less than 0.05. At the same time, the KEGG pathway with p-value less than 0.05 was selected as the enrichment biological pathway.

4.5. Protein—Protein Interaction Network

In this study, the STRING (Available online: http://string-db.org/) database was used to construct the Protein-Protein Interaction (PPI) between proteins encoded by differentially expressed genes. The STRING database is a database that collects protein—protein interactions, gene regulatory relationships, document mining analysis, and protein co-expression analysis, and calculates physical interactions and predicts interaction relationships. The protein interaction threshold was set at

0.4. The protein interaction data obtained from the online STRING database is imported into the Cytoscape software, and the node with the degree greater than 2 and the BC was greater than 0.01 was obtained by using its Network Analysis plug-in tool. The node as a network centre node (Hub). The protein represented by the central node was usually the key protein (Hub gene) [37] with important physiological functions. Then, the MCODE algorithm in Cytoscape was used to further cluster analysis to find the core gene module in the protein interaction network, and to dig the biological function or pathway that was significantly related to the disease.

4.6. Predicted miRNAs

Numerous studies have confirmed that alterations of specific microRNAs (miRNAs) levels are closely related to human pathologies [38]. A small number of miRNA biological functions have been elucidated. Thus, miRNAs were predicted by the TargetScanHuman (Available online: http://www.targetscan.org/vert_71/) tool for differentially expressed genes. The standard for screening predicted miRNAs was 8 mer—a (exact match to positions 2–8 of the mature miRNA followed by an "A") and the percentage of context ++ score (CS) should not be less than 95%. This CS is the cumulative sum of 14 features for a particular site, including the type of site, complementary pairing, minimum distance, length of open reading frame (ORF), conserved target probability (PCT), and so on. To further analyze and explore the pathogenesis of the disease. The miEAA (miRNA Enrichment Analysis and Annotation Tool, Available online: https://ccb-compute2.cs.uni-saarland. de/mieaa_tool/) online tool was used to enrich and annotate the predicted miRNAs by combining the two diseases. The miEAA's *p*-value adjustment method was error detection rate (false discovery rate, FDR), the category of *p*-value less than 0.1 was significantly related.

5. Conclusions

In this article, we first rely on transcriptome data to obtain reliable and useful differentially expressed genes (DEGs) by differential expression analysis. Then, we used the transcriptomics of DEGs to find out and analyze the common pathway of MDD and GBM from three directions. At present, more and more miRNA are the biomarkers of disease, which are related to the pathophysiology of various diseases, including MDD and GBM. However, due to the large number of predicted miRNAs, further studies are needed to find suitable biomarkers for MDD and GBM. Finally, we determine the important biological pathways that are common to MDD and GBM by statistical knowledge. It is worth mentioning that, Chemokine signaling pathway not only found in functional enrichment of coding genes is a common pathways between MDD and GBM, also found in the core gene module of the protein interaction network. Our findings provide the first direct transcriptomic evidence that common pathway in two diseases for the common pathogenesis of the human MDD and GBM. Our results provide a new reference methods and values for the study of the pathogenesis of depression and glioblastoma.

Supplementary Materials: Supplementary materials can be found at www.mdpi.com/1422-0067/19/1/234/s1.

Acknowledgments: This study was financially supported by the Special Project of National Science and Technology Cooperation (2014DFB30010) and National Natural Science Foundation of China (61501071).

Author Contributions: Yongfang Xie composed and finalized the manuscript, and revised final improvement; Ling Wang composed and finalized the manuscript, and revised final improvement, as well as performed data processing and analysis; Zengyan Xie and Chuisheng zeng revised the paper; Kunxian Shu performed primary study design, manuscript editing and final improvement.

Conflicts of Interest: The authors declare no conflict of interest.

References

1. Rooney, A.G.; Carson, A.; Grant, R. Depression in cerebral glioma patients: A systematic review of observational studies. *J. Natl. Cancer Inst.* **2011**, *103*, 61–76. [CrossRef] [PubMed]
2. Rooney, A.G.; Brown, P.D.; Reijneveld, J.C.; Grant, R. Depression in glioma: A primer for clinicians and researchers. *J. Neurol. Neurosurg. Psychiatry* **2014**, *85*, 230–235. [CrossRef] [PubMed]
3. Pranckeviciene, A.; Bunevicius, A. Depression screening in patients with brain tumors: A review. *CNS Oncol.* **2015**, *4*, 71–78. [CrossRef] [PubMed]
4. Seddighi, A.; Seddighi, A.S.; Nikouei, A.; Ashrafi, F.; Nohesara, S. Psychological aspects in brain tumor patients: A prospective study. *Hell. J. Nucl. Med.* **2015**, *18* (Suppl. 1), 63–67. [PubMed]
5. Sah, D.W. Therapeutic potential of rna interference for neurological disorders. *Life Sci.* **2006**, *79*, 1773–1780. [CrossRef] [PubMed]
6. Gautier, L.; Cope, L.; Bolstad, B.M.; Irizarry, R.A. Affy-analysis of affymetrix genechip data at the probe level. *Bioinformatics* **2004**, *20*, 307–315. [CrossRef] [PubMed]
7. Falcon, S.; Gentleman, R. Using gostats to test gene lists for go term association. *Bioinformatics* **2007**, *23*, 257–258. [CrossRef] [PubMed]
8. Ritchie, M.E.; Phipson, B.; Wu, D.; Hu, Y.; Law, C.W.; Shi, W.; Smyth, G.K. Limma powers differential expression analyses for rna-sequencing and microarray studies. *Nucleic Acids Res.* **2015**, *43*, e47. [CrossRef] [PubMed]
9. Carlson, M. KEGG.db: A Set of Annotation Maps for KEGG. Available online: https://bioconductor.org/packages/release/data/annotation/html/KEGG.db.html (accessed on 9 January 2018).
10. Franceschini, A.; Szklarczyk, D.; Frankild, S.; Kuhn, M.; Simonovic, M.; Roth, A.; Lin, J.; Minguez, P.; Bork, P.; von Mering, C.; et al. String v9.1: Protein—protein interaction networks, with increased coverage and integration. *Nucleic Acids Res.* **2013**, *41*, D808–D815. [CrossRef] [PubMed]
11. Shannon, P.; Markiel, A.; Ozier, O.; Baliga, N.S.; Wang, J.T.; Ramage, D.; Amin, N.; Schwikowski, B.; Ideker, T. Cytoscape: A software environment for integrated models of biomolecular interaction networks. *Genome Res.* **2003**, *13*, 2498–2504. [CrossRef] [PubMed]
12. Bader, G.D.; Hogue, C.W. An automated method for finding molecular complexes in large protein interaction networks. *BMC Bioinform.* **2003**, *4*, 2. [CrossRef]
13. Agarwal, V.; Bell, G.W.; Nam, J.W.; Bartel, D.P. Predicting effective microrna target sites in mammalian mrnas. *eLife* **2015**, *4*, 1–38. [CrossRef] [PubMed]
14. Backes, C.; Khaleeq, Q.T.; Meese, E.; Keller, A. Mieaa: Microrna enrichment analysis and annotation. *Nucleic Acids Res.* **2016**, *44*, W110–W116. [CrossRef] [PubMed]
15. Da Silva, R.; Uno, M.; Marie, S.K.; Oba-Shinjo, S.M. Lox expression and functional analysis in astrocytomas and impact of idh1 mutation. *PLoS ONE* **2015**, *10*, e0119781. [CrossRef] [PubMed]
16. Sun, W.; Park, K.W.; Choe, J.; Rhyu, I.J.; Kim, I.H.; Park, S.K.; Choi, B.; Choi, S.H.; Park, S.H.; Kim, H. Identification of novel electroconvulsive shock-induced and activity-dependent genes in the rat brain. *Biochem. Biophys. Res. Commun.* **2005**, *327*, 848–856. [CrossRef] [PubMed]
17. Berent, D.; Macander, M.; Szemraj, J.; Orzechowska, A.; Galecki, P. Vascular endothelial growth factor a gene expression level is higher in patients with major depressive disorder and not affected by cigarette smoking, hyperlipidemia or treatment with statins. *Acta Neurobiol. Exp.* **2014**, *74*, 82–90.
18. Stefano, A.L.D.; Labussiere, M.; Lombardi, G.; Eoli, M.; Bianchessi, D.; Pasqualetti, F.; Farina, P.; Cuzzubbo, S.; Gallego-Perez-Larraya, J.; Boisselier, B. Vegfa snp rs2010963 is associated with vascular toxicity in recurrent glioblastomas and longer response to bevacizumab. *J. Neurooncol.* **2015**, *121*, 499–504. [CrossRef] [PubMed]
19. Lopez-Gines, C.; Cerda-Nicolas, M.; Gil-Benso, R.; Pellin, A.; Lopez-Guerrero, J.A.; Callaghan, R.; Benito, R.; Roldan, P.; Piquer, J.; Llacer, J.; et al. Association of chromosome 7, chromosome 10 and egfr gene amplification in glioblastoma multiforme. *Clin. Neuropathol.* **2005**, *24*, 209–218. [PubMed]
20. Chen, X.; Long, F.; Cai, B.; Chen, X.; Chen, G. A novel relationship for schizophrenia, bipolar and major depressive disorder part 7: A hint from chromosome 7 high density association screen. *Behav. Brain Res.* **2015**, *293*, 241–251. [CrossRef] [PubMed]
21. Fujioka, Y.; Tsuda, M.; Hattori, T.; Sasaki, J.; Sasaki, T.; Miyazaki, T.; Ohba, Y. The ras–pi3k signaling pathway is involved in clathrin-independent endocytosis and the internalization of influenza viruses. *PLoS ONE* **2011**, *6*, e16324. [CrossRef] [PubMed]

22. Li, W.; Li, K.; Zhao, L.; Zou, H. Bioinformatics analysis reveals disturbance mechanism of mapk signaling pathway and cell cycle in glioblastoma multiforme. *Gene* **2014**, *547*, 346–350. [CrossRef] [PubMed]
23. Wang, L.; Peng, D.; Xie, B.; Jiang, K.; Fang, Y. The extracellular signal-regulated kinase pathway may play an important role in mediating antidepressant-stimulated hippocampus neurogenesis in depression. *Med. Hypotheses* **2012**, *79*, 87–91. [CrossRef] [PubMed]
24. Møller, H.G.; Rasmussen, A.P.; Andersen, H.H.; Johnsen, K.B.; Henriksen, M.; Duroux, M. A systematic review of microrna in glioblastoma multiforme: Micro-modulators in the mesenchymal mode of migration and invasion. *Mol. Neurobiol.* **2013**, *47*, 131–144. [CrossRef] [PubMed]
25. Dwivedi, Y. Micrornas as biomarker in depression pathogenesis. *Ann. Psychiatry Ment. Health* **2013**, *1*, 1003. [PubMed]
26. Miguel-Hidalgo, J.J.; Hall, K.O.; Bonner, H.; Roller, A.M.; Syed, M.; Park, C.J.; Ball, J.P.; Rothenberg, M.E.; Stockmeier, C.A.; Romero, D.G. Microrna-21: Expression in oligodendrocytes and correlation with low myelin mrnas in depression and alcoholism. *Prog. Neuropsychopharmacol. Biol. Psychiatry* **2017**, *79*, 503–514. [CrossRef] [PubMed]
27. Masoudi, M.S.; Mehrabian, E.; Mirzaei, H. Mir-21: A key player in glioblastoma pathogenesis. *J. Cell. Biochem.* **2017**. [CrossRef] [PubMed]
28. Irshad, K.; Mohapatra, S.K.; Srivastava, C.; Garg, H.; Mishra, S.; Dikshit, B.; Sarkar, C.; Gupta, D.; Chandra, P.S.; Chattopadhyay, P. A combined gene signature of hypoxia and notch pathway in human glioblastoma and its prognostic relevance. *PLoS ONE* **2015**, *10*, e0118201. [CrossRef] [PubMed]
29. Ning, S.; Lei, L.; Wang, Y.; Yang, C.; Liu, Z.; Li, X.; Zhang, K. Preliminary comparison of plasma notch-associated microRNA-34b and -34c levels in drug naive, first episode depressed patients and healthy controls. *J. Affect. Disord.* **2016**, *194*, 109–114.
30. Salvadore, G.; Veen, J.W.V.D.; Zhang, Y.; Marenco, S.; Machadovieira, R.; Baumann, J.; Ibrahim, L.A.; Luckenbaugh, D.A.; Shen, J.; Drevets, W.C. An investigation of amino-acid neurotransmitters as potential predictors of clinical improvement to ketamine in depression. *Int. J. Neuropsychopharmacol.* **2012**, *15*, 1063–1072. [CrossRef] [PubMed]
31. Panosyan, E.H.; Lin, H.J.; Koster, J.; Lasky, J.L. In search of druggable targets for gbm amino acid metabolism. *BMC Cancer* **2017**, *17*, 162. [CrossRef] [PubMed]
32. Wei, B.; Wang, L.; Zhao, X.; Jin, Y.; Kong, D.; Hu, G.; Sun, Z. Co-mutated pathways analysis highlights the coordination mechanism in glioblastoma multiforme. *Neoplasma* **2014**, *61*, 424–432. [CrossRef] [PubMed]
33. Shintaro, O.; Hiroshi, K. Inhibitors of fatty acid amide hydrolase and monoacylglycerol lipase: New targets for future antidepressants. *Curr. Neuropharmacol.* **2015**, *13*, 760–775.
34. Wu, W.J.; Wang, Q.; Zhang, W.; Li, L. Identification and prognostic value of differentially expressed proteins of patients with platinum resistance epithelial ovarian cancer in serum. *Zhonghua Fu Chan Ke Za Zhi* **2016**, *51*, 515–523. [PubMed]
35. Brazma, A.; Parkinson, H.; Sarkans, U.; Shojatalab, M.; Vilo, J.; Abeygunawardena, N.; Holloway, E.; Kapushesky, M.; Kemmeren, P.; Lara, G.G. Arrayexpress—A public repository for microarray gene expression data at the ebi. *Nucleic Acids Res.* **2003**, *31*, 68–71. [CrossRef] [PubMed]
36. Rice, A.M.; McLysaght, A. Dosage-sensitive genes in evolution and disease. *BMC Biol.* **2017**, *15*, 78. [CrossRef] [PubMed]
37. Jeong, H.; Mason, S.P.; Barabasi, A.L.; Oltvai, Z.N. Lethality and centrality in protein networks. *Nature* **2001**, *411*, 41–42. [CrossRef] [PubMed]
38. Hesse, M.; Arenz, C. Microrna maturation and human disease. *Methods Mol. Biol.* **2014**, *1095*, 11–25. [PubMed]

 © 2018 by the authors. Licensee MDPI, Basel, Switzerland. This article is an open access article distributed under the terms and conditions of the Creative Commons Attribution (CC BY) license (http://creativecommons.org/licenses/by/4.0/).

International Journal of
Molecular Sciences

Article

miR-25-3p, Positively Regulated by Transcription Factor AP-2α, Regulates the Metabolism of C2C12 Cells by Targeting *Akt1*

Feng Zhang [1,2], Kun Chen [2], Hu Tao [1], Tingting Kang [2], Qi Xiong [1], Qianhui Zeng [2], Yang Liu [1], Siwen Jiang [2,*] and Mingxin Chen [1,*]

[1] Hubei Key Laboratory of Animal Embryo Engineering and Molecular Breeding, Institute of Animal Husbandry and Veterinary, Hubei Academy of Agricultural Sciences, Wuhan 430064, China; zhangfeng0130@163.com (F.Z.); taohu00@gmail.com (H.T.); phenixxq@163.com (Q.X.); liuyang430209@163.com (Y.L.)
[2] Key Laboratory of Swine Genetics and Breeding of the Agricultural Ministry and Key Laboratory of Agricultural Animal Genetics, Breeding and Reproduction of the Ministry of Education, College of Animal Science and Technology, Huazhong Agricultural University, Wuhan 430070, China; kunchen1989@163.com (K.C.); 13163228175@126.com (T.K.); zengqianhui.hzau.cn@webmail.hzau.edu.cn (Q.Z.)
* Correspondence: jiangsiwen@mail.hzau.edu.cn (S.J.); chenmingxin18@163.com (M.C.); Tel.: +86+027-8728-1378 (S.J.); +86+027-8768-0959 (M.C.)

Received: 15 January 2018; Accepted: 6 March 2018; Published: 8 March 2018

Abstract: miR-25, a member of the miR-106b-25 cluster, has been reported as playing an important role in many biological processes by numerous studies, while the role of miR-25 in metabolism and its transcriptional regulation mechanism remain unclear. In this study, gain-of-function and loss-of-function assays demonstrated that miR-25-3p positively regulated the metabolism of C2C12 cells by attenuating phosphoinositide 3-kinase (*PI3K*) gene expression and triglyceride (TG) content, and enhancing the content of adenosine triphosphate (ATP) and reactive oxygen species (ROS). Furthermore, the results from bioinformatics analysis, dual luciferase assay, site-directed mutagenesis, qRT-PCR, and Western blotting demonstrated that miR-25-3p directly targeted the AKT serine/threonine kinase 1 (*Akt1*) 3' untranslated region (3'UTR). The core promoter of miR-25-3p was identified, and the transcription factor activator protein-2α (AP-2α) significantly increased the expression of mature miR-25-3p by binding to its core promoter in vivo, as indicated by the chromatin immunoprecipitation (ChIP) assay, and AP-2α binding also downregulated the expression of *Akt1*. Taken together, our findings suggest that miR-25-3p, positively regulated by the transcription factor AP-2α, enhances C2C12 cell metabolism by targeting the *Akt1* gene.

Keywords: mouse; miR-25-3p; *Akt1*; AP-2α; promoter; cell metabolism

1. Introduction

MicroRNAs (miRNAs) are endogenous, small (~22 nucleotides), and single-stranded noncoding RNAs. The role of different miRNAs in biological systems is well established. They are generally regarded as negative regulators of gene expression, as they bind to the 3' untranslated region (3'UTR) of messengerRNAs (mRNAs), leading to mRNA degradation and/or suppression of mRNA translation [1–3]. Currently, thousands of miRNAs have been identified as participating in a number of biological processes, such as cellular growth, proliferation, development, and metabolism [4].

Based on Solexa sequencing, the expression of microRNA-25 (miR-25) was higher in the longissimus dorsi muscle of Large White pigs (a lean type) than in those of Tongcheng pigs (a Chinese indigenous fatty pig). Because skeletal muscle plays a vital role in whole-body metabolism [5], we speculated that miR-25 could play a regulatory role in metabolism.

Previous studies have reported that miR-25 plays an important role in many biological processes. The expression of miR-25-3p was significantly increased in the plasma of thyroid papillary carcinoma, as compared with patients with benign tumors or healthy individuals [6]. miR-25 expression was higher in ovarian epithelial tissue, gastric cancer, lung adenocarcinoma, and many other tumors, and miR-25 expression levels were also closely related to tumor stage and lymph node metastasis [7–10]. Inhibition of miR-25 markedly improved cardiac contractility in the failing heart [11]. miR-25 could protect cardiomyocytes against oxidative damage by downregulating the mitochondrial calcium uniporter (MCU) [12]. Variations in miR-25 expression influenced the severity of diabetic kidney disease [13]. However, to our knowledge, the role of miR-25 in metabolism has not been reported, and its transcriptional regulatory mechanism is not clear.

Thus, in this study, we first investigated whether miR-25 was involved in metabolism by gain-of-function and loss-of-function assays. Then, the target gene of miR-25, AKT serine/threonine kinase 1 (*Akt1*), which is related to metabolism, was predicted and verified using bioinformatics software and experiments. Finally, the core promoter of miR-25 was identified, and the binding of the transcription factor activator protein-2α (AP-2α) to the core promoter was shown to promote the transcriptional activity of miR-25 and downregulate *Akt1* expression.

2. Results

2.1. miR-25 Is Highly Conserved in Mammals

Clustal Omega (Available online: https://www.ebi.ac.uk/Tools/msa/clustalo/) [14] was used to build the phylogenetic tree of pre-miRNA of miR-25. The results show that compared with other species selected in this study, the genetic relationship between mice and humans, cattle and goats, and gorillas and rhesus monkeys is closer (Figure 1A). The mature sequences of miR-25 are highly conserved in mammals, including pigs, mice, humans, goats, rats, hamsters, gorillas, chimpanzees, cattle, and rhesus monkeys. The "seed" sequences of miR-25 are identical, although there is a base deletion at the end of the chimpanzee sequence (ptr) (Figure 1B).

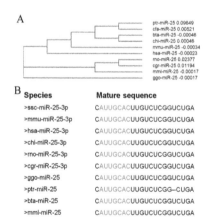

Figure 1. miR-25 is highly conserved in mammals. (**A**) The phylogenetic tree of pre-miRNA of miR-25. pre-miRNA sequences were obtained from NCBI. (**B**) The mature sequences of miR-25 in selected species. These mature sequences were obtained from miRBase. Seed regions are highlighted in red. ssc, sus scrofa; mmu, mus musculus; hsa, homo sapiens; chi, capra hircus; rno, rattus norvegicus; cgr, cricetulus griseus; ggo, gorilla gorilla; ptr, pan troglodytes; bta, bos taurus; mml, macaca mulatta; cfa, canis lupus familiaris.

2.2. Effects of miR-25 on the Metabolism of C2C12 Cells

To investigate the role of miR-25-3p in metabolism, miR-25-3p mimics/negative control (NC) or inhibitors/NC were respectively transfected into growing C2C12 cells (mouse muscle myoblasts). The abundance of miR-25-3p was detected, which was ~3300-fold ($p < 0.01$) higher as compared with another microRNA (Figure S1). The mRNA and protein expression levels of the metabolism-related gene *PI3K* were repressed by miR-25-3p overexpression, while the levels of *PI3K* were upregulated in the inhibitor group, as compared with the negative controls (Figure 2A,B).

In addition, the overexpression of miR-25-3p decreased levels of triglyceride (TG), whereas the knockdown of miR-25-3p increased them (Figure 2C). Conversely, the overexpression of miR-25-3p increased ATP and ROS levels, and the knockdown of miR-25-3p decreased their levels (Figure 2D,E). These data indicate that miR-25-3p plays a role in metabolism.

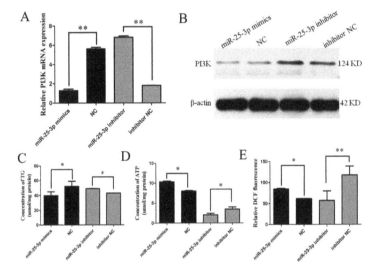

Figure 2. The effect of miR-25 on the metabolism of C2C12 cells. miR-25-3p mimics/NC or inhibitors/NC were respectively transfected into growing C2C12 cells. After 48 h, PI3K expression was detected by qRT-PCR (**A**) and Western blotting (**B**). After 24–48 h transfection, the levels of triglyceride (TG) (**C**), ATP (**D**), and reactive oxygen species (ROS) (**E**) were measured with commercial kits. The fluorescence of DCF represents the content of ROS. NC = negative control (miR-239b-5p of caenorhabditis elegans). β-actin served as the loading control. Data were presented as means ± SD ($n \geq 3$); * $p < 0.05$; ** $p < 0.01$.

2.3. miR-25-3p Directly Targets Akt1

To explore the molecular mechanism of miR-25-3p effects on metabolism, the possible targets for miR-25-3p were predicted using TargetScan, and a putative binding site for miR-25-3p was predicted in the 3′UTR of *Akt1* mRNA. miR-25-3p targeting elements in the *Akt1*-3′UTR were relatively conserved in many mammals, including mice, humans, chimpanzees, rhesus monkeys, and rats (Figure 3A).

To validate whether miR-25-3p directly targets *Akt1*, a luciferase reporter containing a 250 bp fragment from the *Akt1* 3′UTR was tested in vitro. Additionally, we generated a mutated version of the above mentioned reporter, in which five nucleotides of the predicted binding site were changed in order to abolish the putative interaction between miR-25-3p and *Akt1* mRNA (Figure 3B). The *Akt1* 3′UTR and mutant luciferase plasmid were cotransfected with mimics or NC into growing C2C12 cells. Twenty-four hours after transfection, analyses of luciferase activity revealed that miR-25-3p mimics significantly decreased the luciferase activity of the wild reporter plasmid as compared with

NC, while there was no significant effect on the mutant plasmids (Figure 3C). These results revealed that miR-25-3p directly targets the 3′UTR of *Akt1* in vitro.

To directly test the validity of the putative target, we transfected miR-25-3p mimics and miR-25-3p inhibitors into growing C2C12 cells. We found that the overexpression of miR-25-3p repressed *Akt1* expression, as measured by qRT-PCR ($p < 0.01$) and Western blotting, whereas the knockdown of miR-25-3p derepressed it (Figure 3D,E). These results demonstrate that *Akt1* was a target of miR-25-3p.

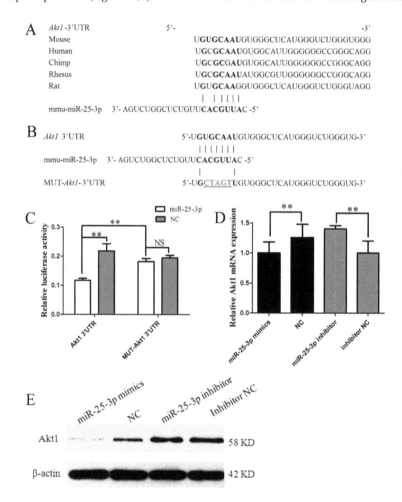

Figure 3. miR-25-3p directly targets the 3′UTR of *Akt1*. (**A**) The sequences of miR-25-3p target elements in the *Akt1* 3′UTR were relatively conserved in many mammals. These sequences were obtained from TargetScan. (**B**) Site-directed mutagenesis of the miR-25-3p target site in the *Akt1* 3′UTR; mutated bases shown in red. (**C**) Dual luciferase reporter assay. The *Akt1* 3′UTR/mutant plasmid was cotransfected with miR-25-3p mimics/NC, respectively, into growing C2C12 cells; dual luciferase activities were measured from cell lysates (24 h after transfection). miR-25-3p mimics/NC or inhibitors/NC were respectively transfected into growing C2C12 cells. After 48 h, *Akt1* expression was detected by qRT-PCR (**D**) and Western blotting (**E**). NC = negative control (miR-239b-5p of caenorhabditis elegans). β-actin served as the loading control. Data were presented as means ± SD ($n \geq 3$). ** $p < 0.01$; NS, not significant.

2.4. Identification and Characterization of the Mouse miR-25-3p Promoter

To further identify the promoter region and regulatory elements of mouse miR-25-3p, we used luciferase assays to analyze a series of deletions in the potential promoter region, as predicted by neural network promoter prediction (NNPP) online software (Figure 4A).The plasmids containing the various lengths of the miR-25-3p promoter were transiently transfected into growing BHK and C2C12 cells. Analyses of luciferase activity revealed that miR-25-3p-P9 (-119/$+144$) showed the greatest transcriptional activity, and the longer fragment showed lower transcriptional activity (Figure 4B), indicating that the region from -1870 to -119 contains one or more *cis*-acting elements that can repress miR-25-3p expression. The result demonstrates that this 263 bp-long sequence was the core promoter of mouse miR-25-3p.

2.5. The Transcription Factor AP-2α Binds to the Core Promoter of Mouse miR-25-3p

To further search the transcription factors that bind to the core promoter of mouse miR-25-3p, AliBaba 2.1 and Genomatix software programs were utilized to analyze the putative transcription factors. As shown in Figure S2, AP-2α was found to be able to bind to the core promoter of mouse miR-25-3p. To examine whether AP-2α influences the activity of the mouse miR-25-3p promoter, an AP-2α overexpression plasmid (pc-AP-2α) was generated and cotransfected with the miR-25-3p-P9 plasmid into growing C2C12 cells. Twenty-four hours after transfection, analyses of luciferase activity showed that pc-AP-2α significantly increased miR-25-3p promoter transcriptional activity (Figure 4C).

To determine the functional importance of the AP-2α binding site, we mutated the AP-2α binding site at -109 to -102, by using the wild-type miR-25-3p-P9 plasmid as the template. The mutant was constructed and transfected into growing C2C12 cells. As shown in Figure 4D, the luciferase activity of the mutant was significantly decreased as compared with the wild-type miR-25-3p-P9 construct. These results indicated that transcription factor AP-2α may induce transcriptional activity by directly binding to the core promoter of mouse miR-25-3p.

To further verify whether transcription factor AP-2α binds to the core promoter of mouse miR-25-3p, ChIP was performed in growing C2C12 cells. Chromatin was immunoprecipitated using the AP-2α antibody, and PCR amplification was performed, using the DNA fragment of the expected size as a template. The ChIP-Q-PCR assay showed that AP-2α interacted with the miR-25-3p promoter within the binding site (Figure 4E). These results confirmed that the transcription factor AP-2αis capable of binding to the AP-2α binding site in the mouse miR-25-3p promoter region, and induces miR-25-3p transcription.

2.6. AP-2α Regulates miR-25-3p and Akt1 Expression

Because *Akt1* was identified as a direct target of miR-25-3p, and the transcription factor AP-2α could upregulate miR-25-3p transcription, the effect of AP-2α on *Akt1* expression was further appraised by the overexpression or knockdown of AP-2α in growing C2C12 cells. As AP-2α mRNA expression was significantly decreased by doublestranded short interfering AP-2α RNA (si-AP-2α-1) and si-AP-2α-2, and the inhibitory effect of si-AP-2α-2 was greater than that of si-AP-2α-1 (Figure S3), si-AP-2α-2 was chosen for subsequent experiments. pc-AP-2α or si-AP-2α was transfected into growing C2C12 cells, respectively. Fourty-eight hours after transfection, RNA and protein were isolated. The overexpression of AP-2α significantly increased miR-25-3p expression, while the knockdown of AP-2α resulted in the significant suppression of miR-25-3p expression (Figure 5A). Conversely, the mRNA and protein expression of Akt1 were significantly suppressed by AP-2α overexpression, and were increased by si-AP-2α (Figure 5B–D). These results indicate that AP-2α activated mature miR-25 expression, and downregulated the expression of *Akt1*.

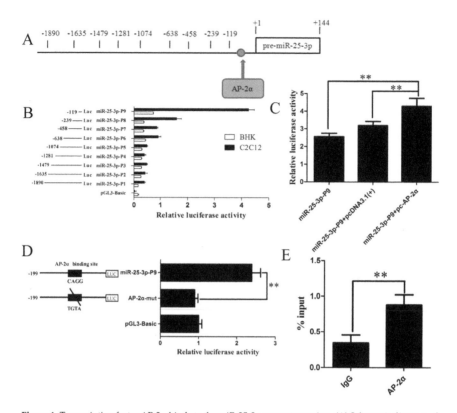

Figure 4. Transcription factor AP-2α binds to the miR-25-3p promoter region. (**A**) Schematic diagram of the AP-2α binding site (arrow, red dot) in the miR-25-3p promoter. The first nucleotide of pre-miR-25-3p was assigned as +1, and the other nucleotides were numbered relative to it. (**B**) A series of progressive deletion mutants were transfected into growing BHK and C2C12 cells, and the promoter activities were analyzed by dual luciferase activity assay. (**C**) miR-25-3p-P9 reporter constructs were cotransfected with pc-AP-2α into growing C2C12 cells. Dual luciferase activity was measured 24 h after transfection. Overexpression of AP-2α upregulated miR-25-3p promoter luciferase activity. pcDNA-3.1(+) was used as a control. (**D**) Site-directed mutagenesis of the AP-2α binding site (CAGG into TGTA) in the miR-25-3p promoter region resulted in the miR-25-3p-P9 luciferase activity being reduced. Data were expressed as the ratio of relative activity, normalized to pRL-TK, and presented as means ± SD ($n \geq 3$). (**E**) Binding of AP-2α to the miR-25-3p promoter region was analyzed by chromatin immunoprecipitation (ChIP). DNA isolated from immunoprecipitated materials was amplified using qRT-PCR. Normal mouse IgG was used as the negative control. Data were normalized by total chromatin (input) and presented as means ± SD ($n = 3$); ** $p < 0.01$.

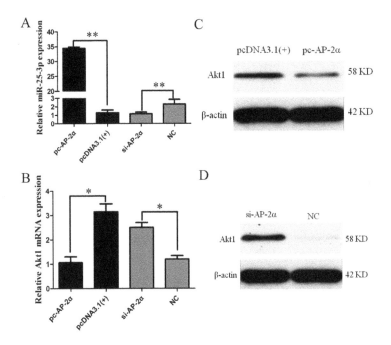

Figure 5. The effects of AP-2α on the expression of miR-25-3p and *Akt1*. The eukaryotic expression plasmid pc-AP-2α or si-AP-2α was transfected into growing C2C12 cells. After 48 h, the expression of miR-25-3p and *Akt1* was detected by qRT-PCR and Western blotting. (**A**) The expression of miR-25-3p was detected by qRT-PCR. (**B**) The mRNA expression of *Akt1* was detected by qRT-PCR. Data were presented as means ± SD ($n = 3$); * $p < 0.05$; ** $p < 0.01$. (**C**) The protein expression of Akt1 was detected by Western blotting after pc-AP-2α transfection. (**D**) The protein expression of Akt1 was detected by Western blotting after si-AP-2α transfection. β-actin served as the loading control.

3. Discussion

Increasing evidence shows that miR-25, a member of the miR-106b-25 cluster, is involved in many biological processes. For instance, miR-25 inhibits human gastric adenocarcinoma cell apoptosis [15], promotes glioblastoma cell proliferation and invasion [16], and regulates human ovarian cancer apoptosis [17]. The miR-106b-25 cluster regulates adult neural stem/progenitor cell proliferation, migration, and differentiation [18,19]. miR-25 plays an important role in heart disease [11,12] and diabetic kidney disease [13]. In addition, numerous studies have demonstrated that miRNAs are implicated in metabolism [20–23]. However, miR-25 has not been functionally related to metabolism until now.

In this study, miR-25 was identified as a novel regulator of metabolism. The gain-of-function and loss-of-function assays showed that miR-25-3p inhibited the expression of *PI3K* and reduced levels of triglyceride (TG), while levels of ATP and ROS were increased. PI3K has been implicated in insulin-regulated glucose metabolism [24], and PI3K signaling has a role in many cellular processes, such as metabolic control, immunity, and cardiovascular homeostasis [25–27]. It is well-known that triglycerides (TG) are a component of lipids, and participate in lipid metabolism. ATP is the most direct source of energy in an organism, and takes part in many metabolic processes. ROS, a class of single electron radicals of oxygen, comprise superoxide anions (O_2^-), hydrogen peroxide (H_2O_2), and hydroxyl radicals (·OH) [28], and are closely related to adipogenesis and myogenesis [28–31]. These data indicate that miR-25-3p indeed participates in metabolism in mice.

To further understand the molecular mechanism by which miR-25-3p regulates metabolism, we searched for potential target genes of miR-25-3p via TargetScan. Fortunately, the 3′UTR of *Akt1* contained a 7 nucleotides perfect match site complementary to the miR-25-3p seed region (Figure 3B). The serine-threonine kinase ATK, also known as protein kinase B (PKB), is an important effector for PI3K signaling as initiated by numerous growth factors and hormones [32]. *Akt* can control glucose uptake by regulating GLUT4 in cells, thereby reducing blood sugar and promoting glycogen synthesis [32–34]. *Akt* usually promotes glycogen synthase kinase-3 alpha (GSK3α) phosphorylation and inhibits its activity [35], and then activates glycogen synthesis [36]. A previous study has demonstrated that overexpression of miR-25-3p downregulates *Akt* expression and inactivates Akt phosphorylation in the tongue squamous cell carcinoma cell line Tca8113 [37]. Consequently, we deduced that the role of miR-25-3p in metabolism may arise from its inhibition of *Akt1*. First, the dual luciferase reporter assay demonstrated that *Akt1* was a direct target of miR-25-3p, shown by the steady decrease luciferase activity of the pmirGLO-Akt1-wt vector; but not the mutant form (Figure 3C). Meanwhile, qRT-PCR and Western blotting results showed that the expression of *Akt1* was inhibited by the miR-25-3p mimics, and that this inhibition was reversed by the miR-25-3p inhibitors (Figure 3D,E). These results suggested that the effect of miR-25-3p in metabolism was due, at least in part, to the suppression of *Akt1*.

An increasing number of studies have shown that transcription factors are capable of binding to miRNA promoter elements and modulating miRNA transcription [38–40]. Therefore, we analyzed the transcriptional mechanism of miR-25-3p in this study. Nine fragments of 5′-flanking sequences of mouse miR-25-3p were isolated. Subsequently, a series of experiments, including dual luciferase, site-directed mutagenesis, and ChIP assays, confirmed that AP-2α bound to the miR-25-3p promoter region and promoted its transcription activity (Figure 4). Moreover, qRT-PCR and Western blotting results showed that overexpression of AP-2α resulted in the upregulation of miR-25-3p and downregulation of *Akt1*, and that the knockdown of AP-2α reversed these results (Figure 5).

The AP-2 family of transcription factors consists of five members, in humans and mice: AP-2α, AP-2β, AP-2γ, AP-2δ, and AP-2ε; which play important roles in several cellular processes, such as apoptosis, migration, and differentiation [41,42]. AP-2α was first identified by its ability to bind to the enhancer regions of SV40 and human metallothionein IIA [43]. Subsequently, numerous studies have demonstrated that AP-2α can regulate gene expression. For instance, AP-2α binding to the *C/EBPα* promoter results in decreased *C/EBPα* expression [44], and AP-2α can bind to the *TACE* promoter and decrease its expression in dendritic cells [45]. Furthermore, Qiao et al. [46] reported that there was an AP-2α binding site in the *DEK* core promoter, and overexpression of AP-2α upregulated *DEK* expression. In this study, we identified that AP-2α binds to the miR-25-3p promoter region and promotes its transcription activity.

In conclusion, our results demonstrate that miR-25-3p acts as a positive regulator of the metabolism of growing C2C12 cells, by affecting *Akt1* gene expression through directly binding to its 3′UTR. Moreover, the transcription factor AP-2α is able to bind to the core promoter of mouse miR-25-3p, activating mature miR-25 expression and downregulating the expression of *Akt1* (Figure 6).

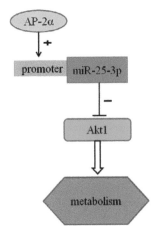

Figure 6. Representation of the proposed mechanism. miR-25-3p is regulated by transcription factor AP-2α, and contributes to C2C12 metabolism by targeting *Akt1*. The arrow-head and "+" represent activation while the blunt-head and "−" represent suppression.

4. Materials and Methods

4.1. miRNA, Small RNA Oligonucleotide Synthesis, and Plasmid Construction

The miR-25-3p oligonucleotides (miR-25-3p mimics, NC, miR-25-3p inhibitors, and inhibitor-NC) and double-stranded short interfering RNAs (siRNAs) targeting AP-2α were designed and synthesized by RiboBio (Guangzhou, China).The oligonucleotides are listed in Table S1.

To construct the AP-2α overexpression vector pc-AP-2α, the AP-2α coding sequence (1314 bp) was amplified from mouse C2C12 cells cDNA using the following primers: forward: 5'-CCC AAGCTTGCCACCATGCTTTGGAAACTGACGGA-3'; reverse: 5'-CCGCTCGAGTCACTTTCTGTG TTTCTCTT-3'. The PCR product was subcloned into the *Hind*III/*Xho*I sites of the pcDNA3.1(+) vector (Invitrogen, Carlsbad, CA, USA).

The potential target site of miR-25-3p, localized in the 3'UTR of *Akt1* mRNA, was predicted by TargetScan (Available online: http://www.targetscan.org/) [47]. The *Akt1* 3'UTR was amplified from C2C12 cell cDNA and inserted into the *Pme*I/*Xho*I sites of the pmirGLO vector (Promega, Madison, WI, USA). Point mutations in the seed region of the predicted miR-25-3p sites within the 3'UTR of *Akt1* were generated using overlap-extension PCR [48]. The corresponding primers are listed in Table S2.

The potential promoter regions of miR-25-3p was predicted by using the neural network promoter prediction (NNPP) software (Available online: http://www.fruitfly.org/seq_tools/promoter.html) [49]. Nine miR-25-3p promoter deletion fragments were amplified from the mouse genome via PCR with the primers listed in Table S3.The nine purified PCR products were ligated into the *Kpn*I/*Hind*III sites of the pGL3-Basic vector (Promega). AliBaba2.1 (Available online: http://www.gene-regulation.com/) [50] and MatInspector (Available online: http://www.genomatix.de/online_help/help_matinspector/ matinspector_help.html) [49] were used to predict the potential transcription factor binding sites. The AP-2α transcription factor binding sites of the miR-25-3p promoter region were also mutated by overlap-extension PCR. The primers are provided in Table S3.

4.2. Cell Culture and Luciferase Reporter Assays

C2C12 (mouse muscle myoblast) and BHK (baby hamster kidney) cells were cultured in DMEM (Gibco, Gaithersburg, MD, USA) containing 10% fetal bovine serum (FBS) (Gibco) at 5% CO_2 and 37 °C.

For luciferase reporter assays, growing C2C12 or BHK cells were seeded in 48-well plates. After 12–16 h, the plated cells were transfected with a recombinant plasmid using Lipofectamine 2000 (Invitrogen). To verify the miR-25-3p targeting *Akt1* 3′UTR, 1 μL miR-25-3p mimics/NC was cotransfected with 0.1 μg *Akt1* 3′UTR/mutant plasmid into C2C12 cells. For the miR-25-3p promoter luciferase reporter assay, 0.4 μg pGL3-Basic or recombinant plasmids and 20 ng pRL-TK vector were transfected. For cotransfection luciferase assays, each well contained 0.2 μg pGL3-(Basic, miR-25-3p-P9 and AP-2α-mut), 20 ng pRL-TK, and 0.2 μg pc-AP-2α. Empty pcDNA-3.1(+) cotransfected with pGL3-(Basic, miR-25-3p-P9 and AP-2α-mut) was used as the control. After 24 h of incubation, luciferase activity was measured using a PerkinElmer 2030 Multilabel Reader (PerkinElmer, Norwalk, CT, USA).

4.3. Triglyceride Content, ATP, and Reactive Oxygen Species (ROS) Assays

For detecting the concentrations of triglyceride (TG), ATP, and ROS, growing C2C12 cells were seeded in 24-well plates the day before transfection. miR-25-3p mimic, NC, miR-25-3p inhibitor, and inhibitor-NC were transfected into confluent (~80%) cells, respectively, at a concentration of 12 nM with Lipofectamine 2000 (Invitrogen). After 24–48 h, the concentrations of TG and ATP in the lysates of cells were measured with commercial kits (Applygen (Beijing, China) and Beyotime (Shanghai, China), respectively) following the manufacturer's instructions, and normalized to the protein content (μmol/mg protein) using the BCA assay kit (Thermo Scientific, Waltham, MA, USA). ROS were measured using the reactive oxygen species assay kit (Beyotime) following the manufacturer's protocol.

4.4. Chromatin Immunoprecipitation (ChIP)

ChIP assays were performed to assess the binding of endogenous AP-2α to the miR-25-3p promoter in C2C12 cells using the EZ-ChIP™ Kit (Millipore, Boston, MA, USA), following a previously described method [49]. Precleared chromatin was incubated with the AP-2α antibody (Santa Cruz Biotechnology, Dallas, TX, USA) or normal mouse IgG (Millipore) antibodies (control) overnight at 4 °C. Purified DNA from the samples and the input controls were analyzed for the presence of miR-25-3p promoter sequences containing putative AP-2α response elements using qPCR. The primers used here are listed in Table S4.

4.5. RNA Isolation and qRT-PCR

For quantifying the mRNA expression of genes, growing C2C12 cells were seeded in 6-well plates. miR-25-3p mimic, NC, miR-25-3p inhibitor, inhibitor-NC, si-AP-2α, and NC were transfected into confluent (~80%) cells, respectively, at a concentration of 50 nM with Lipofectamine 2000 (Invitrogen). After 48 h, total RNA was isolated using a HP Total RNA Kit (Omega, Norcross, GA, USA) according to the manufacturer's protocol. The cDNA was synthesized using a PrimeScript™RT reagent Kit with gDNA Eraser (Takara, Osaka, Japan) according to the manufacturer's protocol. The qRT-PCR was performed in triplicate with iQSYBR green Supermix (Bio-Rad, Hercules, CA, USA) in a LightCycler 480 Realtime PCR machine (Roche, Basel, Switzerland). The mRNA levels of target genes were reported relative to those of the house keeping gene β-actin by using the $2^{-\Delta\Delta Ct}$ method. The qRT-PCR primers are listed in Table S5.

4.6. Protein Isolation and Western Blotting

For detecting the protein expression of PI3K and Akt1, growing C2C12 cells were seeded in 6-well plates. miR-25-3p mimic, NC, miR-25-3p inhibitor, inhibitor-NC, si-AP-2α, and NC were transfected into confluent (~80%) cells, respectively, at a concentration of 50 nM with Lipofectamine 2000 (Invitrogen). After 48 h, total protein was isolated using RIPA Lysis Buffer (Beyotime). The cells were washed briefly with cold phosphate-buffered saline (PBS), 150 μL RIPA Lysis Buffer (containing 1 mM PMSF) was added, incubated for 1 min at room temperature, and then centrifuged at 12,000× *g* for 5 min. The supernatant extract was used for Western blot analysis.

Western blot analysis was performed to analyze the expression levels of Akt1 (Affinity Biosciences, Cincinnati, OH, USA) andPI3K (Abclonal, Wuhan, China) according to the methods of Huang et al. [47]. β-actin (Santa Cruz Biotechnology) served as the loading control.

4.7. Statistical Analysis

All the results are presented as the means ± SD. Student's *t*-test was used for statistical comparisons. A *p* value of < 0.05 was considered to be statistically significant. ** *p* < 0.01; * *p* < 0.05; NS, not significant.

Supplementary Materials: The following are available online at www.mdpi.com/1422-0067/19/3/773/s1.

Acknowledgments: This research wassupported by the China Postdoctoral Science Foundation (2017M610465), the Open Project of Key Laboratory of Animal Embryo Engineering and Molecular Breeding of Hubei Province (KLAEMB201602),the Postdoctoral Innovation Post of Hubei Province (2016), the National Natural Science Foundationof China (31472075, 31402051 and 31501932), and Natural Science Foundation of Hubei Province key projects of technical innovation (2016ABA117).

Author Contributions: Mingxin Chen and Siwen Jiang conceived and supervised the study; Feng Zhang and Kun Chen designed experiments; Tingting Kang and Qianhui Zeng performed experiments; Hu Tao and Qi Xiong analysed data; Feng Zhang wrote the manuscript; Yang Liu made manuscript revisions.

Conflicts of Interest: The authors declare no conflict of interest.

References

1. Bartel, D.P. MicroRNAs: Genomics, biogenesis, mechanism, and function. *Cell* **2004**, *116*, 281–297. [CrossRef]
2. Malan-Muller, S.; Hemmings, S.M.; Seedat, S. Big effects of small RNAs: A review of microRNAs in anxiety. *Mol. Neurobiol.* **2013**, *47*, 726–739. [CrossRef] [PubMed]
3. Carthew, R.; Sontheimer, E. Origins and Mechanisms of miRNAs and siRNAs. *Cell* **2009**, *136*, 642–655. [CrossRef] [PubMed]
4. Ning, B.; Gao, L.; Liu, R.; Liu, Y.; Zhang, N.; Chen, Z. microRNAs in spinal cord injury: Potential roles and therapeutic implications. *Int. J. Biol. Sci.* **2014**, *10*, 997–1006. [CrossRef] [PubMed]
5. Deshmukh, A.; Murgia, M.; Nagaraj, N.; Treebak, J.; Cox, J.; Mann, M. Deep proteomics of mouse skeletal muscle enables quantitation of protein isoforms, metabolic pathways, and transcription factors. *Mol. Cell Proteom.* **2015**, *14*, 841–853. [CrossRef] [PubMed]
6. Li, M.; Song, Q.; Li, H.; Lou, Y.; Wang, L. Correction: Circulating miR-25-3p and miR-451a May Be Potential Biomarkers for the Diagnosis of Papillary Thyroid Carcinoma. *PLoS ONE* **2015**, *10*, e0135549. [CrossRef] [PubMed]
7. Nishida, N.; Nagahara, M.; Sato, T.; Mimori, K.; Sudo, T.; Tanaka, F.; Shibata, K.; Ishii, H.; Sugihara, K.; Doki, Y.; et al. Microarray analysis of colorectal cancer stromal tissue reveals upregulation of two oncogenic miRNA clusters. *Clin. Cancer Res.* **2012**, *18*, 3054–3070. [CrossRef] [PubMed]
8. Razumilava, N.; Bronk, S.F.; Smoot, R.L.; Fingas, C.D.; Werneburg, N.W.; Roberts, L.R.; Mott, J.L. miR-25 targets TNF-related apoptosis inducing ligand (TRAIL) death receptor-4 and promotes apoptosis resistance in cholangiocarcinoma. *Hepatology* **2012**, *55*, 465–475. [CrossRef] [PubMed]
9. Wang, X.; Meng, X.; Li, H.; Liu, W.; Shen, S.; Gao, Z. MicroRNA-25 expression level is an independent prognostic factor in epithelial ovarian cancer. *Clin. Transl. Oncol.* **2014**, *16*, 954–958. [CrossRef] [PubMed]
10. Zhao, H.; Wang, Y.; Yang, L.; Jiang, R.; Li, W. MiR-25 promotes gastric cancer cells growth and motility by targeting RECK. *Mol. Cell. Biochem.* **2014**, *385*, 207–213. [CrossRef] [PubMed]
11. Wahlquist, C.; Jeong, D.; Rojas-Muñoz, A.; Kho, C.; Lee, A.; Mitsuyama, S.; van Mil, A.; Park, W.; Sluijter, J.; Doevendans, P.; et al. Inhibition of miR-25 improves cardiac contractility in the failing heart. *Nature* **2014**, *508*, 531–535. [CrossRef] [PubMed]
12. Pan, L.; Huang, B.; Ma, X.; Wang, S.; Feng, J.; Lv, F.; Liu, Y.; Liu, Y.; Li, C.; Liang, D.; et al. MiR-25 protects cardiomyocytes against oxidative damage by targeting the mitochondrial calcium uniporter. *Int. J. Mol. Sci.* **2015**, *16*, 5420–5433. [CrossRef] [PubMed]

13. Liu, Y.; Li, H.; Liu, J.; Han, P.; Li, X.; Bai, H.; Zhang, C.; Sun, X.; Teng, Y.; Zhang, Y.; et al. Variations in MicroRNA-25 Expression Influence the Severity of Diabetic Kidney Disease. *J. Am. Soc. Nephrol.* **2017**, *28*, 3627–3638. [CrossRef] [PubMed]

14. Sievers, F.; Higgins, D. Clustal Omega, accurate alignment of very large numbers of sequences. *Methods Mol. Biol.* **2014**, *1079*, 105–116. [PubMed]

15. Zhang, Y.; Peng, Z.; Zhao, Y.; Chen, L. microRNA-25 Inhibits Cell Apoptosis of Human Gastric Adenocarcinoma Cell Line AGS via Regulating CCNE1 and MYC. *Med. Sci. Monit.* **2016**, *22*, 1415–1420. [CrossRef] [PubMed]

16. Peng, G.; Yuan, X.; Yuan, J.; Liu, Q.; Dai, M.; Shen, C.; Ma, J.; Liao, Y.; Jiang, W. miR-25 promotes glioblastoma cell proliferation and invasion by directly targeting NEFL. *Mol. Cell. Biochem.* **2015**, *409*, 103–111. [CrossRef] [PubMed]

17. Zhang, H.; Zuo, Z.; Lu, X.; Wang, L.; Wang, H.; Zhu, Z. miR-25 regulates apoptosis by targeting Bim in human ovarian cancer. *Oncol. Rep.* **2012**, *27*, 594–598. [PubMed]

18. Brett, J.O.; Renault, V.M.; Rafalski, V.A.; Webb, A.E.; Brunet, A. The microRNA cluster miR-106b~25 regulates adult neural stem/progenitor cell proliferation and neuronal differentiation. *Aging* **2011**, *3*, 108–124. [CrossRef] [PubMed]

19. Yu, Y.; Lu, X.; Ding, F. microRNA regulatory mechanism by which PLLA aligned nanofibers influence PC12 cell differentiation. *J. Neural Eng.* **2015**, *12*, 046010. [CrossRef] [PubMed]

20. Rottiers, V.; Näär, A. MicroRNAs in metabolism and metabolic disorders. *Nat. Rev. Mol. Cell Biol.* **2012**, *13*, 239–250. [CrossRef] [PubMed]

21. Guo, N.; Zhang, J.; Wu, J.; Xu, Y. Isoflurane promotes glucose metabolism through up-regulation of miR-21 and suppresses mitochondrial oxidative phosphorylation in ovarian cancer cells. *Biosci. Rep.* **2017**, *37*. [CrossRef] [PubMed]

22. Ling, L.; Kokoza, V.; Zhang, C.; Aksoy, E.; Raikhel, A. MicroRNA-277 targets insulin-like peptides 7 and 8 to control lipid metabolism and reproduction in *Aedes aegypti* mosquitoes. *Proc. Natl. Acad. Sci. USA* **2017**, *114*, E8017–E8024. [CrossRef] [PubMed]

23. Gao, P.; Tchernyshyov, I.; Chang, T.; Lee, Y.; Kita, K.; Ochi, T.; Zeller, K.; De Marzo, A.; van Eyk, J.; Mendell, J.; et al. c-Myc suppression of miR-23a/b enhances mitochondrial glutaminase expression and glutamine metabolism. *Nature* **2009**, *458*, 762–765. [CrossRef] [PubMed]

24. Pessin, J.E.; Saltiel, A.R. Signaling pathways in insulin action: Molecular targets of insulin resistance. *J. Clin. Investig.* **2000**, *106*, 165–169. [CrossRef] [PubMed]

25. Vanhaesebroeck, B.; Guillermet-Guibert, J.; Graupera, M.; Bilanges, B. The emerging mechanisms of isoform-specific PI3K signalling. *Nat. Rev. Mol. Cell Biol.* **2010**, *11*, 329–341. [CrossRef] [PubMed]

26. Courtnay, R.; Ngo, D.; Malik, N.; Ververis, K.; Tortorella, S.; Karagiannis, T. Cancer metabolism and the Warburg effect: The role of HIF-1 and PI3K. *Mol. Biol. Rep.* **2015**, *42*, 841–851. [CrossRef] [PubMed]

27. Knight, Z.; Shokat, K. Chemically targeting the PI3K family. *Biochem. Soc. Trans.* **2007**, *35 Pt 2*, 245–249. [CrossRef] [PubMed]

28. Carrière, A.; Fernandez, Y.; Rigoulet, M.; Pénicaud, L.; Casteilla, L. Inhibition of preadipocyte proliferation by mitochondrial reactive oxygen species. *FEBS Lett.* **2003**, *550*, 163–167. [CrossRef]

29. Carrière, A.; Carmona, M.; Fernandez, Y.; Rigoulet, M.; Wenger, R.; Pénicaud, L.; Casteilla, L. Mitochondrial reactive oxygen species control the transcription factor CHOP-10/GADD153 and adipocyte differentiation: A mechanism for hypoxia-dependent effect. *J. Biol. Chem.* **2004**, *279*, 40462–40469. [CrossRef] [PubMed]

30. Ding, Y.; Choi, K.; Kim, J.; Han, X.; Piao, Y.; Jeong, J.; Choe, W.; Kang, I.; Ha, J.; Forman, H.; et al. Endogenous hydrogen peroxide regulates glutathione redox via nuclear factor erythroid 2-related factor 2 downstream of phosphatidylinositol 3-kinase during muscle differentiation. *Am. J. Pathol.* **2008**, *172*, 1529–1541. [CrossRef] [PubMed]

31. Won, H.; Lim, S.; Jang, M.; Kim, Y.; Rashid, M.; Jyothi, K.; Dashdorj, A.; Kang, I.; Ha, J.; Kim, S. Peroxiredoxin-2 upregulated by NF-κB attenuates oxidative stress during the differentiation of muscle-derived C2C12 cells. *Antioxid. Redox Signal.* **2012**, *16*, 245–261. [CrossRef] [PubMed]

32. Cho, H.; Thorvaldsen, J.L.; Chu, Q.; Feng, F.; Birnbaum, M.J. Akt1/PKBα is required for normal growth but dispensable for maintenance of glucose homeostasis in mice. *J. Biol. Chem.* **2001**, *276*, 38349–38352. [CrossRef] [PubMed]

33. Gonzalez, E.; McGraw, T.E. The Akt kinases Isoform specificity in metabolism and cancer. *Cell Cycle* **2009**, *8*, 2502–2508. [CrossRef] [PubMed]

34. Chen, W.S.; Xu, P.-Z.; Gottlob, K.; Chen, M.-L.; Sokol, K.; Shiyanova, T.; Roninson, I.; Weng, W.; Suzuki, R.; Tobe, K.; et al. Growth retardation and increased apoptosis in mice with homozygous disruption of the akt1 gene. *Genes Dev.* **2001**, *15*, 2203–2208. [CrossRef] [PubMed]

35. Beurel, E.; Grieco, S.; Jope, R. Glycogen synthase kinase-3 (GSK3): Regulation, actions, and diseases. *Pharmacol. Ther.* **2015**, *148*, 114–131. [CrossRef] [PubMed]

36. Agius, L. Role of glycogen phosphorylase in liver glycogen metabolism. *Mol. Asp. Med.* **2015**, *46*, 34–45. [CrossRef] [PubMed]

37. Xu, J.; Yang, L.; Ma, C.; Huang, Y.; Zhu, G.; Chen, Q. MiR-25-3p attenuates the proliferation of tongue squamous cell carcinoma cell line Tca8113. *Asian Pac. J. Trop. Med.* **2013**, *6*, 743–747. [CrossRef]

38. Zheng, H.; Dong, X.; Liu, N.; Xia, W.; Zhou, L.; Chen, X.; Yang, Z.; Chen, X. Regulation and mechanism of mouse miR-130a/b in metabolism-related inflammation. *Int. J. Biochem. Cell Biol.* **2016**, *74*, 72–83. [CrossRef] [PubMed]

39. Zhu, B.; Ye, J.; Ashraf, U.; Li, Y.; Chen, H.; Song, Y.; Cao, S. Transcriptional regulation of miR-15b by c-Rel and CREB in Japanese encephalitis virus infection. *Sci. Rep.* **2016**, *6*, 22581. [CrossRef] [PubMed]

40. Bueno, M.; Gómez de Cedrón, M.; Laresgoiti, U.; Fernández-Piqueras, J.; Zubiaga, A.; Malumbres, M. Multiple E2F-induced microRNAs prevent replicative stress in response to mitogenic signaling. *Mol. Cell. Biol.* **2010**, *30*, 2983–2995. [CrossRef] [PubMed]

41. Wenke, A.; Bosserhoff, A. Roles of AP-2 transcription factors in the regulation of cartilage and skeletal development. *FEBS J.* **2010**, *277*, 894–902. [CrossRef] [PubMed]

42. Eckert, D.; Buhl, S.; Weber, S.; Jäger, R.; Schorle, H. The AP-2 family of transcription factors. *Genome Biol.* **2005**, *6*, 246. [CrossRef] [PubMed]

43. Mitchell, P.; Wang, C.; Tjian, R. Positive and negative regulation of transcription in vitro: Enhancer-binding protein AP-2 is inhibited by SV40 T antigen. *Cell* **1987**, *50*, 847–861. [CrossRef]

44. Holt, E.; Lane, M. Downregulation of repressive CUP/AP-2 isoforms during adipocyte differentiation. *Biochem. Biophys. Res. Commun.* **2001**, *288*, 752–756. [CrossRef] [PubMed]

45. Ge, L.; Vujanovic, N. Soluble TNF Regulates TACE via AP-2α Transcription Factor in Mouse Dendritic Cells. *J. Immunol.* **2017**, *198*, 417–427. [CrossRef] [PubMed]

46. Qiao, M.; Li, C.; Zhang, A.; Hou, L.; Yang, J.; Hu, H. Regulation of DEK expression by AP-2α and methylation level of DEK promoter in hepatocellular carcinoma. *Oncol. Rep.* **2016**, *36*, 2382–2390. [CrossRef] [PubMed]

47. Huang, C.; Geng, J.; Wei, X.; Zhang, R.; Jiang, S. MiR-144-3p regulates osteogenic differentiation and proliferation of murine mesenchymal stem cells by specifically targeting Smad4. *FEBS Lett.* **2016**, *590*, 795–807. [CrossRef] [PubMed]

48. Ho, S.N.; Hunt, H.D.; Horton, R.M.; Pullen, J.K.; Pease, L.R. Site-directed mutagenesis by overlap extension using the polymerase chain reaction. *Gene* **1989**, *77*, 51–59. [CrossRef]

49. Deng, B.; Zhang, F.; Chen, K.; Wen, J.; Huang, H.; Liu, W.; Ye, S.; Wang, L.; Yang, Y.; Gong, P.; et al. MyoD promotes porcine PPARgamma gene expression through an E-box and a MyoD-binding site in the PPARgamma promoter region. *Cell Tissue Res.* **2016**, *365*, 381–391. [CrossRef] [PubMed]

50. Wei, X.; Cheng, X.; Peng, Y.; Zheng, R.; Chai, J.; Jiang, S. STAT5a promotes the transcription of mature mmu-miR-135a in 3T3-L1 cells by binding to both miR-135a-1 and miR-135a-2 promoter elements. *Int. J. Biochem. Cell Biol.* **2016**, *77 Pt A*, 109–119. [CrossRef] [PubMed]

 © 2018 by the authors. Licensee MDPI, Basel, Switzerland. This article is an open access article distributed under the terms and conditions of the Creative Commons Attribution (CC BY) license (http://creativecommons.org/licenses/by/4.0/).

International Journal of
Molecular Sciences

Review

Genetic and Epigenetic Control of *CDKN1C* Expression: Importance in Cell Commitment and Differentiation, Tissue Homeostasis and Human Diseases

Emanuela Stampone [1,†], Ilaria Caldarelli [1,†], Alberto Zullo [2,3,†], Debora Bencivenga [1], Francesco Paolo Mancini [2], Fulvio Della Ragione [1,*] and Adriana Borriello [1,*]

[1]   Department of Precision Medicine, University of Campania "Luigi Vanvitelli", 80138 Naples, Italy; ema.stampone@gmail.com (E.S.); ilariacaldarelli@libero.it (I.C.); deborabencivenga@yahoo.it (D.B.)
[2]   Department of Sciences and Technologies, University of Sannio, 82100 Benevento, Italy; albzullo@unisannio.it (A.Z.); mancini@unisannio.it (F.P.M.)
[3]   CEINGE Biotecnologie Avanzate S. C. A R. L., 80145 Naples, Italy
*   Correspondence: fulvio.dellaragione@unicampania.it (F.D.R.); adriana.borriello@unicampania.it (A.B.); Tel.: +39-081-566-5812 (F.D.R.); +39-081-566-7554 (A.B.)
†   These authors equally contributed to the manuscript.

Received: 14 March 2018; Accepted: 31 March 2018; Published: 2 April 2018

Abstract: The *CDKN1C* gene encodes the p57[Kip2] protein which has been identified as the third member of the CIP/Kip family, also including p27[Kip1] and p21[Cip1]. In analogy with these proteins, p57[Kip2] is able to bind tightly and inhibit cyclin/cyclin-dependent kinase complexes and, in turn, modulate cell division cycle progression. For a long time, the main function of p57[Kip2] has been associated only to correct embryogenesis, since *CDKN1C*-ablated mice are not vital. Accordingly, it has been demonstrated that *CDKN1C* alterations cause three human hereditary syndromes, characterized by altered growth rate. Subsequently, the p57[Kip2] role in several cell phenotypes has been clearly assessed as well as its down-regulation in human cancers. *CDKN1C* lies in a genetic locus, 11p15.5, characterized by a remarkable regional imprinting that results in the transcription of only the maternal allele. The control of *CDKN1C* transcription is also linked to additional mechanisms, including DNA methylation and specific histone methylation/acetylation. Finally, long non-coding RNAs and miRNAs appear to play important roles in controlling p57[Kip2] levels. This review mostly represents an appraisal of the available data regarding the control of *CDKN1C* gene expression. In addition, the structure and function of p57[Kip2] protein are briefly described and correlated to human physiology and diseases.

Keywords: p57[Kip2]; CDKN1C; epigenetics; disease; cell differentiation

---

1. Introduction

A well-orchestrated sequence of events allows the transition between the various phases of cell division cycle and the precise control of a perfect execution and accomplishment of each phase. Central actors in this process are heterodimers formed by cyclin/cyclin-dependent kinase complexes (CDK) whose activity is strictly regulated by a number of factors, including their amount, localization, and post-synthetic modifications (mainly phosphorylations). A further important modulation is due to the interaction with additional inhibitory proteins resulting in the formation of heterotrimers, generally lacking the kinase activity. These proteins are defined CDK inhibitors (CKI) or, alternatively, CDK regulator. One family of CKI, established on the basis of sequence homology and specificity of action, is named CIP/Kip and includes three members, namely p21[Cip1/WAF1], p27[Kip1] and p57[Kip2]. Due to their broad inhibitory effect on cyclin-CDK complexes, CIP/Kip members have been mainly considered as antiproliferative proteins and their encoding genes as potential tumor suppressor genes. However, strong emerging pieces of evidence have demonstrated that the activities of CIP/Kip members are well beyond that of modulators of cell division [1]. Indeed, in function of their localization and interactors, CIP/Kip members might regulate a plethora of events including cell differentiation, cell movement, apoptosis, autophagy and all the major steps of carcinogenesis [1]. In addition, the tissue-specific functions of p57^{Kip2} cannot be substituted by other CIP/Kip family members, suggesting that each of them has peculiar roles in cell physiology.

In this brief review, we provide an appraisal of the published data on the p57^{Kip2} protein, that represents the least studied member within the CIP/Kip family. Our attention will be mainly focused on the regulation of *CDKN1C* (the p57^{Kip2} encoding gene) expression and its relevance in human diseases, including overgrowth and undergrowth syndromes.

1.1. p57^{Kip2} Protein

Human *CDKN1C* encodes a 316-amino-acid protein that migrates at 57 kDa by SDS-PAGE electrophoresis, hence the name p57^{Kip2}. p57^{Kip2} is the last identified member of the CIP/Kip family of the cyclin-dependent kinase inhibitors, also including p21$^{Cip1/WAF1}$ and p27^{Kip1} [2,3]. The CIP/Kip proteins share structural similarity mainly related to the common activity of cell cycle regulators. The most characterized Cyclin/CDK inhibitory activity relies on two common features: a CDK binding/inhibitory domain (KID) located at the amino-term and the nuclear localization signal (NLS) at the carboxy terminal of the protein [4]. The KID includes three short peculiar motifs: a cyclin-binding domain, a CDK-binding site and a 3$_{10}$ helix that, due to a specific pair of amino acids (phenylalanine-tyrosine), is able to mimic the adenine component of ATP, therefore blocking the catalytic site of CDKs [5]. Similarly to other CIP/Kip members, KID is necessary and sufficient to bind and inhibit CDK activity. Specifically, it has been reported that p57^{Kip2} inhibits the kinase activity of cyclin-CDK complexes in vitro, including cyclin E (A)/CDK2 and cyclin D1,2/CDK4 [2,3,6]. Besides CDKs, several other proteins have been reported to interact with the p57^{Kip2} amino-terminal domain. Particularly, known interactors of p57^{Kip2} at its N-terminal domain are the basic helix-loop-helix transcription factors, such as MyoD, Mash1, NeuroD, and Nex/Math2 [6–8]. Furthermore, p57^{Kip2} interacts, both in vivo and in vitro, with the transcription factor B-Myb, which plays an important role during early embryonic development. Particularly, B-Myb competes with cyclin A2 for binding to p57^{Kip2}, thus determining the release of active cyclin/CDK2 [9].

The carboxy-terminal region of p57Kip2 contains a QT box domain, rich in glutamine and threonine residues, which is homologous to the corresponding QT domain of p27^{Kip1} and can be responsible for further interactions of the protein. It has been reported that the QT box directly binds to c-Jun NH2-terminal kinase/stress-activated protein kinase, determining its inhibition [10,11]. In the QT domain, a consensus sequence for a putative nuclear localization signal (NLS) has been identified [2,3]. Proceeding towards the C-terminal, p57^{Kip2} presents, in homology with p21$^{Cip1/WAF1}$, a binding domain for the proliferating cell nuclear antigen (PCNA), a cofactor of DNA polymerase delta. Thus, it is able to bind and inhibit PCNA, even though with much lower affinity than p21$^{Cip1/WAF1}$ [12].

Whereas the p57^{Kip2} amino- and carboxy-terminal domains are similar in sequence in mammals, the internal domain, consisting of proline/alanine-rich motifs, results as a peculiarity of human p57^{Kip2}: the PAPA region, a sort of hinge between the N- and the C-end of the protein. It is absent in p21$^{Cip1/WAF1}$ and p27^{Kip1} and is responsible for the difference between the sequence-derived molecular weight and the SDS-PAGE observed molecular weight. The PAPA region is scarcely conserved in mouse and rat, where it is substituted by a proline-rich region followed by an acidic repeat in which glutamic or aspartic acid occur every four amino acids [3]. However, the functional meaning of the PAPA region is still unknown, even though some authors retain that it is important for protein interactions.

A peculiar characteristic of p57^{Kip2} protein is a limited degree of stable secondary and tertiary structures under physiological conditions. Specifically, the protein belongs to the so-called intrinsically unstructured proteins (IUPs), which can adopt different conformations upon binding to distinct and specific interactors. This property is shared with its siblings p21$^{Cip1/WAF1}$ and p27^{Kip1}, and with numerous proteins involved in the control of cell proliferation. As a matter of fact, more than 70% of human cancer-associated proteins are IUPs. This conformational flexibility allows a considerable versatility in terms of biomolecular interactors, expanding the range of their functions and, in turn, their involvement in numerous cellular processes [1]. On the other hand, post-synthetic changes of an IUP, like (but not only) phosphorylations, might play a fundamental role in guiding the protein towards specific interactions and specific functions. So far, only few phosphorylation sites have been identified in human p57^{Kip2} protein, such as threonine 310 (T310). Particularly, T310 phosphorylation has been suggested as being important for human protein degradation [1] and level control. Specifically, in analogy with p27^{Kip1} threonine 187 phosphorylation [13], the phosphorylation on T310 determines a phosphodegron which functions as a recognition site for the substrate recognition subunit (S-phase kinase-associated protein 2, Skp2) of the E3 ubiquitin ligase SCF complex (Skp1/Cul1/F-box protein). The Skp2-SCF complex guides target proteins to proteasomal degradation in a cell cycle-dependent manner (from late G1 to early M phase) and its activity appears strongly deregulated in human cancers [12]. Furthermore, besides the Skp2-SCF complex, the activity of the SCF-FBL12 complex, whose substrate recognition subunit (FBL12) is different from Skp2, has been reported to be involved in TGFβ1-induced p57^{Kip2} ubiquitin-dependent proteasomal degradation in osteoblast cells [14].

1.2. p57^{Kip2} in Embryonic and Adult Tissues

p57^{Kip2}, unlike the other two CKIs, shows a fine-tuned temporal and spatial expression from embryogenesis up to the adult life. p27^{Kip1} and p57^{Kip2} are widely expressed during embryogenesis. p27^{Kip1} is more abundant in ovary, testis, thymus, spleen and developing retina, instead, p57^{Kip2} is mostly localized in cartilage, skeletal muscle, palate, pancreas, and intestine. Interestingly, the CKIs show a complementary expression pattern in several embryonic areas. Indeed, in the adrenal gland, p27^{Kip1} is only expressed in the medulla, while p57^{Kip2} is exclusively found in the cortex [2,3,15,16]. In contrast, p21$^{Cip1/WAF1}$ is highly expressed in terminally differentiated cells of adult tissues rather than in embryonic cells [17,18], except for the embryonic carcass where there is an extensive muscle differentiation [18–20].

In adult tissues, p21$^{Cip1/WAF1}$ and p27^{Kip1} are widely expressed, whereas p57^{Kip2} is detectable only in a restricted subset of mouse and human tissues/organs, such as placenta, fat, kidney, ovary, adrenal gland, endometrium, lung, prostate, brain, kidney, pancreas, testis, heart and skeletal muscle [2,3,16]. In most tissues, p57^{Kip2} is expressed at a low level. This may reflect the heterogeneity of some of these tissues where only certain cell types express the protein. The highest expression level is found in human placenta, particularly in the villus section of placenta, together with other genes involved in growth and tissue remodeling, like IGF2 and GPC3 [21]. In mice, during placental development, p57^{kip2} is expressed in giant trophoblast cells. Therefore, p57^{Kip2} has been postulated to be involved in the allocation of maternal nutrients through the placenta [22,23]. Since human placenta lacks a cell type equivalent to the giant trophoblast cell, the function of p57^{Kip2} in human placenta might be different and further investigations appear necessary [24].

p57^{Kip2} level declines, in most organs, before birth, whereas p27^{Kip1} expression persists after birth and throughout adult life, suggesting that p57^{Kip2} is important during early organogenesis [15]. The crucial role of p57^{Kip2} in embryogenesis is corroborated by the finding that *CDKN1C* knockout mice (p57^{KO}) die after birth with only less than 10% of the mutant mice surviving until weaning. p57^{KO} mice show severe defects such as macroglossia, cleft palate, omphalocele and gastrointestinal abnormalities, skeletal muscle and endochondral ossification defects, adrenocortical hyperplasia, lens cell hyperproliferation and apoptosis [25,26]. p57^{KO} mice also present several placental abnormalities, including trophoblastic dysplasia [27,28]. Conversely, p21^{KO} mice develop normally [29,30] and p27^{KO} mice do not present gross developmental defects, even though the protein is expressed during embryogenesis and it is required for development. However, p27^{KO} mice display organ hyperplasia and tumorigenesis, consistent with the expected function of inhibitor of cell proliferation [31–33]. The importance of a proper control of p57^{Kip2} dosage is also evident in mice that express a twofold level of p57^{Kip2}. They show an increase of embryonic lethality and a decreased body size [25,28]. Furthermore, the replacement of *Cdkn1c* with *Cdkn1b* (p57^{KO}p27^{KI}) cannot completely compensate for the specific role of p57^{Kip2}. In general, p27^{Kip1} knock-in corrected many of the abnormalities observed in p57^{KO} mice, except for omphalocele, dysplasia of placenta and renal papilla [34]. This evidence supported the opinion that most of the functions performed by both p27^{Kip1} and p57^{Kip2} proteins during development are attributable to the CKI role through their conserved N-terminal KID domain. Thus, the phenotypic differences noticed in p27^{KO} and p57^{KO} mice most probably reflect both their different spatiotemporal expression patterns and the diverse cellular behavior towards an incomplete cell cycle inhibition [34]. However, it is also possible that the C-terminus domain of both CKIs plays similar functions or affects superimposable pathways. In addition, it should be taken into consideration that p57^{KO}p27^{KI} mice express non-physiological levels of p27^{Kip1}.

In adult tissues, all the three CIP/Kip proteins are specifically expressed in terminally differentiated cells, but, of great interest, also in certain undifferentiated quiescent stem cells, probably because of their CKI activity. The finding that most p57^{KO} mice die soon after birth represented an obstacle for the characterization of p57^{Kip2} function in adult tissues. This issue has been overcome by the generation of conditional KO mice. So far, the tissue-specific deletion of *Cdkn1c* has been performed only in adult hematopoietic stem cells (HSCs) and in neural stem cells (NSCs), evidencing the pivotal role of p57^{Kip2} in the quiescence and maintenance of adult stem cells [35,36].

Among hematopoietic cell populations, p57^{Kip2} is the only CKI to be prevalent in a pool of cells with long-term repopulating capability [37] and hematopoietic-specific ablation of p57^{Kip2} in adult mice determines a clear depletion of the HSC population [35]. On the contrary, p21$^{Cip1/WAF1}$ seems to be mainly important in regulating HSC cell cycle during stress condition when DNA is damaged [38,39]; instead, p27^{Kip1} has limited activities, but becomes more effective in later committed progenitors [38]. In vitro experiments partially confirmed the in vivo observations. High p57^{Kip2} mRNA and protein expression have been reported in the HSC side population, especially in c-kit(+)/Sca-1(+)/Lineage-SP cells and p57^{Kip2} has been designated as responsible for the cell cycle blockage since its downregulation is required for S phase entry [37,40]. Moreover, RNA-sequencing analysis of HSC populations derived from a mouse model with a lacZ knock-in at *Mds1* and *Evi1* complex locus, which eliminates the ME domain, has revealed the silencing of p57^{Kip2} expression and it is correlated with the reduction in the number of HSCs and a complete loss of long-term repopulation capacity [41]. Similar pieces of evidence have been obtained later by analyzing CKIs activities in NSCs. p21^{KO} and p27^{KO} mice show an increased proliferation of intermediate progenitor cells rather than of NSCs in the dentate gyrus of the hippocampus, where the two CKIs are barely expressed [42,43]. In contrast, p57^{Kip2} is abundant in NSCs and its expression decreases when these cells become committed and proliferative. Conditional deletion of p57^{Kip2} resulted initially in a transient recruitment of NSCs into the cell cycle, thus activating neurogenesis in brain of both young and aged mice, and later in an excessive depletion of the quiescent NSC population and impairment of hippocampal neurogenesis [36]. The new "disposable stem cell model" proposed recently by Encinas, might explain this phenomenon. During youth, the generation of new neurons is abundant in brain and progressively decreases with age. NSCs, upon activation, asymmetrically divide for limited rounds and then terminally differentiate into astrocytes, thus, dramatically reducing the pool of NSCs [44].

In vitro experiments reveal a dual role of p57^{Kip2}: one is related to the division capability of adult stem cells and the other one to differentiation. Indeed, p57^{Kip2} mRNA and protein have been reported to be increased during differentiation of cerebral cortical precursor [45], oligodendrocytes [46], keratinocytes [47,48], podocytes [49] and skeletal myoblasts [50–52]. Skeletal muscle has a certain regenerative potential, given the presence of the satellite cells, which are muscle progenitor cells that become activated following muscle injury, thus progressing through self-renewal, proliferation, differentiation, and fusion with pre-existing mature muscle fibers to replenish the lost muscle tissue [53].

In skeletal muscle cells, p57[Kip2] participates in the balancing of progenitor cell maintenance with muscle differentiation [54]. Indeed, *Cdkn1c* is upregulated in murine G_0 muscle satellite cells, and its inhibition is needed for satellite cell proliferation [55,56].

Data supporting a possible functional repair of the cardiac tissue have been accumulated over the last decades [57]. Indeed, this hypothesis relies also on the presence of cardiac progenitor cells, named cardiac stem cells (CSCs). The block of cell cycle progression in murine c-kit+ CSCs is due to a complex signaling which involves also the upregulation of *Cdkn1c* [58]. Moreover, experimental evidence in mice demonstrated that cell cycle withdrawal in neonatal cardiomyocytes is associated with an increased expression of p57[Kip2], p21[Cip1/WAF1], and p27[Kip1] and that in adult cardiomyocytes, silencing CDK inhibitors, including p57[Kip2], induces cell cycle re-entry [59–61]. In addition, studies in transgenic mouse reported a cardioprotective effect of ventricular-specific overexpression of p57[Kip2] with no side-effects on heart development [62]. Interestingly, also studies on zebrafish demonstrated that the repression of p57[Kip2] expression promotes heart regeneration [63].

Importantly, several pathways have been reported to modulate the expression of p57[Kip2] [64]. TGFβ/Smad signaling upregulates p57[Kip2] expression in HSCs, mediating the maintenance of hematopoietic stem cells [65], while it has been reported to induce p57[Kip2] degradation in osteoblasts [14]. On the contrary, Wnt/β-catenin and Notch/Hes pathways are reported to reduce p57[Kip2] expression in several cell types. For example, in midbrain dopaminergic neurons, Wnt1 downregulates p57[Kip2] [66], in lens epithelium [67] and in pancreas [68] Notch effectors suppress p57[Kip2] expression. However, the general picture is complex and difficult to understand due to the cross-talk and overlapping of different signal pathways.

1.3. CDKN1C Mapping and Structure

CDKN1C is localized, in humans, at the 11p15.5 locus and includes four exons and three introns (Figure 1). *CDKN1C* alternative splicing results in the formation of three mature mRNAs that have the same open reading frame, but different untranslated regions [69,70]. Human 11p15.5 locus contains numerous genes subjected to an imprinting modulation (Figure 1).

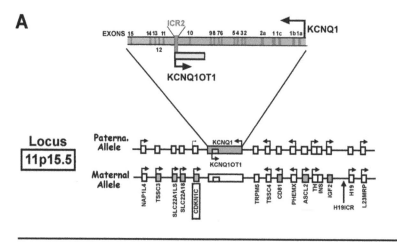

Figure 1. *Cont.*

B

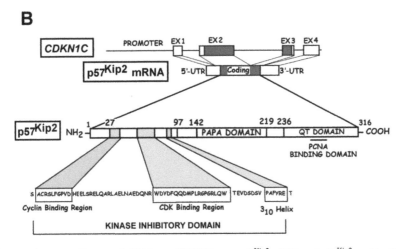

Figure 1. Structure of human 11p15.5 locus, *CDKN1C* gene, p57^Kip2 mRNA and p57^Kip2 protein. Panel (**A**) The panel shows the structure of the 11p15.5 locus with details of the *KCNQ1* exon organization (in blue boxes). *KCNQ1OT1* gene is included in the *KCNQ1* gene and transcribed in a different direction. The ICR2 region is shown in orange; Panel (**B**) The figure shows the structure of *CDKN1C* gene and p57^Kip2 mRNA. In addition, at the bottom of the figure, it is represented the domain organization of p57^Kip2 protein and the sequence of the KID (kinase inhibitory domain).

Importantly, the homolog region in mouse (i.e., the distal region of chromosome 7) shows an equal cluster of linked genes, arguing for the significance of their coordinate regulation and for the presence of maintained regulatory mechanisms [71,72].

The human 11p15.5 gene cluster might be divided into two distinct domains, both presenting a specific "*in cis*" acting ICR (Imprinted Control Region). The centromeric domain of the cluster is 800 kb long and is controlled by ICR2. The domain includes, in addition to *CDKN1C*, *KCNQ1* (*KvLQT1* or potassium voltage-gated channel, KQT-like subfamily member 1), *KCNQ1OT1* (also known as *LIT1*, *KCNQ1*-overlapping transcript 1 or long QT intronic transcript 1), *PHLDA2* (Pleckstrin homology-like domain family A member 2) and *SLC22A18* (Solute carrier family 22 member 18).

Structurally, ICR2 maps inside *KCNQ1* intron 10 and is methylated on the maternal chromosome; it encompasses the promoter for the non-coding RNA Kcnq1ot1 (antisense to *KCNQ1*) (Figure 1) [73].

ICR1 is telomeric and regulates the imprinting of *H19* (a gene for a long noncoding RNA) and *IGF2* (encoding for insulin-like growth factor 2) by restricting the access to the enhancers (i.e., ICR1 is a chromatin insulator) [74]. Interestingly several of these genes have distinct imprinting. Indeed, *IGF2* is paternally expressed, *H19* is maternally transcribed, and *CDKN1C* is maternally expressed, even though a weak expression of the paternal allele has been demonstrated in some human tissues [75].

Two main promoter elements have been identified in mouse *Cdkn1c* that are similar in humans. First, a proximal promoter element (−165 to +15 from the transcriptional start site) contains several *consensus* sequences for Egr1 and Sp1 [52,76]. Intriguingly, both transcription factors are ubiquitously expressed and have been reported to regulate other members of the CIP/Kip family of CDK inhibitors [77,78]. Furthermore, a binding site for GATA2, a transcription factor playing a pivotal role in hematopoiesis, particularly in early and late stages of erythropoiesis, and in the TGF-β-response has also been described [79]. Finally, this promoter region also contains recognition sequences for the transcriptional repressors CTIP2/Bcl11b, implicated in the developmental process and carcinogenesis, and the T-box transcription factor TBX3, which is involved in the tissue patterning and differentiation during embryonic development and is up-regulated in a plethora of cancers [80]. Importantly, the accessibility of the reported transcriptional modulators to the *CDKN1C* promoter is strongly influenced by the high presence of CpG islands, located upstream and downstream of the transcription start site, responsible for genomic imprinting and epigenetic gene silencing. This is achieved by CpG dinucleotide methylation and/or through chromatin remodeling by histone covalent modifications (histones H3 and H4 methylation and acetylation) [81]. More distal promoters have also been identified. They embrace E-boxes or E-box-like motifs for the interaction with basic-HLH proteins, including activators, like TCF4/E2-2 [82], E47 [83], Smad1/Atf2 complex [84], repressors, as Hes1 (a Notch effector) that, in intestinal crypt progenitor cells, inhibits *Cdkn1c* transcription by binding to a site located at −3300 [85], or Hes-related repressor protein Herp2 that acts as transcriptional repressor of *CDKN1C* in proliferating lens epithelial cells [67]. Furthermore, a glucocorticoid response element is located 5076 to 5062 bases upstream of the transcription start site of the human *CDKN1C* gene and is responsible for the glucocorticoid inducibility of the *CDKN1C* gene [86], thus explaining, at least in part, the antiproliferative effect of dexamethasone in human tumor cells such as Hela cell line [87].

In mouse, additional key elements of *Cdkn1c* transcription are located distantly from the gene. As a matter of facts, enhancers for its expression lie more than 25 kb downstream of the gene. Experiments with artificial chromosome also suggest the existence of enhancer(s) located very distantly from *CDKN1C* [88]. Accordingly, in humans, it has been suggested the presence of numerous *CDKN1C* enhancer elements localized in a region between 255–387 kb [89].

2. Control of *CDKN1C* Transcription

CDKN1C lies in humans and mice in a very complicated cluster of imprinted genes, controlled through superimposed *cis*-acting mechanisms. Genomic imprinting is an epigenetic process that results in parent-of-origin specific allelic expression [90]. A relatively small subset of genes within the mammalian genome (0.4%) is imprinted [91,92] showing a mono-allelic expression either in specific phenotypes of the whole organism or in peculiar tissues that favors the maternal (e.g., *CDKN1C* and *UBE3A*) or the paternal allele (e.g., *DLK1* and *NNAT*) [93]. Imprinted expression is initially determined by differential DNA methylation that is established in the germline [94].

Regarding *CDKN1C*, its transcription is regulated by the imprinting center KvDMR1 that acquires DNA methylation in the maternal germline [69,95,96]. This differentially methylated region spans the promoter of the paternally expressed long non-coding RNA Kcnq1ot1 required for continuous domain-wide imprinting. The *CDKN1C* promoter and gene body are also directly methylated on the paternal allele post-fertilization, after allelic silencing has been established [97].

Besides *cis*-acting mechanisms responsible for the imprinted silencing of the paternal allele (briefly summarized in the previous paragraph), *trans*-acting mechanisms participate in the epigenetic modulation of *CDKN1C* gene expression [98]. Indeed, a complex interplay among DNA methylation and post-translational modifications of histones contributes to the chromatin dynamics at the promoter and in *CDKN1C* gene body.

2.1. DNA CpG Island Methylation

CDKN1C gene is included in a CpG island extended about −600 bp from the transcriptional start site up into the gene body. This CpG island presents, in mice but not in humans, a differential methylation between the two inherited alleles, being the paternal one hypermethylated and the maternal one hypomethylated. This methylation pattern seems to be acquired successively to the ICR2-dependent DNA-modifications and is involved in the maintenance and reinforcement of the imprinted repression of the paternal allele. Several regulators have been involved in this process. One of them is Lsh (lymphoid-specific helicase) a protein belonging to the family of SNF2/helicases that act as chromatin remodeler and regulate DNA methylation. Lsh directly binds to *CDKN1C* promoter and allows the maintenance of hypermethylation of the paternal allele [99].

Complete biallelic hypermethylation occurs in human tumors and tumor cell lines [100,101], as well as in some undifferentiated tissues and cell types such as skeletal myoblasts. In this cell model, the activation of the transcription factor MyoD drives DNA demethylation on the maternal allele, therefore allowing the Myo-D-dependent expression of p57^{Kip2} [102].

Most interesting is also the role played by different members of the DNA methyltransferase (DNMT) family, the enzymes catalyzing the transfer of methyl groups to cytosines. Results from genetic ablation studies support the notion that not only DNMT1, mostly in charge of maintaining the methylation pattern of CpG islands during DNA replication, but also DNMT3a which is generally involved (together with DNMT3b) in de novo methylation of most imprinting control regions in the germline, are involved in *CDKN1C* promoter methylation. As matter of fact, both DNMT1 [56,103], and DNMT3a [56] have been found associated with *CDKN1C* promoter. Consistently with the importance of DNA methylation not only in paternal allele imprinting but also in p57^{Kip2} expression modulation in specific cellular and cell cycle phase contexts, treatment of many human tumor cell lines with demethylating agents such as 5-azacytidine and 5-aza-2′-deoxycytidine results generally in p57^{Kip2} expression activation [102,104].

2.2. Histone Marks

Histone modifications represent fundamental factors involved in chromatin plasticity, controlling gene promoter accessibility and gene expression activation [105,106].

Acetylation and methylation of core histone tails, in addition to DNA methylation, are key mechanisms for regulating *CDKN1C* transcription. Accordingly, the level of H3 and H4 acetylation directly correlates with the gene expression and, in turn, with several phenotypes including differentiation and carcinogenesis.

Specifically, a decrease of H3 lysine 4 dimethylation and histone H3 lysine 9 and 14 acetylation is observed on the paternal allele respect to the maternal one, facilitating its inactivation [107,108]. On the other hand, histone acetylation results to be increased on the paternal locus at the level of KvDMR1, following the expression of the long non-coding RNA and the corresponding *CDKN1C* inactivation [81].

Under various conditions, a direct correlation between *CDKN1C* expression and H3K9/K14 acetylation has been demonstrated. For example, cancer cells with low (or absent) p57^{Kip2} present histone hypoacetylation and, vice versa, tumors with a high level of the CKI show hyperacetylation [109,110]. These findings are confirmed by the re-expression of p57^{Kip2} after histone deacetylase (HDAC) inhibitor treatment [76]. Mechanistically, these changes involve the binding of HDACs, mostly HDAC1 and HDAC2, to the *CDKN1C* promoter region. We must underline that HDAC1 is highly expressed in many cancers including gastric [111], colorectal [112], hepatic [113], breast [114], and pancreatic cancer [115]. HDAC2 has been found mutated in colon cancer [116] and is overexpressed in esophageal [117], prostate [118], and gastrointestinal carcinomas [119].

An additional recognized histone epigenetic mark includes a lysine trimethylation, specifically H3K27me3 (trimethylation of lysine 27 of histone H3). Such a modification, also responsible for the paternal allele exclusion, is involved in the maturation of glial cells [120].

This histone trimethylation mark is due to the Polycomb repressive complex 2 (PRC2). The increase of H3K27me3 reduces *CDKN1C* expression, while its decrease, due to a reduction in the levels or activity of EZH, a specific promoter-binding PRC2 subunit, up-regulates gene transcription [121].

Di- and trimethylation of lysine 9 of the histone H3, an additional histone modification, also appears to control the expression of *KvDMR1* on the maternal locus, while it is not present on the paternal allele, in accord with the imprinted silencing of the paternally-derived allele [107,108].

In the same context, it is important to stress the role of histone modifications in MyoD control of *CDKN1C* expression. As matter of fact, an altered accumulation of H3K9me2 on the maternal KvDMR1 allele results in the lack of response to MyoD in that it reduces the accessibility of the transcription factor to the DNA [102].

In brief, the regulation of *KvDMR1* due to epigenetic factors (methylation of DNA, acetylation/methylation of histones) appears a key mechanism in the control of *CDKN1C* expression. The last several years have identified the existence of a strict crosstalk between all the epigenetic modifications, including the binding of modifying enzymes to the specific sites of action. In this complex interplay, an important role for non-coding RNAs has emerged.

2.3. LncRNA Involvement in Epigenetic Regulation

As for many genes playing fundamental roles in development, *CDKN1C* gene expression is controlled by lncRNAs, which act in strict crosstalk with signal pathway-induced transcription factors and chromatin modifiers, accounting for spatial- and temporal-specific gene activation during development or cell commitment and differentiation in adult life. Specifically, the macro lncRNA Kcnq1ot1, first discovered both in humans and mice as a KvDMR1-associated RNA, has emerged as a critical regulator of the chromatin status of the gene, at least in relation to the imprinting control [122]. *KCNQ1* and *KCNQ1OT1* share a region of overlapping DNA and are transcribed in opposite directions. *KCNQ1* encodes the potassium voltage-gated channel subfamily Q member 1, a protein required for the repolarization phase of the cardiac action potential. Differently from *KCNQ1* that allows the synthesis of a protein, *KCNQ1OT1* codifies a long-noncoding RNA that regulates the expression of several genes. Its promoter is normally hypermethylated in the maternal allele, thus hampering its expression. On the contrary, the paternal allele is normally transcribed, being hypomethylated [72,123]. When expressed, the long non-coding RNA remains in the nucleus where it is able to act on its own chromosome (i.e., it acts on chromatin *in cis*). Mechanistically, it is able to interact with histone methyltransferase complexes (like G9a, Suz12, and Ezh2) causing the enrichment of repressive histone modifications. This activity results in the epigenetic inactivation of paternally inherited *CDKN1C* [124]. Kcnq1ot1 is also able to bind DNMT1, allowing the hypermethylation of *CDKN1C* promoter and therefore implementing the repression of gene expression [103].

Additional long non-coding RNAs, like Tug1 (taurine upregulated gene 1) [125], Linc00668 [126] and HEIH-coding RNA [127] have been reported to modulate *CDKN1C* expression. Tug1 is a long non-coding RNA mostly occurring in the retina and in the brain. It has been proposed to control cell growth by epigenetically down-regulating *CDKN1C*. In addition, Tug1 seems to predict a negative prognosis in gastric cancer [125]. These data indicate that lncRNAs regulate p57^{Kip2} at the cellular level probably acting in a phenotype-specific manner.

3. *CDKN1C* Expression and Human Diseases

Genetic and epigenetic disorders in the imprinted region 11p15 and *CDKN1C* mutations can lead to embryonic abnormalities, such as those occurring in Beckwith-Wiedmann syndrome (BWS; OMIM 130650), IMAGe syndrome (OMIM 614732) and Russell–Silver syndrome (RSS-OMIM 180860) and acquired diseases such as cancer.

3.1. Human Developmental Disorders

BWS, IMAGe syndrome and RSS are genetic diseases with different features, belonging to the group of congenital imprinting disorders. BWS has a prevalence of 1–5:10,000 live births and is characterized by overgrowth, tumor predisposition, abdominal wall defects and congenital malformations such as macroglossia, hemihyperplasia, hyperinsulinaemic hypoglycemia, ear anomalies, nephrologic and capillary malformations and organomegaly [128]. Epigenetic and genetic alterations in the imprinting cluster on chromosome 11p15.5 are responsible for up to 80% of BWS cases.

They include epigenetic alterations such as methylation defects, specifically loss of methylation at IC2 regulatory region (IC2-LoM) and gain of methylation at IC1 (IC1-GoM), as well as genetic alterations including uniparental paternal disomy of the 11p15.5 locus (UPD), followed by a lower percentage of cases with microdeletion/duplications or point mutations involving either one of the two ICRs responsible for the locus imprinting region [129–131]. Of interest, in BWS patients without methylation defects, *CDKN1C* gene mutations are frequently noticed [132], reaching the 50–70% in familial BWS cases. Although with a lower occurrence, *CDKN1C* mutations are also reported in sporadic BWS cases and they have been identified as the causative genetic alterations [128,133]. Missense/nonsense mutations are reported along the entire sequence of the gene, leading to increased proliferation and risk of cancer [134,135]. Interestingly, these (epi) genotypes have been associated with specific phenotypes which discriminate mainly between overgrowth in pre- or postnatal age (Figure 2) [129].

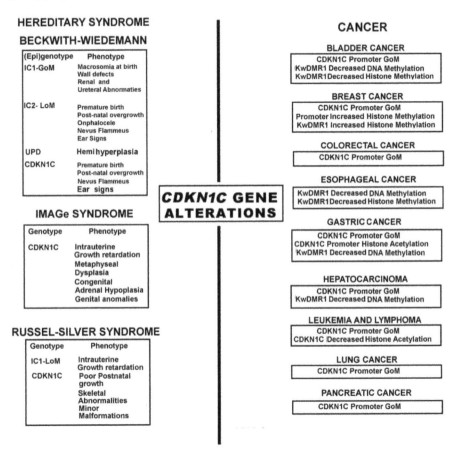

Figure 2. *CDKN1C* gene alterations in hereditary syndromes and human cancers. The figure reports on the left the Syndromes in which the *CDKN1C* gene is altered with the description of genotype alterations and main phenotypic features. On the right are reported the cancers showing *CDKN1C* genetic changes. IC1, ICR1 region; IC2, ICR2 region; GoM, gain of methylation; LoM, loss of methylation; UPD, Uniparental disomy.

Opposite to BWS, IMAGe (Intrauterine growth retardation, Metaphyseal dysplasia, congenital Adrenal hypoplasia, and Genital anomalies) syndrome is a rare condition (www.orpha.net) in which a cluster of CDKN1C missense mutations in the PCNA binding domain, result in growth inhibition) [136]. Interestingly, BWS and IMAGe, are characterized by loss of function and gain of function mutations of *CDKN1C*, respectively [136,137].

Imprinting alterations in the 11p15 region are also described in RSS, a disease (prevalence of 1–30:100,000 live birth) characterized by intrauterine growth retardation, very limited postnatal growth, skeletal abnormalities such as peculiar craniofacial characteristics and body asymmetry, and several minor malformations. Moreover, in RSS is also reported a maternal duplication in this region [138,139] and methylation alterations of imprinted genes on chromosome 7 [140]. Moreover, in one case of RSS a *CDKN1C* mutation affecting the PCNA binding domain has been found [141].

3.2. Human Cancers

Gene encoding cyclin-dependent inhibitors are frequently altered in human tumors. Among them, INK4 family represents the most clear paradigm [142,143]. The discovery of the involvement of p57^{Kip2} in BWS and in some human tumors suggests that p57^{Kip2}, like p27^{Kip1}, might also have a role in the process of carcinogenesis (Figure 2) [144]. Based on the roles played by the protein in the nuclear compartment, currently, it is considered a tumor suppressor; however, differently from p27^{Kip1} [143], somatic mutations have been rarely reported in tumors [145,146], underlining the importance of p57^{Kip2} expression control as the main cause of its altered levels in cancer. Particularly, a downregulation of *CDKN1C* is generally reported in cancer including gastric [147] and urothelial cancer [148,149], pancreatic adenocarcinomas [150], adrenocortical [151], lung [152], and breast cancer [153] as well as several leukemias [154]. Moreover, many authors have attributed to the p57^{Kip2} levels a value of prognostic marker since a decrease of its expression has been correlated to a poor prognosis [153,155] As described above, different epigenetic and genetic mechanisms can modulate the expression of *CDKN1C*. Essentially, loss of imprinting, DNA methylation and post-translational modifications of histones in the promoter region as well microRNAs downregulate *CDKN1C* in human cancers. Among them, the main cause of reduction of p57^{Kip2} in cancers is generally the increased methylation of the large CpG islands localized in the *CDKN1C* promoter [81,150,153,156]. Particularly, the promoter methylation of *CDKN1C* has been found critical in hematological malignancies such as acute lymphoblastic leukemia [157,158] and large B cell lymphoma [104]. Recently, Zohny and colleagues have proposed as diagnostic markers of breast cancer the expression levels of p21$^{CIP1/Waf1}$ and p57^{Kip2} combined with the promoter methylation of *CDKN1C*, since they found a silenced expression of the two CKIs and a hypermethylation of *CDKN1C* promoter in more than 50% of the breast cancer specimens analyzed, together with no hypermethylation at promoter of p21$^{CIP1/Waf1}$-coding gene [159]. Moreover, several miRNAs have been reported to control p57^{Kip2} mRNA levels. miR21 downregulates *CDKN1C* in prostate cancer [160], miR25 in gastric cancer and glioma [161,162] and miR92b in brain tumors [163]. Furthermore, miR221/222 are reported to reduce p57^{Kip2} and p27^{Kip1} expression in hepatocarcinoma [164], in glioblastoma [165], in oral cancer [166], in colorectal cancer [167] and B-cell malignancies EBV-associated [168] Experimental data confirmed, at least in ovarian cancers, the specific action of miR-221/222 on *CDKN1C* [169]. A further putative mechanism at the basis of *CDKN1C* down-regulation might be related to an increased rate of protein degradation, mainly due to the Skp2 overexpression, as frequently observed in human cancers. However, the relevance of Skp2-dependent degradation of p57^{Kip2} in carcinogenesis is still debated.

3.3. Other CDKN1C-Related Human Diseases

Gestational diseases, such as pre-eclampsia and intra-uterine growth restriction (IUGR), are also associated with altered p57^{Kip2} expression. Pre-eclampsia and IUGR are associated respectively with downregulation [170] and upregulation [171] of *CDKN1C*, underlining that a correct control of gene transcription is required for the proper development and progression of the pregnancy. p57^{Kip2}, as reported above, is abundantly expressed in placental tissues, and, therefore, its dysregulated expression, in humans, is associated with placental mesenchymal and vascular proliferative disorders, such as placental mesenchymal dysplasia (PMD), and complete and partial hydatidiform moles [172,173].

Importantly, placentomegaly due to abnormal proliferation of extravillous trophoblasts, and accumulation of intervillous fibrinoid can be observed also in BWS syndrome [172,174].

PMD is a rare condition (0.02% pregnancies) associated with different fetal outcomes, ranging from structurally normal fetus/newborn (in most cases) to fetal and neonatal abnormalities, including those present in BWS, and mortality [175]. Indeed, PMD and BWS are associated in one-third of cases and paternal uniparental disomy at *IGF2* and *CDKN1C* locus has been proposed as the genetic link between them [176–178].

The identification of p57^{Kip2} as an important player in these placental diseases has led to the development of diagnostic procedures based on immunohistochemistry using anti-p57^{Kip2} antibodies and histological analysis for the characterization of hydatidiform moles and PMD, and their differential diagnosis [179–181].

Although some human diseases may represent very different clinical entities, some common pathways may be identified in their etiopathogenesis. This could be the case for cancer, metabolic diseases and related cardiovascular disease [182–184]. Indeed, p57^{Kip2}, among others, may provide a similar origin for neoplastic proliferation and metabolic disorders. p57^{Kip2} is specifically expressed in the endocrine portion of the pancreas and particularly in β-cells where it is paternally imprinted [185]. In focal hyperinsulinism of infancy, a syndrome characterized by hyperinsulinemic hypoglycemia, p57^{Kip2} is not expressed within the focal adenomatous hyperplastic lesions. This missing expression is caused by somatic loss of heterozygosity and associated with increased proliferation of β-cells [185]. On this basis, p57^{Kip2} negative human pancreatic islets restored proper glucose control when transplanted into hyperglycemic, immunodeficient mice [186]. Interestingly, a gain-of-function mutant of p57^{Kip2} is associated with early-adulthood-onset diabetes, in addition to intrauterine growth restriction and short stature [187]. Not only glucose metabolism but also lipid metabolism could be affected by the genetic dosage of *CDKN1C*. In fact, a double dose of *Cdkn1c* promoted the brown adipose tissue development in a mouse model of the RSS [188]. Conversely, a loss-of-function mutation of p57^{Kip2}, in the same model, hindered completely the formation of brown adipocytes [188].

The expression of p57^{Kip2} may also change according to genetic variations that do not pertain to *CDKN1C*, the p57^{Kip2}-coding gene but are located in neighboring genomic sites. In particular, a mutation has been identified at the *KCNQ1* locus that increases the expression of p57^{Kip2} in mouse pancreatic islets by epigenetically modifying *Cdkn1c* [189]. Differently from the maternally transmitted *CDKN1C* diseases, this mutation is effective only when inherited from the paternal allele and, no matter which is the mechanism increasing the levels of p57^{Kip2} in the endocrine pancreas, β-cell mass is reduced [189].

Thus, it is conceivable that altering p57^{Kip2} expression could be a promising therapeutic strategy also in humans with type 2 diabetes and/or obesity. Indeed, oral administration of FTY720, a sphingosine 1-phosphate receptor agonist, normalizes glycemia in diabetic db/db mice by downregulating islet p57^{Kip2} and promoting β-cell regeneration [190]. However, body weight significantly increased in treated animals. Probably, tissue-specific targeting of p57^{Kip2} should be pursued, in order to avoid undesired effect in either pancreatic or adipose tissue when trying to manipulate p57^{Kip2} in one of the two tissues.

4. Future Directions

It has been clearly established that p57^{Kip2} plays pivotal and specific roles in human physiology that cannot be replaced by the other two members of its protein family, i.e., p27^{Kip1} and p21$^{Cip/WAF1}$. As a matter of fact, *CDKN1C*-deleted mice show a very high percentage of mortality demonstrating that the protein is necessary for correct embryogenesis. Accordingly, in humans, three important syndromes, all showing an altered growth, are due to *CDKN1C* alterations. Particularly, the BWS is characterized by signs of overgrowth with infants considerably larger than normal and the IMAGe and RSS are both characterized by slow growth before and after birth and growth retardation.

Int. J. Mol. Sci. **2018**, *19*, 1055

It is well known that adult stem cells settle specific niches in different tissues and organs, thus supporting their repair/regeneration [57,191–193]. In this regard, p57^{Kip2} has been discovered as a very important factor [36,56,61,63,194].

The molecular mechanisms by which p57^{Kip2} is so important for a normal growth and tissue differentiation, are still not well understood. They refer, only in part, to the capability of the protein to modulate cyclin/CDK activity, a function that is played approximately by the first 100 residues of p57^{Kip2} where the KID is localized. On the other hand, numerous activities of CKI have been associated with the C-end region that includes the PAPA domain and the QT domain. These protein domains seem to be involved in the control of cell movement and the organization of mitotic spindle (via interaction with the cytoskeleton). Numerous pieces of evidence also suggest that p57^{Kip2} C-terminus participates in endoreduplication, apoptosis, autophagy, and senescence. Altogether these observations point to the definite identification of p57^{Kip2} interactors as a pivotal issue in the studies on CKI. Unfortunately, the protein belongs to the family of IUP, namely, proteins lacking a tertiary structure that fold upon binding. This structural feature allows p57^{Kip2} to have a large degree of plasticity and to interact with several different proteins. Although this is certainly a great advantage in terms of function, it results in remarkable difficulties in the precise definition of the mechanism of p57^{Kip2} action. An additional important aspect of p57^{Kip2} studies is the knowledge of processes that regulate its level. *CDKN1C* is subject to an epigenetic control and only the maternal allele is expressed. The gene lies in a locus, 11p15.5, that represents a major example of regional imprinting and that is strongly regulated by mechanisms acting in cis (i.e., directly on DNA structure) as well as in trans (namely via DNA methylation and histone acetylation). An additional level of intracellular control of p57^{Kip2} amount could be related to the protein degradation that involves, a not completely clarified, ubiquitination/proteasomal mechanism.

In conclusion, while the important role of p57^{Kip2} is clear, the details of its regulation and interactors appear enigmatic and intensive research and the development of novel cellular and animal models are required. This is particularly relevant in view of the plethora of p57^{Kip2} functions played in different tissues and distinct phenotypes both in normal and pathological conditions. It is conceivable that the elucidation of these aspects will provide important directions for human physiologic research and for the development of novel strategies for targeted therapy of several relevant human diseases.

Acknowledgments: This work was supported in part by Grant No. 11653 from the Associazione Italiana per la Ricerca sul Cancro (AIRC).

Author Contributions: Emanuela Stampone, Ilaria Caldarelli, Alberto Zullo, and Debora Bencivenga drafted the manuscript. Francesco Paolo Mancini Fulvio Della Ragione, and Adriana Borriello drafted and revised the manuscript. All authors read and approved the final manuscript.

Conflicts of Interest: The authors declare no conflict of interest.

Abbreviations

BWS	Beckwith-Wiedmann syndrome
CSC	Cardia stem cell
CDK	Cyclin-dependent kinase
CKI	CDK inhibitor
DNMT	DNA methyltransferase
HSC	Hematopoietic stem cell
ICR	Imprinted control region
IGF2	Insulin-like gorwth factor
HDAC	histone deacetylase
IUGR	Intra-uterine growth restriction
IUP	Intrinsically unstructurated protein
KID	CDK binding/inhibitory domain
LSH	lymphoid-specific helicase
NLS	Nuclear localization signal

NSC	Neural stem cell
PCNA	Proliferating cell nuclear antigen
PMD	Placental mesenchymal dysplasia
PRC2	Polycomb repressive complex 2
$p57^{KO}$	CDKN1C knockout
$p57^{KO}p27^{KI}$	CDKN1C knockout + CDKN1B knockin
RSS	Russell-Silver syndrome
Skp2	S-phase kinase-associated protein 2
SCF	Skp1/Cul1/F-box
TUG	Taurine upregulated gene 1
T310	Threonin 310

References

1. Borriello, A.; Caldarelli, I.; Bencivenga, D.; Criscuolo, M.; Cucciolla, V.; Tramontano, A.; Oliva, A.; Perrotta, S.; Della Ragione, F. p57(Kip2) and cancer: Time for a critical appraisal. *Mol. Cancer Res.* **2011**, *9*, 1269–1284. [CrossRef] [PubMed]

2. Lee, M.H.; Reynisdottir, I.; Massague, J. Cloning of p57KIP2, a cyclin-dependent kinase inhibitor with unique domain structure and tissue distribution. *Genes Dev.* **1995**, *9*, 639–649. [CrossRef] [PubMed]

3. Matsuoka, S.; Edwards, M.C.; Bai, C.; Parker, S.; Zhang, P.; Baldini, A.; Harper, J.W.; Elledge, S.J. p57KIP2, a structurally distinct member of the p21CIP1 Cdk inhibitor family, is a candidate tumor suppressor gene. *Genes Dev.* **1995**, *9*, 650–662. [CrossRef] [PubMed]

4. Fotedar, R.; Fitzgerald, P.; Rousselle, T.; Cannella, D.; Dorée, M.; Messier, H.; Fotedar, A. p21 contains independent binding sites for cyclin and cdk2: Both sites are required to inhibit cdk2 kinase activity. *Oncogene* **1996**, *12*, 2155–2164. [PubMed]

5. Hashimoto, Y.; Kohri, K.; Kaneko, Y.; Morisaki, H.; Kato, T.; Ikeda, K.; Nakanishi, M. Critical role for the 310 helix region of p57(Kip2) in cyclin-dependent kinase 2 inhibition and growth suppression. *J. Biol. Chem.* **1998**, *273*, 16544–16550. [CrossRef] [PubMed]

6. Reynaud, E.G.; Guillier, M.; Leibovitch, M.P.; Leibovitch, S.A. Dimerization of the amino terminal domain of p57Kip2 inhibits cyclin D1-cdk4 kinase activity. *Oncogene* **2000**, *19*, 1147–1152. [CrossRef] [PubMed]

7. Vaccarello, G.; Figliola, R.; Cramerotti, S.; Novelli, F.; Maione, R. p57Kip2 is induced by MyoD through a p73-dependent pathway. *J. Mol. Biol.* **2006**, *356*, 578–588. [CrossRef] [PubMed]

8. Joseph, B.; Andersson, E.R.; Vlachos, P.; Södersten, E.; Liu, L.; Teixeira, A.I.; Hermanson, O. p57Kip2 is a repressor of Mash1 activity and neuronal differentiation in neural stem cells. *Cell Death Differ.* **2009**, *16*, 1256–1265. [CrossRef] [PubMed]

9. Joaquin, M.; Watson, R.J. The cell cycle-regulated B-Myb transcription factor overcomes cyclin-dependent kinase inhibitory activity of p57(KIP2) by interacting with its cyclin-binding domain. *J. Biol. Chem.* **2003**, *278*, 44255–44264. [CrossRef] [PubMed]

10. Chang, T.S.; Kim, M.J.; Ryoo, K.; Park, J.; Eom, S.J.; Shim, J.; Nakayama, K.I.; Nakayama, K.; Tomita, M.; Takahashi, K.; et al. p57KIP2 modulates stress-activated signaling by inhibiting c-Jun NH2-terminal kinase/stress-activated protein Kinase. *J. Biol. Chem.* **2003**, *278*, 48092–48098. [CrossRef] [PubMed]

11. Yamamoto, T.; Digumarthi, H.; Aranbayeva, Z.; Wataha, J.; Lewis, J.; Messer, R.; Qin, H.; Dickinson, D.; Osaki, T.; Schuster, G.S.; et al. EGCG-targeted p57/KIP2 reduces tumorigenicity of oral carcinoma cells: Role of c-Jun N-terminal kinase. *Toxicol. Appl. Pharmacol.* **2007**, *224*, 318–325. [CrossRef] [PubMed]

12. Watanabe, H.; Pan, Z.Q.; Schreiber-Agus, N.; DePinho, R.A.; Hurwitz, J.; Xiong, Y. Suppression of cell transformation by the cyclin-dependent kinase inhibitor p57KIP2 requires binding to proliferating cell nuclear antigen. *Proc. Natl. Acad. Sci. USA* **1998**, *95*, 1392–1397. [CrossRef] [PubMed]

13. Borriello, A.; Cucciolla, V.; Criscuolo, M.; Indaco, S.; Oliva, A.; Giovane, A.; Bencivenga, D.; Iolascon, A.; Zappia, V.; Della Ragione, F. Retinoic acid induces p27Kip1 nuclear accumulation by modulating its phosphorylation. *Cancer Res.* **2006**, *66*, 4240–4248. [CrossRef] [PubMed]

14. Kim, M.; Nakamoto, T.; Nishimori, S.; Tanaka, K.; Chiba, T. A new ubiquitin ligase involved in p57KIP2 proteolysis regulates osteoblast cell differentiation. *EMBO Rep.* **2008**, *9*, 878–884. [CrossRef] [PubMed]

15. Nagahama, H.; Hatakeyama, S.; Nakayama, K.; Nagata, M.; Tomita, K.; Nakayama, K. Spatial and temporal expression patterns of the cyclin-dependent kinase (CDK) inhibitors p27Kip1 and p57Kip2 during mouse development. *Anat. Embryol.* **2001**, *203*, 77–87. [CrossRef] [PubMed]

16. Fagerberg, L.; Hallström, B.M.; Oksvold, P.; Kampf, C.; Djureinovic, D.; Odeberg, J.; Habuka, M.; Tahmasebpoor, S.; Danielsson, A.; Edlund, K.; et al. Analysis of the human tissue-specific expression by genome-wide integration of transcriptomics and antibody-based proteomics. *Mol. Cell. Proteom.* **2014**, *13*, 397–406. [CrossRef] [PubMed]

17. Macleod, K.F.; Sherry, N.; Hannon, G.; Beach, D.; Tokino, T.; Kinzler, K.; Vogelstein, B.; Jacks, T. p53-dependent and independent expression of p21 during cell growth, differentiation, and DNA damage. *Genes Dev.* **1995**, *9*, 935–944. [CrossRef] [PubMed]

18. Parker, S.B.; Eichele, G.; Zhang, P.; Rawls, A.; Sands, A.T.; Bradley, A.; Olson, E.N.; Harper, J.W.; Elledge, S.J. p53-independent expression of p21Cip1 in muscle and other terminally differentiating cells. *Science* **1995**, *267*, 1024–1027. [CrossRef] [PubMed]

19. Halevy, O.; Novitch, B.G.; Spicer, D.B.; Skapek, S.X.; Rhee, J.; Hannon, G.J.; Beach, D.; Lassar, A.B. Correlation of terminal cell cycle arrest of skeletal muscle with induction of p21 by MyoD. *Science* **1995**, *267*, 1018–1021. [CrossRef] [PubMed]

20. Fredersdorf, S.; Milne, A.W.; Hall, P.A.; Lu, X. Characterization of a panel of novel anti-p21Waf1/Cip1 monoclonal antibodies and immunochemical analysis of p21Waf1/Cip1 expression in normal human tissues. *Am. J. Pathol.* **1996**, *148*, 825–835. [PubMed]

21. Sood, R.; Zehnder, J.L.; Druzin, M.L.; Brown, P.O. Gene expression patterns in human placenta. *Proc. Natl. Acad. Sci. USA* **2006**, *103*, 5478–5483. [CrossRef] [PubMed]

22. Tunster, S.J.; Van de Pette, M.; John, R.M. Fetal overgrowth in the Cdkn1c mouse model of Beckwith-Wiedemann syndrome. *Dis. Models Mech.* **2011**, *4*, 814–821. [CrossRef] [PubMed]

23. Tunster, S.J.; van de Pette, M.; John, R.M. Impact of genetic background on placental glycogen storage in mice. *Placenta* **2012**, *33*, 124–127. [CrossRef] [PubMed]

24. Enders, A.C.; Carter, A.M. What can comparative studies of placental structure tell us?—A review. *Placenta* **2004**, *25* (Suppl. A), S3–S9. [CrossRef] [PubMed]

25. Yan, Y.; Frisén, J.; Lee, M.H.; Massagué, J.; Barbacid, M. Ablation of the CDK inhibitor p57Kip2 results in increased apoptosis and delayed differentiation during mouse development. *Genes Dev.* **1997**, *11*, 973–983. [CrossRef] [PubMed]

26. Zhang, P.; Liégeois, N.J.; Wong, C.; Finegold, M.; Hou, H.; Thompson, J.C.; Silverman, A.; Harper, J.W.; DePinho, R.A.; Elledge, S.J. Altered cell differentiation and proliferation in mice lacking p57KIP2 indicates a role in Beckwith-Wiedemann syndrome. *Nature* **1997**, *387*, 151–158. [CrossRef] [PubMed]

27. Zhang, P.; Wong, C.; DePinho, R.A.; Harper, J.W.; Elledge, S.J. Cooperation between the Cdk inhibitors p27(KIP1) and p57(KIP2) in the control of tissue growth and development. *Genes Dev.* **1998**, *12*, 3162–3167. [CrossRef] [PubMed]

28. Takahashi, K.; Nakayama, K.; Nakayama, K. Mice lacking a CDK inhibitor, p57Kip2, exhibit skeletal abnormalities and growth retardation. *J. Biochem.* **2000**, *127*, 73–83. [CrossRef] [PubMed]

29. Deng, C.; Zhang, P.; Harper, J.W.; Elledge, S.J.; Leder, P. Mice lacking p21CIP1/WAF1 undergo normal development, but are defective in G1 checkpoint control. *Cell* **1995**, *82*, 675–684. [CrossRef]

30. Brugarolas, J.; Chandrasekaran, C.; Gordon, J.I.; Beach, D.; Jacks, T.; Hannon, G.J. Radiation-induced cell cycle arrest compromised by p21 deficiency. *Nature* **1995**, *377*, 552–557. [CrossRef] [PubMed]

31. Fero, M.L.; Rivkin, M.; Tasch, M.; Porter, P.; Carow, C.E.; Firpo, E.; Polyak, K.; Tsai, L.H.; Broudy, V.; Perlmutter, R.M.; et al. A syndrome of multiorgan hyperplasia with features of gigantism, tumorigenesis, and female sterility in p27(Kip1)-deficient mice. *Cell* **1996**, *85*, 733–744. [CrossRef]

32. Kiyokawa, H.; Kineman, R.D.; Manova-Todorova, K.O.; Soares, V.C.; Hoffman, E.S.; Ono, M.; Khanam, D.; Hayday, A.C.; Frohman, L.A.; Koff, A. Enhanced growth of mice lacking the cyclin-dependent kinase inhibitor function of p27(Kip1). *Cell* **1996**, *85*, 721–732. [CrossRef]

33. Nakayama, K.; Ishida, N.; Shirane, M.; Inomata, A.; Inoue, T.; Shishido, N.; Horii, I.; Loh, D.Y.; Nakayama, K. Mice lacking p27(Kip1) display increased body size, multiple organ hyperplasia, retinal dysplasia, and pituitary tumors. *Cell* **1996**, *85*, 707–720. [CrossRef]

34. Susaki, E.; Nakayama, K.I. Functional similarities and uniqueness of p27 and p57: Insight from a knock-in mouse model. *Cell Cycle* **2009**, *8*, 2497–2501. [CrossRef] [PubMed]

35. Matsumoto, A.; Takeishi, S.; Kanie, T.; Susaki, E.; Onoyama, I.; Tateishi, Y.; Nakayama, K.; Nakayama, K.I. p57 is required for quiescence and maintenance of adult hematopoietic stem cells. *Cell Stem Cell* **2011**, *9*, 262–271. [CrossRef] [PubMed]

36. Furutachi, S.; Matsumoto, A.; Nakayama, K.I.; Gotoh, Y. p57 controls adult neural stem cell quiescence and modulates the pace of lifelong neurogenesis. *EMBO J.* **2013**, *32*, 970–981. [CrossRef] [PubMed]

37. Yamazaki, S.; Iwama, A.; Takayanagi, S.; Morita, Y.; Eto, K.; Ema, H.; Nakauchi, H. Cytokine signals modulated via lipid rafts mimic niche signals and induce hibernation in hematopoietic stem cells. *EMBO J.* **2006**, *25*, 3515–3523. [CrossRef] [PubMed]

38. Cheng, T.; Rodrigues, N.; Shen, H.; Yang, Y.; Dombkowski, D.; Sykes, M.; Scadden, D.T. Hematopoietic stem cell quiescence maintained by p21cip1/waf1. *Science* **2000**, *287*, 1804–1808. [CrossRef] [PubMed]

39. van Os, R.; Kamminga, L.M.; Ausema, A.; Bystrykh, L.V.; Draijer, D.P.; van Pelt, K.; Dontje, B.; de Haan, G. A Limited role for p21Cip1/Waf1 in maintaining normal hematopoietic stem cell functioning. *Stem Cells* **2007**, *25*, 836–843. [CrossRef] [PubMed]

40. Umemoto, T.; Yamato, M.; Nishida, K.; Yang, J.; Tano, Y.; Okano, T. p57Kip2 is expressed in quiescent mouse bone marrow side population cells. *Biochem. Biophys. Res. Commun.* **2005**, *337*, 14–21. [CrossRef] [PubMed]

41. Zhang, Y.; Stehling-Sun, S.; Lezon-Geyda, K.; Juneja, S.C.; Coillard, L.; Chatterjee, G.; Wuertzer, C.A.; Camargo, F.; Perkins, A.S. PR-domain-containing Mds1-Evi1 is critical for long-term hematopoietic stem cell function. *Blood* **2011**, *118*, 3853–3861. [CrossRef] [PubMed]

42. Pechnick, R.N.; Zonis, S.; Wawrowsky, K.; Pourmorady, J.; Chesnokova, V. p21Cip1 restricts neuronal proliferation in the subgranular zone of the dentate gyrus of the hippocampus. *Proc. Natl. Acad. Sci. USA* **2008**, *105*, 1358–1363. [CrossRef] [PubMed]

43. Qiu, J.; Takagi, Y.; Harada, J.; Topalkara, K.; Wang, Y.; Sims, J.R.; Zheng, G.; Huang, P.; Ling, Y.; Scadden, D.T.; et al. p27Kip1 constrains proliferation of neural progenitor cells in adult brain under homeostatic and ischemic conditions. *Stem Cells* **2009**, *27*, 920–927. [CrossRef] [PubMed]

44. Encinas, J.M.; Michurina, T.V.; Peunova, N.; Park, J.H.; Tordo, J.; Peterson, D.A.; Fishell, G.; Koulakov, A.; Enikolopov, G. Division-coupled astrocytic differentiation and age-related depletion of neural stem cells in the adult hippocampus. *Cell Stem Cell* **2011**, *8*, 566–579. [CrossRef] [PubMed]

45. Tury, A.; Mairet-Coello, G.; DiCicco-Bloom, E. The cyclin-dependent kinase inhibitor p57Kip2 regulates cell cycle exit, differentiation, and migration of embryonic cerebral cortical precursors. *Cereb. Cortex* **2011**, *21*, 1840–1856. [CrossRef] [PubMed]

46. Dugas, J.C.; Ibrahim, A.; Barres, B.A. A crucial role for p57(Kip2) in the intracellular timer that controls oligodendrocyte differentiation. *J. Neurosci.* **2007**, *27*, 6185–6196. [CrossRef] [PubMed]

47. Martinez, L.A.; Chen, Y.; Fischer, S.M.; Conti, C.J. Coordinated changes in cell cycle machinery occur during keratinocyte terminal differentiation. *Oncogene* **1999**, *18*, 397–406. [CrossRef] [PubMed]

48. Gosselet, F.P.; Magnaldo, T.; Culerrier, R.M.; Sarasin, A.; Ehrhart, J.C. BMP2 and BMP6 control p57(Kip2) expression and cell growth arrest/terminal differentiation in normal primary human epidermal keratinocytes. *Cell Signal.* **2007**, *19*, 731–739. [CrossRef] [PubMed]

49. Hiromura, K.; Haseley, L.A.; Zhang, P.; Monkawa, T.; Durvasula, R.; Petermann, A.T.; Alpers, C.E.; Mundel, P.; Shankland, S.J. Podocyte expression of the CDK-inhibitor p57 during development and disease. *Kidney Int.* **2001**, *60*, 2235–2246. [CrossRef] [PubMed]

50. Figliola, R.; Maione, R. MyoD induces the expression of p57Kip2 in cells lacking p21Cip1/Waf1: Overlapping and distinct functions of the two cdk inhibitors. *J. Cell. Physiol.* **2004**, *200*, 468–475. [CrossRef] [PubMed]

51. Reynaud, E.G.; Pelpel, K.; Guillier, M.; Leibovitch, M.P.; Leibovitch, S.A. p57(Kip2) stabilizes the MyoD protein by inhibiting cyclin E-Cdk2 kinase activity in growing myoblasts. *Mol. Cell. Biol.* **1999**, *19*, 7621–7629. [CrossRef] [PubMed]

52. Figliola, R.; Busanello, A.; Vaccarello, G.; Maione, R. Regulation of p57(KIP2) during muscle differentiation: Role of Egr1, Sp1 and DNA hypomethylation. *J. Mol. Biol.* **2008**, *380*, 265–277. [CrossRef] [PubMed]

53. Zullo, A.; Mancini, F.P.; Schleip, R.; Wearing, S.; Yahia, L.; Klingler, W. The interplay between fascia, skeletl muscle, nerves, adipose tissue, inflammation and mechanical stress in musculo-fascial regeneration. *J. Gerontol. Geriatr.* **2017**, *65*, 271–283.

54. Zalc, A.; Hayashi, S.; Auradé, F.; Bröhl, D.; Chang, T.; Mademtzoglou, D.; Mourikis, P.; Yao, Z.; Cao, Y.; Birchmeier, C. Antagonistic regulation of p57kip2 by Hes/Hey downstream of Notch signaling and muscle regulatory factors regulates skeletal muscle growth arrest. *Development* **2014**, *141*, 2780–2790. [CrossRef] [PubMed]

55. Fukada, S.; Uezumi, A.; Ikemoto, M.; Masuda, S.; Segawa, M.; Tanimura, N.; Yamamoto, H.; Miyagoe-Suzuki, Y.; Takeda, S. Molecular signature of quiescent satellite cells in adult skeletal muscle. *Stem Cells* **2007**, *25*, 2448–2459. [CrossRef] [PubMed]

56. Naito, M.; Mori, M.; Inagawa, M.; Miyata, K.; Hashimoto, N.; Tanaka, S.; Asahara, H. Dnmt3a Regulates Proliferation of Muscle Satellite Cells via p57Kip2. *PLoS Genet.* **2016**, *12*, e1006167. [CrossRef] [PubMed]

57. Sommese, L.; Zullo, A.; Schiano, C.; Mancini, F.P.; Napoli, C. Possible Muscle Repair in the Human Cardiovascular System. *Stem Cell Rev* **2017**, *13*, 170–191. [CrossRef] [PubMed]

58. Johnson, A.M.; Kartha, C.C. Proliferation of murine c-kit(pos) cardiac stem cells stimulated with IGF-1 is associated with Akt-1 mediated phosphorylation and nuclear export of FoxO3a and its effect on downstream cell cycle regulators. *Growth Factors* **2014**, *32*, 53–62. [CrossRef] [PubMed]

59. Zhang, Y.; Mignone, J.; MacLellan, W.R. Cardiac Regeneration and Stem Cells. *Physiol. Rev.* **2015**, *95*, 1189–1204. [CrossRef] [PubMed]

60. Evans-Anderson, H.J.; Alfieri, C.M.; Yutzey, K.E. Regulation of cardiomyocyte proliferation and myocardial growth during development by FOXO transcription factors. *Circ. Res.* **2008**, *102*, 686–694. [CrossRef] [PubMed]

61. Di Stefano, V.; Martelli, F. Removing the brakes to cardiomyocyte cell cycle. *Cell Cycle* **2011**, *10*, 1176–1177. [CrossRef] [PubMed]

62. Haley, S.A.; Zhao, T.; Zou, L.; Klysik, J.E.; Padbury, J.F.; Kochilas, L.K. Forced expression of the cell cycle inhibitor p57Kip2 in cardiomyocytes attenuates ischemia-reperfusion injury in the mouse heart. *BMC Physiol.* **2008**, *8*, 4. [CrossRef] [PubMed]

63. Xiao, C.; Gao, L.; Hou, Y.; Xu, C.; Chang, N.; Wang, F.; Hu, K.; He, A.; Luo, Y.; Wang, J.; et al. Chromatin-remodelling factor Brg1 regulates myocardial proliferation and regeneration in zebrafish. *Nat. Commun.* **2016**, *7*, 13787. [CrossRef] [PubMed]

64. Borriello, A.; Caldarelli, I.; Bencivenga, D.; Cucciolla, V.; Oliva, A.; Usala, E.; Danise, P.; Ronzoni, L.; Perrotta, S.; Della Ragione, F. p57Kip2 is a downstream effector of BCR-ABL kinase inhibitors in chronic myelogenous leukemia cells. *Carcinogenesis* **2011**, *32*, 10–18. [CrossRef] [PubMed]

65. Scandura, J.M.; Boccuni, P.; Massagué, J.; Nimer, S.D. Transforming growth factor beta-induced cell cycle arrest of human hematopoietic cells requires p57KIP2 up-regulation. *Proc. Natl. Acad. Sci. USA* **2004**, *101*, 15231–15236. [CrossRef] [PubMed]

66. Castelo-Branco, G.; Wagner, J.; Rodriguez, F.J.; Kele, J.; Sousa, K.; Rawal, N.; Pasolli, H.A.; Fuchs, E.; Kitajewski, J.; Arenas, E. Differential regulation of midbrain dopaminergic neuron development by Wnt-1, Wnt-3a, and Wnt-5a. *Proc. Natl. Acad. Sci. USA* **2003**, *100*, 12747–12752. [CrossRef] [PubMed]

67. Jia, J.; Lin, M.; Zhang, L.; York, J.P.; Zhang, P. The Notch signaling pathway controls the size of the ocular lens by directly suppressing p57Kip2 expression. *Mol. Cell. Biol.* **2007**, *27*, 7236–7247. [CrossRef] [PubMed]

68. Georgia, S.; Soliz, R.; Li, M.; Zhang, P.; Bhushan, A. p57 and Hes1 coordinate cell cycle exit with self-renewal of pancreatic progenitors. *Dev. Biol.* **2006**, *298*, 22–31. [CrossRef] [PubMed]

69. Hatada, I.; Mukai, T. Genomic imprinting of p57KIP2, a cyclin-dependent kinase inhibitor, in mouse. *Nat. Genet.* **1995**, *11*, 204–206. [CrossRef] [PubMed]

70. Potikha, T.; Kassem, S.; Haber, E.P.; Ariel, I.; Glaser, B. p57Kip2 (cdkn1c): Sequence, splice variants and unique temporal and spatial expression pattern in the rat pancreas. *Lab. Investig.* **2005**, *85*, 364–375. [CrossRef] [PubMed]

71. Lefebvre, L.; Mar, L.; Bogutz, A.; Oh-McGinnis, R.; Mandegar, M.A.; Paderova, J.; Gertsenstein, M.; Squire, JA.; Nagy, A. The interval between Ins2 and Ascl2 is dispensable for imprinting centre function in the murine Beckwith-Wiedemann region. *Hum. Mol. Genet.* **2009**, *18*, 4255–4267. [CrossRef] [PubMed]

72. Mancini-DiNardo, D.; Steele, S.J.; Ingram, R.S.; Tilghman, S.M. A differentially methylated region within the gene Kcnq1 functions as an imprinted promoter and silencer. *Hum. Mol. Genet.* **2003**, *12*, 283–294. [CrossRef] [PubMed]

73. Smith, A.C.; Choufani, S.; Ferreira, J.C.; Weksberg, R. Growth regulation, imprinted genes, and chromosome 11p15.5. *Pediatr. Res.* **2007**, *61 Pt 2*, 43r–47r. [CrossRef] [PubMed]

74. Lee, M.P.; DeBaun, M.R.; Mitsuya, K.; Galonek, H.L.; Brandenburg, S.; Oshimura, M.; Feinberg, A.P. Loss of imprinting of a paternally expressed transcript, with antisense orientation to KVLQT1, occurs frequently in Beckwith-Wiedemann syndrome and is independent of insulin-like growth factor II imprinting. *Proc. Natl. Acad. Sci. USA* **1999**, *96*, 5203–5208. [CrossRef] [PubMed]

75. Hark, A.T.; Schoenherr, C.J.; Katz, D.J.; Ingram, R.S.; Levorse, J.M.; Tilghman, S.M. CTCF mediates methylation-sensitive enhancer-blocking activity at the H19/Igf2 locus. *Nature* **2000**, *405*, 486–489. [CrossRef] [PubMed]

76. Cucciolla, V.; Borriello, A.; Criscuolo, M.; Sinisi, A.A.; Bencivenga, D.; Tramontano, A.; Scudieri, A.C.; Oliva, A.; Zappia, V.; Della Ragione, F. Histone deacetylase inhibitors upregulate p57Kip2 level by enhancing its expression through Sp1 transcription factor. *Carcinogenesis* **2008**, *29*, 560–567. [CrossRef] [PubMed]

77. Gartel, A.L.; Goufman, E.; Najmabadi, F.; Tyner, A.L. Sp1 and Sp3 activate p21 (WAF1/CIP1) gene transcription in the Caco-2 colon adenocarcinoma cell line. *Oncogene* **2000**, *19*, 5182–5188. [CrossRef] [PubMed]

78. Yokota, T.; Matsuzaki, Y.; Miyazawa, K.; Zindy, F.; Roussel, M.F.; Sakai, T. Histone deacetylase inhibitors activate INK4d gene through Sp1 site in its promoter. *Oncogene* **2004**, *23*, 5340–5349. [CrossRef] [PubMed]

79. Billing, M.; Rörby, E.; May, G.; Tipping, A.J.; Soneji, S.; Brown, J.; Salminen, M.; Karlsson, G.; Enver, T.; Karlsson, S. A network including TGFbeta/Smad4, Gata2, and p57 regulates proliferation of mouse hematopoietic progenitor cells. *Exp. Hematol.* **2016**, *44*, 399–409. [CrossRef] [PubMed]

80. Li, X.; Ruan, X.; Zhang, P.; Yu, Y.; Gao, M.; Yuan, S.; Zhao, Z.; Yang, J.; Zhao, L. TBX3 promotes proliferation of papillary thyroid carcinoma cells through facilitating PRC2-mediated p57(KIP2) repression. *Oncogene* **2018**. [CrossRef] [PubMed]

81. Kikuchi, T.; Toyota, M.; Itoh, F.; Suzuki, H.; Obata, T.; Yamamoto, H.; Kakiuchi, H.; Kusano, M.; Issa, J.P.; Tokino, T.; et al. Inactivation of p57KIP2 by regional promoter hypermethylation and histone deacetylation in human tumors. *Oncogene* **2002**, *21*, 2741–2749. [CrossRef] [PubMed]

82. Schmidt-Edelkraut, U.; Daniel, G.; Hoffmann, A.; Spengler, D. Zac1 regulates cell cycle arrest in neuronal progenitors via Tcf4. *Mol. Cell. Biol.* **2014**, *34*, 1020–1030. [CrossRef] [PubMed]

83. Pfurr, S.; Chu, Y.H.; Bohrer, C.; Greulich, F.; Beattie, R.; Mammadzada, K.; Hils, M.; Arnold, S.J.; Taylor, V.; Schachtrup, K.; et al. The E2A splice variant E47 regulates the differentiation of projection neurons via p57(KIP2) during cortical development. *Development* **2017**, *144*, 3917–3931. [CrossRef] [PubMed]

84. Jia, H.; Cong, Q.; Chua, J.F.; Liu, H.; Xia, X.; Zhang, X.; Lin, J.; Habib, S.L.; Ao, J.; Zuo, Q.; et al. p57Kip2 is an unrecognized DNA damage response effector molecule that functions in tumor suppression and chemoresistance. *Oncogene* **2015**, *34*, 3568–3581. [CrossRef] [PubMed]

85. Kim, T.H.; Shivdasani, R.A. Genetic evidence that intestinal Notch functions vary regionally and operate through a common mechanism of Math1 repression. *J. Biol. Chem.* **2011**, *286*, 11427–11433. [CrossRef] [PubMed]

86. Alheim, K.; Corness, J.; Samuelsson, M.K.; Bladh, L.G.; Murata, T.; Nilsson, T.; Okret, S. Identification of a functional glucocorticoid response element in the promoter of the cyclin-dependent kinase inhibitor p57Kip2. *J. Mol. Endocrinol.* **2003**, *30*, 359–368. [CrossRef] [PubMed]

87. Lin, K.T.; Wang, L.H. New dimension of glucocorticoids in cancer treatment. *Steroids* **2016**, *111*, 84–88. [CrossRef] [PubMed]

88. John, R.M.; Ainscough, J.F.; Barton, S.C.; Surani, M.A. Distant cis-elements regulate imprinted expression of the mouse p57(Kip2) (Cdkn1c) gene: Implications for the human disorder, Beckwith-Wiedemann syndrome. *Hum. Mol. Genet.* **2001**, *10*, 1601–1609. [CrossRef] [PubMed]

89. Gurrieri, F.; Zollino, M.; Oliva, A.; Pascali, V.; Orteschi, D.; Pietrobono, R.; Camporeale, A.; Coll Vidal, M.; Partemi, S.; Brugada, R.; et al. Mild Beckwith-Wiedemann and severe long-QT syndrome due to deletion of the imprinting center 2 on chromosome 11p. *Eur. J. Hum. Genet.* **2013**, *21*, 965–969. [CrossRef] [PubMed]

90. John, R.M.; Surani, M.A. Genomic imprinting, mammalian evolution, and the mystery of egg-laying mammals. *Cell* **2000**, *101*, 585–588. [CrossRef]

91. Surani, M.A.; Barton, S.C.; Norris, M.L. Development of reconstituted mouse eggs suggests imprinting of the genome during gametogenesis. *Nature* **1984**, *308*, 548–550. [CrossRef] [PubMed]

92. McGrath, J.; Solter, D. Completion of mouse embryogenesis requires both the maternal and paternal genomes. *Cell* **1984**, *37*, 179–183. [CrossRef]

93. Monk, D.; Sanches, R.; Arnaud, P.; Apostolidou, S.; Hills, F.A.; Abu-Amero, S.; Murrell, A.; Friess, H.; Reik, W.; Stanier, P.; et al. Imprinting of IGF2 P0 transcript and novel alternatively spliced INS-IGF2 isoforms show differences between mouse and human. *Hum. Mol. Genet.* **2006**, *15*, 1259–1269. [CrossRef] [PubMed]

94. Surani, M.A. Imprinting and the initiation of gene silencing in the germ line. *Cell* **1998**, *93*, 309–312. [CrossRef]

95. Hatada, I.; Inazawa, J.; Abe, T.; Nakayama, M.; Kaneko, Y.; Jinno, Y.; Niikawa, N.; Ohashi, H.; Fukushima, Y.; Iida, K.; et al. Genomic imprinting of human p57KIP2 and its reduced expression in Wilms' tumors. *Hum. Mol. Genet.* **1996**, *5*, 783–788. [CrossRef] [PubMed]

96. John, R.M.; Lefebvre, L. Developmental regulation of somatic imprints. *Differentiation* **2011**, *81*, 270–280. [CrossRef] [PubMed]

97. Bhogal, B.; Arnaudo, A.; Dymkowski, A.; Best, A.; Davis, T.L. Methylation at mouse Cdkn1c is acquired during postimplantation development and functions to maintain imprinted expression. *Genomics* **2004**, *84*, 961–970. [CrossRef] [PubMed]

98. Rossi, M.N.; Andresini, O.; Matteini, F.; Maione, R. Transcriptional regulation of p57(kip2) expression during development, differentiation and disease. *Front. Biosci.* **2018**, *23*, 83–108.

99. Fan, T.; Hagan, J.P.; Kozlov, S.V.; Stewart, C.L.; Muegge, K. Lsh controls silencing of the imprinted Cdkn1c gene. *Development* **2005**, *132*, 635–644. [CrossRef] [PubMed]

100. Riemenschneider, M.J.; Reifenberger, J.; Reifenberger, G. Frequent biallelic inactivation and transcriptional silencing of the DIRAS3 gene at 1p31 in oligodendroglial tumors with 1p loss. *Int. J. Cancer* **2008**, *122*, 2503–2510. [CrossRef] [PubMed]

101. Neumann, L.C.; Weinhäusel, A.; Thomas, S.; Horsthemke, B.; Lohmann, D.R.; Zeschnigk, M. EFS shows biallelic methylation in uveal melanoma with poor prognosis as well as tissue-specific methylation. *BMC Cancer* **2011**, *11*, 380. [CrossRef] [PubMed]

102. Andresini, O.; Ciotti, A.; Rossi, M.N.; Battistelli, C.; Carbone, M.; Maione, R. A cross-talk between DNA methylation and H3 lysine 9 dimethylation at the KvDMR1 region controls the induction of Cdkn1c in muscle cells. *Epigenetics* **2016**, *11*, 791–803. [CrossRef] [PubMed]

103. Mohammad, F.; Mondal, T.; Guseva, N.; Pandey, G.K.; Kanduri, C. Kcnq1ot1 noncoding RNA mediates transcriptional gene silencing by interacting with Dnmt1. *Development* **2010**, *137*, 2493–2499. [CrossRef] [PubMed]

104. Li, Y.; Nagai, H.; Ohno, T.; Yuge, M.; Hatano, S.; Ito, E.; Mori, N.; Saito, H.; Kinoshita, T. Aberrant DNA methylation of p57(KIP2) gene in the promoter region in lymphoid malignancies of B-cell phenotype. *Blood* **2002**, *100*, 2572–2577. [CrossRef] [PubMed]

105. Li, C.; Choi, H.P.; Wang, X.; Wu, F.; Chen, X.; Lü, X.; Jing, R.; Ryu, H.; Wang, X.; Azadzoi, K.M.; et al. Post-Translational Modification of Human Histone by Wide Tolerance of Acetylation. *Cells* **2017**, *6*, E34. [CrossRef] [PubMed]

106. Faundes, V.; Newman, W.G.; Bernardini, L.; Canham, N.; Clayton-Smith, J.; Dallapiccola, B.; Davies, S.J.; Demos, M.K.; Goldman, A.; Gill, H.; et al. Histone Lysine Methylases and Demethylases in the Landscape of Human Developmental Disorders. *Am. J. Hum. Genet.* **2018**, *102*, 175–187. [CrossRef] [PubMed]

107. Umlauf, D.; Goto, Y.; Cao, R.; Cerqueira, F.; Wagscha, A.; Zhang, Y.; Feil, R. Imprinting along the Kcnq1 domain on mouse chromosome 7 involves repressive histone methylation and recruitment of Polycomb group complexes. *Nat. Genet.* **2004**, *36*, 1296–1300. [CrossRef] [PubMed]

108. Lewis, A.; Mitsuya, K.; Umlauf, D.; Smith, P.; Dean, W.; Walter, J.; Higgins, M.; Feil, R.; Reik, W. Imprinting on distal chromosome 7 in the placenta involves repressive histone methylation independent of DNA methylation. *Nat. Genet.* **2004**, *36*, 1291–1295. [CrossRef] [PubMed]

109. Algar, E.M.; Muscat, A.; Dagar, V.; Rickert, C.; Chow, C.W.; Biegel, J.A.; Ekert, P.G.; Saffery, R.; Craig, J.; Johnstone, R.W.; et al. Imprinted CDKN1C is a tumor suppressor in rhabdoid tumor and activated by restoration of SMARCB1 and histone deacetylase inhibitors. *PLoS ONE* **2009**, *4*, e4482. [CrossRef] [PubMed]

110. Attia, M.; Rachez, C.; De Pauw, A.; Avner, P.; Rogner, U.C. Nap1l2 promotes histone acetylation activity during neuronal differentiation. *Mol. Cell. Biol.* **2007**, *27*, 6093–6102. [CrossRef] [PubMed]

111. Cao, L.L.; Yue, Z.; Liu, L.; Pei, L.; Yin, Y.; Qin, L.; Zhao, J.; Liu, H.; Wang, H.; Jia, M. The expression of histone deacetylase HDAC1 correlates with the progression and prognosis of gastrointestinal malignancy. *Oncotarget* **2017**, *8*, 39241–39253. [CrossRef] [PubMed]

112. Jo, Y.K.; Park, N.Y.; Shin, J.H.; Jo, D.S.; Bae, J.E.; Choi, E.S.; Maeng, S.; Jeon, H.B.; Roh, S.A.; Chang, J.W.; et al. Up-regulation of UVRAG by HDAC1 Inhibition Attenuates 5FU-induced Cell Death in HCT116 Colorectal Cancer Cells. *Anticancer Res.* **2018**, *38*, 271–277. [PubMed]

113. Zhou, H.; Cai, Y.; Liu, D.; Li, M.; Sha, Y.; Zhang, W.; Wang, K.; Gong, J.; Tang, N.; Huang, A.; et al. Pharmacological or transcriptional inhibition of both HDAC1 and 2 leads to cell cycle blockage and apoptosis via p21(Waf1/Cip1) and p19(INK4d) upregulation in hepatocellular carcinoma. *Cell Prolif.* **2018**. [CrossRef] [PubMed]

114. Wakahara, M.; Sakabe, T.; Kubouchi, Y.; Hosoya, K.; Hirooka, Y.; Yurugi, Y.; Nosaka, K.; Shiomi, T.; Nakamura, H.; Umekita, Y. Subcellular Localization of maspin correlates with histone deacetylase 1 expression in human breast cancer. *Anticancer Res.* **2017**, *37*, 5071–5077. [PubMed]

115. Cai, M.H.; Xu, X.G.; Yan, S.L.; Sun, Z.; Ying, Y.; Wang, B.K.; Tu, Y.X. Depletion of HDAC1, 7 and 8 by histone deacetylase inhibition confers elimination of pancreatic cancer stem cells in combination with gemcitabine. *Sci. Rep.* **2018**, *8*, 1621. [CrossRef] [PubMed]

116. Ropero, S.; Ballestar, E.; Alaminos, M.; Arango, D.; Schwartz, S.Jr.; Esteller, M. Transforming pathways unleashed by a HDAC2 mutation in human cancer. *Oncogene* **2008**, *27*, 4008–4012. [CrossRef] [PubMed]

117. Li, S.; Wang, F.; Qu, Y.; Chen, X.; Gao, M.; Yang, J.; Zhang, D.; Zhang, N.; Li, W.; Liu, H. HDAC2 regulates cell proliferation, cell cycle progression and cell apoptosis in esophageal squamous cell carcinoma EC9706 cells. *Oncol. Lett.* **2017**, *13*, 403–409. [CrossRef] [PubMed]

118. Hulsurkar, M.; Li, Z.; Zhang, Y.; Li, X.; Zheng, D.; Li, W. Beta-adrenergic signaling promotes tumor angiogenesis and prostate cancer progression through HDAC2-mediated suppression of thrombospondin-1. *Oncogene* **2017**, *36*, 1525–1536. [CrossRef] [PubMed]

119. Song, J.; Noh, J.H.; Lee, J.H.; Eun, J.W.; Ahn, Y.M.; Kim, S.Y.; Lee, S.H.; Park, W.S.; Yoo, N.J.; Lee, J.Y.; et al. Increased expression of histone deacetylase 2 is found in human gastric cancer. *APMIS* **2005**, *113*, 264–268. [CrossRef] [PubMed]

120. Iida, A.; Iwagawa, T.; Baba, Y.; Satoh, S.; Mochizuki, Y.; Nakauchi, H.; Furukawa, T.; Koseki, H.; Murakami, A.; Watanabe, S. Roles of histone H3K27 trimethylase Ezh2 in retinal proliferation and differentiation. *Dev. Neurobiol.* **2015**, *75*, 947–960. [CrossRef] [PubMed]

121. Heinen, A.; Tzekova, N.; Graffmann, N.; Torres, K.J.; Uhrberg, M.; Hartung, H.P.; Küry, P. Histone methyltransferase enhancer of zeste homolog 2 regulates Schwann cell differentiation. *Glia* **2012**, *60*, 1696–1708. [CrossRef] [PubMed]

122. Mitsuya, K.; Meguro, M.; Lee, M.P.; Katoh, M.; Schulz, T.C.; Kugoh, H.; Yoshida, M.A.; Niikawa, N.; Feinberg, A.P.; Oshimura, M. LIT1, an imprinted antisense RNA in the human KvLQT1 locus identified by screening for differentially expressed transcripts using monochromosomal hybrids. *Hum. Mol. Genet.* **1999**, *8*, 1209–1217. [CrossRef] [PubMed]

123. Du, M.; Zhou, W.; Beatty, L.G.; Weksberg, R.; Sadowski, P.D. The KCNQ1OT1 promoter, a key regulator of genomic imprinting in human chromosome 11p15.5. *Genomics* **2004**, *84*, 288–300. [CrossRef] [PubMed]

124. Pandey, R.R.; Mondal, T.; Mohammad, F.; Enroth, S.; Redrup, L.; Komorowski, J.; Nagano, T.; Mancini-Dinardo, D.; Kanduri, C. Kcnq1ot1 antisense noncoding RNA mediates lineage-specific transcriptional silencing through chromatin-level regulation. *Mol. Cell* **2008**, *32*, 232–246. [CrossRef] [PubMed]

125. Zhang, E.; He, X.; Yin, D.; Han, L.; Qiu, M.; Xu, T.; Xia, R.; Xu, L.; Yin, R.; De, W. Increased expression of long noncoding RNA TUG1 predicts a poor prognosis of gastric cancer and regulates cell proliferation by epigenetically silencing of p57. *Cell Death Dis.* **2016**, *7*, e2109. [CrossRef] [PubMed]

126. Zhang, E.; Yin, D.; Han, L.; He, X.; Si, X.; Chen, W.; Xia, R.; Xu, T.; Gu, D.; De, W.; et al. E2F1-induced upregulation of long noncoding RNA LINC00668 predicts a poor prognosis of gastric cancer and promotes cell proliferation through epigenetically silencing of CKIs. *Oncotarget* **2016**, *7*, 23212–23226. [CrossRef] [PubMed]

127. Yang, F.; Zhang, L.; Huo, X.S.; Yuan, J.H.; Xu, D.; Yuan, S.X.; Zhu, N.; Zhou, W.P.; Yang, G.S.; Wang, Y.Z.; et al. Long noncoding RNA high expression in hepatocellular carcinoma facilitates tumor growth through enhancer of zeste homolog 2 in humans. *Hepatology* **2011**, *54*, 1679–1689. [CrossRef] [PubMed]

128. Brioude, F.; Kalish, J.M.; Mussa, A.; Foster, A.C.; Bliek, J.; Ferrero, G.B.; Boonen, S.E.; Cole, T.; Baker, R.; Bertoletti, M.; et al. Expert consensus document: Clinical and molecular diagnosis, screening and management of Beckwith-Wiedemann syndrome: An international consensus statement. *Nat. Rev. Endocrinol.* **2018**, *14*, 229–249. [CrossRef] [PubMed]

129. Mussa, A.; Russo, S.; De Crescenzo, A.; Freschi, A.; Calzari, L.; Maitz, S.; Macchiaiolo, M.; Molinatto, C.; Baldassarre, G.; Mariani, M.; et al. (Epi)genotype-phenotype correlations in Beckwith-Wiedemann syndrome. *Eur. J. Hum. Genet.* **2016**, *24*, 183–190. [CrossRef] [PubMed]

130. Heide, S.; Chantot-Bastaraud, S.; Keren, B.; Harbison, M.D.; Azzi, S.; Rossignol, S.; Michot, C.; Lackmy-Port Lys, M.; Demeer, B.; Heinrichs, C.; et al. Chromosomal rearrangements in the 11p15 imprinted region: 17 new 11p15.5 duplications with associated phenotypes and putative functional consequences. *J. Med. Genet.* **2018**, *55*, 205–213. [CrossRef] [PubMed]

131. Mussa, A.; Russo, S.; de Crescenzo, A.; Freschi, A.; Calzari, L.; Maitz, S.; Macchiaiolo, M.; Molinatto, C.; Baldassarre, G.; Mariani, M.; et al. Fetal growth patterns in Beckwith-Wiedemann syndrome. *Clin. Genet.* **2016**, *90*, 21–27. [CrossRef] [PubMed]

132. Cooper, W.N.; Schofield, P.N.; Reik, W.; Macdonald, F.; Maher, E.R. Molecular subtypes and phenotypic expression of Beckwith-Wiedemann syndrome. *Eur. J. Hum. Genet.* **2005**, *13*, 1025–1032. [CrossRef] [PubMed]

133. Brioude, F.; Luharia, A.; Evans, G.A.; Raza, H.; Haire, A.C.; Grundy, R.; Bowdin, S.C.; Riccio, A.; Sebastio, G.; Bliek, J.; et al. Mutations of the Imprinted CDKN1C Gene as a Cause of the Overgrowth Beckwith-Wiedemann Syndrome: Clinical Spectrum and Functional Characterization. *Hum. Mutat.* **2015**, *36*, 894–902. [CrossRef] [PubMed]

134. Bourcigaux, N.; Gaston, V.; Logié, A.; Bertagna, X.; Le Bouc, Y.; Gicquel, C. High expression of cyclin E and G1 CDK and loss of function of p57KIP2 are involved in proliferation of malignant sporadic adrenocortical tumors. *J. Clin. Endocrinol. Metab.* **2000**, *85*, 322–330. [CrossRef] [PubMed]

135. Duquesnes, N.; Callot, C.; Jeannot, P.; Daburon, V.; Nakayama, K.I.; Manenti, S.; Davy, A.; Besson, A. p57(Kip2) knock-in mouse reveals CDK-independent contribution in the development of Beckwith-Wiedemann syndrome. *J. Pathol.* **2016**, *239*, 250–261. [CrossRef] [PubMed]

136. Arboleda, V.A.; Lee, H.; Parnaik, R.; Fleming, A.; Banerjee, A.; Ferraz-de-Souza, B.; Délot, E.C.; Rodriguez-Fernandez, I.A.; Braslavsky, D.; Bergadá, I.; et al. Mutations in the PCNA-binding domain of CDKN1C cause IMAGe syndrome. *Nat. Genet.* **2012**, *44*, 788–792. [CrossRef] [PubMed]

137. Dias, R.P.; Maher, E.R. An imprinted IMAGe: Insights into growth regulation through genomic analysis of a rare disease. *Genome Med.* **2012**, *4*, 60. [CrossRef] [PubMed]

138. Begemann, M.; Spengler, S.; Kanber, D.; Haake, A.; Baudis, M.; Leisten, I.; Binder, G.; Markus, S.; Rupprecht, T.; Segerer, H.; et al. Silver-Russell patients showing a broad range of ICR1 and ICR2 hypomethylation in different tissues. *Clin. Genet.* **2011**, *80*, 83–88. [CrossRef] [PubMed]

139. Eggermann, K.; Bliek, J.; Brioude, F.; Algar, E.; Buiting, K.; Russo, S.; Tümer, Z.; Monk, D.; Moore, G.; Antoniadi, T.; et al. EMQN best practice guidelines for the molecular genetic testing and reporting of chromosome 11p15 imprinting disorders: Silver-Russell and Beckwith-Wiedemann syndrome. *Eur. J. Hum. Genet.* **2016**, *24*, 1377–1387. [CrossRef] [PubMed]

140. Monk, D.; Wakeling, E.L.; Proud, V.; Hitchins, M.; Abu-Amero, S.N.; Stanier, P.; Preece, M.A.; Moore, G.E. Duplication of 7p11.2-p13, including GRB10, in Silver-Russell syndrome. *Am. J. Hum. Genet.* **2000**, *66*, 36–46. [CrossRef] [PubMed]

141. Brioude, F.; Oliver-Petit, I.; Blaise, A.; Praz, F.; Rossignol, S.; Le Jule, M.; Thibaud, N.; Faussat, A.M.; Tauber, M.; Le Bouc, Y.; et al. CDKN1C mutation affecting the PCNA-binding domain as a cause of familial Russell Silver syndrome. *J. Med. Genet.* **2013**, *50*, 823–830. [CrossRef] [PubMed]

142. Iolascon, A.; Giordani, L.; Moretti, A.; Basso, G.; Borriello, A.; Della Ragione, F. Analysis of CDKN2A, CDKN2B, CDKN2C, and cyclin Ds gene status in hepatoblastoma. *Hepatology* **1998**, *27*, 989–995. [CrossRef] [PubMed]

143. Iolascon, A.; Giordani, L.; Moretti, A.; Tonini, G.P.; Lo Cunsolo, C.; Mastropietro, S.; Borriello, A.; Della Ragione, F. Structural and functional analysis of cyclin-dependent kinase inhibitor genes (CDKN2A, CDKN2B, and CDKN2C) in neuroblastoma. *Pediatr. Res.* **1998**, *43*, 139–144. [CrossRef] [PubMed]

144. Pateras, I.S.; Apostolopoulou, K.; Niforou, K.; Kotsinas, A.; Gorgoulis, V.G. p57KIP2: "Kip"ing the cell under control. *Mol. Cancer Res.* **2009**, *7*, 1902–1919. [CrossRef] [PubMed]

145. Bencivenga, D.; Caldarelli, I.; Stampone, E.; Mancini, F.P.; Balestrieri, M.L.; Della Ragione, F.; Borriello, A. p27(Kip1) and human cancers: A reappraisal of a still enigmatic protein. *Cancer Lett.* **2017**, *403*, 354–365. [CrossRef] [PubMed]

146. Bonilla, F.; Orlow, I.; Cordon-Cardo, C. Mutational study of p16CDKN2/MTS1/INK4A and p57KIP2 genes in hepatocellular carcinoma. *Int. J. Oncol.* **1998**, *12*, 583–588. [CrossRef] [PubMed]

147. Shin, J.Y.; Kim, H.S.; Lee, K.S.; Kim, J.; Park, J.B.; Won, M.H.; Chae, S.W.; Choi, Y.H.; Choi, K.C.; Park, Y.E.; et al. Mutation and expression of the p27KIP1 and p57KIP2 genes in human gastric cancer. *Exp. Mol. Med.* **2000**, *32*, 79–83. [CrossRef] [PubMed]

148. Oya, M.; Schulz, W.A. Decreased expression of p57(KIP2)mRNA in human bladder cancer. *Br. J. Cancer* **2000**, *83*, 626–631. [CrossRef] [PubMed]

149. Hoffmann, M.J.; Flor, A.R.; Seifert, H.H.; Schulz, W.A. Multiple mechanisms downregulate CDKN1C in human bladder cancer. *Int. J. Cancer* **2005**, *114*, 406–413. [CrossRef] [PubMed]

150. Sato, N.; Matsubayashi, H.; Abe, T.; Fukushima, N.; Goggins, M. Epigenetic down-regulation of CDKN1C/p57KIP2 in pancreatic ductal neoplasms identified by gene expression profiling. *Clin. Cancer Res.* **2005**, *11*, 4681–4688. [CrossRef] [PubMed]

151. Giovannoni, I.; Boldrini, R.; Benedetti, M.C.; Inserra, A.; De Pasquale, M.D.; Francalanci, P. Pediatric adrenocortical neoplasms: Immunohistochemical expression of p57 identifies loss of heterozygosity and abnormal imprinting of the 11p15.5. *Pediatr. Res.* **2017**, *81*, 468–472. [CrossRef] [PubMed]

152. Sun, Y.; Jin, S.D.; Zhu, Q.; Han, L.; Feng, J.; Lu, X.Y.; Wang, W.; Wang, F.; Guo, R.H. Long non-coding RNA LUCAT1 is associated with poor prognosis in human non-small lung cancer and regulates cell proliferation via epigenetically repressing p21 and p57 expression. *Oncotarget* **2017**, *8*, 28297–28311. [CrossRef] [PubMed]

153. Qiu, Z.; Li, Y.; Zeng, B.; Guan, X.; Li, H. Downregulated CDKN1C/p57(kip2) drives tumorigenesis and associates with poor overall survival in breast cancer. *Biochem. Biophys. Res. Commun.* **2018**, *497*, 187–193. [CrossRef] [PubMed]

154. Radujkovic, A.; Dietrich, S.; Andrulis, M.; Benner, A.; Longerich, T.; Pellagatti, A.; Nanda, K.; Giese, T.; Germing, U.; Baldus, S.; et al. Expression of CDKN1C in the bone marrow of patients with myelodysplastic syndrome and secondary acute myeloid leukemia is associated with poor survival after conventional chemotherapy. *Int. J. Cancer* **2016**, *139*, 1402–1413. [CrossRef] [PubMed]

155. Guo, H.; Nan, K.; Hu, T.; Meng, J.; Hui, W.; Zhang, X.; Qin, H.; Sui, C. Prognostic significance of co-expression of nm23 and p57 protein in hepatocellular carcinoma. *Hepatol. Res.* **2010**, *40*, 1107–1116. [CrossRef] [PubMed]

156. Kobatake, T.; Yano, M.; Toyooka, S.; Tsukuda, K.; Dote, H.; Kikuchi, T.; Toyota, M.; Ouchida, M.; Aoe, M.; Date, H.; et al. Aberrant methylation of p57KIP2 gene in lung and breast cancers and malignant mesotheliomas. *Oncol. Rep.* **2004**, *12*, 1087–1092. [CrossRef] [PubMed]

157. Shen, L.; Toyota, M.; Kondo, Y.; Obata, T.; Daniel, S.; Pierce, S.; Imai, K.; Kantarjian, H.M.; Issa, J.P.; Garcia-Manero, G. Aberrant DNA methylation of p57KIP2 identifies a cell-cycle regulatory pathway with prognostic impact in adult acute lymphocytic leukemia. *Blood* **2003**, *101*, 4131–4136. [CrossRef] [PubMed]

158. Roman-Gomez, J.; Jimenez-Velasco, A.; Agirre, X.; Prosper, F.; Heiniger, A.; Torres, A. Lack of CpG island methylator phenotype defines a clinical subtype of T-cell acute lymphoblastic leukemia associated with good prognosis. *J. Clin. Oncol.* **2005**, *23*, 7043–7049. [CrossRef] [PubMed]

159. Zohny, S.F.; Baothman, O.A.; El-Shinawi, M.; Al-Malki, A.L.; Zamzami, M.A.; Choudhry, H. The KIP/CIP family members p21^{Waf1/Cip1} and p57^{Kip2} as diagnostic markers for breast cancer. *Cancer Biomark.* **2017**, *18*, 413–423. [CrossRef] [PubMed]

160. Mishra, S.; Lin, C.L.; Huang, T.H.; Bouamar, H.; Sun, L.Z. MicroRNA-21 inhibits p57Kip2 expression in prostate cancer. *Mol. Cancer* **2014**, *13*, 212. [CrossRef] [PubMed]

161. Kim, Y.K.; Yu, J.; Han, T.S.; Park, S.Y.; Namkoong, B.; Kim, D.H.; Hur, K.; Yoo, M.W.; Lee, H.J.; Yang, H.K.; et al. Functional links between clustered microRNAs: Suppression of cell-cycle inhibitors by microRNA clusters in gastric cancer. *Nucleic Acids Res.* **2009**, *37*, 1672–1681. [CrossRef] [PubMed]

162. Zhang, J.; Gong, X.; Tian, K.; Chen, D.; Sun, J.; Wang, G.; Guo, M. miR-25 promotes glioma cell proliferation by targeting CDKN1C. *Biomed. Pharmacother.* **2015**, *71*, 7–14. [CrossRef] [PubMed]

163. Nass, D.; Rosenwald, S.; Meiri, E.; Gilad, S.; Tabibian-Keissar, H.; Schlosberg, A.; Kuker, H.; Sion-Vardy, N.; Tobar, A.; Kharenko, O.; et al. MiR-92b and miR-9/9* are specifically expressed in brain primary tumors and can be used to differentiate primary from metastatic brain tumors. *Brain Pathol.* **2009**, *19*, 375–383. [CrossRef] [PubMed]

164. Fornari, F.; Gramantieri, L.; Ferracin, M.; Veronese, A.; Sabbioni, S.; Calin, G.A.; Grazi, G.L.; Giovannini, C.; Croce, C.M.; Bolondi, L.; et al. MiR-221 controls CDKN1C/p57 and CDKN1B/p27 expression in human hepatocellular carcinoma. *Oncogene* **2008**, *27*, 5651–5661. [CrossRef] [PubMed]

165. Fu, X.; Wang, Q.; Chen, J.; Huang, X.; Chen, X.; Cao, L.; Tan, H.; Li, W.; Zhang, L.; Bi, J.; et al. Clinical significance of miR-221 and its inverse correlation with p27Kip[1] in hepatocellular carcinoma. *Mol. Biol. Rep.* **2011**, *38*, 3029–3035. [CrossRef] [PubMed]

166. Yang, C.J.; Shen, W.G.; Liu, C.J.; Chen, Y.W.; Lu, H.H.; Tsai, M.M.; Lin, S.C. miR-221 and miR-222 expression increased the growth and tumorigenesis of oral carcinoma cells. *J. Oral Pathol. Med.* **2011**, *40*, 560–566. [CrossRef] [PubMed]

167. Sun, K.; Wang, W.; Zeng, J.J.; Wu, C.T.; Lei, S.T.; Li, G.X. MicroRNA-221 inhibits CDKN1C/p57 expression in human colorectal carcinoma. *Acta Pharmacol. Sin.* **2011**, *32*, 375–384. [CrossRef] [PubMed]

168. Bazot, Q.; Paschos, K.; Skalska, L.; Kalchschmidt, J.S.; Parker, G.A.; Allday, M.J. Epstein-Barr Virus Proteins EBNA3A and EBNA3C Together Induce Expression of the Oncogenic MicroRNA Cluster miR-221/miR-222 and Ablate Expression of Its Target p57KIP2. *PLoS Pathog.* **2015**, *11*, e1005031. [CrossRef] [PubMed]

169. Wurz, K.; Garcia, R.L.; Goff, B.A.; Mitchell, P.S.; Lee, J.H.; Tewari, M.; Swisher, E.M. MiR-221 and MiR-222 alterations in sporadic ovarian carcinoma: Relationship to CDKN1B, CDKN1C and overall survival. *Genes Chromosomes Cancer* **2010**, *49*, 577–584. [CrossRef] [PubMed]

170. Unek, G.; Ozmen, A.; Mendilcioglu, I.; Simsek, M.; Korgun, E.T. The expression of cell cycle related proteins PCNA, Ki67, p27 and p57 in normal and preeclamptic human placentas. *Tissue Cell* **2014**, *46*, 198–205. [CrossRef] [PubMed]

171. McMinn, J.; Wei, M.; Schupf, N.; Cusmai, J.; Johnson, E.B.; Smith, A.C.; Weksberg, R.; Thaker, H.M.; Tycko, B. Unbalanced placental expression of imprinted genes in human intrauterine growth restriction. *Placenta* **2006**, *27*, 540–549. [CrossRef] [PubMed]

172. Allias, F.; Lebreton, F.; Collardeau-Frachon, S.; Azziza, J.; Pasquier, C.J.; Arcin-Thoury, F.; Patrier, S.; Devouassoux-Shisheboran, M. Immunohistochemical expression of p57 in placental vascular proliferative disorders of preterm and term placentas. *Fetal Pediatr. Pathol.* **2009**, *28*, 9–23. [CrossRef] [PubMed]

173. Fukunaga, M. Immunohistochemical characterization of p57(KIP2) expression in early hydatidiform moles. *Hum. Pathol.* **2002**, *33*, 1188–1192. [CrossRef] [PubMed]

174. McCowan, L.M.; Becroft, D.M. Beckwith-Wiedemann syndrome, placental abnormalities, and gestational proteinuric hypertension. *Obstet. Gynecol.* **1994**, *83 Pt 2*, 813–817. [PubMed]

175. Linn, R.L.; Minturn, L.; Yee, L.M.; Maniar, K.; Zhang, Y.; Fritsch, M.K.; Kashireddy, P.; Kapur, R.; Ernst, L.M. Placental mesenchymal dysplasia without fetal development in a twin gestation: A case report and review of the spectrum of androgenetic biparental mosaicism. *Pediatr. Dev. Pathol.* **2015**, *18*, 146–154. [CrossRef] [PubMed]

176. H'Mida, D.; Gribaa, M.; Yacoubi, T.; Chaieb, A.; Adala, L.; Elghezal, H.; Saad, A. Placental mesenchymal dysplasia with beckwith-wiedemann syndrome fetus in the context of biparental and androgenic cell lines. *Placenta* **2008**, *29*, 454–460. [CrossRef] [PubMed]

177. Kaiser-Rogers, K.A.; McFadden, D.E.; Livasy, C.A.; Dansereau, J.; Jiang, R.; Knops, J.F.; Lefebvre, L.; Rao, K.W.; Robinson, W.P. Androgenetic/biparental mosaicism causes placental mesenchymal dysplasia. *J. Med. Genet.* **2006**, *43*, 187–192. [CrossRef] [PubMed]

178. Kalish, J.M.; Conlin, L.K.; Bhatti, T.R.; Dubbs, H.A.; Harris, M.C.; Izumi, K.; Mostoufi-Moab, S.; Mulchandani, S.; Saitta, S.; States, L.J.; et al. Clinical features of three girls with mosaic genome-wide paternal uniparental isodisomy. *Am. J. Med. Genet. A* **2013**, *161a*, 1929–1939. [CrossRef] [PubMed]

179. Samadder, A.; Kar, R. Utility of p57 immunohistochemistry in differentiating between complete mole, partial mole & non-molar or hydropic abortus. *Indian J. Med. Res.* **2017**, *145*, 133–137. [PubMed]

180. Sasaki, S.; Sasaki, Y.; Kunimura, T.; Sekizawa, A.; Kojima, Y.; Iino, K. Clinical Usefulness of Immunohistochemical Staining of p57 kip2 for the Differential Diagnosis of Complete Mole. *Biomed. Res. Int.* **2015**, *2015*, 905648. [CrossRef] [PubMed]

181. Castrillon, D.H.; Sun, D.; Weremowicz, S.; Fisher, R.A.; Crum, C.P.; Genest, D.R. Discrimination of complete hydatidiform mole from its mimics by immunohistochemistry of the paternally imprinted gene product p57KIP2. *Am. J. Surg. Pathol.* **2001**, *25*, 1225–1230. [CrossRef] [PubMed]

182. De Nigris, F.; Balestrieri, M.L.; Napoli, C. Targeting c-Myc, Ras and IGF cascade to treat cancer and vascular disorders. *Cell Cycle* **2006**, *5*, 1621–1628. [CrossRef] [PubMed]

183. Balestrieri, M.L.; Fiorito, C.; Crimi, E.; Felice, F.; Schiano, C.; Milone, L.; Casamassimi, A.; Giovane, A.; Grimaldi, V.; del Giudice, V.; et al. Effect of red wine antioxidants and minor polyphenolic constituents on endothelial progenitor cells after physical training in mice. *Int. J. Cardiol.* **2008**, *126*, 295–297. [CrossRef] [PubMed]

184. Vitiello, M.; Zullo, A.; Servillo, L.; Mancini, F.P.; Borriello, A.; Giovane, A.; Della Ragione, F.; D'Onofrio, N.; Balestrieri, M.L. Multiple pathways of SIRT6 at the crossroads in the control of longevity, cancer, and cardiovascular diseases. *Ageing Res. Rev.* **2017**, *35*, 301–311. [CrossRef] [PubMed]

185. Kassem, S.A.; Ariel, I.; Thornton, P.S.; Hussain, K.; Smith, V.; Lindley, K.J.; Aynsley-Green, A.; Glaser, B. p57(KIP2) expression in normal islet cells and in hyperinsulinism of infancy. *Diabetes* **2001**, *50*, 2763–2769. [CrossRef] [PubMed]

186. Avrahami, D.; Li, C.; Yu, M.; Jiao, Y.; Zhang, J.; Naji, A.; Ziaie, S.; Glaser, B.; Kaestner, K.H. Targeting the cell cycle inhibitor p57Kip2 promotes adult human beta cell replication. *J. Clin. Investig.* **2014**, *124*, 670–674. [CrossRef] [PubMed]

187. Kerns, S.L.; Guevara-Aguirre, J.; Andrew, S.; Geng, J.; Guevara, C.; Guevara-Aguirre, M.; Guo, M.; Oddoux, C.; Shen, Y.; Zurita, A. A novel variant in CDKN1C is associated with intrauterine growth restriction, short stature, and early-adulthood-onset diabetes. *J. Clin. Endocrinol. Metab.* **2014**, *99*, E2117–E2122. [CrossRef] [PubMed]

188. Van De Pette, M.; Tunster, S.J.; McNamara, G.I.; Shelkovnikova, T.; Millership, S.; Benson, L.; Peirson, S.; Christian, M.; Vidal-Puig, A.; John, R.M. Cdkn1c Boosts the Development of Brown Adipose Tissue in a Murine Model of Silver Russell Syndrome. *PLoS Genet.* **2016**, *12*, e1005916. [CrossRef] [PubMed]

189. Asahara, S.; Etoh, H.; Inoue, H.; Teruyama, K.; Shibutani, Y.; Ihara, Y.; Kawada, Y.; Bartolome, A.; Hashimoto, N.; Matsuda, T. Paternal allelic mutation at the Kcnq1 locus reduces pancreatic beta-cell mass by epigenetic modification of Cdkn1c. *Proc. Natl. Acad. Sci. USA* **2015**, *112*, 8332–8337. [CrossRef] [PubMed]

190. Zhao, Z.; Choi, J.; Zhao, C.; Ma, Z.A. FTY720 normalizes hyperglycemia by stimulating beta-cell in vivo regeneration in db/db mice through regulation of cyclin D3 and p57(KIP2). *J. Biol. Chem.* **2012**, *287*, 5562–5573. [CrossRef] [PubMed]

191. Grimaldi, V.; Schiano, C.; Casamassimi, A.; Zullo, A.; Soricelli, A.; Mancini, F.P.; Napoli, C. Imaging Techniques to Evaluate Cell Therapy in Peripheral Artery Disease: State of the Art and Clinical Trials. *Clin. Physiol. Funct. Imaging* **2016**, *36*, 165–178. [CrossRef] [PubMed]

192. Nödl, M.-T.; Fossati, S.M.; Domingues, P.; Sánchez, F.J.; Zullo, L. The Making of an Octopus Arm. *Evodevo* **2015**, *6*, 19. [CrossRef] [PubMed]

193. Fossati, S.M.; Candiani, S.; Nödl, M.-T.; Maragliano, L.; Pennuto, M.; Domingues, P.; Benfenati, F.; Pestarino, M.; Zullo, L. Identification and Expression of Acetylcholinesterase in Octopus Vulgaris Arm Development and Regeneration: A Conserved Role for ACHE? *Mol. Neurobiol.* **2015**, *52*, 45–56. [CrossRef] [PubMed]

194. Zou, P.; Yoshihara, H.; Hosokawa, K.; Tai, I.; Shinmyozu, K.; Tsukahara, F.; Maru, Y.; Nakayama, K.; Nakayama, K.I.; Suda, T. p57Kip2 and p27Kip1 Cooperate to Maintain Hematopoietic Stem Cell Quiescence through Interactions with Hsc70. *Cell Stem Cell* **2011**, *9*, 247–261. [CrossRef] [PubMed]

 © 2018 by the authors. Licensee MDPI, Basel, Switzerland. This article is an open access article distributed under the terms and conditions of the Creative Commons Attribution (CC BY) license (http://creativecommons.org/licenses/by/4.0/).

International Journal of
Molecular Sciences

Article

Preliminary RNA-Seq Analysis of Long Non-Coding RNAs Expressed in Human Term Placenta

Marta Majewska [1,*], Aleksandra Lipka [2], Lukasz Paukszto [3], Jan Pawel Jastrzebski [3], Marek Gowkielewicz [2], Marcin Jozwik [2] and Mariusz Krzysztof Majewski [1]

[1] Department of Human Physiology, School of Medicine, Collegium Medicum, University of Warmia and Mazury in Olsztyn, 10-082 Olsztyn, Poland; mariusz.majewski@uwm.edu.pl
[2] Department of Gynecology and Obstetrics, School of Medicine, Collegium Medicum, University of Warmia and Mazury in Olsztyn, 10-045 Olsztyn, Poland; aleksandra.lipka@uwm.edu.pl (A.L.); marekgowkielewicz@gmail.com (M.G.); prof.jozwik@gmail.com (M.J.)
[3] Department of Plant Physiology, Genetics and Biotechnology, Faculty of Biology and Biotechnology, University of Warmia and Mazury in Olsztyn, 10-719 Olsztyn, Poland; pauk24@gmail.com (L.P.); bioinformatyka@gmail.com (J.P.J.)
* Correspondence: marta.majewska@uwm.edu.pl; Tel.: +48-89-524-53-34; Fax: +48-89-524-53-07

Received: 12 June 2018; Accepted: 24 June 2018; Published: 27 June 2018

Abstract: Development of particular structures and proper functioning of the placenta are under the influence of sophisticated pathways, controlled by the expression of substantial genes that are additionally regulated by long non-coding RNAs (lncRNAs). To date, the expression profile of lncRNA in human term placenta has not been fully established. This study was conducted to characterize the lncRNA expression profile in human term placenta and to verify whether there are differences in the transcriptomic profile between the sex of the fetus and pregnancy multiplicity. RNA-Seq data were used to profile, quantify, and classify lncRNAs in human term placenta. The applied methodology enabled detection of the expression of 4463 isoforms from 2899 annotated lncRNA loci, plus 990 putative lncRNA transcripts from 607 intergenic regions. Those placentally expressed lncRNAs displayed features such as shorter transcript length, longer exon length, fewer exons, and lower expression levels compared to messenger RNAs (mRNAs). Among all placental transcripts, 175,268 were classified as mRNAs and 15,819 as lncRNAs, and 56,727 variants were discovered within unannotated regions. Five differentially expressed lncRNAs (*HAND2-AS1, XIST, RP1-97J1.2, AC010084.1, TTTY15*) were identified by a sex-bias comparison. Splicing events were detected within 37 genes and 4 lncRNA loci. Functional analysis of *cis*-related potential targets for lncRNAs identified 2021 enriched genes. It is presumed that the obtained data will expand the current knowledge of lncRNAs in placenta and human non-coding catalogs, making them more contemporary and specific.

Keywords: placenta; long non-coding RNA (lncRNA); human; pregnancy; high-throughput RNA sequencing (RNA-Seq); transcriptome

1. Introduction

The placenta serves as a metabolic, respiratory, excretory, and endocrine organ, whose proper functioning is required for adequate embryonic development during pregnancy [1]. Fetal growth is a multifactorial and complex process modulated simultaneously by maternal, fetal, placental, and environmental factors predetermined by genetic potential. A properly functioning placenta fine-tunes the expression of various genes essential in pregnancy maintenance and fetal development [2,3].

Spatiotemporal expression is a huge impediment in any transcriptome analysis, especially in the placenta, an organ that constantly adapts to feto-maternal environmental alterations.

Comprehensive analysis of messenger RNA (mRNA) expression in first- and second-trimester placentas compared to term placentas by microarray assay revealed more genes with increasing than decreasing expression [4]. Furthermore, a source of variability in the placental transcriptome is embryo sex-dependent bias connected with the expression of genes located on the sex chromosomes, which can also affect the expression level of autosomal genes [5]. However, sex-dependent biases in respect to growth, development [6], and predisposition to pregnancy complications are interesting, and placental gene expression regulation remains unclear [5].

Previous studies implied that risk factor profiles for various pathologies are different between singleton and twin births [7]. Furthermore, without any obvious pattern, twin pregnancies are more likely to involve disorders than single pregnancies. Higher perinatal risk, dangerous for both mother and fetus, is associated with the number of embryos in utero. The main risks of multiple pregnancies are early and late miscarriage, preeclampsia (PE), antepartum bleeding, postpartum hemorrhage, preterm delivery, intrauterine growth restriction (IUGR), placental abruption, and stillbirth [8–10]. To date, the correlations between gene expression profile and multiplicity of gestation have been studied only in the beaver, and it is suggested that a greater number of fetuses might have a negative influence on pregnancy outcome [2].

The great progress of RNA sequencing (RNA-Seq) and the capabilities of numerous bioinformatic approaches make it a powerful technology for thorough transcriptome analysis, which enables characterization of gene expression, alternative splicing events, large-scale discovery of novel transcripts, Single Nucleotide Variant (SNV) prediction, and functional annotation [2,3,11,12]. It is not surprising that in a complex organ like the placenta, there are various distinct transcripts, including mRNA, microRNA (miRNA), and long non-coding RNA (lncRNA), that are not present in other tissues [13]. The lncRNAs are still unexplored ncRNAs characterized by a small number of exons and a sequence length >200 nt that are highly diverse and species-specific with tissue-specific expression [14–18]. LncRNAs act by a range of mechanisms and molecular functions [19], with expression restricted to particular developmental stages [20], and they participate in important biological processes such as embryogenesis [18,21,22], tissue development [23], genomic imprinting [24], and different disease courses [25–27]. Given the complex nature of physiological pregnancy, it is important to elucidate possible molecular mechanisms underlying the placental development of male and female fetuses during single and twin pregnancy.

Disruptions to adaptive changes in the placental transcriptome as a response to altering the feto-maternal environment may be associated with pregnancy complications and compromised fetal outcomes. In this context, defining differences in placenta-specific gene expression regarding the sex of the fetus and the multiplicity of gestation could contribute to the understanding of placental development and function. Since revealing factors that influence the placental expression profile is necessary, this study was conducted to examine whether there are differences in the transcriptomic profile of the human placenta compared for sex of the fetus and number of fetuses. A stringent pathway was applied to identify, analyze, and compare placental transcriptome from male and female fetuses during single and twin pregnancies. This study focused on the lncRNA profile to investigate possible mechanisms regulating the expression profile of the human placenta.

2. Results

2.1. Characteristics of RNA-Seq Data

In total, $2 \times 119{,}560{,}140$ raw paired-end reads were generated, and subsequently $2 \times 109{,}363{,}183$ reads were acquired after trimming. The 218,726,366 clean reads were mapped to a reference human genome, and an average of 86.14% reads were mapped uniquely. Among all mapped transcripts (258,353; Figure 1), 67.84% were classified as mRNA, 6.12% were classified as lncRNA, and 2.60% were classified as pseudogenes.

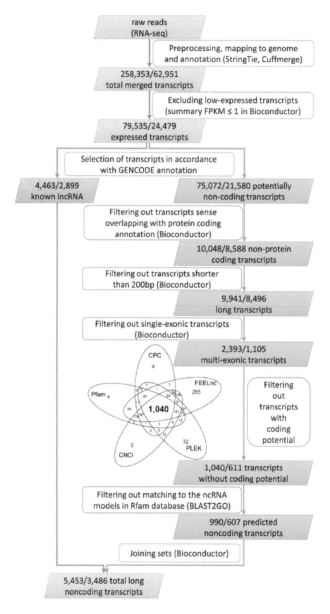

Figure 1. Identification and classification pathway of known (green) and novel (orange) long non-coding RNAs (lncRNAs) expressed in human term placenta; common stages of analysis (blue) join both paths. Numbers in parallelograms refer to amount of lnc transcripts/lncRNA loci. Rectangles show processes and applied tools. Venn diagram presents a number of transcripts without coding potential assigned by Coding-Non-Coding Index (CNCI), Coding Potential Calculator (CPC), FEELnc, Pfam, and PLEK software.

Moreover, 1.48% of the expressed transcripts were derived from other RNAs (e.g., processed transcript, Ig genes, or misc RNA) and 21.96% originated from unannotated regions, which included potentially new lncRNA transcripts (Figure 2a). After excluding low expressed transcripts (fragments per kilobase of transcript per million mapped reads (FPKM) ≤1), 79,535 of the identified transcripts (TCONs) were directed for further analysis (Figures 1 and 2b).

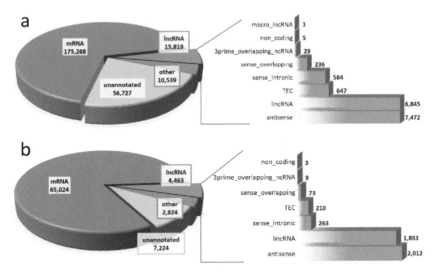

Figure 2. Classification of the assembled human placental transcripts according to their Ensembl code class (pie graphs) detailing lncRNA distribution (bar graphs) of: (**a**) all expressed loci; (**b**) transcripts with expression value (fragments per kilobase of transcript per million mapped reads, FPKM) higher than 1.

The dynamic range of the expression values was calculated and is presented as a box plot of logarithmic transformed FPKM values for each sample separately (Figure 3a), and the FPKM density distribution is shown in Figure 3b.

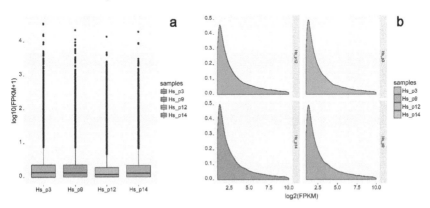

Figure 3. Transcript expression level distribution of each human term placenta sample. (**a**) Box plot of FPKM distribution with different samples on the horizontal axis and logarithmic values of FPKM on the vertical axis; (**b**) density plot of expression distribution with logarithmic values of FPKM on the horizontal axis and density on the vertical axis.

2.2. Identification and Profiling of lncRNAs

An lncRNA profile of human term placenta was identified and characterized by applying a stringent pathway (Figure 1). GENCODE enabled selection of 4463 known lncRNAs and 75,072 other than annotated lncRNAs (including 7224 unannotated transcripts; Figure 2b) that qualified for verification of their coding potential and small RNA features (Figure 1). Filtering out sense-overlapping transcripts with protein coding annotation resulted in 10,048 non-protein coding transcripts, corresponding to 8588 potentially non-coding regions. After excluding sequences shorter than 200 nt, 9941 transcripts were obtained. Next, filtering of single-exon variants enabled identification of 2393 multi-exon transcripts. An assessment of coding potential with Coding Potential Calculator (CPC), Coding-Non-Coding Index (CNCI), FEELnc, Pfam, and PLEK generated 1340, 1790, 2222, 1767, and 1439, respectively, for each method (Supplementary Table S1). Intersecting the aforementioned methods allowed determination of the set of 1040 potentially non-coding transcripts (Venn diagram, Figure 1). The remaining transcripts were devoid of non-mRNA sequences, and as a result, 990 variants, corresponding to 607 regions, were classified as predicted lncRNAs. The set of known lncRNAs was composed of 4463 lnc transcripts corresponding to 2899 lncRNA loci. Among them, 2012 were antisense lncRNAs, 1893 lincRNAs, 263 sense intronic transcripts, and 73 sense overlapping (Figure 2b). The classification of the final set of 5453 lncRNA transcripts, according to genomic localization and relation to nearest annotated genes, is shown in Table 1. The 5252 and 201 lncRNA transcripts were distributed within autosomes and sex chromosomes, respectively. Among all 990 predicted lncRNA transcripts, 395 unknown transcripts (Table 1) have not been annotated so far and were deposited (BankIt accession nos. MG828427–MG828821; Supplementary Table S2).

Table 1. Classification of 5453 lncRNA transcripts (class code module in Cuffcompare).

Class-Code	Description	Isoform (TCONS)	Locus (XLOC)
"_"	unknown, intergenic region	395	344
"o"	overlapped with existed gene with a dramatic difference in gene structures	208	170
"x"	overlapped with existed gene in an opposite direction	160	150
"i"	located in introns	2	2
"="	complete match (of known lncRNA)	3747	2698
"j"	potentially novel isoform (of known lncRNA)	941	579

Expression levels of antisense, lincRNA biotype classes, and newly discovered lncRNAs were comparable (Figure 4).

2.3. Feature Comparison of lncRNA and mRNA

In the current study, 5453 lncRNA and 65,024 mRNA transcripts with FPKM were identified >1. The lncRNA and mRNA transcripts were compared for their total length, exon length, exon number, and expression level (Figure 5). The average length of identified lncRNAs was 1906 nt, while that of mRNAs was 2917 nt (Figure 5a). More than 30% of lncRNAs were in the range of 500–1000 nt, and more than 50% of mRNAs were longer than 2000 nt. Distant length distribution between lncRNA (19.51%) and mRNA (5.18%) was observed in the range 200–500 nt (Figure 5a). The mean exon length of lncRNAs was 737 nt, which was much shorter than mRNAs (337 nt; Figure 5b). Most of the mRNA exons (44%) ranged between 100 and 200 nt, whereas most of the lncRNAs (28.52%) had exon lengths above 500 nt (Figure 5b). The most numerous group of lncRNAs (38.40%) comprised two exons, while only 0.31% of lncRNAs had more than 10 exons, versus mRNAs constituting the largest group (26.86%; Figure 5c). The expression profiles of lncRNA and mRNA biotypes are presented as logarithmic distributions (Figure 5d). The average mRNA expression level was higher than that of the lncRNAs (0.43 vs. 0.31).

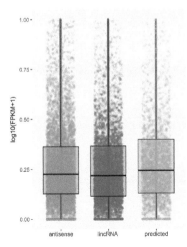

Figure 4. Dispersion of normalized FPKM values presented for the two most numerous lncRNA biotypes: antisense (red), lincRNA (blue), and transcripts predicted as lncRNA (green). Each point represents an individual transcript.

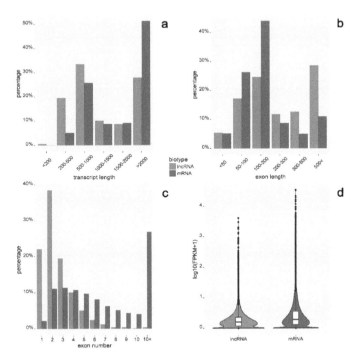

Figure 5. Global summary of comparison between lncRNA (red) and messenger RNA (mRNA) (blue) structural features. lncRNA and mRNA transcripts compared by (**a**) length; (**b**) exon length; (**c**) exon number; (**d**) expression level presented by log10(FPKM + 1); boxes inside each violin plot depict interquartile ranges and individual medians. The differences of average values were statistically significant in each comparison (*p*-value < 2×10^{-16} using Welch two-sample *t*-test).

2.4. Sex Biases in lncRNA Expression Levels

The expression level (FPKM) of long non-coding transcripts was estimated for both sex and multiplicity biases. A sex-bias comparison revealed five differentially expressed lncRNAs (Table 2; p-adjusted < 0.05) and 21 protein-coding genes (Supplementary Table S3). Among the lncRNAs, two loci, XLOC_042918 (chromosome 4) and XLOC_061548 (chromosome X), revealed higher expression levels in female libraries. However, three lncRNA loci, XLOC_050164 (chromosome 6), XLOC_062450, and XLOC_062528 (chromosome Y), were expressed only in male libraries. For protein-coding genes, 11 were upregulated, while 10 were downregulated in female–male comparison (Table S3). The multiplicity-bias comparison did not detect any significant changes in the expression levels of lncRNA and protein-coding genes transcripts.

2.5. Splicing Alterations in Placental Transcriptome

JunctionSeq allows detection of alternative isoform regulation (AIR) genes, also known as differential transcript usage (DTU). As a result, differentially expressed exons and altered spliced patterns of placental transcripts were detected (male vs. female). Comparing the placental transcriptome from male and female samples revealed 37 AIR/DTU genes displaying 38 and 8 statistically significant differential exon and splice-junction usages, respectively. The use of the JunctionSeq analysis tool led to the detection of new splice junctions in the gene encoding pregnancy-specific β-1-glycoprotein 4 (*PSG4*). Three transcripts with multiple distinct exonic regions, Rho GTPase activating protein 45 (*ARHGAP45*); GATA binding protein 2 (*GATA2*), and long non-coding RNA (*RP11-440I14.3*), were also indicated. Four genes, peptidylprolyl isomerase G (*PPIG*), HLA class II histocompatibility antigen DRB5 beta chain (*HLA-DRB5*), torsin 1A interacting protein 1 (*TOR1AIP1*), and cysteine and serine rich nuclear protein 1 (*CSRNP1*), displayed simultaneous differential exon and splice-junction usage. Among all AIR/DTU events, four significant differential usages of exons were localized within lncRNA loci: *H19*, *AC132217.4*, *RP11-440I14.3*, and *AC005154.6* (Figure 6; Supplementary Table S4). Within *H19*, exon 27 was upregulated in female samples. In female placentas, variable expression of exons 5 and 7 of *RP11-440I14.3* was also observed, although in male placentas, exon 15 of *AC132217.4* and exon 13 of *AC005154.6* were upregulated (Figure 6; Supplementary Table S4).

Table 2. Differences in expression level of lncRNAs in sex-bias comparison.

Gene_ID	lncRNA Variant ID	Ensembl Gene ID	HGNC Symbol	Gene Name	Biotype	Locus	Samples		Expression Level [FPKM]		log2fc
							Male	Female	Male	Female	
XLOC_042918	11	ENSG00000237125	HAND2-AS1	HAND2-AS1	antisense	4:173524692-173591465	Hs_p3, Hs_p9	Hs_p12, Hs_p14	2.416	23.984	3.312
XLOC_050164	1	ENSG00000227012	LINC02527	RP1-97J1.2	lincRNA	6:111900309-111909386	Hs_p3, Hs_p9	Hs_p12, Hs_p14	2.095	0.000	"-Inf"
XLOC_061548	9	ENSG00000229807	XIST	XIST	lincRNA	X:73792204-73852753	Hs_p3, Hs_p9	Hs_p12, Hs_p14	1.185	90.740	6.259
XLOC_062450	1	ENSG00000229308	NA	AC010084.1	lincRNA	Y:4036485-4106081	Hs_p3, Hs_p9	Hs_p12, Hs_p14	1.444	0.000	"-Inf"
XLOC_062528	3	ENSG00000233864	TTTY15	TTTY15	lincRNA	Y:12662333-12692233	Hs_p3, Hs_p9	Hs_p12, Hs_p14	5.355	0.000	"-Inf"

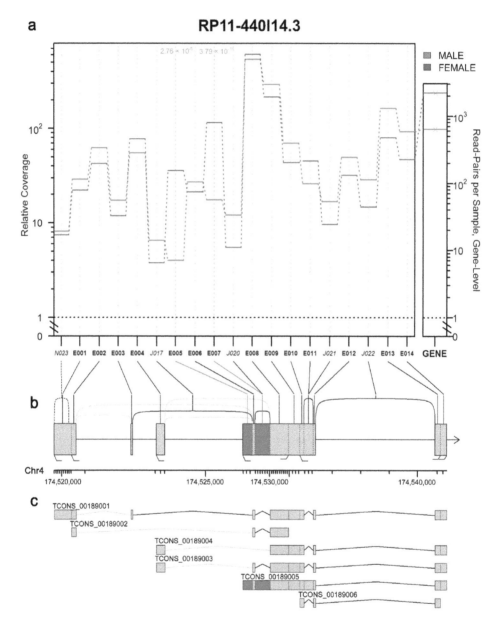

Figure 6. Presentation of differential transcript usage: (**a**) JunctionSeq gene profile plot for *RP11-440I14.3* lncRNA identified in male (red) and female (blue) placental samples. This plot displays estimates for the mean normalized read-pair coverage count for each exon and splice junction. The small panel on the far right displays the total mean normalized read-pair count based on gene level. (**b**) Gene diagram displaying the exonic regions (boxes, labeled E001–E014), known splice junctions (solid lines, labeled J017–J022), and novel splice junction (dashed line, labeled N023) for *RP11-440I14.3* lncRNA localized on chromosome 4 (Chr4). (**c**) The panel shows exon-intron structures of *RP11-440I14.3* variants. Statistically significant differences (*p*-adjust < 0.05) in exon usage are marked in pink.

2.6. Functional Analysis of Nearest Neighbor Genes to lncRNAs

Potential *cis*-target genes were predicted, revealing possible lncRNA regulation functions in term placental tissues. The genes located within 2000 nt distance (upstream and downstream) from the identified lncRNAs were considered as target genes, and the approach produced 2021 genes. Those genes closely related to lncRNAs were analyzed for Gene Ontology (GO) enrichment, as shown in Figure 7. The majority of *cis*-target genes were enriched ($p < 0.05$) to biological process (148 terms), cellular component (56 terms), and molecular function (20 terms) according to GO classification. GO annotation showed that 61 and 107 protein-coding *cis*-target genes were enriched in in utero embryonic development and vasculature development, respectively. Within the cellular component GO category, 1772 and 1684 *cis*-targets were assigned according to cell and intracellular compartments (Supplementary Table S5).

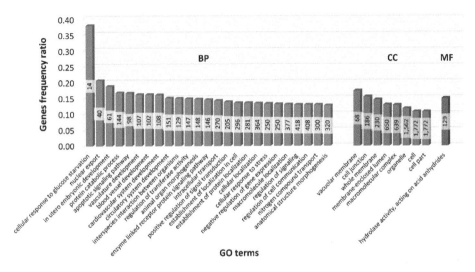

Figure 7. Gene Ontology (GO) annotations (level 1) of *cis* lncRNA target protein-coding genes presenting enriched terms in biological process (BP), cellular component (CC), and molecular function (MF). The height of each bar represents the ratio of target protein-coding genes involved in the particular process relative to all genes associated with a given process in the GO database. The numbers in bars represent the amount of genes involved in a particular GO term.

2.7. Validation of RNA-Seq Results Using External Transcriptomic Datasets

Validation with external data confirmed the presence and expression tendencies of the majority of novel (607) and known (2899) lncRNA loci predicted in this study (Figure 8). For external data, mean expression values in logarithmic scale ranged between 0.31 and 0.44 for newly discovered lncRNAs, and between 0.31 and 0.42 for known lncRNA loci (Figure 8, Table 3). Mean expression values for our data ranged from 0.43 to 0.51 for new loci and from 0.37 to 0.42 for known lncRNA regions (Figure 8, Table 3; Supplementary Table S6). Expression levels for 1276 highly expressed lncRNA loci (with FPKM > 2 in at least half the samples) showed that 142 novel and 610 known lncRNA loci had the same high expression profile in external data and our data. As the results obtained for external data were largely consistent with our results, it may further indicate the reliability of the results obtained in this study.

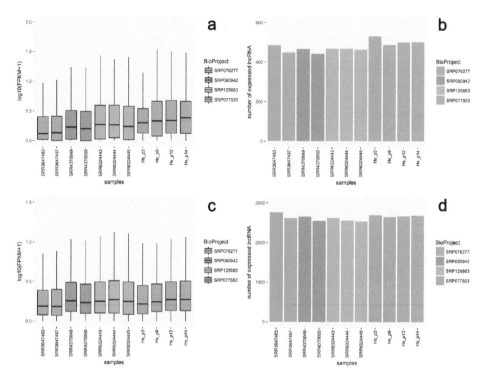

Figure 8. Comparison of transcript expression level distribution between external datasets downloaded from BioProjects SRP076277 (BioSamples SRR3647483 and SRR3647497), SRP090942 (BioSamples SRR4370049 and SRR4370050), SRP125683 (BioSamples SRR6324443, SRR6324444, and SRR6324445), and our dataset (Hs_p3, Hs_p9, Hs_p12, and Hs_p14). (**a**,**c**) Normalized FPKM distribution (box plots) and (**b**,**d**) sum of expression loci (bar graphs) for novel (**upper panel**) and known (**lower panel**) lncRNAs.

Table 3. Summary statistics of logarithm FPKM values for novel and known lncRNAs in external data and our datasets.

		SRR3647483	SRR3647497	SRR4370049	SRR4370050	SRR6324443	SRR6324444	SRR6324445	Hs_p3	Hs_p9	Hs_p12	Hs_p14
Novel	Min.	0.0000	0.0000	0.0000	0.0000	0.0000	0.0000	0.0000	0.0000	0.0000	0.0000	0.0000
	1st Qu.	0.0200	0.0000	0.0272	0.0000	0.0343	0.0527	0.0288	0.1313	0.0843	0.1331	0.1333
	Median	0.1178	0.1345	0.2295	0.2124	0.2744	0.2801	0.2483	0.3100	0.3456	0.3519	0.3937
	Mean	0.3073	0.3158	0.3851	0.3723	0.4278	0.4400	0.4262	0.4341	0.4693	0.4878	0.5055
	3rd Qu.	0.4167	0.4262	0.5216	0.5149	0.6155	0.6120	0.6010	0.5563	0.6927	0.7024	0.7068
	Max.	3.5141	3.5084	3.1126	3.0138	2.8940	2.8243	2.7285	3.5662	3.0406	3.6352	2.8402
Known	Min.	0.0000	0.0000	0.0000	0.0000	0.0000	0.0000	0.0000	0.0000	0.0000	0.0000	0.0000
	1st Qu.	0.0810	0.0732	0.1281	0.1034	0.1093	0.1085	0.0909	0.0972	0.1283	0.1410	0.1412
	Median	0.1932	0.1882	0.2657	0.2461	0.2649	0.2865	0.2619	0.2273	0.2574	0.2870	0.2867
	Mean	0.3159	0.3127	0.3951	0.3755	0.4016	0.4227	0.4067	0.3698	0.3945	0.4150	0.4228
	3rd Qu.	0.4016	0.4150	0.5165	0.4978	0.5357	0.5665	0.5464	0.4826	0.5086	0.5361	0.5451
	Max.	3.4224	3.3734	3.4381	3.3992	4.5228	5.1047	5.1712	3.7855	3.5651	3.3986	3.5616

Min., minimum; 1st Qu., first quantile; 3rd Qu., third quantile; Max., maximum of log10(FPKM) expression value.

3. Discussion

Placenta fine-tunes the expression of various genes involved in major molecular mechanisms essential in pregnancy maintenance and fetal development [2,3,28]. For this reason, any alterations in expression and further processing of specific genes may be correlated with impaired placental function and may directly affect pregnancy outcome [29]. Additionally, the expression mechanisms at both the transcriptional and post-transcriptional level are regulated by numerous lncRNA and lncRNA–RNA interactions [30]. Different expression of genes and their regulatory elements can potentially impact many biological processes and might constitute one of the main regulators of molecular pathways within the placenta [28]. To the best of the authors' knowledge, the expression profile of lncRNA in the human term placental transcriptome has not yet been studied. Therefore, in this study, lncRNA landscape analysis of human term placenta was performed.

Among the placental transcripts obtained in our study, 67.84% (175,268) were classified as mRNA, 6.12% (15,819) as lncRNA, and 2.60% (6726) as pseudogenes (Figure 2a). In all, 21.96% of variants (56,727) originated from unannotated regions. According to the current data, 4463 known and 990 previously unknown (predicted) lncRNAs are expressed in human term placental tissue (Figure 1). In comparison, RNA-Seq analysis of first-trimester human placenta transcriptome revealed transcript biotypes in the following classes: 77% protein-coding genes, 9.8% long non-coding genes, and 6.5% pseudogenes [31]. However, the current analysis allowed identification of 21 genes with significantly different expression between males and females, compared to 58 genes discovered by Gonzalez et al. [31]. Further, in placentas of severe preeclampsia cases (~27 weeks of gestation), Gormley et al. [32] classified 15,060 transcripts as mRNA, 823 as lncRNA, and 547 as pseudogenes. Moreover, among the 15,646 dysregulated lncRNAs in early-onset preeclampsia placental tissue, 12,195 were categorized as intergenic, 5182 as antisense, and 1352 as intron sense–overlapping sequences [33]. The present study indicates that among 5453 lncRNA transcripts, the set of 2012 (36.90%) antisense placental transcripts was the largest group, together with 1893 lincRNAs (34.71%) located within the intergenic regions (Figure 2b). Similarly, the class of sense-overlapping sequences (73) was among the smallest groups. The results regarding the lncRNA expression profile in human placenta extend and complement the present transcriptomic databases, which enables genome-wide analysis across tissues and conditions [34]. Moreover, validation with external datasets confirmed the obtained results regarding known and novel lncRNA transcript expression in human term placenta. A general comparison of mRNA and lncRNA features, indicating shorter transcript lengths, longer exon lengths, fewer exons, and lower expression levels for lncRNAs, was consistent with studies in other mammals [17,18,35–39]. Nevertheless, the differences between this study and various transcriptomic experiments result from a strict tissue-specificity pattern of lncRNA expression, restricted spatiotemporal specificity, and differences in adopted pathways.

The expression of mammalian lncRNAs is strictly associated with their regulatory role in a tissue-specific manner. Among various tissues, the testis and ovary were indicated as the most enriched in lncRNAs [40], suggesting their huge regulatory potential within the reproductive system. The expression level analysis in this study revealed five differentially expressed lncRNAs enriched within human term placenta only in sex-bias comparison. It was found that the multiplicity-bias comparison revealed no significant changes in lncRNA expression level. Two lncRNA loci, *HAND2-AS1* and X chromosome inactive–specific transcript (*XIST*), displayed higher expression levels in female libraries and three others, *RP1-97J1.2*, *AC010084.1*, and *TTTY15*, were expressed solely in male libraries. *XIST* as X chromosome–specific was highly enriched in the female libraries. *XIST* is a kind of functional lncRNA uniquely involved in the formation of repressive chromatin and regulation of the X chromosome inactivation process by *cis* action [41–44]. *XIST*'s expression occurs in a spatiotemporal manner, regulating and influencing female development [45]. *HAND2-AS*, as antisense to *HAND2*, may regulate its expression. *HAND2* is a kind of transcription factor that plays a key role, e.g., in vascularization, development, and differentiation of sympathetic neurons [46,47]. Moreover, *HAND2* fosters a level of fibulin-1, which contributes to progesterone action during implantation [48–50]. Usually, the majority of lncRNAs exist as single variants [17], but *HAND2* and *XIST* exhibit more variants: 11 and 9, respectively. Therefore, fetal sex-specific expression of the aforementioned lncRNAs and their variants in the placenta might impact proper placental development and function. That is why further molecular insights into their function must be gained to fully discover their implication in pregnancy outcome. There were 21 protein-coding genes differentially expressed in female and male term placentas. Among them, microsomal glutathione transferase 1 (*MGST1*) was identified to have a confirmed role in oxidative stress protection [51]; relaxin family peptide receptor 1 (*RXFP1*) a receptor for relaxin, a key hormone in mammalian pregnancy [52], and semaphorin 3A (*SEMA3A*) play essential roles in preventing nerve fiber growth in the placenta to protect the fetus from external stress [53].

Previous transcriptomic studies performed on beaver discoid placenta revealed that there are differences in gene expression between twin and triple pregnancies and that the number of fetuses may affect pregnancy outcome [2]. It was found that a multiplicity-bias comparison revealed significant changes of lncRNA expression level in human term placenta. It should be mentioned that such changes may appear in earlier pregnancy stages. Additionally, it cannot be excluded that a similar analysis performed on a greater number of samples would reveal multiplicity as a significant factor affecting the placental transcriptome. The present study should be considered as a pilot screen that may be a good starting point for future functional analysis of more groups of samples. A better understanding of the molecular factors and specific biomarkers during single and twin pregnancies that are predisposed to pathology might be helpful in determining effective prevention strategies. Given the complex nature of physiological pregnancy, such studies are needed to continue to elucidate possible molecular mechanisms underlying placental development during single and twin pregnancies.

Alternative isoform regulation (AIR) can enhance transcriptome diversity and gain another biological function of a single gene by events such as alternative splice sites, alternative transcription start sites, methylation, nucleosome occupancy, internal promoters, nonsense-mediated decay, and/or transcript switching [54]. Alternative splicing events, besides increasing transcriptome complexity, may also disrupt processes or generate pathologies [55]. In the present study, 37 genes and 4 lncRNA loci were identified with AIR/DTU between female and male placental samples. This study enabled detection of a novel splice junction in the gene encoding pregnancy-specific beta-1-glycoprotein 4 (*PSG4*). Pregnancy-specific glycoproteins (*PSGs*) are a specific group of highly expressed trophoblast genes crucial for placentation, acting as regulators of trophoblast cell migration, cytokine secretion, and the establishment of uteroplacental circulation [56]. PSGs are the most abundant proteins in the maternal blood in late pregnancy [57]. A decreased PSG level in maternal serum may be associated with spontaneous abortion, intrauterine growth retardation, or preeclampsia [58–60]. Human *PSG* loci (*PSG1–PSG11*) are enriched with various types of copy number variations, which may be linked with impaired fertility and pregnancy complications such as preeclampsia [61].

Multiple distinct exonic regions were detected in *ArhGAP45* (also named *HMHA1/HA-1*), which functions as a Rho GTPase [62,63]. Rho GTPases are engaged in the proper functioning of the endothelial barrier [64], embryogenesis [65], neural development [66], cytokinesis, and differentiation [67]. *ArhGAP45* mRNA expression is elevated in preeclamptic placentas and is under the control of oxygen accessibility [68]. GATA binding protein 2 (*GATA2*) regulates stage-specific trophoblastic gene expression of the preimplantation human embryo [69–71].

A substantial contribution of lncRNAs in placental formation and function is well known; an evident example is *H19*, a placenta-specific lncRNA highly expressed during mammalian embryonic development [72–74]. *H19* is implicated in the regulation of human placenta trophoblast proliferation, placental development [75,76], and fetal growth [77,78]. Moreover, the dynamic profile of *H19* expression may support normal pregnancy, while its impaired regulation might promote preeclampsia, early-onset preeclampsia (EOPE), and IUGR [77,79,80]. *AC132217.4* lncRNA, because it affects 3′UTR and enhances expression level, fosters mRNA stability and upregulates expression of IGF2 circulating growth factor, which acts during pregnancy to promote both fetal and placental growth [81].

Differential usage of exons was also detected in lincRNA *RP11-440I14.3*, localized in *cis* position to hydroxyprostaglandin dehydrogenase (*HPGD*). Hydroxyprostaglandin dehydrogenase inactivates prostaglandins E2 (*PGE2*) and D2 (*PGD2*), which affect several biological processes, such as reproduction, differentiation, and inflammation [82]. In the uterus, PGs play a key role in infection-induced pregnancy loss, in which the concentration of this mediator is increased. As AIR/DTU was detected in genes and lncRNAs, whose functions are related to placental and embryonic development, it should be further investigated to indicate whether the expression profile of specific isoforms can affect the proper or pathological pregnancy course.

GO analysis was applied to explore the function of the *cis*-target genes. A variety of subclasses of ncRNAs, like piRNA, miRNA, siRNA, and lncRNA, have regulatory roles in gene expression [83–87]. In the present research, enrichment analysis of *cis*-related potential targets for lncRNAs identified 2021 genes. The 61 protein-coding genes were found to be regulated by lncRNA transcripts, and GO enrichment showed that they were enriched in in utero embryonic development (GO:0001701), suggesting that predicted lncRNA functions during pregnancy are linked with developmental, growth, and regulation related processes. Generally, annotation with GO terms displays many of the placentally expressed lncRNA transcripts involved in the regulation of various biological processes also implicated in the gestation course.

Taken together, since the functions of the majority of lncRNAs have yet to be uncovered, tremendous effort should be made to decipher their implication in the course of gestation, placental development, and reproductive disorders. The present research may be used as a resource for functional studies, which is a huge challenge in determining the influence of lncRNAs on reproductive processes. The authors' previous study [3] established the placental gene expression landscape of human term placenta during uncomplicated single and twin pregnancies. Therefore, it is hoped that the results of this study will broaden the placenta-specific transcriptome database, which will be useful in a functional field of future research.

4. Materials and Methods

4.1. Research Material

The lncRNA expression profile of human term placenta was compared between the sex of the fetus (*n* = 2) and pregnancy multiplicity (*n* = 2). All procedures regarding tissue collection, the characteristics of placental samples (*n* = 4), RNA extraction, and RNA-Seq were described previously [3]. Briefly, Hs_p3 (male) and Hs_p14 (female) originated from single pregnancies, whereas Hs_p9 (male) and Hs_12 (female) were from twin pregnancies. To identify lncRNAs expressed in human term placentas, cDNA libraries were constructed and sequenced on the HiSeq 2500 Illumina platform (Illumina, San Diego, CA, USA). The raw data were submitted to the National Center for Biotechnology Information (NCBI) Sequence Read Archive (SRA) under accession No. SRP077553. The experimental protocol was approved by the Bioethics Committee of the Warmia-Mazury Medical Chamber (OIL.164/15/Bioet; 2 April 2015) in Olsztyn, Poland.

4.2. Transcriptome Assembly and Identification of Novel Transcripts

The quality of reads was checked using the FastQC tool. Preprocessing using a Trimmomatic tool v. 0.32 [88] included the following: removal of Illumina adaptors and poly(A) stretches, exclusion of low-quality reads (Phred cutoff = 20), and trimming of reads to equal 90 nt in length. Next, paired-end clean reads were aligned to the reference human genome (Homo_sapiens.GRCh38.dna.primary_assembly.fa) with annotation (Homo_sapiens.GRCh38.87.gtf) applying the STAR (v. 2.4, https://github.com/alexdobin/STAR) mapper. As a result, a BAM file alignment of the trimmed reads to the reference genome was obtained for each sample. StringTie v. 1.0.4 (https://ccb.jhu.edu/software/stringtie) [89] and Cuffmerge, as part of the Cufflinks tool v. 2.2.1 (http://cole-trapnell-lab.github.io/cufflinks) [90], were applied to expand gene and transcript annotations based on Ensembl human reference (release 90, August 2017).

This approach enabled the identification of unannotated regions and novel splice variants expressed in the placenta. An expanded annotation file (merge.gtf) was used for expression calculation (Cuffquant), normalization (Cuffnorm), and differential analysis (Cuffdiff). All transcript sequences were extracted to a FASTA file using a gffread script (Figure 1).

4.3. Classification, Characterization, and Validation of lncRNAs

Low-expressed transcripts with FPKM values $\leq$ 1 (expression sum of 4 libraries) were excluded from the set of merged transcripts. Next, all transcripts longer than 200 nt were passed for further analysis (Figure 1). Selected transcripts were divided into 2 main datasets: (1) known lncRNA transcripts (biotypes of GENCODE (https://www.gencodegenes.org/) "lincRNA", "antisense", "sense_intronic", "sense_overlapping", "bidirectional_promoter_lncRNA", "non_coding", "macro_lncRNA", "TEC" (to be experimentally confirmed), and "3prime_overlapping_ncRNA"); and (2) potentially non-coding RNA (unannotated transcripts and other than lncRNAs). The second dataset, including unknown non-coding sequences, was reduced by removing a transcript assigned to the "protein_coding" Ensembl class code. Next, these transcripts were subjected to multi-exon filtering. Transcript coding potential was assessed by several tools: The Coding Potential Calculator (CPC) (http://cpc.cbi.pku.edu.cn) [91], Pfam (https://pfam.xfam.org) [92], CPAT (http://rna-cpat.sourceforge.net) [93], Coding-Non-Coding Index (CNCI) (https://github.com/www-bioinfo-org/CNCI) [94], and PLEK (https://sourceforge.net/projects/plek/files) [95]. CPC (score < 0) enabled the assessment of ORF occurrence (Figure 1). Transcripts that encoded any conserved protein domains were removed, applying the following parameters: CPC (cutoff < 0), Pfam database (*e*-value 10^{-5}; release 27), CPAT (cutoff < 0.43), CNCI (cutoff < 0), and PLEK (cutoff < 0). Further, surviving transcripts were searched in Rfam using Blast2GO software (https://www.blast2go.com) [96], to exclude small ncRNAs (rRNAs, tRNAs, snRNAs, snoRNAs, and miRNAs). Sequences of both known and unknown datasets were denoted as the final set of lncRNAs (Figure 1). Obtained data regarding known and novel lncRNAs were validated by comparison with external data generated in similar studies. SRA resources were searched to find projects focused on RNA-Seq of term placental tissues from normal pregnancies ended by cesarean section. Data from the 3 most accurate BioProjects, SRP076277 (BioSamples SRR3647483 and SRR3647497), SRP090942 (BioSamples SRR4370049 and SRR4370050), and SRP125683 (BioSamples SRR6324443, SRR6324444 and SRR6324445), were chosen for further analysis. Then the raw data were processed with the same approach and parameters that were applied to our data analysis. Downloaded data were aligned to the reference human genome (Homo_sapiens.GRCh38.dna.primary_assembly.fa) with a previously generated merged.gtf annotation file. Then, BAM files were sorted by coordinates and used to calculate FPKM values. Expression values for 607 lncRNA loci predicted as novel and 2899 known lncRNA regions were merged and compared with FPKM values obtained for datasets from the aforementioned BioProjects.

4.4. Different Expression and Splicing Analysis

The reads assembled to mRNA and lncRNA sequences were normalized to FPKM values using Cuffnorm. Applying Cuffdiff, the corresponding *p*-values were determined for 2 comparisons: sex and multiplicity bias in placental tissue. Thresholds for significantly different expression were set as follows: *p*-adjusted < 0.05 and log2 fold change (log2FC) $\geq$ 1.0. A structural comparison between lncRNA and mRNA transcripts was performed by custom R bioconductor scripts. The QoRTs/JunctionSeq pipeline [54] was adopted for differentially expressed exons and splice junction analysis (*p*-adjusted < 0.05).

4.5. LncRNA Target cis Gene Prediction

Based on the localization of lncRNA in relation to mRNA, *cis* interactions were predicted, since the *cis* role refers to the influence of lncRNAs on vicinity target genes localized within 2000 nt upstream or downstream of each protein coding gene on the same chromosome. Functional enrichment analysis (*p*-adjusted < 0.05) of the potential *cis* target genes was performed by Kobas 3.0 software (http://www.kobas.cbi.pku.edu.cn) [97] including Gene Ontology (GO) (http://www.geneontology.org).

Supplementary Materials: Supplementary materials can be found at http://www.mdpi.com/1422-0067/19/7/1894/s1.

Author Contributions: M.M. conceived the study design, performed experimental work, and drafted the manuscript; A.L. assisted in the study design and performed experimental work; L.P. and J.P.J. analyzed the data; M.G. provided placental samples; M.J. and M.K.M. helped in writing the final version of the manuscript. All authors have seen and approved the final version.

Funding: This study was supported by the School of Medicine, Collegium Medicum (61.610.001-300), University of Warmia and Mazury in Olsztyn.

Conflicts of Interest: The authors declare no conflict of interest.

Availability of Supporting Data: The sequencing data from this study have been submitted (http://www.ncbi.nlm.nih.gov/sra) to the NCBI Sequence Read Archive under accession No. SRP077553. In addition, all identified lncRNA sequences of the novel transcripts have been deposited in the GenBank (BankIt accession Nos. MG828427–MG828821).

References

1. Murthi, P. Review: Placental homeobox genes and their role in regulating human fetal growth. *Placenta* **2014**, *28*, S46–S50. [CrossRef] [PubMed]

2. Lipka, A.; Paukszto, L.; Majewska, M.; Jastrzebski, J.P.; Myszczynski, K.; Panasiewicz, G.; Szafranska, B. Identification of differentially expressed placental transcripts during multiple gestations in the Eurasian beaver (*Castor fiber* L.). *Reprod. Fertil. Dev.* **2017**, *29*, 2073–2084. [CrossRef] [PubMed]

3. Majewska, M.; Lipka, A.; Paukszto, L.; Jastrzebski, J.P.; Myszczynski, K.; Gowkielewicz, M.; Jozwik, M.; Majewski, M.K. Transcriptome profile of the human placenta. *Funct. Integr. Genom.* **2017**, *17*, 551–563. [CrossRef] [PubMed]

4. Uusküla, L.; Männik, J.; Rull, K.; Minajeva, A.; Kõks, S.; Vaas, P.; Teesalu, P.; Reimand, J.; Laan, M. Mid-gestational gene expression profile in placenta and link to pregnancy complications. *PLoS ONE* **2012**, *7*, e49248. [CrossRef] [PubMed]

5. Buckberry, S.; Bianco-Miotto, T.; Bent, S.J.; Dekker, G.A.; Roberts, C.T. Integrative transcriptome meta-analysis reveals widespread sex-biased gene expression at the human fetal-maternal interface. *Mol. Hum. Reprod.* **2014**, *20*, 810–819. [CrossRef] [PubMed]

6. Misra, D.P.; Salafia, C.M.; Miller, R.K.; Charles, A.K. Non-linear and gender-specific relationships among placental growth measures and the fetoplacental weight ratio. *Placenta* **2009**, *30*, 1052–1057. [CrossRef] [PubMed]

7. Gupta, S.; Fox, N.S.; Feinberg, J.; Klauser, C.K.; Rebarber, A. Outcomes in twin pregnancies reduced to singleton pregnancies compared with ongoing twin pregnancies. *Am. J. Obstet. Gynecol.* **2015**, *213*, 580.e1–580.e5. [CrossRef] [PubMed]

8. Chauhan, S.P.; Scardo, J.A.; Hayes, E.; Abuhamad, A.Z.; Berghella, V. Twins: Prevalence, problems, and preterm births. *Am. J. Obstet. Gynecol.* **2010**, *203*, 305–315. [CrossRef] [PubMed]

9. Ananth, C.V.; Demissie, K.; Smulian, J.C.; Vintzileos, A.M. Relationship among placenta previa, fetal growth restriction, and preterm delivery: A population-based study. *Obstet. Gynecol.* **2001**, *98*, 299–306. [CrossRef] [PubMed]

10. Huppertz, B.; Ghosh, D.; Sengupta, J. An integrative view on the physiology of human early placental villi. *Prog. Biophys. Mol. Biol.* **2014**, *114*, 33–48. [CrossRef] [PubMed]

11. Wang, Z.; Gerstein, M.; Snyder, M. RNA-Seq: A revolutionary tool for transcriptomics. *Nat. Rev. Genet.* **2009**, *10*, 57–63. [CrossRef] [PubMed]

12. Hong, Y.B.; Jung, S.C.; Lee, J.; Moon, H.S.; Chung, K.W.; Choi, B.O. Dynamic transcriptional events in distal sural nerve revealed by transcriptome analysis. *Exp. Neurobiol.* **2014**, *23*, 169–172. [CrossRef] [PubMed]

13. Gu, Y.; Sun, J.; Groome, L.J.; Wang, Y. Differential miRNA expression profiles between the first and third trimester human placentas. *Am. J. Physiol. Endocrinol. Metab.* **2013**, *304*, 836–843. [CrossRef] [PubMed]

14. Mercer, T.R.; Dinger, M.E.; Mattick, J.S. Long non-coding RNAs: Insights into functions. *Nat. Rev. Genet.* **2009**, *10*, 155–159. [CrossRef] [PubMed]

Int. J. Mol. Sci. **2018**, *19*, 1894

15. Jiang, C.; Ding, N.; Li, J.; Jin, X.; Li, L.; Pan, T.; Huo, C.; Li, Y.; Xu, J.; Li, X. Landscape of the long non-coding RNA transcriptome in human heart. *Brief. Bioinform.* **2018**. [CrossRef] [PubMed]

16. Paralkar, V.R.; Mishra, T.; Luan, J.; Yao, Y.; Kossenkov, A.V.; Anderson, S.M.; Dunagin, M.; Pimkin, M.; Gore, M.; Sun, D.; et al. Lineage and species-specific long noncoding RNAs during erythro-megakaryocytic development. *Blood* **2014**, *123*, 1927–1937. [CrossRef] [PubMed]

17. Mattick, J.S.; Rinn, J.L. Discovery and annotation of long noncoding RNAs. *Nat. Struct. Mol. Biol.* **2015**, *22*, 5–7. [CrossRef] [PubMed]

18. Pauli, A.; Eivind, V.; Lin, M.F.; Garber, M.; Vastenhouw, N.L.; Levin, J.Z.; Fan, L.; Sandelin, A.; Rinn, J.L.; Regev, A.; et al. Systematic identification of long noncoding RNAs expressed during zebrafish embryogenesis. *Genome Res.* **2012**, *22*, 577–591. [CrossRef] [PubMed]

19. Wang, K.C.; Chang, H.Y. Molecular mechanisms of long noncoding RNAs. *Mol. Cell* **2011**, *43*, 904–914. [CrossRef] [PubMed]

20. Amaral, P.P.; Mattick, J.S. Noncoding RNA in development. *Mamm. Genome* **2008**, *19*, 454–492. [CrossRef] [PubMed]

21. Pauli, A.; Rinn, J.L.; Schie, A.F. Non-coding RNAs as regulators of embryogenesis. *Nat. Rev. Genet.* **2011**, *12*, 136–149. [CrossRef] [PubMed]

22. Bouckenheimer, J.; Assou, S.; Riquier, S.; Hou, C.; Philippe, N.; Sansac, C.; Lavabre-Bertrand, T.; Commes, T.; Lemaître, J.M.; Boureux, A.; et al. Long non-coding RNAs in human early embryonic development and their potential in ART. *Hum. Reprod. Update* **2016**, *23*, 19–40. [CrossRef] [PubMed]

23. Zhao, W.; Mu, Y.; Ma, L.; Wang, C.; Tang, Z.; Yang, S.; Zhou, R.; Hu, X.; Li, M.H.; Li, K. Systematic identification and characterization of long intergenic non-coding RNAs in fetal porcine skeletal muscle development. *Sci. Rep.* **2015**, *5*, 8957. [CrossRef] [PubMed]

24. Sleutels, F.; Zwart, R.; Barlow, D.P. The non-coding Air RNA is required for silencing autosomal imprinted genes. *Nature* **2002**, *415*, 810–813. [CrossRef] [PubMed]

25. Bawa, P.; Zackaria, S.; Verma, M.; Gupta, S.; Srivatsan, R.; Chaudhary, B.; Srinivasan, S. Integrative Analysis of Normal Long Intergenic Non-Coding RNAs in Prostate Cancer. *PLoS ONE* **2015**, *10*, e0122143. [CrossRef] [PubMed]

26. Zhou, Y.; Wu, K.; Jiang, J.; Huang, J.; Zhang, P.; Zhu, Y.; Hu, G.; Lang, J.; Shi, Y.; Hu, L.; et al. Integrative Analysis Reveals Enhanced Regulatory Effects of Human Long Intergenic Non-Coding RNAs in Lung Adenocarcinoma. *J. Genet. Genom.* **2015**, *42*, 423–436. [CrossRef] [PubMed]

27. Cui, W.; Qian, Y.; Zhou, X.; Lin, Y.; Jiang, J.; Chen, J.; Zhao, Z.; Shen, B. Discovery and characterization of long intergenic non-coding RNAs (lincRNA) module biomarkers in prostate cancer: An integrative analysis of RNA-Seq data. *BMC Genom.* **2015**, *16*, S3. [CrossRef] [PubMed]

28. McAninch, D.; Roberts, C.T.; Bianco-Miotto, T. Mechanistic Insight into Long Noncoding RNAs and the Placenta. *Int. J. Mol. Sci.* **2017**, *18*, 1371. [CrossRef] [PubMed]

29. Kaartokallio, T.; Cervera, A.; Kyllönen, A.; Laivuori, K.; Kere, J.; Laivuori, H.; FINNPEC Core Investigator Group. Gene expression profiling of pre-eclamptic placentae by RNA sequencing. *Sci. Rep.* **2015**, *5*, 14107. [CrossRef] [PubMed]

30. Szcześniak, M.W.; Makałowska, I. lncRNA-RNA Interactions across the Human transcriptome. *PLoS ONE* **2016**, *11*, e0150353. [CrossRef] [PubMed]

31. Gonzalez, T.L.; Sun, T.; Koeppel, A.F.; Lee, B.; Wang, E.T.; Farber, C.R.; Rich, S.S.; Sundheimer, L.W.; Buttle, R.A.; Chen, Y.I.; et al. Sex differences in the late first trimester human placenta transcriptome. *Biol. Sex Differ.* **2018**, *9*, 4. [CrossRef] [PubMed]

32. Gormley, M.; Ona, K.; Kapidzic, M.; Garrido-Gomez, T.; Zdravkovic, T.; Fisher, S.J. Preeclampsia: Novel insights from global RNA profiling of trophoblast subpopulations. *Am. J. Obstet. Gynecol.* **2017**, *217*, 200.e1–200.e17. [CrossRef] [PubMed]

33. Long, W.; Rui, C.; Song, X.; Dai, X.; Xue, X.; Lu, Y.; Shen, R.; Li, J.; Li, J.; Ding, H. Distinct expression profiles of lncRNAs between early-onset preeclampsia and preterm controls. *Clin. Chim. Acta* **2016**, *463*, 193–199. [CrossRef] [PubMed]

34. Iyer, M.K.; Niknafs, Y.S.; Malik, R.; Singhal, U.; Sahu, A.; Hosono, Y.; Barrette, T.R.; Prensner, J.R.; Evans, J.R.; Zhao, S.; et al. The Landscape of Long Noncoding RNAs in the Human Transcriptome. *Nat. Genet.* **2015**, *47*, 199–208. [CrossRef] [PubMed]

35. Cabili, M.N.; Trapnell, C.; Goff, L.; Koziol, M.; Tazon-Vega, B.; Regev, A.; Rinn, J.L. Integrative annotation of human large intergenic noncoding RNAs reveals global properties and specific subclasses. *Genes Dev.* **2011**, *25*, 1915–1927. [CrossRef] [PubMed]

36. Ulitsky, I.; Shkumatava, A.; Jan, C.H.; Sive, H.; Bartel, D.P. Conserved function of lincRNAs in vertebrate embryonic development despite rapid sequence evolution. *Cell* **2011**, *147*, 1537–1550. [CrossRef] [PubMed]

37. Zhang, K.; Huang, K.; Luo, Y.; Li, S. Identification and functional analysis of long non-coding RNAs in mouse cleavage stage embryonic development based on single cell transcriptome data. *BMC Genom.* **2014**, *15*, 845. [CrossRef] [PubMed]

38. Derrien, T.; Johnson, R.; Bussotti, G.; Tanzer, A.; Djebali, S.; Tilgner, H.; Guernec, G.; Martin, D.; Merkel, A.; Knowles, D.G.; et al. The GENCODE v7 catalog of human long noncoding RNAs: Analysis of their gene structure, evolution, and expression. *Genome Res.* **2012**, *22*, 1775–1789. [CrossRef] [PubMed]

39. Xia, J.; Xin, L.; Zhu, W.; Li, L.; Li, C.; Wang, Y.; Mu, Y.; Yang, S.; Li, K. Characterization of long non-coding RNA transcriptome in high-energy diet induced nonalcoholic steatohepatitis minipigs. *Sci. Rep.* **2016**, *6*, 30709. [CrossRef] [PubMed]

40. Tang, Z.; Wu, Y.; Yang, Y.; Yang, Y.T.; Wang, Z.; Yuan, J.; Yang, Y.; Hua, C.; Fan, X.; Niu, G.; et al. Comprehensive analysis of long non-coding RNAs highlights their spatio-temporal expression patterns and evolutional conservation in *Sus scrofa*. *Sci. Rep.* **2017**, *7*, 43166. [CrossRef] [PubMed]

41. Brown, C.J.; Ballabio, A.; Rupert, J.L.; Lafreniere, R.G.; Grompe, M.; Tonlorenzi, R.; Willard, H.F. A gene from the region of the human X inactivation centre is expressed exclusively from the inactive X chromosome. *Nature* **1991**, *349*, 38–44. [CrossRef] [PubMed]

42. Pontier, D.B.; Gribnau, J. Xist regulation and function explored. *Hum. Genet.* **2011**, *130*, 223–236. [CrossRef] [PubMed]

43. Wutz, A. Gene silencing in X-chromosome inactivation: Advances in understanding facultative heterochromatin formation. *Nat. Rev. Genet.* **2011**, *12*, 542–553. [CrossRef] [PubMed]

44. Lee, J.T.; Bartolomei, M.S. X-inactivation, imprinting, and long noncoding RNAs in health and disease. *Cell* **2013**, *152*, 1308–1323. [CrossRef] [PubMed]

45. Lee, J.T. Lessons from X-chromosome inactivation: Long ncRNA as guides and tethers to the epigenome. *Genes Dev.* **2009**, *23*, 1831–1842. [CrossRef] [PubMed]

46. Firulli, A.B. A HANDful of questions: The molecular biology of the heart and neural crest derivatives (HAND)-subclass of basic helix-loop-helix transcription factors. *Gene* **2003**, *312*, 27–40. [CrossRef]

47. Hendershot, T.J.; Liu, H.; Clouthier, D.E.; Shepherd, I.T.; Coppola, E.; Studer, M.; Firulli, A.B.; Pittman, D.L.; Howard, M.J. Conditional deletion of Hand2 reveals critical functions in neurogenesis and cell type-specific gene expression for development of neural crest-derived noradrenergic sympathetic ganglion neurons. *Dev. Biol.* **2008**, *319*, 179–191. [CrossRef] [PubMed]

48. Huyen, D.V.; Bany, B.M. Evidence for a conserved function of heart and neural crest derivatives expressed transcript 2 in mouse and human decidualization. *Reproduction* **2011**, *142*, 353–368. [CrossRef] [PubMed]

49. Cho, H.; Okada, H.; Tsuzuki, T.; Nishigaki, A.; Yasuda, K.; Kanzaki, H. Progestin-induced heart and neural crest derivatives expressed transcript 2 is associated with fibulin-1 expression in human endometrial stromal cells. *Fertil. Steril.* **2013**, *99*, 248–255. [CrossRef] [PubMed]

50. Okada, H.; Tsuzuki, T.; Shindoh, H.; Nishigaki, A.; Yasuda, K.; Kanzaki, H. Regulation of decidualization and angiogenesis in the human endometrium: Mini review. *J. Obstet. Gynaecol. Res.* **2014**, *40*, 1180–1187. [CrossRef] [PubMed]

51. Morgenstern, R.; Zhang, J.; Johansson, K. Microsomal glutathione transferase 1: Mechanism and functional roles. *Drug Metab. Rev.* **2011**, *43*, 300–306. [CrossRef] [PubMed]

52. Nowak, M.; Gram, A.; Boos, A.; Aslan, S.; Ay, S.S.; Önyay, F.; Kowalewski, M.P. Functional implications of the utero-placental relaxin (RLN) system in the dog throughout pregnancy and at term. *Reproduction* **2017**, *154*, 415–431. [CrossRef] [PubMed]

53. Marzioni, D.; Tamagnone, L.; Capparuccia, L.; Marchini, C.; Amici, A.; Todros, T.; Bischof, P.; Neidhart, S.; Grenningloh, G.; Castellucci, M. Restricted innervation of uterus and placenta during pregnancy: Evidence for a role of the repelling signal Semaphorin 3A. *Dev. Dyn.* **2004**, *231*, 839–848. [CrossRef] [PubMed]

54. Hartley, S.W.; Mullikin, J.C. Detection and visualization of differential splicing in RNA-Seq data with JunctionSeq. *Nucleic Acids. Res.* **2016**, *44*, e127. [CrossRef] [PubMed]

55. Mourier, T.; Jeffares, D.C. Eukaryotic intron loss. *Science* **2003**, *300*, 1393. [CrossRef] [PubMed]

56. Ha, C.T.; Wu, J.A.; Irmak, S.; Lisboa, F.A.; Dizon, A.M.; Warren, J.W.; Ergun, S.; Dveksler, G.S. Human pregnancy specific beta-1-glycoprotein 1 (PSG1) has a potential role in placental vascular morphogenesis. *Biol. Reprod.* **2010**, *83*, 27–35. [CrossRef] [PubMed]

57. Horne, C.H.; Towler, C.M. Pregnancy-specific beta1-glycoprotein: A review. *Obstet. Gynecol. Surv.* **1978**, *33*, 761–768. [CrossRef] [PubMed]

58. Towler, C.M.; Horne, C.H.; Jandial, V.; Campbell, D.M.; MacGillivray, I. Plasma levels of pregnancy-specific beta 1-glycoprotein in complicated pregnancies. *Br. J. Obstet. Gynaecol.* **1977**, *84*, 258–263. [CrossRef] [PubMed]

59. Silver, R.M.; Heyborne, K.D.; Leslie, K.K. Pregnancy specific beta 1 glycoprotein (SP-1) in maternal serum and amniotic fluid; pre-eclampsia, small for gestational age fetus and fetal distress. *Placenta* **1993**, *14*, 583–589. [CrossRef]

60. Arnold, L.L.; Doherty, T.M.; Flor, A.W.; Simon, J.A.; Chou, J.Y.; Chan, W.Y.; Mansfield, B.C. Pregnancy-specific glycoprotein gene expression in recurrent aborters: A potential correlation to interleukin-10 expression. *Am. J. Reprod. Immunol.* **1999**, *41*, 174–182. [CrossRef] [PubMed]

61. Chang, C.L.; Semyonov, J.; Cheng, P.J.; Huang, S.Y.; Park, J.I.; Tsai, H.J.; Lin, C.Y.; Grützner, F.; Soong, Y.K.; Cai, J.J.; et al. Widespread divergence of the CEACAM/PSG genes in vertebrates and humans suggests sensitivity to selection. *PLoS ONE* **2013**, *8*, e61701. [CrossRef] [PubMed]

62. De Kreuk, B.J.; Schaefer, A.; Anthony, E.C.; Tol, S.; Fernandez-Borja, M.; Geerts, D.; Pool, J.; Hambach, L.; Goulmy, E.; Hordijk, P.L. The human minor Histocompatibility Antigen1 is a RhoGAP. *PLoS ONE* **2013**, *8*, e73962. [CrossRef] [PubMed]

63. Holland, O.J.; Linscheid, C.; Hodes, H.C.; Nauser, T.L.; Gilliam, M.; Stone, P.; Chamley, L.W.; Petroff, M.G. Minor histocompatibility antigens are expressed in syncytiotrophoblast and trophoblast debris: Implications for maternal alloreactivity to the fetus. *Am. J. Pathol.* **2012**, *180*, 256–266. [CrossRef] [PubMed]

64. Amado-Azevedo, J.; Reinhard, N.R.; van Bezu, J.; van Nieuw Amerongen, G.P.; van Hinsbergh, V.W.M.; Hordijk, P.L. The minor histocompatibility antigen 1 (HMHA1)/ArhGAP45 is a RacGAP and a novel regulator of endothelial integrity. *Vascul. Pharmacol.* **2018**, *101*, 38–47. [CrossRef] [PubMed]

65. Boissel, L.; Houssin, N.; Chikh, A.; Rynditch, A.; Van Hove, L.; Moreau, J. Recruitment of Cdc42 through the GAP domain of RLIP participates in remodeling of the actin cytoskeleton and is involved in Xenopus gastrulation. *Dev. Biol.* **2007**, *312*, 331–343. [CrossRef] [PubMed]

66. Ligeti, E.; Welti, S.; Scheffzek, K. Inhibition and termination of physiological responses by GTPase activating proteins. *Physiol. Rev.* **2012**, *92*, 237–272. [CrossRef] [PubMed]

67. Mishima, M.; Glotzer, M. Cytokinesis: A logical GAP. *Curr. Biol.* **2003**, *13*, 589–591. [CrossRef]

68. Linscheid, C.; Heitmann, E.; Singh, P.; Wickstrom, E.; Qiu, L.; Hodes, H.; Nauser, T.; Petroff, M.G. Trophoblast expression of the minor histocompatibility antigen HA-1 is regulated by oxygen and is increased in placentas from preeclamptic women. *Placenta* **2015**, *36*, 832–838. [CrossRef] [PubMed]

69. Assou, S.; Boumela, I.; Haouzi, D.; Monzo, C.; Dechaud, H.; Kadoch, I.J.; Hamamah, S. Transcriptome analysis during human trophectoderm specification suggests new roles of metabolic and epigenetic genes. *PLoS ONE* **2012**, *7*, e39306. [CrossRef] [PubMed]

70. Blakeley, P.; Fogarty, N.M.; del Valle, I.; Wamaitha, S.E.; Hu, T.X.; Elder, K.; Snell, P.; Christie, L.; Robson, P.; Niakan, K.K. Defining the three cell lineages of the human blastocyst by single-cell RNA-seq. *Development* **2015**, *142*, 3151–3165. [CrossRef] [PubMed]

71. Home, P.; Kumar, R.P.; Ganguly, A.; Saha, B.; Milano-Foster, J.; Bhattacharya, B.; Ray, S.; Gunewardena, S.; Paul, A.; Camper, S.A.; et al. Genetic redundancy of GATA factors in the extraembryonic trophoblast lineage ensures the progression of preimplantation and postimplantation mammalian development. *Development* **2017**, *144*, 876–888. [CrossRef] [PubMed]

72. Keniry, A.; Oxley, D.; Monnier, P.; Kyba, M.; Dandolo, L.; Smits, G.; Reik, W. The H19 lincRNA is a developmental reservoir of miR-675 that suppresses growth and Igf1r. *Nat. Cell Biol.* **2012**, *14*, 659–665. [CrossRef] [PubMed]

73. Saben, J.; Zhong, Y.; McKelvey, S.; Dajani, N.K.; Andres, A.; Badger, T.M.; Gomez-Acevedo, H.; Shankar, K. A comprehensive analysis of the human placenta transcriptome. *Placenta* **2014**, *35*, 125–131. [CrossRef] [PubMed]

74. Taylor, D.H.; Chu, E.T.; Spektor, R.; Soloway, P.D. Long Non-Coding RNA Regulation of Reproduction and Development. *Mol. Reprod. Dev.* **2015**, *82*, 932–956. [CrossRef] [PubMed]

75. Gao, W.L.; Liu, M.; Yang, Y.; Yang, H.; Liao, Q.; Bai, Y.; Li, Y.X.; Li, D.; Peng, C.; Wang, Y.L. The imprinted H19 gene regulates human placental trophoblast cell proliferation via encoding miR-675 that targets Nodal Modulator 1 (NOMO1). *RNA Biol.* **2012**, *9*, 1002–1010. [CrossRef] [PubMed]

76. Song, X.; Luo, X.; Gao, Q.; Wang, Y.; Gao, Q.; Long, W. Dysregulation of LncRNAs in Placenta and Pathogenesis of Preeclampsia. *Curr. Drug Targets* **2017**, *10*, 1165–1170. [CrossRef] [PubMed]

77. Yu, L.; Chen, M.; Zhao, D.; Yi, P.; Lu, L.; Han, J.; Zheng, X.; Zhou, Y.; Li, L. The H19 gene imprinting in normal pregnancy and pre-eclampsia. *Placenta* **2009**, *30*, 443–447. [CrossRef] [PubMed]

78. Giabicani, É.; Brioude, F.; Le Bouc, Y.; Netchine, I. Imprinted disorders and growth. *Ann. Endocrinol. (Paris)* **2017**, *78*, 112–113. [CrossRef] [PubMed]

79. Zuckerwise, L.; Li, J.; Lu, L.; Men, Y.; Geng, T.; Buhimschi, C.S.; Buhimschi, I.A.; Bukowski, R.; Guller, S.; Paidas, M.; et al. H19 long noncoding RNA alters trophoblast cell migration and invasion by regulating TβR3 in placentae with fetal growth restriction. *Oncotarget* **2016**, *7*, 38398–38407. [CrossRef] [PubMed]

80. Ying, W.; Jingli, F.; Wei, S.W.; Li, W.L. Genomic imprinting status of IGF-II and H19 in placentas of fetal growth restriction patients. *J. Genet.* **2010**, *89*, 213–216. [PubMed]

81. Li, X.; Ma, C.; Zhang, L.; Li, N.; Zhang, X.; He, J.; He, R.; Shao, M.; Wang, J.; Kang, L.; et al. LncRNAAC132217.4, a KLF8-regulated long non-coding RNA, facilitates oral squamous cell carcinoma metastasis by upregulating IGF2 expression. *Cancer Lett.* **2017**, *407*, 45–56. [CrossRef] [PubMed]

82. Aisemberg, J.; Bariani, M.V.; Vercelli, C.A.; Wolfson, M.L.; Franchi, A.M. Lipopolysaccharide-induced murine embryonic resorption involves nitric oxide-mediatedinhibition of the NAD+-dependent 15-hydroxyprostaglandin dehydrogenase. *Reproduction* **2012**, *144*, 447–454. [CrossRef] [PubMed]

83. Girard, A.; Sachidanandam, R.; Hannon, G.J.; Carmell, M.A. A germline-specific class of small RNAs binds mammalian Piwi proteins. *Nature* **2006**, *442*, 199–202. [CrossRef] [PubMed]

84. Aravin, A.; Gaidatzis, D.; Pfeffer, S.; Lagos-Quintana, M.; Landgraf, P.; Iovino, N.; Morris, P.; Brownstein, M.J.; Kuramochi-Miyagawa, S.; Nakano, T.; et al. A novel class of small RNAs bind to MILI protein in mouse testes. *Nature* **2006**, *442*, 203–207. [CrossRef] [PubMed]

85. Fu, X.D. Non-coding RNA: A new frontier in regulatory biology. *Natl. Sci. Rev.* **2014**, *1*, 190–204. [CrossRef] [PubMed]

86. Gomes, A.Q.; Nolasco, S.; Soares, H. Non-coding RNAs: Multi-tasking molecules in the cell. *Int. J. Mol. Sci.* **2013**, *14*, 16010–16039. [CrossRef] [PubMed]

87. Ulitsky, I.; Bartel, D.P. lincRNAs: Genomics, evolution, and mechanisms. *Cell* **2013**, *154*, 26–46. [CrossRef] [PubMed]

88. Bolger, A.M.; Lohse, M.; Usadel, B. Trimmomatic: A flexible trimmer for Illumina sequence data. *Bioinformatics* **2014**, *30*, 2114–2120. [CrossRef] [PubMed]

89. Pertea, M.; Pertea, G.M.; Antonescu, C.M.; Chang, T.C.; Mendell, J.T.; Salzberg, S.L. StringTie enables improved reconstruction of a transcriptome from RNA-seq reads. *Nat. Biotechnol.* **2015**, *33*, 290–295. [CrossRef] [PubMed]

90. Trapnell, C.; Roberts, A.; Goff, L.; Pertea, G.; Kim, D.; Kelley, D.R.; Pimentel, H.; Salzberg, S.L.; Rinn, J.L.; Pachter, L. Differential gene and transcript expression analysis of RNA-seq experiments with TopHat and Cufflinks. *Nat. Protoc.* **2012**, *7*, 562–578. [CrossRef] [PubMed]

91. Kong, L.; Zhang, Y.; Ye, Z.Q.; Liu, X.Q.; Zhao, S.Q.; Wei, L.; Gao, G. CPC: Assess the protein-coding potential of transcripts using sequence features and support vector machine. *Nucleic Acids Res.* **2007**, *35*, 345–349. [CrossRef] [PubMed]

92. Mistry, J.; Bateman, A.; Finn, R.D. Predicting active site residue annotations in the Pfam database. *BMC Bioinform.* **2007**, *8*, 298. [CrossRef] [PubMed]

93. Wang, L.; Park, H.J.; Dasari, S.; Wang, S.; Kocher, J.P.; Li, W. CPAT: Coding-Potential Assessment Tool using an alignment-free logistic regression model. *Nucleic Acids Res.* **2013**, *41*, e74. [CrossRef] [PubMed]

94. Sun, L.; Luo, H.; Bu, D.; Zhao, G.; Yu, K.; Zhang, C.; Liu, Y.; Chen, R.; Zhao, Y. Utilizing sequence intrinsic composition to classify protein-coding and long non-coding transcripts. *Nucleic Acids Res.* **2013**, *41*, e166. [CrossRef] [PubMed]

95. Li, A.; Zhang, J.; Zhou, Z. PLEK: A tool for predicting long non-coding RNAs and messenger RNAs based on an improved k-mer scheme. *BMC Bioinform.* **2014**, *15*, 311. [CrossRef] [PubMed]

96. Conesa, A.; Götz, S.; Garcia-Gomez, J.M.; Terol, J.; Talon, M.; Robles, M. Blast2GO: A universal tool for annotation, visualization and analysis in functional genomics research. *Bioinformatics* **2005**, *21*, 3674–3676. [CrossRef] [PubMed]

97. Wu, J.; Mao, X.; Cai, T.; Luo, J.; Wei, L. KOBAS server: A web-based platform for automated annotation and pathway identification. *Nucleic Acids Res.* **2006**, *34*, 720–724. [CrossRef] [PubMed]

© 2018 by the authors. Licensee MDPI, Basel, Switzerland. This article is an open access article distributed under the terms and conditions of the Creative Commons Attribution (CC BY) license (http://creativecommons.org/licenses/by/4.0/).

International Journal of
Molecular Sciences

Article

Transcriptome Analysis of *Novosphingobium pentaromativorans* US6-1 Reveals the Rsh Regulon and Potential Molecular Mechanisms of *N*-acyl-L-homoserine Lactone Accumulation

Hang Lu and Yili Huang *

Zhejiang Provincial Key Laboratory of Organic Pollution Process and Control,
Department of Environmental Science, College of Environmental and Resource Sciences,
Zhejiang University, Hangzhou 310027, China; 21514007@zju.edu.cn
* Correspondence: yilihuang@zju.edu.cn; Tel./Fax: +86-571-8898-2592

Received: 4 August 2018; Accepted: 2 September 2018; Published: 5 September 2018

Abstract: In most bacteria, a bifunctional Rsh responsible for (p)ppGpp metabolism is the key player in stringent response. To date, no transcriptome-wide study has been conducted to investigate the Rsh regulon, and the molecular mechanism of how Rsh affects the accumulation of *N*-acyl-L-homoserine lactone (AHL) remains unknown in sphingomonads. In this study, we identified an rsh_{US6-1} gene by sequence analysis in *Novosphingobium pentaromativorans* US6-1, a member of the sphingomonads. RNA-seq was used to determine transcription profiles of the wild type and the ppGpp-deficient rsh_{US6-1} deletion mutant (Δrsh). There were 1540 genes in the Rsh_{US6-1} regulon, including those involved in common traits of sphingomonads such as exopolysaccharide biosynthesis. Furthermore, both RNA-seq and quantitative real-time polymerase chain reaction (qRT-PCR) showed essential genes for AHL production (*novI* and *novR*) were positively regulated by Rsh_{US6-1} during the exponential growth phase. A degradation experiment indicated the reason for the AHL absence in Δrsh was unrelated to the AHL degradation. According to RNA-seq, we proposed σ^E, DksA, Lon protease and RNA degradation enzymes might be involved in the Rsh_{US6-1}-dependent expression of *novI* and *novR*. Here, we report the first transcriptome-wide analysis of the Rsh regulon in sphingomonads and investigate the potential mechanisms regulating AHL accumulation, which is an important step towards understanding the regulatory system of stringent response in sphingomonads.

Keywords: Rsh regulon; *Novosphingobium pentaromativorans* US6-1; sphingomonads; RNA-seq; *N*-acyl-L-homoserine lactone; ppGpp

1. Introduction

Bacteria need to co-ordinate cellular responses to unfavorable environmental conditions [1]. One major strategy to cope with environmental stress is the activation of stringent response, a global regulatory system [2]. The stringent response is activated by (p)ppGpp (guanosine tetraphosphate and guanosine pentaphosphate) [3]. The proteins from the RelA/SpoT (Rsh) family are the key players, synthesizing (p)ppGpp from ATP and either GTP or GDP, and degrading (p)ppGpp to pyrophosphate and either GTP or GDP. Most species in *Betaproteobacteria* and *Gammaproteobacteria* contain two multi-domain Rsh enzymes, RelA and SpoT, while the majority of bacteria contain only a single Rsh protein [3]. A number of studies have demonstrated that Rsh affected the expression of a wide range of genes involved in physiological processes in bacteria such as *Escherichia coli*, *Staphylococcus aureus* and *Rhizobium etli* [4–6]. In *R. etli*, there were 834 genes in the Rsh regulon involved in various cellular processes such as transcriptional regulation, signal transduction, production of sigma factors and non-coding RNAs [6]. Sphingomonads, a group of *Alphaproteobacteria*, are widely distributed in polluted and oligotrophic environments [7,8]. Sphingomonads have drawn much attention for their traits, including the pronounced abilities of degrading a wide range of recalcitrant natural and xenobiotic compounds such as polycyclic aromatic hydrocarbons (PAH) [7,9], the substitution of sphingolipids for lipopolysaccharide in their outer membrane [10,11] as well as their production of exopolysaccharides (EPS) [12,13]. The genomes of sphingomonads contain one single *rsh* gene, which responds to environmental stress as in other bacteria [14]. However, to date, no transcriptome-wide study has been conducted to investigate Rsh target genes in any strain of sphingomonads.

One of the physiological activities regulated by Rsh is quorum sensing (QS), a mechanism of intercellular communication. In Gram-negative bacteria, the main QS signal is N-acyl-L-homoserine lactone (AHL), which is produced by LuxI-type synthases. LuxR-type receptors can bind to AHL and then stimulate the expression of *luxI* homologs [15]. AHL accumulation is dependent on RelA/SpoT homologs in various bacteria such as *Pseudomonas aeruginosa* [16–18] and *R. etli* [19], and AHL degradation is regulated by (p)ppGpp via AttM in *Agrobacterium tumefaciens* [20,21]. In *P. aeruginosa*, deletion of both *relA* and *spoT* resulted in increased levels of 4-hydroxyl-2-heptylquinoline and 3,4-dihydroxy-2-heptylquinoline via up-regulated *pqsA* and *pqsR* expression and decreased levels of butanoyl-homoserine lactone and 3-oxo-dodecanoyl-homoserine lactone via down-regulated *rhlI*, *rhlR*, *lasI*, and *lasR* expression [18]. In recent years, an increasing number of strains in sphingomonads which produced AHLs have been isolated [22]. The comparative genomic analyses of 62 sphingomonads genomes showed that the canonical *luxI/R*-type QS network was widespread within sphingomonads [23]. In the previous study, a Tn5 mutant of *Novosphingobium* sp. Rr 2-17, deficient in AHL accumulation, was found to have an insertion in an *rsh* gene, suggesting that QS was under the regulation of Rsh in *Novosphingobium*, a member of sphingomonads. However, the potential molecular mechanism remains unknown.

Novosphingobium pentaromativorans US6-1, which has been shown to have potential in aromatic hydrocarbons bioremediation, is a type strain belonging to sphingomonads [24]. Its genome sequencing has been completed and the genome database is accessible from the public NCBI database [25]. In this study, we identified an *rsh* gene in *N. pentaromativorans* US6-1 (annotated as rsh_{US6-1}) by sequence analysis. The wild-type strain produced ppGpp in static culture medium while rsh_{US6-1} deletion mutant (Δrsh) did not. Therefore, the transcription profiles of the wild type and Δrsh grown in static medium was determined by RNA-seq to identify the Rsh_{US6-1} regulon. Furthermore, we determined whether AHL accumulation was affected by Rsh_{US6-1} and investigated the potential molecular mechanisms via quantitative real-time polymerase chain reaction (qRT-PCR), transcriptome analysis and an AHL degradation experiment. These results are useful to understand the regulatory system of stringent response in sphingomonads.

2. Results and Discussion

2.1. Sequence Analysis of Rsh_{US6-1} Protein

The open reading frame of the full-length rsh_{US6-1} gene (accession number WP_007011921) was 2094 nucleotides in length. Rsh_{US6-1} contained the nitrogen-terminal metal-dependent hydrolase domain (HD) (amino acids 26–177) and synthetase domain (Syn) (amino acids 236–347). The carboxy-terminal domain of Rsh_{US6-1} consisted of the TGS domain (for: Thr-tRNA synthetase, GTPase and SpoT) (amino acids 385–444) and ACT domain (for: aspartate kinase, chorismate mutase and T protein) (amino acids 625–687) (Figure 1a). These four domains are commonly present in Rsh [3].

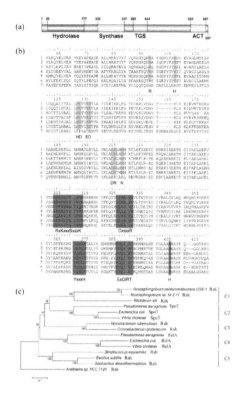

Figure 1. Sequence analysis of Rsh$_{US6-1}$ protein. (**a**) Rsh$_{US6-1}$ domain; (**b**) amino acid alignment of RelA/SpoT homologs. Line 1: Rsh$_{US6-1}$. Line 2: Rsh$_{Rr2-17}$. Line 3: Rsh$_{Mtb}$. Line 4: Rsh$_{Seq}$. Line 5: Rsh$_{Ret}$. Line 6: SpoT$_{Ecoli}$. Line 7: RelA$_{Ecoli}$ (see in Materials and Methods). Light grey boxes represent six conserved motifs in HD domain (HD 1: R. HD 2: H. HD 3: HD. HD 4: ED. HD 5: DR. HD 6: N) while dark grey boxes represent five conserved motifs in Syn domain (Syn 1: RxKxxxSxxxK. Syn 2: DxxxxR. Syn 3: YxxxH. Syn 4: ExQIRT. Syn 5: H) [26]; (**c**) phylogenetic tree based on RelA/SpoT homologs. Sequences, available at NCBI GenBank, were aligned by using the ClustalW algorithm [27] in MEGA version 6.0 [28]. Then a phylogenetic tree was constructed by the neighbor-joining method [29] with 1000 bootstrap replicates [30] after cutting off the redundant sequences at the end, with the use of MEGA version 6.0. The Rsh protein from *Anabaena sp.* PCC 7120 (accession number BAB77915) was used as outgroup [31]. There were five clusters: C1 for Rsh in *Alphaproteobacteria*, C2 for SpoT in *Gammaproteobacteria*, C3 for Rsh in *Actinobacteria*, C4 for RelA in *Gammaproteobacteria*, C5 for Rsh in *Firmicutes*.

Rsh$_{US6-1}$ displayed high sequence similarity to Rsh proteins whose functions had been identified in other strains [32–36]. The identities of Rsh$_{US6-1}$ with Rsh$_{Rr2-17}$, Rsh$_{Mtb}$, Rsh$_{Seq}$, Rsh$_{Ret}$, SpoT$_{Ecoli}$ and RelA$_{Ecoli}$ were 88%, 37%, 38%, 43%, 38% and 30% respectively. In the HD domain (Figure 1b), Rsh$_{US6-1}$ contained six motifs, which are predicted to be essential for (p)ppGpp hydrolysis [26]. RelA$_{Ecoli}$ was more divergent in these motifs, which was consistent with its loss of hydrolase activity. In the Syn domain (Figure 1b), Rsh$_{US6-1}$ also contained five conserved motifs that were proved structurally and biochemically important for (p)ppGpp synthetase activity [26].

The phylogenic analysis of RelA/SpoT homologs was performed (Figure 1c). The result showed that there was a clear separation of the RelA/SpoT families, forming five clusters: C1 for Rsh in *Alphaproteobacteria*, C2 for SpoT in *Gammaproteobacteria*, C3 for Rsh in *Actinobacteria*, C4 for RelA in *Gammaproteobacteria*, C5 for Rsh in *Firmicutes*. Rsh$_{US6-1}$ was grouped with Rsh$_{Rr2-17}$ in C1. These data suggested that Rsh$_{US6-1}$ possessed the sequence required for the (p)ppGpp synthetase and hydrolase activities.

2.2. N-Acyl-L-homoserine Lactone (AHL) Accumulation in the Cross Feeding Assay and Extract Assay

To determine the effect of Rsh$_{US6-1}$ on the AHL accumulation, cross-feeding and extract assays were performed. The results of the cross-feeding and extract assays were consistent (Figure 2). In the cross-feeding assay, strains were grown as biofilm and only US6-1 showed AHL accumulation. In the extract assay, US6-1 accumulated AHL signals in static cultures while no AHL was detected in Δ*rsh* during the whole growth period. To verify that the *rsh*$_{US6-1}$ deletion was responsible for the absence of AHL, a *rsh*$_{US6-1}$ complementation strain Δ*rsh* (pCM62-rsh) was created. Δ*rsh* (pCM62-rsh) restored the same phenotype as the wild-type strain, suggesting that Rsh$_{US6-1}$ was required for the AHL accumulation. In the extract assay, the AHL molecules in the extract of culture would diffuse through the soft agar to the reporter strain *A. tumefaciens* A136, and activate the *traI-lacZ* fusion. Therefore, we could determine the quantity of AHL molecules by the diffusion area indicated by the diameter of the blue coloration in the presence of 5-bromo-4-chloro-3-indolyl-β-D-galactopyranoside (X-Gal) [37]. The relationship between the growth of US6-1 and AHL accumulation was analyzed carefully. The bacteria cells grew exponentially for about 48 h and then entered the stationary phase. Similarly, the AHL accumulation was maximized at 48 h and then declined, indicating US6-1 could degrade AHL signals.

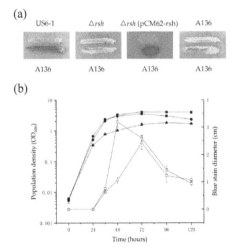

Figure 2. AHL accumulation in US6-1, Δ*rsh* and Δ*rsh* (pCM62-rsh). (**a**) AHL accumulation assay by cross-feeding. P5Y3 agar plate was covered with 50 μL X-gal. The AHL reporter strain A136 and the tested strains were streaked side by side on the agar plates. Plates were incubated for 24 h, when the activation of the reporter was recorded. Strain A136 versus A136 was used as negative control. One representative experiment out of three independent biological replicates is shown. (**b**) Time course of population density of US6-1 (filled squares), Δ*rsh* (filled circles) and Δ*rsh* (pCM62-rsh) (filled triangles) and AHL accumulation of US6-1 (open squares) and Δ*rsh* (pCM62-rsh) (open triangles) in the extract assay. Fresh colonies of strains were first inoculated into P5Y3 broth at the shaking speed of 200 rpm to an OD$_{600}$ value of 1. Then 250 μL of this seed culture was re-inoculated into 50 mL of fresh P5Y3 broth and incubated statically. The growth curve was drawn from measuring the OD$_{600}$ values. AHL signals in cultures were then extracted by ethyl acetate (EA). The extracts were spotted onto LB soft agar plates plus X-gal mixed with A136. The plates were incubated for 12 h at 30 °C and the diameters of blue stains which represented the quantity of AHL signals in bacterial cultures were measured. No AHL was detected in Δ*rsh*. Values shown are the average of biological triplicate experiments with standard deviations marked with error bars.

2.3. ppGpp Accumulation in Strain US6-1 and Its Derivatives

We monitored the endogenous ppGpp levels in US6-1, Δ*rsh* and Δ*rsh* (pCM62-rsh) (Figure 3). The presence of ppGpp in US6-1 culture at 72 h of incubation (stationary phase) was determined while no ppGpp was detected in Δ*rsh*. When we analyzed ppGpp accumulation in Δ*rsh* (pCM62-rsh), the complementation of ppGpp production was observed, confirming that Rsh$_{US6-1}$ was responsible for ppGpp synthesis. These results showed that the stringent response in US6-1 and the complementary strain was induced under the current culture conditions. Nutrient limitation is one of the conditions that can induce the stringent response. However, the medium used in this study was nutrient-rich P5Y3 medium. Several studies has recently found that antibiotics, acid stress and oxidative stress could also induce the stringent response [38]. Therefore, it seems that many factors which can induce the stringent response of bacteria remain unknown. At 36 h of incubation (exponential growth phase), we could not detect the ppGpp accumulation in US6-1 culture. This was probably because the ppGpp level was below the detectable range, as ppGpp is generally thought to be present at basal levels in the exponential growth phase [3]. However, previous studies showed that ppGpp could still regulate the expression of a wide range of genes during this growth period [3,6].

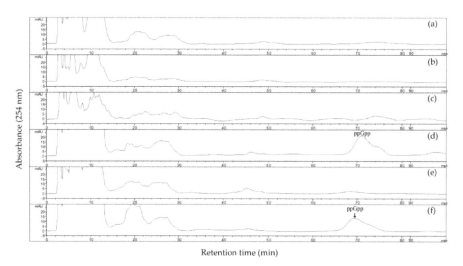

Figure 3. ppGpp accumulation in strain US6-1, Δ*rsh* and Δ*rsh* (pCM62-rsh) in different growth phases. Levels of ppGpp in cultures of US6-1 (**a,d**), Δ*rsh* (**b,e**) and Δ*rsh* (pCM62-rsh) (**c,f**) in the exponential growth phase (36 h) and in the stationary phase (72 h) were monitored by reverse phase high-performance liquid chromatography (HPLC) analysis. The eluted nucleotides were monitored at 254 nm and identified by comparison with the retention time of 100 μM ppGpp standard. ppGpp standard was eluted at 71 min under the current conditions and the position is indicated by an arrow.

2.4. Global Overview of the Rsh$_{US6-1}$ Regulon

The RNA-seq of strain US6-1 and Δ*rsh* cultivated in static conditions, in which US6-1 accumulated AHL signals, was conducted to determine the Rsh$_{US6-1}$ regulon. An average of 24 million clean reads in one sample were obtained. The total mapping radios and uniquely mapping ratios of clean reads to reference genome of US6-1 were up to 98% and 95%, indicating that the sequencing was deep enough to cover almost all kinds of transcripts in the cells (Table S1). The genome of US6-1 contains 5110 annotated protein-encoding genes, 59 pseudo genes and 82 RNA genes [25]. The comparative transcriptome analysis between strain US6-1 and Δ*rsh* revealed 1540 Rsh$_{US6-1}$-dependent differentially expressed genes (DEGs), which were defined according to the combination of the absolute value of a fold change ≥2 and an adjust p value ≤ 0.05 (Table S2). Compared with the wild type, 911 of these genes were up-regulated (17.83%) and 629 genes (12.31%) were down-regulated in Δ*rsh* (Δ*rsh* vs. US6-1). The differential expression varied between an 8.82-fold up-regulation of a gene encoding a protein of ferrisiderophore receptor (WP_007014902) and a 40.5-fold down-regulation of an autoinducer synthesis protein (WP_007013362). The transcriptome data were validated by analyzing the expression levels of 39 representative genes using qRT-PCR (Table S3). The result showed that 29 (74.36%) of the tested genes were regulated in the same direction (up or down) in both RNA-seq data and qRT-PCR (Δ*rsh* vs. US6-1), and the fold change of each gene in qRT-PCR was $\log_2(x) > 1$ or $\log_2(1/x) < -1$ (representing a plain fold-change >2) with statistical differences. Among these 29 genes, the expression of 24 genes by qRT-PCR (the complementary strain vs. US6-1) was not different significantly or was regulated in the opposite direction in the RNA-seq data. Therefore, the transcriptome data were in good agreement with the qRT-PCR data, which proved the reliability of the transcriptome data.

These genes were further grouped based on Kyoto Encyclopedia of Genes and Genomes (KEGG) annotation and pathway enrichment analysis (Figure 4a). The Fisher's exact test identified 12 significantly over-represented KEGG pathways. The result suggested that Rsh_{US6-1} controlled a variety of metabolic pathways. The most of the DEGs involved in ribosomal protein production, porphyrin and chlorophyll metabolism, amino acyl-tRNA biosynthesis, oxidative phosphorylation, TCA cycle, phenylalanine, tyrosine and tryptophan biosynthesis were up-regulated in the mutant, which suggested that the role of Rsh_{US6-1}, similar to that in *E. coli*, was to repress the majority of cellular processes to reallocate cellular resources [39]. Although one of the hallmarks of stringent response is the induction of amino acid biosynthesis [39,40], unexpectedly, in Δ*rsh* most genes involved in amino acid biosynthesis were up-regulated. Similarly, in *relA* mutant of *Pectobacterium atrosepticum*, several unlinked clusters of genes involved in branched chain amino acid metabolism (*ilvGMEDA*, *ilvIH*, *ilvBN* and *leuABCD*) were also up-regulated [41]. There may be other mechanisms which can induce amino acid biosynthesis when US6-1 grows under environmental stress. According to the validation of the transcriptome data, DEGs involved in AHL production, sigma factor synthesis, RNA degradation, EPS biosynthesis, PAH degradation, sphingolipid production, cell division and shape and GTP synthesis were chosen, and their expression profiles were visualized in the heat map (Figure 4b, Table S3). Except the genes related to AHL production, the most of these genes were under negative control of Rsh_{US6-1}. These DEGs will be discussed in detail in the following paragraphs.

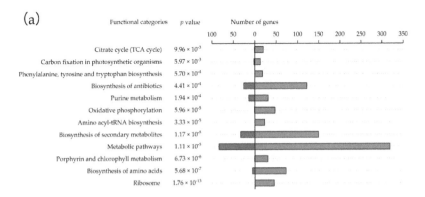

Figure 4. *Cont.*

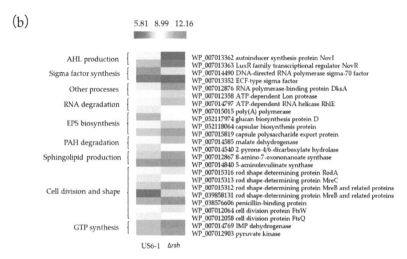

Figure 4. Differentially expressed genes (DEGs) overview (Δ*rsh* vs. US6-1) in the RNA-seq data. (**a**) Over-represented Kyoto Encyclopedia of Genes and Genomes (KEGG) categories. KEGG annotation and pathway enrichment analysis was performed using the KEGG pathway database (http://www. genome.jp/kegg/). The *p* values denote the enriched levels in a KEGG pathway, which was calculated using a Fisher's exact test [42]. Up- and down-regulated genes (Δ*rsh* vs. US6-1) are represented by red and blue bars respectively, representing the number of genes per functional category. (**b**) Heat map of log₂ expression ratios of specific DEGs involved in AHL production, sigma factor synthesis, RNA degradation, EPS biosynthesis, PAH degradation, sphingolipid production, cell division and shape, GTP synthesis and other functions in strain US6-1 and Δ*rsh*. Expression values are reflected by red-blue coloring as indicated. The heat map was drawn by the software HemI [43].

2.5. Essential Genes for AHL Production were Positively Regulated by Rsh$_{US6-1}$ in the Exponential Growth Phase

There are only one gene *novI*, encoding an autoinducer synthase (WP_007013362) and one gene *novR* encoding a LuxR family transcriptional regulator containing an autoinducer binding domain (WP_007013363) in the genome of US6-1. These two genes are essential for AHL production. The transcriptome data showed that in strain Δ*rsh*, *novI* and *novR* were down-regulated 40.5 and 2.35-fold at 36 h of incubation, respectively. The qRT-PCR analysis of culture samples from different growth periods confirmed the decreased expression levels of *novI* and *novR* in Δ*rsh* (Table 1). The *novI* and *novR* were both down-regulated significantly at 36 h and 48 h (exponential growth phase) in Δ*rsh* while Δ*rsh* (pCM62-rsh) could complemented their expression.

Although the essential genes for AHL production were down-regulated in the exponential growth phase due to the lack of ppGpp, Δ*rsh* still possibly produced low concentrations of AHL signals. Since US6-1 could degrade AHL signals, we analyzed whether Δ*rsh* degraded AHL in the exponential growth phase. When provided with 2 μM medium-chain AHL (C8-AHL and 3-OH-C8-AHL) as cosubstrates in culture medium, Δ*rsh* did not show the capacity to degrade the medium-chain AHL at 48 h (Figure 5). However, at 72 h of incubation, there were fewer AHL signals in Δ*rsh* cultures. These results suggested that Δ*rsh* might not degrade AHL signals in the exponential growth phase. The activation of its capacity to degrade AHL was likely related to the entrance to the stationary phase. Overall, the reason why Rsh$_{US6-1}$ was required for the AHL accumulation was that Rsh$_{US6-1}$ positively regulated essential genes for AHL production in the exponential growth phase, and might be unrelated to the AHL degradation.

Table 1. Relative expression levels of *novI* and *novR* by quantitative reverse transcription polymerase chain reaction (qRT-PCR). RNA samples were extracted from bacterial cultures after 24 h, 36 h, 48 h and 72 h. The levels of gene expression in US6-1, Δ*rsh*, Δ*rsh* (pCM62-rsh) were normalized to 16S rRNA gene, and the relative levels in Δ*rsh* and Δ*rsh* (pCM62-rsh) to that in the wild-type strain (set as value of 1) were reported. Values shown are the average of biological triplicate experiments with standard deviations (mean ± standard deviation (SD)). "*" represents $p < 0.05$; "**" represents $p < 0.01$; "***" represents $p < 0.001$.

Genes	Strains	24 h	36 h	48 h	96 h
novI	US6-1	1.166 ± 0.773	1.060 ± 0.466	1.115 ± 0.647	1.039 ± 0.356
	Δ*rsh*	0.261 ± 0.083	0.011 ± 0.002 *	0.020 ± 0.004 *	0.026 ± 0.004 **
	Δ*rsh* (pCM62-rsh)	0.501 ± 0.085 ***	0.648 ± 0.216	0.753 ± 0.097	3.975 ± 1.547 *
novR	US6-1	1.012 ± 0.185	1.006 ± 0.129	1.001 ± 0.055	1.029 ± 0.314
	Δ*rsh*	0.119 ± 0.018 **	0.068 ± 0.005 ***	0.547 ± 0.017 ***	0.881 ± 0.354
	Δ*rsh* (pCM62-rsh)	0.601 ± 0.050 **	0.463 ± 0.076	0.543 ± 0.054	2.486 ± 0.972

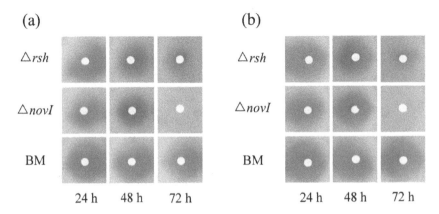

Figure 5. Identification of degradation of the medium-chain AHL by Δ*rsh*. Strain Δ*rsh* was incubated with 2 µM C8-AHL (**a**) and 3-OH-C8-AHL (**b**) in culture medium for 24 h (early exponential growth phase), 48 h (late exponential growth phase) and 72 h (stationary phase), after which the EA extracts were spotted onto AHL reporter plates mixed with A136. The diameters of the blue stains represent the quantity of residual AHL signals. To avoid the influence of the AHL produced by strain US6-1 in the exponential growth phase, *novI* deletion mutant (Δ*novI*) which could not produce AHL was used as positive control. Blank medium (BM) with AHL was used as negative control.

2.6. Potential Molecular Mechanisms of Rsh$_{US6-1}$-Dependent Expression of novI and novR

In the case of *E. coli*, (p)ppGpp regulates the expression of genes through binding to β'-subunit of RNA polymerase to destabilize the short-lived open complexes that form at certain promoters [3]. However, genes in the Rsh$_{US6-1}$ regulon were numerous and the proteins which can modulate the expression of *luxI/R* homologs vary in bacteria [44]. Therefore, the transcriptome data were used to investigate the potential molecular mechanisms of the decreased expression of *novI* and *novR* in Δ*rsh*.

During exponential growth, transcription is widely under control of the housekeeping sigma factor σ^{70}. ppGpp can regulate the expression of genes via the regulation of alternative sigma factor competition [45]. Previous studies showed that QS was influenced by alternative sigma factors. In *P. aeruginosa* PAO1 and *P. fluorescens* UK4, ppGpp up-regulates the expression of *rpoS* encoding a stationary phase sigma factor and RpoS increases the expression level of AHL-related genes as well as the AHL production [16,46]. However, the *Pseudomonas* model might not be applicable in strain US6-1, since its genome does not encode a corresponding RpoS. US6-1 possesses 22 putative sigma factors. Only one housekeeping sigma factor σ^{70} (WP_007014490) was up-regulated 2.2-fold while a extracytoplasmic function (ECF) sigma factor (WP_007013352) was down-regulated 2.73-fold in Δrsh. The link between ECF sigma factor and QS has recently given rise to controversy. ECF sigma factors can bind directly to -10 and -35 elements in the *luxR* promoter, thus inducing *luxR* expression. The binding sites of this kind of sigma factors are also identified in the upstream region of *luxS* [47,48]. However, the effect of ECF sigma factors on QS was not observed in another study [49].

Our data showed that *dksA* (WP_007012876) was up-regulated significantly (6.41-fold) in Δrsh. DksA, a (p)ppGpp cofactor, can bind the secondary channel and sensitize RNAP to (p)ppGpp at many promoters [3]. The previous study showed that DksA could also function without ppGpp [50]. In *P. aeruginosa*, DksA represses the expression of *rhlI* during the exponential growth [51].

The gene encoding an ATP-dependent Lon protease (WP_007012358) was up-regulated 2.01-fold due to the Rsh$_{US6-1}$ absence. The gene is involved in the toxin–antitoxin (TA) module, a (p)ppGpp-dependent mechanism related to antibiotic tolerance and resistance [3]. Lon protease is a powerful negative regulator of two HSL-mediated QS systems in *P. aeruginosa* [52]. In the TA model, (p)ppGpp activates the Lon protease [3]. Interestingly, our data indicated an opposite type of the TA module regulation of ppGpp in strain US6-1.

Moreover, the expression levels of two genes involved in the putative functions of RNA degradation (WP_007014797 and WP_007015015) were increased 2.62 and 2.19-fold respectively in Δrsh. The enzyme responsible for RNA degradation has been reported to degrade mRNA of AHL synthase genes and thus decrease the AHL production rapidly [53]. Overall, the above transcriptome data showed that ppGpp could regulate several factors including the ECF sigma factor, DksA, ATP-dependent Lon protease and RNA degradation enzymes. These factors have been proved to affect the expression of *luxI/R* homologs in other bacteria. However, the relationship between these factors and *luxI/R* homologs in sphingomonads remains unknown and needs further studies. We proposed these potential molecular mechanisms of ppGpp-dependent expression of *novI* and *novR*, which might contribute to giving insight into the complicated cross-talk between stringent response and QS in sphingomonads.

2.7. Repressed Exopolysaccharides (EPS) Biosynthesis

Many strains in sphingomonads can produce EPS [12,13]. The transcriptome data showed the gene encoding a putative glucan biosynthesis protein (WP_052117974) and two genes involved in capsule biosynthesis and export (WP_052118064, WP_007015819) were up-regulated 3.03, 2.07 and 3.01-fold respectively in strain Δrsh. Glucans are major constituents of capsular materials [54]. Thus, we analyzed Δrsh for Congo red binding. Congo red binds certain polysaccharides as well as polymers that display amyloid-like properties [12]. On Congo red plates, the colony biofilm of Δrsh was light red, while the wild-type strain appeared slightly orange. The complemented strain showed a phenotype similar to that of the wild-type strain (Figure 6a). Furthermore, the culture of Δrsh bound more Congo red than the wild type, leaving less Congo red in the cell-free supernatant. The Congo red bounding of Δrsh (pCM62-rsh) showed no significant difference with that of strain US6-1 (Figure 6b). The results suggested that Rsh$_{US6-1}$ affected EPS biosynthesis as a repressor, consistent with the changes of the expression levels of the above genes. This phenomenon was also observed in *Sinorhizobium meliloti* [55]. When the local environment becomes unfavorable (e.g., nutrient or oxygen deficiency), the motile behavior is important for bacteria to move towards a better environment to survive. In several strains of sphingomonads, the presence of EPS can result in a phenotypic shift from planktonic cells to non-motile cells [13]. Therefore, the repression of Rsh$_{US6-1}$ on the EPS biosynthesis might be a survival strategy of strain US6-1.

(a)

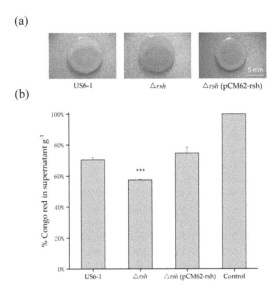

(b)

Figure 6. Assessment of EPS production of strain US6-1, Δ*rsh* and Δ*rsh* (pCM62-rsh). (**a**) Colony biofilm morphology of strains on the P5Y3 plates containing 40 μg/mL Congo red. One representative experiment out of three independent biological replicates is shown. The white bar represents 5 mm. (**b**) Congo red binding assay. The biomass of strains grown in static culture for 96 h was determined gravimetrically as cell wet weigh. The bacterial mass was then mixed with 1 mL of Congo red (100 μg/mL) in 0.9% saline for binding. After removing the biomass, the OD$_{490}$ of unbound Congo red in supernatant was measured and the percentage of Congo red left in supernatant g^{-1} was shown. The OD$_{490}$ of 100 μg/mL Congo red in 0.9% saline was used as the control. The results are averages of three replicates, and the error bars indicate standard deviations. "***" represents $p < 0.001$.

2.8. Rsh$_{US6-1}$ Regulon Involved in Other Processes

The large plasmids present in US6-1 possess putative biodegradation genes that play a key role in PAH degradation [56]. These genes were investigated in the transcriptome data. Two genes (WP_007014585, WP_007014540) located in the large plasmid pLA3 showed up-regulation in Δ*rsh*, which was similar to the situation in *Sphingomonas* sp. LH128 [14]. For strain LH128, the expression of phenanthrene catabolic genes were decreased while *rsh* was up-regulated after 6 months of starvation. This indicated that Rsh$_{US6-1}$ might affect the degradation of PAH. Many researches on sphingomonads mainly focused on enzymes which could degrade PAH directly [56]. Considering sphingomonads always degrade aromatic compounds in contaminated soil with limited nutrient availability [7], studying the relationship between Rsh and genes involved in the degradation of aromatic compounds would help understand a comprehensive degradation mechanism. Future studies carried out in an environment contaminated with PAH using Δ*rsh* are needed to investigate the relationship.

All sphingomonads contain in their outer membranes sphingolipids which replace lipopolysaccharides [7]. Serine palmitoyltransferase (SPT) is a key enzyme in sphingolipid biosynthesis [57]. Two up-regulated genes (WP_007012867 and WP_007014840) were annotated as α-oxoamine synthases, a family of enzymes including SPT. Their identification of amino acid sequences with SPT from *Sphingomonas paucimobilis* and *Sphingomonas wittichii* were more than 30% [57,58], indicating they were responsible for encoding SPT. Sphingolipids play an important role in bacterial survival under stress [59]. However, our results indicated that when spingomonads encountered environmental stress, sphingolipid biosynthesis was repressed via stringent response instead of being induced, probably for energy and resource saving.

The expression level of shape determination genes (*rodA*, *mreC* and *mreB*), cell-division genes (*ftsW*, *ftsQ*) and genes encoding penicillin-binding proteins increased as a result of the deletion of rsh_{US6-1}. These genes were involved in cell size and shape, indicating that the morphology of US6-1 was regulated by Rsh_{US6-1}. Starvation stress can result in the changes of cell morphology of *Sphingomonas* [14,60]. The reason might be that the change of cell size and shape improves the uptake of scarce nutrients due to the increase in cell surface area to volume ratio [61]. Our data suggested that US6-1 might adapt to environment stress via the change in cell size or shape regulated by Rsh_{US6-1}.

Genes responsible for putative regulatory functions were investigated. Several transcriptional regulators including the cold shock protein, an UspA and a PadR family transcriptional regulator were down-regulated. The balance of available GTP might be altered due to the up-regulation of two genes involved in GTP synthesis, a gene encoding an IMP dehydrogenase (WP_007014769) and a pyruvate kinase (WP_007012903). In *Bacillus subtilis*, Rsh regulates transcription mainly by altering the balance of GTP [3]. Whether Rsh also affects the expression of genes through the changes of GTP concentration in sphingomonads remains unknown. In addition, the genome of US6-1 contains 16 chemotaxis-related genes in a "che" cluster, in which several genes were up-regulated. Many genes involved in iron homeostasis were also in the Rsh_{US6-1} regulon. There was an up-regulation in the expression level of genes encoding transporters ExbD and a ferrous iron transporter B and a 3.48-fold down-regulation in the gene related to bacterioferritin. Among proteins annotated as TonB-dependent receptors, 10 were up-regulated and 13 were down-regulated. The absence of Rsh_{US6-1} seemed to result in a disruption in iron homeostasis.

3. Materials and Methods

3.1. Bacterial Strains, Plasmids and Growth Conditions

Bacterial strains and plasmids used in this study are listed in Table 2. Strain US6-1 and its derivatives were grown in P5Y3 (5 g/L peptone, 3 g/L yeast exact and 25 g/L sea salt, pH = 6.5 ± 0.2) at 30 °C. Unless noticed otherwise, in all experiments fresh colonies of US6-1 and its derivatives were first inoculated into P5Y3 broth at the shaking speed of 200 rpm to an OD_{600} value of 1 and 250 µL of this seed culture was then re-inoculated into 50 mL of fresh P5Y3 broth and incubated statically. *E. coli* and *A. tumefaciens* A136 were grown in Luria Broth (10 g/L tryptone, 5 g/L yeast extract, 10 g/L NaCl) at 37 °C or 30 °C, respectively. When appropriate, the following antibiotics were added to the medium: kanamycin (100 µg/mL), streptomycin (100 µg/mL), rifampicin (50 µg/mL) and tetracycline (9 µg/mL) for strain US6-1 and its derived strains, kanamycin (50 µg/mL), tetracycline (9 µg/mL) and ampicillin (100 µg/mL) for *E. coli*, tetracycline (4.5 µg/mL) and spectinomycin (50 µg/mL) for A136.

Table 2. Bacterial strains and plasmids used in this study.

Strains or Plasmids	Relevant Traits	Source or Reference
Strains		
N. pentaromativorans		
US6-1	Wild type (JCM 12182)	Microbe division JCM
Δ*rsh*	US6-1, rsh_{US6-1} deletion mutant, Rif^r	This study
Δ*rsh* (pCM62-rsh)	Δ*rsh* with the plasmid pCM62-rsh, Tc^r	This study
Δ*novI*	US6-1, *novI* deletion mutant, Rif^r	Lab stock
E. coli		
DH5α	F- *hsdR17 endA1 thi-1 gyrA96 relA1 supE44ΔlacU169 (ψ80dlacZΔM15)*	TransGen Biotech

Table 2. *Cont.*

Strains or Plasmids	Relevant Traits	Source or Reference
S17-1 (λpir)	*E. coli* K-12 Tp^r Sm^r *recA thi hsdRM*$^+$ RP4::2-Tc::Mu::Km Tn7, λ*pir* phage lysogen	[62]
A. tumefaciens A136 (pCF218) (pCF372)	*traI-lacZ* fusion; AHL biosensor; Tc^r Sp^r	[63]
Plasmids		
pMD19-T	T-vector, Amp^r	TaKaRa, Dalian, China
pAK405	Plasmid for allelic exchange and markerless gene deletions; Km^r	[64]
pAK405-rsh	pAK405 with fusion of up- and downstream regions of *rsh*$_{US6-1}$; Km^r	This study
pCM62	Broad-host-range cloning vector; IncP origin of replication; Tc^r	[65]
pCM62-rsh	pCM62 with *rsh*$_{US6-1}$ ORF and 502 bp of upstream region; Tc^r	This study

3.2. Sequence Analysis of Rsh$_{US6-1}$

The nucleotide sequence of *rsh*$_{US6-1}$ (accession number WP_007011921) and its deduced amino acid sequences were obtained from NCBI GenBank (https://www.ncbi.nlm.nih.gov, 15 November 2016). Domain prediction was carried out using Conserved Domain Database in NCBI. Multiple-sequence alignments were accomplished using BIOEDIT version 7.0.9.0 (Ibis Therapeutics, Carlsbad, CA, USA) [66]. The identities of Rsh$_{US6-1}$ with Rsh$_{Rr2-17}$ (Rsh from *Novosphingobium* sp. Rr 2-17, accession number ACH57394) [33], Rsh$_{Mtb}$ (Rsh from *Mycobacterium tuberculosis*, accession number CAB01260) [34], Rsh$_{Seq}$ (Rsh from *Streptococcus equisimilis*, accession number CAA51353) [35], Rsh$_{Ret}$ (Rsh from *Rhizobium etli*, accession number ABC90188) [36], SpoT$_{Ecoli}$ (SpoT from *E. coli*, accession number AAC76674) [32] and RelA$_{Ecoli}$ (RelA from *E. coli*, accession number AAC75826) were calculated by BLAST+ program [67]. For construction of phylogenetic trees, sequences were aligned by using the ClustalW algorithm [27] in MEGA version 6.0 [28]. Then a phylogenetic tree was constructed by the neighbor-joining method [29] with 1000 bootstrap replicates [30] after cutting off the redundant sequences at the end, with the use of MEGA version 6.0. The Rsh protein from *Anabaena sp.* PCC 7120 (accession number BAB77915) was used as outgroup [31].

3.3. Strain Construction

Primers used in the present study are listed in Table S4. The in-frame gene deletion mutant was constructed as described previously [64]. Briefly, up- and down-stream flanking regions of *rsh*$_{US6-1}$, approximately 400 bp each, were amplified by PCR with the primer pairs rsh-5O/rsh-5I and rsh-3O/rsh-3I and joined using overlap PCR with rsh-5O/rsh-3O. The resulting fragments were cloned into the plasmid pAK405 via the *Bam*HI/*Hind*III restriction site. The pAK405 derivative was subsequently delivered into US6-1 via conjugal transfer from *E. coli* S17-1 (λpir). After the conjugal transfer, bacteria were plated on P5Y3 supplemented with kanamycin (100 μg/mL) and rifampicin (50 μg/mL). Individual colonies were restreaked once on the same medium and then plated on P5Y3 supplemented with 100 μg/mL streptomycin to select for the second homologous recombination event. The resulting colonies were restreaked on both P5Y3 supplemented with 100 μg/mL streptomycin and P5Y3 containing 100 μg/mL kanamycin, and kanamycin-sensitive clones were analyzed by colony PCR using primers rsh-FO/rsh-RO.

For complementation studies, fragments containing the *rsh*$_{US6-1}$ ORFs and upstream regions were amplified by PCR using the primer pairs PrshF/PrshR. The PCR products were cloned into the plasmid pCM62 [65] to generate pCM62-rsh via the EcoRI/HindIII restriction site. The plasmid pCM62-rsh was transformed into Δ*rsh* by electroporation (2 kV) using a MicroPulse electroporator (Bio-Rad, Hercules, CA, USA). The complementary strain was verified by qRT-PCR and the tetracycline resistance (9 µg/mL). A Bacterial Genomic DNA Extraction Kit, a Gel Extraction Kit and a Plasmid Miniprep Kit were purchased from TransGen Biotech (Beijing, China) and used according to the manufacturer's instruction.

3.4. Growth Curve

Fresh colonies of US6-1 and its derivatives were first inoculated into P5Y3 broth at the shaking speed of 200 rpm to an OD_{600} value of 1. Then, 250 µL of this seed culture was re-inoculated into 50 mL of fresh P5Y3 broth and incubated statically. The growth curve was drawn from measuring the OD_{600} values. The OD_{600} values were monitored using the Spectrophotometer (Metash, Shanghai, China).

3.5. Detection of AHL Accumulation

Strains were tested for AHL accumulation by cross-feeding and extract assays according to the previous study [37]. In the cross-feeding assay, a P5Y3 agar plate was covered with 50 µL X-gal (20 mg/mL stock solution in dimethylformamide). The AHL reporter strain A136 and the tested strains were streaked side by side on the agar plates. Plates were incubated for 24 h, when the activation of the reporter was recorded. A136 versus A136 was used as negative control.

In the extract assay, AHL signals in culture broths were extracted three times by equal volume of ethyl acetate (EA). The EA extracts were combined and evaporated at 55 °C to dry under reduced pressure. The extracts were re-dissolved in 1 mL of EA and passed through a sodium sulfate column to remove water. The extracts were spotted onto LB soft agar plates plus X-gal mixed with A136. The plates were incubated for 12 h at 30 °C and the diameters of the blue stains which represented the quantity of AHL signals in bacterial culture were recorded.

3.6. ppGpp Analysis

ppGpp was extracted with formic acid [68]. At 36 h (exponential growth phase) and 72 h (stationary phase) of incubation, bacterial cells were collected by centrifugation and suspended in 1 mL of 0.9% saline. Then 100 µL of 11 M formic acid was added to the suspension. The sample was vigorously mixed and incubated on ice for 30 min. These samples were centrifuged at $10,000 \times g$ for 10 min at 4 °C. The supernatant was filtered through 0.2 µm filters and stored at −20 °C until use in HPLC analysis.

To assay ppGpp, 100 µL of the extract was subjected to 1100 Series HPLC (Agilent, Santa Clara, CA, USA) by using a ZORBAX SB-C18 column (4.6 × 250 mm, 5 µm, Agilent) at a flow rate of 1.0 mL/min. The mobile phase (pH 6.0) consisted of 125 mM KH_2PO_4, 10 mM tetrabutyl ammonium dihydrogen phosphate, 60 mL/L methanol, and 1 g/L KOH [68]. The eluted nucleotides were monitored at 254 nm and identified by comparison with the retention time of 100 µM ppGpp standards (TriLink Biosciences, San Diego, CA, USA). The ppGpp standard was eluted at 71 min under the current conditions.

3.7. Identification of AHL Degradation of Δrsh in the Exponential Growth Phase

To determine whether AHL was degraded by Δ*rsh* in the exponential growth phase, Δ*rsh* was incubated with 2 µM medium-chain AHL (C8-AHL and 3-OH-C8-AHL) for 24 h, 48 h and 72 h, after which the EA extracts were spotted onto AHL reporter plates [69]. The residual AHL was detected by A136. To avoid the influence of the AHL produced by US6-1 in the exponential growth phase, *novl* deletion mutant, which could not produce AHL, was used as positive control. Blank medium with AHL was used as negative control.

3.8. RNA Extraction and RNA-seq

Cultures of US6-1, Δ*rsh* and Δ*rsh* (pCM62-rsh) after 24 h, 36 h, 48 h and 72 h of static incubation were frozen in liquid nitrogen and stored at −80 °C. Total RNA extraction was performed with RNAprep pure Cell/Bacteria Kit (Tiangen biotech, Beijing, China) according to the manufacturer's instructions. RNA concentration and purity were determined at 230 nm using a NanoDrop ND-1000 (Thermo Scientific, Waltham, MA, USA). Genomic DNA contamination was removed and cDNA was synthesized using FastKing gDNA Dispelling RT SuperMix (Tiangen biotech, Beijing, China).

Samples of cDNA of US6-1 and Δ*rsh* at 36 h of incubation (exponential phase) were chosen for RNA-seq analysis. Library construction and sequencing were performed on a BGISEQ-500 platform by Beijing Genomic Institution (Shenzhen, China). All the raw sequencing reads were filtered to remove low quality reads and reads with adaptors, reads in which unknown bases are more than 10%. Clean reads were then obtained. These reads were stored as FASTQ format and then deposited in the NCBI Sequence Read Archive (http://trace.ncbi.nlm.nih.gov/Traces/sra, 22 May 2018) under the accession number SRP148564. HISAT [70] was used to map clean reads to genome of US6-1. For gene expression analysis, the matched reads were calculated and then normalized to reads per kilobaseper million mapped reads using RESM software [71]. The significance of the differential expression of genes was defined according to the combination of the absolute value of the fold change ≥2 and an adjusted *p* value ≤ 0.05 using DESeq2 algorithm [72]. KEGG-based annotation and pathway enrichment analysis was performed using the KEGG pathway database (http://www.genome.jp/kegg/, 25 September 2017). All these KEGG terms were decided by a Fisher's exact test with phyper function in R language [42]. An adjust *p* value of 0.01 was used as the threshold to obtain significantly over-represented KEGG terms.

3.9. Quantitative Real-Time Polymerase Chain Reaction (qRT-PCR)

Quantitative real-time PCR (qRT-PCR) was performed according to protocols described previously [73]. cDNA samples were used as the templates for amplification by using SYBR Premix Ex TaqTM Kit (Tli RNaseH Plus) (TaKaRa, Dalian, China) in Applied Biosystems StepOnePlus™ Real-Time PCR System (Thermo Scientific, Waltham, MA, USA). Primers were designed on the Sangon website (http://www.sangon.com/, 14 January 2018) and listed in Table S4. The following PCR program was used: 95 °C for 30 s, followed by 40 cycles of 95 °C for 5 s, 60 °C for 30 s. A heat dissociation curve (60–95 °C) was checked after the final PCR cycle to determine the specificity of the PCR amplification. The gene expression levels of derivatives of US6-1 relative to the wild-type strain were analyzed using the $2^{-\Delta\Delta Ct}$ analysis method [74] (Δ*rsh* vs. US6-1 or the complementary strain vs. US6-1). The 16S rRNA gene was used as internal references to normalize cDNA templates. Negative controls with nuclease-free water as templates for each primer set were included in each run. The expression profiles obtained from the RNA-seq were validated by qRT-PCR of 39 representative genes involved in functional groups including AHL production, sigma factor synthesis, RNA degradation, EPS biosynthesis, PAH degradation, sphingolipid production, cell division and shape, GTP synthesis and other functions. The RNA-seq data were considered valid if they met both two requirements. One was that the gene expression (Δ*rsh* vs. US6-1) was regulated in the same direction (up or down) in both RNA-seq data and qRT-PCR, and the fold change of genes tested by qRT-PCR was $\log_2(x) > 1$ or $\log_2(1/x) < -1$ (i.e., a plain fold change >2) with statistical differences (*p* value < 0.05) [75]. Another was that the gene expression (the complementary strain vs. US6-1) by qRT-PCR was not different significantly or was regulated in the opposite direction in the RNA-seq data. The above valid DEGs were chosen to be clustered. The heat map of their $\log_2$ expression ratios in the RNA-seq data in strain US6-1 and Δ*rsh* was drawn by the software HemI [43].

3.10. Congo Red Binding Assay

The assay to evaluate the EPS production followed the method described previously with minor modification [76]. Cultures were inoculated from a single colony and grown to an OD_{600} of 1 in P5Y3 broth; 10 µL of suspensions were spotted on P5Y3 plates containing 40 µg/mL Congo red (Solarbio, Beijing, China). The plates were grown at 30 °C to assess the morphology of colony biofilm. For quantitative analysis, strains were first cultured in static P5Y3 broth. Then, the bacterial mass was collected at 96 h by centrifuging at 8000 rpm for 5 min and the supernatant was discarded. Determination of biomass was performed gravimetrically as cell wet weight (g). The precipitate was suspended in 1 mL of 100 µg/mL Congo red in 0.9% saline and then incubated for 90 min with shaking at 30 rpm at room temperature. Centrifugation was performed again to separate the mass and solution, and then the OD_{490} of the supernatant was measured. The percentage of Congo red left in the supernatant was calculated by relating to the OD_{490} of 100 µg/mL Congo red in 0.9% saline.

3.11. Statistical Analysis

All assays were performed in triplicate and data were expressed as the means and standard deviations (mean ± SD). Analysis of statistical differences was conducted with Student's *t*-test.

4. Conclusions

This is the first transcriptome-wide study of the Rsh regulon in sphingomonads, which is an important step towards understanding the regulatory system of stringent response in sphingomonads. Our study showed that there was a wide range of genes in the Rsh_{US6-1} regulon, including those involved in common traits of sphingomonads such as EPS biosynthesis, PAH degradation and sphingolipid biosynthesis. Furthermore, we focused on the potential molecular mechanisms of Rsh_{US6-1}-dependent AHL accumulation. Essential genes for AHL production (*novI* and *novR*) were positively regulated by Rsh_{US6-1} in the exponential growth phase. Several factors including the ECF sigma factor, DksA, ATP-dependent Lon protease and RNA degradation enzymes might be involved in the ppGpp-dependent expression of *novI* and *novR*. Future studies will focus on the validation of the above proposed mechanisms, which might provide an insight into the complicated cross-talk between stringent response and QS in sphingomonads.

Supplementary Materials: The following are available online at http://www.mdpi.com/1422-0067/19/9/2631/s1, Table S1: Results of mapping clean reads of RNA-Seq to the reference genome of US6-1; Table S2: Complete list of DEGs according to RNA-seq (Δ*rsh* vs. US6-1); Table S3: Validation of the RNA-seq data by qRT-PCR of 39 representative genes; Table S4: Primers used in this study.

Author Contributions: The authors' responsibilities were as follows: Conceptualization, Y.H.; Formal analysis, H.L.; Supervision, Y.H.; Writing—original draft, H.L.

Acknowledgments: This work was supported financially by the National Natural Science Foundation of China 21577121 and the 973 Program (2014CB441103). We thanked Yanhua Zeng for her advices for the study.

Conflicts of Interest: The authors declare no conflict of interest.

Abbreviations

AHL	*N*-acyl-L-homoserine Lactone
QS	Quorum Sensing
DEGs	Differentially Expressed Genes
PAH	Polycyclic Aromatic Hydrocarbon
ECF	Extracytoplasmic Function
SPT	Serine Palmitoyltransferase
EPS	Exopolysaccharide

References

1. Balsalobre, C. Concentration matters!! ppGpp, from a whispering to a strident alarmone. *Mol. Microbiol.* **2011**, *79*, 827–829. [CrossRef] [PubMed]

2. Potrykus, K.; Cashel, M. (p)ppGpp: Still magical? *Annu. Rev. Microbiol.* **2008**, *62*, 35–51. [CrossRef] [PubMed]

3. Hauryliuk, V.; Atkinson, G.C.; Murakami, K.S.; Tenson, T.; Gerdes, K. Recent functional insights into the role of (p)ppGpp in bacterial physiology. *Nat. Rev. Microbiol.* **2015**, *13*, 298–309. [CrossRef] [PubMed]

4. Durfee, T.; Hansen, A.M.; Zhi, H.; Blattner, F.R.; Jin, D.J. Transcription profiling of the stringent response in *Escherichia coli*. *J. Bacteriol.* **2008**, *190*, 1084–1096. [CrossRef] [PubMed]

5. Geiger, T.; Francois, P.; Liebeke, M.; Fraunholz, M.; Goerke, C.; Krismer, B.; Schrenzel, J.; Lalk, M.; Wolz, C. The stringent response of *Staphylococcus aureus* and its impact on survival after phagocytosis through the induction of intracellular PSMs expression. *PLoS Pathog.* **2012**, *8*. [CrossRef] [PubMed]

6. Vercruysse, M.; Fauvart, M.; Jans, A.; Beullens, S.; Braeken, K.; Cloots, L.; Engelen, K.; Marchal, K.; Michiels, J. Stress response regulators identified through genome-wide transcriptome analysis of the (p)ppGpp-dependent response in *Rhizobium etli*. *Genome Biol.* **2011**, *12*. [CrossRef] [PubMed]

7. Stolz, A. Molecular characteristics of xenobiotic-degrading sphingomonads. *Appl. Microbiol. Biotechnol.* **2009**, *81*, 793–811. [CrossRef] [PubMed]

8. Huang, Y.L.; Feng, H.; Lu, H.; Zeng, Y.H. Novel 16S rDNA primers revealed the diversity and habitats-related community structure of sphingomonads in 10 different niches. *Anton. Leeuw.* **2017**, *110*, 877–889. [CrossRef] [PubMed]

9. Alvarez, A.; Benimeli, C.S.; Saez, J.M.; Fuentes, M.S.; Cuozzo, S.A.; Polti, M.A.; Amoroso, M.J. Bacterial bio-resources for remediation of hexachlorocyclohexane. *Int. J. Mol. Sci.* **2012**, *13*, 15086–15106. [CrossRef] [PubMed]

10. Yamamoto, A.; Yano, I.; Masui, M.; Yabuuchi, E. Isolation of a novel sphingoglycolipid containing glucuronic acid and 2-hydroxy fatty acid from *Flavobacterium devorans* ATCC 10829. *J. Biochem.* **1978**, *83*, 1213–1216. [CrossRef] [PubMed]

11. Kawasaki, S.; Moriguchi, R.; Sekiya, K.; Nakai, T.; Ono, E.; Kume, K.; Kawahara, K. The cell envelope structure of the lipopolysaccharide-lacking gram-negative bacterium *Sphingomonas paucimobilis*. *J. Bacteriol.* **1994**, *176*, 284–290. [CrossRef] [PubMed]

12. Francez-Charlot, A.; Kaczmarczyk, A.; Vorholt, J.A. The branched CcsA/CckA-ChpT-CtrA phosphorelay of *Sphingomonas melonis* controls motility and biofilm formation. *Mol. Microbiol.* **2015**, *97*, 47–63. [CrossRef] [PubMed]

13. Pollock, T.J.; Armentrout, R.W. Planktonic/sessile dimorphism of polysaccharide-encapsulated sphingomonads. *J. Ind. Microbiol. Biotechnol.* **1999**, *23*, 436–441. [CrossRef] [PubMed]

14. Fida, T.T.; Moreno-Forero, S.K.; Heipieper, H.J.; Springael, D. Physiology and transcriptome of the polycyclic aromatic hydrocarbon-degrading *Sphingomonas* sp. LH128 after long-term starvation. *Microbiology* **2013**, *159*, 1807–1817. [CrossRef] [PubMed]

15. Tay, S.B.; Yew, W.S. Development of quorum-based anti-virulence therapeutics targeting Gram-negative bacterial pathogens. *Int. J. Mol. Sci.* **2013**, *14*, 16570–16599. [CrossRef] [PubMed]

16. Van Delden, C.; Comte, R.; Bally, M. Stringent response activates quorum sensing and modulates cell density-dependent gene expression in *Pseudomonas aeruginosa*. *J. Bacteriol.* **2001**, *183*, 5376–5384. [CrossRef] [PubMed]

17. Erickson, D.L.; Lines, J.L.; Pesci, E.C.; Venturi, V.; Storey, D.G. *Pseudomonas aeruginosa relA* contributes to virulence in *Drosophila melanogaster*. *Infect. Immun.* **2004**, *72*, 5638–5645. [CrossRef] [PubMed]

18. Schafhauser, J.; Lepine, F.; McKay, G.; Ahlgren, H.G.; Khakimova, M.; Nguyen, D. The stringent response modulates 4-hydroxy-2-alkylquinoline biosynthesis and quorum-sensing hierarchy in *Pseudomonas aeruginosa*. *J. Bacteriol.* **2014**, *196*, 1641–1650. [CrossRef] [PubMed]

19. Moris, M.; Braeken, K.; Schoeters, E.; Verreth, C.; Beullens, S.; Vanderleyden, J.; Michiels, J. Effective symbiosis between *Rhizobium etli* and *Phaseolus vulgatis* requires the alarmone ppGpp. *J. Bacteriol.* **2005**, *187*, 5460–5469. [CrossRef] [PubMed]

20. Zhang, H.B.; Wang, C.; Zhang, L.H. The quormone degradation system of *Agrobacterium tumefaciens* is regulated by starvation signal and stress alarmone (p)ppGpp. *Mol. Microbiol.* **2004**, *52*, 1389–1401. [CrossRef] [PubMed]

21. Wang, C.; Zhang, H.B.; Wang, L.H.; Zhang, L.H. Succinic semialdehyde couples stress response to quorum-sensing signal decay in *Agrobacterium tumefaciens*. *Mol. Microbiol.* **2006**, *62*, 45–56. [CrossRef] [PubMed]

22. Huang, Y.L.; Zeng, Y.H.; Yu, Z.L.; Zhang, J.; Feng, H.; Lin, X.C. *In silico* and experimental methods revealed highly diverse bacteria with quorum sensing and aromatics biodegradation systems—A potential broad application on bioremediation. *Bioresour. Technol.* **2013**, *148*, 311–316. [CrossRef] [PubMed]

23. Gan, H.M.; Gan, H.Y.; Ahmad, N.H.; Aziz, N.A.; Hudson, A.O.; Savka, M.A. Whole genome sequencing and analysis reveal insights into the genetic structure, diversity and evolutionary relatedness of *luxI* and *luxR* hornologs in bacteria belonging to the *Sphingomonadaceae* family. *Front. Cell. Infect. Microbiol.* **2015**, *4*, 188. [CrossRef] [PubMed]

24. Lyu, Y.H.; Zheng, W.; Zheng, T.L.; Tian, Y. Biodegradation of polycyclic aromatic hydrocarbons by *Novosphingobium pentaromativorans* US6-1. *PLoS ONE* **2014**, *9*, e101438. [CrossRef] [PubMed]

25. Choi, D.H.; Kwon, Y.M.; Kwon, K.K.; Kim, S.J. Complete genome sequence of *Novosphingobium pentaromativorans* US6-1(T). *Stand. Genomic Sci.* **2015**, *10*, 107. [CrossRef] [PubMed]

26. Steinchen, W.; Bange, G. The magic dance of the alarmones (p)ppGpp. *Mol. Microbiol.* **2016**, *101*, 531–544. [CrossRef] [PubMed]

27. Yuan, J.; Amend, A.; Borkowski, J.; Demarco, R.; Bailey, W.; Liu, Y.; Xie, G.C.; Blevins, R. Multiclustal: A systematic method for surveying Clustal W alignment parameters. *Bioinformatics* **1999**, *15*, 862–863. [CrossRef] [PubMed]

28. Hall, B.G. Building phylogenetic trees from molecular data with MEGA. *Mol. Biol. Evol.* **2013**, *30*, 1229–1235. [CrossRef] [PubMed]

29. Saitou, N.; Nei, M. The neighbor-joining method—A new method for reconstructing phylogenetic trees. *Mol. Biol. Evol.* **1987**, *4*, 406–425. [PubMed]

30. Felsenstein, J. Confidence-limits on phylogenies—An approach using the bootstrap. *Evolution* **1985**, *39*, 783–791. [CrossRef] [PubMed]

31. Zhang, S.R.; Lin, G.M.; Chen, W.L.; Wang, L.; Zhang, C.C. ppGpp metabolism is involved in heterocyst development in the cyanobacterium *Anabaena* sp. strain PCC 7120. *J. Bacteriol.* **2013**, *195*, 4536–4544. [CrossRef] [PubMed]

32. Cashel, M.; Gallant, J. Two compounds implicated in the function of the *RC* gene of *Escherichia coli*. *Nature* **1969**, *221*, 838–841. [CrossRef] [PubMed]

33. Gan, H.M.; Buckley, L.; Szegedi, E.; Hudson, A.O.; Savka, M.A. Identification of an *rsh* Gene from a *Novosphingobium* sp. necessary for quorum-sensing signal accumulation. *J. Bacteriol.* **2009**, *191*, 2551–2560. [CrossRef] [PubMed]

34. Primm, T.P.; Andersen, S.J.; Mizrahi, V.; Avarbock, D.; Rubin, H.; Barry, C.E. The stringent response of *Mycobacterium tuberculosis* is required for long-term survival. *J. Bacteriol.* **2000**, *182*, 4889–4898. [CrossRef] [PubMed]

35. Mechold, U.; Cashel, M.; Steiner, K.; Gentry, D.; Malke, H. Functional analysis of a *relA/spoT* gene homolog from *Streptococcus equisimilis*. *J. Bacteriol.* **1996**, *178*, 1401–1411. [CrossRef] [PubMed]

36. Calderón-Flores, A.; Du Pont, G.; Huerta-Saquero, A.; Merchant-Larios, H.; Servín-González, L.; Durán, S. The stringent response is required for amino acid and nitrate utilization, nod factor regulation, nodulation, and nitrogen fixation in *Rhizobium etli*. *J. Bacteriol.* **2005**, *187*, 5075–5083. [CrossRef] [PubMed]

37. Huang, Y.L.; Ki, J.S.; Case, R.J.; Qian, P.Y. Diversity and acyl-homoserine lactone production among subtidal biofilm-forming bacteria. *Aquat. Microb. Ecol.* **2008**, *52*, 185–193. [CrossRef]

38. Liu, K.Q.; Bittner, A.N.; Wang, J.D. Diversity in (p)ppGpp metabolism and effectors. *Curr. Opin. Microbiol.* **2015**, *24*, 72–79. [CrossRef] [PubMed]

39. Magnusson, L.U.; Farewell, A.; Nystrom, T. ppGpp: a global regulator in *Escherichia coli*. *Trends Microbiol.* **2005**, *13*, 236–242. [CrossRef] [PubMed]

40. Eymann, C.; Homuth, G.; Scharf, C.; Hecker, M. *Bacillus subtilis* functional genomics: global characterization of the stringent response by proteome and transcriptome analysis. *J. Bacteriol.* **2002**, *184*, 2500–2520. [CrossRef] [PubMed]

41. Bowden, S.D.; Eyres, A.; Chung, J.C.S.; Monson, R.E.; Thompson, A.; Salmond, G.P.C.; Spring, D.R.; Welch, M. Virulence in *Pectobacterium atrosepticum* is regulated by a coincidence circuit involving quorum sensing and the stress alarmone, (p)ppGpp. *Mol. Microbiol.* **2013**, *90*, 457–471. [CrossRef] [PubMed]

42. Karunakaran, D.K.P.; Al Seesi, S.; Banday, A.R.; Baumgartner, M.; Olthof, A.; Lemoine, C.; Măndoiu, I.I.; Kanadia, R.N. Network-based bioinformatics analysis of spatio-temporal RNA-Seq data reveals transcriptional programs underpinning normal and aberrant retinal development. *BMC Genomics* **2016**, *17*, 495. [CrossRef] [PubMed]

43. Deng, W.K.; Wang, Y.B.; Liu, Z.X.; Cheng, H.; Xue, Y. HemI: A toolkit for illustrating heatmaps. *PLoS ONE* **2014**, *9*, e111988. [CrossRef] [PubMed]

44. Galloway, W.R.J.D.; Hodgkinson, J.T.; Bowden, S.D.; Welch, M.; Spring, D.R. Quorum sensing in Gram-negative bacteria: Small-molecule modulation of AHL and AI-2 quorum sensing pathways. *Chem. Rev.* **2011**, *111*, 28–67. [CrossRef] [PubMed]

45. Dalebroux, Z.D.; Swanson, M.S. ppGpp: Magic beyond RNA polymerase. *Nat. Rev. Microbiol.* **2012**, *10*, 203–212. [CrossRef] [PubMed]

46. Liu, X.X.; Ji, L.; Wang, X.; Li, J.R.; Zhu, J.L.; Sun, A.H. Role of RpoS in stress resistance, quorum sensing and spoilage potential of *Pseudomonas fluorescens*. *Int. J. Food Microbiol.* **2018**, *270*, 31–38. [CrossRef] [PubMed]

47. Gu, D.; Guo, M.; Yang, M.J.; Zhang, Y.X.; Zhou, X.H.; Wang, Q.Y. A sigma(E)-mediated temperature gauge controls a switch from LuxR-mediated virulence gene expression to thermal stress adaptation in *Vibrio alginolyticus*. *PLoS Pathog.* **2016**, *12*, e1005645. [CrossRef] [PubMed]

48. Li, J.; Overall, C.C.; Johnson, R.C.; Jones, M.B.; McDermott, J.E.; Heffron, F.; Adkins, J.N.; Cambronne, E.D. ChIP-Seq Analysis of the sigma(E) regulon of *Salmonella enterica* serovar Typhimurium reveals new genes implicated in heat shock and oxidative stress response. *PLoS ONE* **2015**, *10*, e0138466. [CrossRef]

49. Devescovi, G.; Venturi, V. The *Burkholderia cepacia rpoE* gene is not involved in exopolysaccharide production and onion pathogenicity. *Can. J. Microbiol.* **2006**, *52*, 260–265. [CrossRef] [PubMed]

50. Magnusson, L.U.; Gummesson, B.; Joksimović, P.; Farewell, A.; Nyström, T. Identical, independent, and opposing roles of ppGpp and DksA in *Escherichia coli*. *J. Bacteriol.* **2007**, *189*, 5193–5202. [CrossRef] [PubMed]

51. Jude, F.; Köhler, T.; Branny, P.; Perron, K.; Mayer, M.P.; Comte, R.; van Delden, C. Posttranscriptional control of quorum-sensing-dependent virulence genes by DksA in *Pseudomonas aeruginosa*. *J. Bacteriol.* **2003**, *185*, 3558–3566. [CrossRef] [PubMed]

52. Takaya, A.; Tabuchi, F.; Tsuchiya, H.; Isogai, E.; Yamamoto, T. Negative regulation of quorum-sensing systems in *Pseudomonas aeruginosa* by ATP-dependent Lon protease. *J. Bacteriol.* **2008**, *190*, 4181–4188. [CrossRef] [PubMed]

53. Baumgardt, K.; Charoenpanich, P.; Mcintosh, M.; Schikora, A.; Stein, E.; Thalmann, S.; Kogel, K.H.; Klug, G.; Becker, A.; Evguenieva-Hackenberg, E. RNase E affects the expression of the acyl-homoserine lactone synthase gene *sinI* in *Sinorhizobium meliloti*. *J. Bacteriol.* **2014**, *196*, 1435–1447. [CrossRef] [PubMed]

54. McIntosh, M.; Stone, B.A.; Stanisich, V.A. Curdlan and other bacterial (1 → 3)-β-D-glucans. *Appl. Microbiol. Biotechnol.* **2005**, *68*, 163–173. [CrossRef] [PubMed]

55. Wells, D.H.; Long, S.R. The *Sinorhizobium meliloti* stringent response affects multiple aspects of symbiosis. *Mol. Microbiol.* **2002**, *43*, 1115–1127. [CrossRef] [PubMed]

56. Yun, S.H.; Choi, C.W.; Lee, S.Y.; Lee, Y.G.; Kwon, J.; Leem, S.H.; Chung, Y.H.; Kahng, H.Y.; Kim, S.J.; Kwon, K.K.; et al. Proteomic characterization of plasmid pLA1 for biodegradation of polycyclic aromatic hydrocarbons in the marine bacterium, *Novosphingobium pentaromativorans* US6-1. *PLoS ONE* **2014**, *9*, e90812. [CrossRef] [PubMed]

57. Raman, M.C.C.; Johnson, K.A.; Clarke, D.J.; Naismith, J.H.; Campopiano, D.J. The serine palmitoyltransferase from *Sphingomonas wittichii* RW1: An interesting link to an unusual acyl carrier protein. *Biopolymers* **2010**, *93*, 811–822. [CrossRef] [PubMed]

58. Ikushiro, H.; Hayashi, H.; Kagamiyama, H. A water-soluble homodimeric serine palmitoyltransferase from *Sphingomonas paucimobilis* EY2395T Strain. *J. Biol. Chem.* **2001**, *276*, 18249–18256. [CrossRef] [PubMed]

59. An, D.D.; Na, C.Z.; Bielawski, J.; Hannun, Y.A.; Kasper, D.L. Membrane sphingolipids as essential molecular signals for *Bacteroides* survival in the intestine. *Proc. Natl. Acad. Sci. USA* **2011**, *108*, 4666–4671. [CrossRef] [PubMed]

60. Eguchi, M.; Nishikawa, T.; MacDonald, K.; Cavicchioli, R.; Gottschal, J.C.; Kjelleberg, S. Responses to stress and nutrient availability by the marine ultramicrobacterium *Sphingomonas* sp. strain RB2256. *Appl. Environ. Microbiol.* **1996**, *62*, 1287–1294. [PubMed]

61. Vanoverbeek, L.S.; Eberl, L.; Givskov, M.; Molin, S.; Vanelsas, J.D. Survival of, and induced stress resistance in, carbon-starved *Pseudomonas-fluorescens* cells residing in soil. *Appl. Environ. Microbiol.* **1995**, *61*, 4202–4208.

62. Simon, R.; Priefer, U.; Pühler, A. A broad host range mobilization system for in vivo genetic engineering: Transposon mutagenesis in Gram negative bacteria. *Nat. Biotechnol.* **1983**, *1*, 784–791. [CrossRef]

63. Mclean, R.J.C.; Whiteley, M.; Stickler, D.J.; Fuqua, W.C. Evidence of autoinducer activity in naturally occurring biofilms. *FEMS Microbiol. Lett.* **1997**, *154*, 259–263. [CrossRef] [PubMed]

64. Kaczmarczyk, A.; Vorholt, J.A.; Francez-Charlot, A. Markerless gene deletion system for sphingomonads. *Appl. Environ. Microbiol.* **2012**, *78*, 3774–3777. [CrossRef] [PubMed]

65. Marx, C.J.; Lidstrom, M.E. Development of improved versatile broad-host-range vectors for use in methylotrophs and other Gram-negative bacteria. *Microbiology* **2001**, *147*, 2065–2075. [CrossRef] [PubMed]

66. Hall, T.A. BioEdit: A user-friendly biological sequence alignment editor and analysis program for Windows 95/98/NT. *Nucleic Acids Symp. Ser.* **1999**, *41*, 95–98.

67. Camacho, C.; Coulouris, G.; Avagyan, V.; Ma, N.; Papadopoulos, J.; Bealer, K.; Madden, T.L. BLAST plus: Architecture and applications. *BMC Bioinform.* **2009**, *10*, 421. [CrossRef] [PubMed]

68. Washio, K.; Lim, S.P.; Roongsawang, N.; Morikawa, M. Identification and characterization of the genes responsible for the production of the cyclic lipopeptide arthrofactin by *Pseudomonas* sp. MIS38. *Biosci. Biotechnol. Biochem.* **2010**, *74*, 992–999. [CrossRef] [PubMed]

69. Gao, J.; Ma, A.Z.; Zhuang, X.L.; Zhuang, G.Q. An *N*-acyl homoserine lactones synthase in the ammonia-oxidizing bacterium *Nitrosospira multiformis*. *Appl. Environ. Microbiol.* **2014**, *80*, 951–958. [CrossRef] [PubMed]

70. Kim, D.; Langmead, B.; Salzberg, S.L. HISAT: A fast spliced aligner with low memory requirements. *Nat. Methods* **2015**, *12*, 357–360. [CrossRef] [PubMed]

71. Li, B.; Dewey, C.N. RSEM: Accurate transcript quantification from RNA-Seq data with or without a reference genome. *BMC Bioinform.* **2011**, *12*, 323. [CrossRef] [PubMed]

72. Love, M.I.; Huber, W.; Anders, S. Moderated estimation of fold change and dispersion for RNA-seq data with DESeq2. *Genome Biol.* **2014**, *15*, 550. [CrossRef] [PubMed]

73. Zeng, Y.H.; Wang, Y.L.; Yu, Z.L.; Huang, Y.L. Hypersensitive response of plasmid-encoded ahl synthase gene to lifestyle and nutrient by *Ensifer adhaerens* X097. *Front. Microbiol.* **2017**, *8*, 1160. [CrossRef] [PubMed]

74. Schmittgen, T.D.; Livak, K.J. Analyzing real-time PCR data by the comparative C_T method. *Nat. Protoc.* **2008**, *3*, 1101–1108. [CrossRef] [PubMed]

75. Hanna, N.; Ouahrani-Bettache, S.; Drake, K.L.; Adams, L.G.; Köhler, S.; Occhialini, A. Global Rsh-dependent transcription profile of *Brucella suis* during stringent response unravels adaptation to nutrient starvation and cross-talk with other stress responses. *BMC Genomics* **2013**, *14*, 459. [CrossRef] [PubMed]

76. Liu, H.Z.; Xiao, Y.J.; Nie, H.L.; Huang, Q.Y.; Chen, W.L. Influence of (p)ppGpp on biofilm regulation in *Pseudomonas putida* KT2440. *Microbiol. Res.* **2017**, *204*, 1–8. [CrossRef] [PubMed]

 © 2018 by the authors. Licensee MDPI, Basel, Switzerland. This article is an open access article distributed under the terms and conditions of the Creative Commons Attribution (CC BY) license (http://creativecommons.org/licenses/by/4.0/).

Review

Selenium-Related Transcriptional Regulation of Gene Expression

Mikko J. Lammi [1,2,]* and Chengjuan Qu [2]

1 Key Laboratory of Trace Elements and Endemic Diseases, National Health and Family Planning, Institute of Endemic Diseases, School of Public Health of Health Science Center, Xi'an Jiaotong University, Xi'an 710061, China
2 Department of Integrative Medical Biology, University of Umeå, 901 87 Umeå, Sweden; chengjuan.qu@umu.se
* Correspondence: mikko.lammi@umu.se; Tel.: +358-40-5870601

Received: 3 August 2018; Accepted: 5 September 2018; Published: 8 September 2018

Abstract: The selenium content of the body is known to control the expression levels of numerous genes, both so-called selenoproteins and non-selenoproteins. Selenium is a trace element essential to human health, and its deficiency is related to, for instance, cardiovascular and myodegenerative diseases, infertility and osteochondropathy called Kashin–Beck disease. It is incorporated as selenocysteine to the selenoproteins, which protect against reactive oxygen and nitrogen species. They also participate in the activation of the thyroid hormone, and play a role in immune system functioning. The synthesis and incorporation of selenocysteine occurs via a special mechanism, which differs from the one used for standard amino acids. The codon for selenocysteine is a regular in-frame stop codon, which can be passed by a specific complex machinery participating in translation elongation and termination. This includes a presence of selenocysteine insertion sequence (SECIS) in the 3′-untranslated part of the selenoprotein mRNAs. Nonsense-mediated decay is involved in the regulation of the selenoprotein mRNA levels, but other mechanisms are also possible. Recent transcriptional analyses of messenger RNAs, microRNAs and long non-coding RNAs combined with proteomic data of samples from Keshan and Kashin–Beck disease patients have identified new possible cellular pathways related to transcriptional regulation by selenium.

Keywords: selenium; selenocysteine; selenoproteins; selenocysteine insertion sequence; nonsense-mediated decay

1. Introduction

Selenium is a trace element and a vital nutrient component. It is present in various forms, such as inorganic sodium selenate and sodium selenite and, for instance, also as selenomethionine, selenocysteine, y-glutamyl-selenium-methylselenocysteine, selenium-methyl selenocysteine and methylselenol [1,2]. These various forms have different oxidation states: +6 in selenates, +4 in selenites, 0 in elemental selenium, and −2 in inorganic and organic selenides [3], which affect their bioavailability and properties. In food, selenium is mainly associated with protein in animal tissue, particularly in meat and seafood, but also in bread and cereals [1,2]. In plants, selenium is converted to methylated selenium components, selenomethione and selenocysteine (Sec) [4,5]. There are also many enteral formulas supplemented with selenium, too [2].

The major bioavailable forms of selenium are organic forms, such as Sec and selenomethionine, and inorganic selenate and selenite. A recommended average daily intake for individuals over 14 years of age is 55 μg according to the Office of Dietary Supplements of the National Institutes of Health (Bethesda, MD, USA), while the need in younger children ranges between 15–40 μg/day. Globally, the dietary intake can often be below that recommendation [6], and there may be even one billion people affected by selenium deficiency, mainly due to a low selenium content in the soil. Geographical variation in the soil selenium contents occurs globally, and especially in China the soil shows highly variable contents [7].

The first findings related to selenium were those noted on its toxicity already in 19th century. The symptoms of selenium toxicity include fatigue, and disturbances in connective tissue, cardiovascular, nervous, gastrointestinal and respiratory systems [8]. Selenosis due to unusually high concentrations of dietary selenium leads also to poor dental health, brittle hair and nails, nausea, vomiting and pulmonary oedema, the symptom severity depending on the level of poisoning [9]. In addition, selenium can even interact with arsenic, increasing the toxicity [10].

Many chronic diseases are also related to decreased selenium status. Increased incidences of myocardial infarction and death have been noticed to be associated with selenium deficiency [11]. Low selenium has also been related to cancer, renal diseases and viral infection [4]. Chinese endemic diseases, such as the Keshan and Kashin–Beck diseases are mainly observed in a geographic belt located from northeast to southwest in China with a very low content of water-soluble selenium [12]. Selenium deficiency is also associated with fibrosis of various organs, such as heart, liver, kidney, thyroid and pancreas, and both to fibrotic cystic and oral submucosa [13].

The supplementation of selenium in foodstuff or fertilizers to achieve adequate supply has been practiced in variable ways. In the 1970s, Finland was among those countries that had the lowest selenium levels in the population. Since Finland also had a high incidence of cardiovascular diseases, an association of these factors was hypothesized. A large-scale fortification of fertilizers supplemented with selenium increased the selenium contents of bread and milk in Finland so that the selenium levels in the serum almost doubled in the population [3]. Patients with viral or bacterial infections may benefit from dietary selenium supplementation [14].

The preventive and therapeutic efficiency of selenium in cancer depends on the form of selenium [15]. It has been noticed that four-valent sodium selenite can specifically oxidize the vicinal sulfhydryl groups [16,17], in contrast to six-valent selenate [16]. It has been found that hydrophobic polymers coating the plasma membrane can make the cells resistant to the recognition and destruction by the innate immune system, while redox-active sodium selenite can inhibit this process by unmasking specific tumor antigens, which upon immune recognition allow natural killer cells to eliminate the tumor cells [18]. Importantly, sodium selenite can also directly activate the natural killer cells [19]. Besides solid tumors [20], selenite appears to have an anti-leukemic effect [21], and to be potentially useful for the treatment of multidrug-resistant acute myeloid leukemia [22].

In this review, we briefly discuss the functions of the selenoproteins, the events related to selenoprotein biosynthesis, and how selenium deficiency affects the gene regulation of selenoproteins, but also other genes. The role of long non-coding RNAs (lncRNAs) is also handled, while the focus of this review is on human and other mammalian observations.

2. Selenoprotein Functions and Selenoprotein-Related Disorders

Selenium is incorporated into proteins mainly as Sec, the 21st amino acid [23]. There are 24 selenoproteins identified in rodents and 25 in humans [24]. These include five glutathione peroxidases, three iodothyronine deiodinases, two thioredoxin reductases, thioredoxin-glutathione reductase 3, selenophosphate synthetase 2, methionine sulfoxide reductase B1, and selenoproteins F, H, I, K, M, N, O, P, S, T, V and W [25]. The best-known selenoproteins are glutathione peroxidases, thioredoxin reductases, iodothyronine deiodinases and selenoprotein P.

Selenoprotein biosynthesis is essential for life, since deletion of the murine gene encoding selenocysteine tRNA, which inserts selenocysteine to a growing polypeptide during translation, led to an early embryonic lethality [26]. The functions of many of the selenoproteins have been identified so far, although there are still those with unknown functions [27]. Almost all of them are oxidoreductases, and they have been localized to plasma membrane, endoplasmic reticulum, cytosol, mitochondria, and nucleus, and even two secreted extracellular selenoproteins are present [27].

They also have diverse patterns of tissue distribution, which can vary from ubiquitous to very tissue-specific locations [28]. The selenoproteins involve antioxidant and redox reactions by the detoxification of peroxides, a regeneration of reduced thioredoxin and a reduction of oxidized methionine residues, and iodothyronine deiodinases regulate the activity of thyroid hormones [28]. Some selenoproteins are also involved in calcium mobilization, selenium transport and endoplasmic reticulum stress [28].

Universal genetic code has three codons reserved for translation termination. Of those, the UGA codon also signals for the incorporation of Sec into the selenoproteins [29]. A Sec insertion sequence (SECIS), which is a specific stem–loop structure in the 3′-untranslated regions (UTRs) of the messenger RNAs (mRNAs) in eukaryotes, makes the cotranslational incorporation of Sec into nascent polypeptides possible [30]. The mammalian selenoproteins usually contain a single Sec residue located in the enzymes' active site, with the exception of the selenoprotein P, which has multiple residues [31]. The presence of Sec in the active site of the selenoproteins is important for its activity, and a misinterpretation of the UGA codon can lead to a significant loss of function of the selenoprotein. Still, many selenoproteins have functional orthologs, which have cysteine occupying the position of Sec [32], although they usually have poor activity in comparison to the Sec-containing selenoprotein.

Mutations and inborn errors in the genes involved in selenoprotein biosynthesis have been identified [33]. The first selenoprotein-related mutations were observed in the selenoprotein N [34], and it has been noticed that its mutations led to a spectrum of myopathies. Other disorders include neurologic phenotypes due to, for instance, the selenium deficiency in the brain caused by impaired selenium transport in the selenoprotein P-deficient mice [35,36]. The mutations in the glutathione peroxidase 4 cause Sedaghatian-type spondylometaphyseal chondrodysplasia [37], and those in thioredoxin system may affect the heart and the adrenals [38,39]. There are no known mutations of iodothyronine deiodinases in humans, but in mice the deletion of iodothyronine deiodinase 2 impaired bone stability [40]. The mutations in SECIS-binding protein 2 gene affect thyroid hormone regulation [41], but can have variable other features as well. The first identified human mutation in selenocysteine transfer RNA (tRNA[Ser] Sec) manifested a similar phenotype as mutations in SECIS-binding protein 2 gene [42]. In mice, deletion of the gene encoding the tRNA[Ser] Sec in osteochondroprogenitor yielded a phenotype similar to Kashin–Beck disease [43], although the phenotype is clearly stronger than the one observed in human Kashin–Beck disease [44].

3. Biosynthesis of Selenocysteine and Selenoproteins

There are two principal elements in the biosynthesis of selenoproteins and recoding of the UGA stop codon [45]. These are the involvement of (1) tRNA[Ser] Sec, which has an anticodon to the UGA codon, and (2) SECIS in the 3′-UTR. tRNA[Ser] Sec, first described in 1970 [46,47], is a unique tRNA in being able to regulate a whole class of proteins, namely the selenoproteins.

The aminoacylation of tRNA[Ser] Sec differs remarkably from the other tRNAs, since it requires the action of four different enzymes instead of only the one needed for the others [45]. In eukaryotes, the synthesis of Sec is started so that serine aminoacyl tRNA synthetase charges tRNA[Ser] Sec with serine. In the next step, tRNA[Ser] Sec serine is phosphorylated by a specific phosphoseryl tRNA kinase and, finally, the phosphate of *O*-phosphoserine of the tRNA[Ser] Sec is substituted by selenium atom donated by selenophosphate in a reaction catalyzed by selenocysteine synthase [45]. Selenophosphate synthetase 2 is essential in mammals to generate selenophosphate, while a highly homologous selenophosphate synthetase 1 has only little or no effect on selenoprotein synthesis [48,49].

Every selenoprotein mRNA contains a stem–loop–stem–loop structure in the 3′-end of their mRNA, which is an approximately 200 nucleotides long sequence. They are essential for the recoding of the UGA codons during selenoprotein translation. It has been noticed that the SECIS motif is in fact important for the modulation of the efficiency of Sec insertion, so that several thousand-fold differences can be observed between the strongest and weakest SECIS elements [50]. The minimum distance of the SECIS motif from the UGA codon is evaluated to be 51–111 nucleotides [51].

There are many proteins involved in the recoding of Sec. The best characterized of those is SECIS-binding protein 2 (SBP2) [52], which is found in all eukaryotes expressing selenoproteins. It is one of the limiting factors for Sec insertion [53], and its silencing caused a selective down-regulation of the selenoprotein expression in the mammalian cells [54]. It has been shown that SBP2 binding affinity also has a major importance for differential selenoprotein mRNA translation and sensitivity to so-called nonsense-mediated decay (NMD) [55]. The cells use a complex NMD pathway to destroy mRNAs with the premature stop codon to prevent their translation, which would lead to a synthesis of incorrect proteins [56]. Translation initiation factor 4A3 (eIF4A3) is a SECIS-binding protein, which takes part in a selective translational control of a subset of the selenoproteins [57]. Its binding to a subset of SECIS elements physically limits the binding of SBP2 there, thus preventing Sec insertion [57], and leads to the premature stop of translation. It has been shown that approximately half of the selenoprotein transcriptome has a SECIS element susceptible to NMD, which leads to remarkable changes of the transcript levels under selenium deficiency [58].

Ribosomal protein L30 is a SECIS-binding protein as a component of the large unit of eukaryotic ribosome, and can apparently recruit SBP2 to the Sec recoding machinery [59], but its exact function in Sec insertion needs further investigation. A multifunctional protein, nucleolin, is another SECIS-binding protein, which is likely to functions in Sec translation complex as a link to the ribosome [60].

4. Regulation of Selenoprotein Transcriptome by Selenium

Selenium intake has been observed to significantly affect the selenoproteome. During selenium deficiency, the gene expression and production of certain obviously essential selenoproteins is prioritized over others [61,62]. In a one-day-old layer of chicken liver, a prolonged deficiency of selenium in the diet continued for up to 65 days decreased 19 selenoprotein mRNA expressions of 21 investigated ones [63]. However, the sensitivities of various selenoprotein gene expressions to decrease was differential [63]. In general, interpretations of the results accumulated from the published studies are somewhat complicated, since the tissue distribution of the specific selenoproteins and their various intra- and extracellular locations is variable [27,64].

A considerable number of the selenoproteins participate in glutathione and thioperoxin systems, which play important roles in maintaining cellular redox balance [65]. Glutathione peroxidases are catalysts for the reduction of H_2O_2 or organic hydroxyperoxides to water or corresponding alcohols, most often by using glutathione as a reductant. Their dysregulation has been associated with diabetes, cancer, inflammation, and obesity [66]. Glutathione is one of the most important intracellular antioxidant systems, although it is also present in extracellular environments [67]. Thus, it acts as the major redox buffer in the majority of the cells. Glutathione in its reduced form can scavenge reactive oxygen and nitrogen species, a process which yields oxidized glutathione disulfide [65]. An enzyme glutathione reductase recycles oxidized glutathione back to its reduced form using nicotinamide adenine dinucleotide phosphate as a universal reducing agent [68]. Besides this system, glutathione metabolism involves numerous other proteins and electrophiles [69].

During aging, the redox balance of our body is jeopardized by oxidative stress, which is indicated by the oxidation of, for instance, glutathione, which is also related to our health and disease [70]. Selenoglutathione is a selenium analogue of glutathione, and a biologically potent redox substrate. Its oxidized form was recently shown to have the potential to repair misfolded disulfide-bond-containing proteins [71]. Since diselenide bonds can be expected to be more stable than disulfide ones, selenoglutathione analogues have also raised medical interest as possibly effective antioxidative agents [72].

Thioredoxin is a highly conserved antioxidant protein, which also participates in the maintenance of the redox balance by thiol-disulfide exchange reactions [73]. The thioredoxin system acts together with glutathione, and they can provide the electrons that cross-control the cellular redox environment [74].

4.1. Glutathione Peroxidases

Five of the glutathione peroxidases in humans are selenoproteins. Ubiquitously present glutathione peroxidase 1 was the first identified selenoprotein, which was then found to protect hemoglobin from oxidative breakdown [75]. Its gene expression and enzyme activity was almost totally lost in the liver and the heart under the depletion of selenium in the diet of rats, while in glutathione peroxidase 4, also known as phospholipid hydroperoxide glutathione reductase, gene expression was practically unchanged despite the 75% and 60% decrease of enzymatic activity, respectively [64].

In rats, glutathione peroxidase 2 was observed to be expressed only in the epithelium of the gastrointestinal tract, therefore, it is also called gastrointestinal glutathione peroxidase [76]. Its expression has also been reported for the human liver [77]. The selenium deficiency even appeared to increase the gene expression of glutathione peroxidase 2, while glutathione peroxidase 1 levels dropped and glutathione peroxidase 4 levels remained unchanged [78].

Glutathione peroxidase 3 is an abundant plasma protein, which is mostly secreted by kidney proximal tubules [79]. Studies with double knock-out mice for the glutathione peroxidase 3 and the selenoprotein P showed that together these two proteins account for the majority (>97%) of all plasma selenium in mice [80].

Besides H_2O_2, glutathione peroxidase 4 can also reduce the hydroperoxides present in phospholipids and cholesterol [66]. It is present in three different isoforms, namely the cytosolic, mitochondrial and sperm nuclear forms [81]. Their expression levels have been shown to be relatively resistant to the selenium deficiency [54,64], and therefore they are considered to be among the essential selenoproteins [82].

4.2. Thioredoxin Reductases

Humans and higher eukaryotes have three thioredoxin reductases, namely cytosolic, mitochondrial and testis-specific [83], which all contain the in-frame UGA codon to encode Sec residue. Although the mutation of Sec to Cys leads to a major decrease in their catalytic activity, Sec is not catalytically essential to reduce thioredoxin due to orthologs, which can perform identical reactions with approximately the same efficiencies [84]. Thioredoxin reductase 1 has been shown to be different from thioredoxin reductase 3 by containing a C-terminal structure, which restricts the motion of the C-terminal tail containing Gly-Cys-Sec-Gly tetrapeptide so that only that C-terminal redox center can participate with Se-dependent reduction by the N-terminal redox center, while thioredoxin reductase 3 does not have that same structure [85]. Therefore, thioredoxin reductase 3 has a greater access to the substrate and an alternative mechanistic pathway [85].

The selenium deficiency was shown to decrease the expression levels of both thioredoxin reductases 1 and 2 in rat heart, as well as glutathione reductases 1 and 4, which is known to impair the recovery from ischemia-reperfusion, while selenium supplementation significantly increased their expression levels [86]. Although glutathione peroxidases and thioredoxin reductases have been shown to be similarly modulated by the selenium availability in cardiac tissue [86], and to cooperate in the protection against free-radical-mediated cell death [87], they were oppositely regulated by selenium in response to replicative senescence in human embryonic fibroblasts [88].

Thioredoxin reductase 1 is overexpressed in many malignant cells and thus it has been considered as a target for anticancer approaches [89]. The transcriptional regulation of human thioredoxin reductase is very complex [90]. It has at least 21 different transcripts, which encode five isoforms differing in the alternative N-terminal domains [91]. It has been observed that selenium overdose can have selective cytotoxic effects on tumor cells [92].

4.3. Iodothyronine Deiodinases

Type 1 iodothyronine deiodinase produces biologically active 3,5,3'-triiodothyronine (T3) from 3,5,3',5'-tetraiodothyronine (T4, or thyroxin) for the plasma, while type II enzyme provides a local conversion. Type 3 enzyme finally converts T3 to inactive T2. All three deiodinases are expressed in a number of fetal and adult tissues [93]. The finding that selenium deficiency inhibits the activity of type I and II iodothyronine deiodinases led to the proposal that they are seleno-enzymes or selenium-containing cofactors [94], and type I iodothyronine deiodinase was soon identified as a selenoenzyme [95]. However, the gene expressions of type I iodothyronine deiodinase and selenoprotein P were more resistant to selenium deficiency-related decline than glutathione reductase [96]. In chicken, type I deiodinase expression was the most sensitive to prolonged selenium deficiency, while the type III enzyme did not change significantly even after 65 days deprivation [63]. Thyroid is much more resistant to dietary deficiency than, for instance, the liver, indicating the dependency on the organ in terms of how selenium deprivation affects the gene expression of the selenoproteins [97]. In general, the thyroideal conversion of T4 to T3 by iodothyronine deiodinases can be considered to be their major function. However, they have importance in other tissues as well, since type I iodothyronine deiodinase transcripts could be found at even higher levels than the thyroid in the ovine skeletal muscle, kidney and heart [98].

4.4. Other Selenoproteins

The selenium availability changes the gene expressions of a number of other selenoproteins, too. In chicken, selenoprotein I was the only one, which did not show a significant decrease at mRNA levels even after 65 days of selenium deficiency, while all the other investigated ones had decreased levels already after 15 days of treatment [63].

Selenoprotein P is unique among the selenoproteins, since it incorporates several Sec residues instead of the normal content of one Sec per one protein molecule. The human selenoprotein contains 10 Sec residues [99]. The liver takes up selenium and incorporates it into selenoprotein P, which is then secreted into the plasma for transport to other tissues [35]. In selenoprotein P knockout mice, a low selenium diet decreased the selenium content most severely in the brain and the testis, while the heart was not affected [36]. Due to its selenium transport characteristics, selenoprotein P also has a special role in the regulation of other selenoproteins.

Selenophosphate-synthetase 2 is an enzyme that participates the biosynthesis of selenoproteins by providing the selenophosphate needed in the biosynthesis of Sec. Thus, it can regulate its own production besides the production of other selenoproteins [100]. In selenoprotein P-depleted mice, all the other selenoproteins were down-regulated in the brain and the testis, in contrast to elevated levels of selenophosphate-synthetase 2 [101], which can obviously provide a compensatory mechanism during low selenium conditions.

A recent selenoproteome transcriptome study showed that the selenium deficiency in an ATDC chondrocyte cell line most dramatically dropped the expression levels of glutathione peroxidase 1 and selenoproteins H, I and W, while selenophosphate synthetase 2 and selenoproteins O, R and S were in fact upregulated [102]. The gene expression levels of glutathione peroxidase 1, thioredoxin reductase 1 and selenoproteins H, P, R and W were the most efficiently increased after the repletion of selenium to selenium-deficient medium [103]. Similar results were obtained for another chondrocyte cell line C28/I2 [103]. A microarray analysis showed that nine selenoproteins were significantly down-regulated in the selenium-deficient liver, while in mice with marginal selenium status none of these transcripts remained decreased relative to the levels in selenium-adequate mice [104]. This was concluded to mean that the selenium-related regulation of the selenoprotein gene expression is mediated by one underlying mechanism [62].

Microarray data of more than 30,000 transcripts from the livers of rats fed with a 0–5 µg selenium/g diet revealed that selenium deficiency down-regulated four selenoprotein genes (glutathione peroxidase 1, thioredoxin reductase 3, and selenoproteins H and T). In mouse, thioredoxin reductase 2 and selenoproteins K and W, muscle 1 were also down-regulated in the liver, and selenoprotein W and muscle 1 in the kidney [102].

5. Regulation of Non-Selenoprotein Transcriptome by Selenium

Selenium deficiency also regulates the expression of genes other than those encoding for the selenoproteins. Although limited in number, the recent large-scale transcriptome and proteome analyses have given new insights into the general level effects of selenium deprivation.

A study investigating the effects of various levels of selenium in the diet was continued for 35 days in mice and 28 days in rats [102]. The microarray analysis of the livers of animals fed with 0–5 µg selenium/g diet revealed only two transcripts (uridine diphosphate-glucuronosyltransferase 2 family, polypeptide B7 and glutathione S-transferase Yc2 subunit), which were up-regulated in the rat liver [102], whereas carbonyl reductase 3 and heat shock protein 1 in the mouse liver and glutathione 5-transferase, alpha 3 in the mouse kidney were also induced [102]. Real-time polymerase chain reaction analyses of the rat liver also indicated that adenosine triphosphate-binding cassette, subfamily C (cystic fibrosis transmembrane conductance regulator/multidrug-resistance-related protein), member 3 and nicotinamide dinucleotide phosphate dehydrogenase, quinone 1 were upregulated genes [102], which are the targets of the nuclear factor erythroid-2 like factor 2 (Nrf2) [105]. Since Nrf2 has a major role in the regulation of the cellular antioxidant response [106], and plays a critical role in cancer prevention, this pathway may be associated with the anti-cancer properties of selenium.

In the range of 0.08 to 0.8 µg selenium/g diet there were no remarkable differences in the gene expressions, while 2 µg selenium/g diet upregulated the following transcripts: regulator of G-protein signaling 4, coiled-coil domain containing 80, RGD1560666, 3-oxoacid CoA transferase 1, and expressed sequence tag UI-R-BS1-ayq-f-06-0-UI.s1. The only down-regulated one was cold-inducible RNA binding protein in comparison to a selenium adequate diet [102].

The biggest number of changes in gene expression became evident at selenium levels considered already toxic. There were 1193 transcripts, which had the significantly changed expressions [102]. The rats fed with a 5 µg selenium/g diet had retarded growth and signs of liver damage markers. Therefore, in order to more reliably reveal the selenium-related changes, the transcripts were filtered against the changed transcripts in calorie-restricted rats [105] and those overlapping with Affymetrix's Rat ToxFX 1.0 array to remove the transcripts probably related to growth and toxicity. Duplicate transcripts were also removed. This reduced the number of the transcripts regarded to be most likely selenium-specific to 667 unique transcripts [102]. Gene ontology analyses revealed a number of biological processes related to cellular movement and morphogenesis, extracellular matrix and cytoskeleton organization, and development and angiogenesis affected by the toxic level of selenium [102].

6. Selenium Regulation of MicroRNAs and Long Non-Coding RNAs

Besides mRNAs, the cells express groups of non-protein-coding transcripts, which can regulate a wide array of protein coding genes. MicroRNAs (miRNAs) and lncRNAs are two major families of these. In an intestinal cell line, selenium deficiency changed the expression of 50 genes and twelve miRNAs. Pathways related to, for instance, arachidonic acid metabolism, glutathione metabolism, oxidative stress, and mitochondrial respiration were found to be selenium-sensitive. Besides, thirteen transcripts were predicted to be targets for selenium-sensitive miRNAs, three of which were recognized by miR-185. More importantly, the silencing of miR-185 increased glutathione peroxidase 2 and selenophosphate synthetase 2 [107]. In a recent study, several miRNAs were predicted as putative regulators of various glutathione peroxidases [108].

In hepatocarcinoma cell line, it was noticed that miRNA-544a interacted with selenoprotein K, suppressing its expression, and the selenium treatment was able to modulate miR-544a expression [109]. On the other hand, selenium increased the expression level of miR-125a, and the overexpression of miR-125a could inhibit cadmium-induced apoptosis in LLC-PK1 cells [110]. In selenium-deficient rats, cardiac dysfunction was mainly associated with five up-regulated miRNAs (miR-374, miR-16, miR-199a-5p, miR-195 and miR-30e) and three down-regulated ones (miR-3571, miR-675 and miR-450a) [111]. In Caco-2 human adenocarcinoma cells, ten selenium-sensitive miRNAs were identified, which are involved in the regulation of 3588 mRNAs. Pathway analysis indicated that they participate in pathways such as the cell cycle, the cellular response to stress, the canonical Wnt/β-catenin, p53 and mitogen-activated kinase signaling pathways [112].

Only two publications that investigated the regulation of lncRNAs by selenium could be found. The first one investigated lncRNAs in the vascular exudative diathesis of chicken, induced by selenium deficiency. A total of 635 differentially-expressed lncRNAs of 15412 detected ones were identified, and gene ontology analysis suggested an importance of redox in particular. The study could verify that 19 target mRNAs of 23 lncRNAs were related to the redox process [113].

The other study investigated the role of lncRNAs in the selenium deficiency-induced muscle injury in chicken. The study identified 38 lncRNAs and 687 mRNAs that were affected by selenium deficiency. Pathway analyses revealed dysregulated pathways associated with phagosomes, cardiac muscle contraction, and peroxisome proliferator-activated receptors (PPAR) in the selenium-deficient group. In particular the study showed a relationship between lncRNA ALDBGAL0000005049 and stearoyl-CoA desaturase in the PPAR pathway. The down-regulation of that lncRNA led to inflammation by regulating stearoyl-CoA desaturase gene expression [114].

There is still a limited number of studies on non-protein-coding RNAs and selenium, and future investigations will certainly shed new light on how they are involved in selenium-related gene expression. The present studies already show that they are potential regulators of cellular responses and gene expressions.

7. Selenium-Related Gene Expressions in Some Pathological Conditions

Selenium has been associated with a number of pathological conditions. Although the mechanism for how selenium is related to pathogenesis is not always very precisely known, new research results may provide new ideas to understand selenium-related diseases.

Keshan disease is an endemic cardiomyopathy related to the selenium deficiency leading to a high mortality rate. To understand better the molecular mechanisms of Keshan disease pathogenesis, a proteomic screening of peripheral blood sera was performed with mass spectrometry to identify differentially-expressed selenium- and zinc-related proteins and their pathway and networks associated with Keshan disease [115]. In total, nineteen selenium- and three zinc-associated proteins were identified among 105 differentially-expressed proteins [115]. Pathway analysis revealed 52 pathways, of which hypoxia-inducible factor-1α and apoptosis pathways were likely to play a role in the selenium-associated functions [115]. A study combining a custom-made microarray, containing 78 probes previously differentially expressed genes in Keshan disease, analyses of peripheral blood mononuclear cell RNA samples obtained from 100 Keshan disease patients and 100 normal controls, and mass spectrometric proteomic analyses of peripheral blood sera of Keshan patients and controls indicated that there are numerous functional pathways and cellular systems associated with the differentially expressed genes and proteins [116]. The limitations of these studies are the fact that the selenium status of the individuals was not known.

Selenium deficiency has also been associated to Kashin–Beck disease, an endemic osteochondropathy. The differentially expressed genes in the transcriptome analyses comparing cartilage samples from Kashin–Beck disease versus normal control [117] or versus osteoarthritis [118] were other than the selenoproteins. Pathways related to reactive oxygen species and vascular endothelial growth factor were significantly elevated in osteoarthritis compared with Kashin–Beck disease, while the expression of genes of collagen- and nitric oxide-related pathways were elevated in Kashin–Beck disease [119]. Analysis of mitochondria-related genes in Kashin–Beck disease chondrocytes versus normal control identified nine up-regulated genes, which involved three canonical pathways: (1) oxidative phosphorylation; (2) apoptosis signaling; and (3) pyruvate metabolism [120]. In addition, microarray expression profiles of the cultured control and the Kashin–Beck disease chondrocytes revealed 232 up- and 427 down-regulated mRNAs and 316 up- and 631 down-regulated lncRNAs in Kashin–Beck disease patients' chondrocytes [121]. A lncRNA-mRNA correlation analysis yielded 509 coding–noncoding gene co-expression networks, and eleven lncRNAs were predicted to have *cis*-regulated target genes, and co-expressed mRNAs and lncRNAs formed a large network associated with biological events of the extracellular matrix [121].

The problem with investigations of Kashin–Beck disease is that the disease often develops already in early childhood. Due to the poor regeneration of the articular cartilage, the collection of biopsies from juvenile cartilage is not normally justified. The diet in Kashin–Beck endemic areas has also changed due to selenium supplementation [122]. Thus, the samples collected in later life most likely do not reflect the physiological or selenium status present at the onset of the disease. Nevertheless, a recent study, which estimated the nutrient intakes of children, showed that the daily intakes of multiple nutrients, not only selenium, were lower in the selenium-supplemented Kashin–Beck area in comparison to the non-selenium-supplemented areas [123]. There were 116 nutrient-related differentially-expressed genes in the peripheral blood of Kashin–Beck disease children (51 up- and 65 down-regulated ones), and 10 significant pathways were also recognized using the KEGG and REACTOME network databases [123]. Overlapping genes with various functions were also associated with different nutrients [123].

8. Conclusions

The transcription and translation of selenoproteins require a specific and complex cellular machinery. Selenium status can define the expression levels of the selenoprotein mRNA transcripts, although there is some degree of organ- and species-dependent variation. Expressions of the non-selenoprotein genes are also affected by the selenium status, but more research is warranted to gain a deeper understanding of the interacting pathways and functions.

Recent studies, which have analysed the miRNA and lncRNA expressions and their relations, have revealed a number of them to be differentially expressed by selenium status. However, the present rather limited knowledge of the functional regulation of the cellular processes controlled by miRNAs and lncRNAs prevents strong interpretations from being made, and further studies are definitely warranted to better understand their role in the transcriptional regulation by selenium.

The Keshan and Kashin–Beck diseases have been associated with selenium deficiency, although the actual mechanisms that are behind these diseases are still not precisely understood. New findings combining microarray and proteomic analyses have revealed new ideas regarding the pathways possibly involved in these diseases.

Author Contributions: M.J.L. and C.Q. jointly selected and analysed the literature for this review, and together wrote the manuscript.

Funding: This research was funded by Reumatikerfonden, Sweden, the Department of Integrative Medical Biology, Umeå University, and Insamlingsstiftelsen för medicinsk forskning, Umeå University.

Conflicts of Interest: The authors declare no conflict of interest.

Abbreviations

eIF4A3	Translation initiation factor 4A3
lncRNA	Long non-coding RNA
miRNA	Micro RNA
mRNA	Messenger RNA
NMD	Nonsense-mediated decay
Nrf2	Nuclear factor, erythroid-2 like factor 2
PPAR	Peroxisome proliferator-activated receptor
SBP2	SECIS-binding protein 2
Sec	Selenocysteine
SECIS	Selenocysteine insertion sequence
tRNA	Transfer RNA
tRNA[Ser] Sec	Selenocysteine transfer RNA
UTR	3′-Untranslated region of messenger RNA
T3	3,5,3′-Triiodothyronine
T4	3,5,3,5″-Tetraiodothyronine

References

1. Kieliszek, M.; Blazejak, S. Current knowledge on the importance of selenium in food for living organisms: A review. *Molecules* **2016**, *21*, 609. [CrossRef] [PubMed]
2. Whanger, P.D. Selenocompounds in plants and animals and their biological significance. *J. Am. Coll. Nutr.* **2002**, *21*, 223–232. [CrossRef] [PubMed]
3. Hartikainen, H.L. Biochemistry of selenium and its impact on food chain quality and human health. *J. Trace Elem. Med. Biol.* **2005**, *18*, 309–318. [CrossRef] [PubMed]
4. Holben, D.H.; Smith, A.M. The diverse role of selenium within selenoproteins. *J. Am. Diat. Assoc.* **1999**, *99*, 836–843. [CrossRef]
5. Papp, L.V.; Lu, J.; Holmgren, A.; Khanna, K.K. From selenium to selenoproteins: Synthesis, identity, and their role in human health. *Antioxid. Redox Signal.* **2007**, *9*, 775–806. [CrossRef] [PubMed]
6. European Food Safety Authority. Scientific opinion on dietary reference values for selenium. *Eur. Food Saf. Auth. J.* **2014**, *12*, 3846.
7. Steinnes, E. Soils and geochemistry. *Environ. Geochem. Health* **2009**, *31*, 523–535. [CrossRef] [PubMed]
8. Kielczykowska, M.; Kocot, J.; Pazdzior, M.; Musik, I. Selenium—A fascinating antioxidant of protective properties. *Adv. Clin. Exp. Med.* **2018**, *27*, 245–255. [CrossRef] [PubMed]
9. Fairweather-Tait, S.J.; Bao, Y.; Broadley, M.R.; Collings, R.; Ford, D.; Heseth, J.E.; Hurst, R. Selenium in human health and disease. *Antioxid. Redox Signal.* **2011**, *14*, 1337–1383. [CrossRef] [PubMed]
10. Sun, H.-J.; Rathinasabapathi, B.; Wu, B.; Luo, J.; Pu, L.-P.; Ma, L.Q. Arsenic and selenium toxicity and their interactive effects in human. *Environ. Int.* **2014**, *69*, 148–158. [CrossRef] [PubMed]
11. Virtamo, J.; Valkeila, E.; Alftan, G.; Punsar, S.; Huttunen, J.K.; Karvonen, M.J. Serum selenium and the risk of coronary heart disease and stroke. *Am. J. Epidemiol.* **1985**, *122*, 276–282. [CrossRef] [PubMed]
12. Tan, J.; Zhu, W.; Wang, W.; Li, R.; Hou, S.; Wang, D.; Yang, L. Selenium in soil and endemic diseases in China. *Sci. Total Environ.* **2002**, *284*, 227–235. [CrossRef]
13. Han, J.; Guo, X.; Wang, L.; Chilufya, M.M.; Lim, P.N.; Qu, C. Selenium deficiency and selenium supplements: Biological effects on fibrosis in chronic diseases, from animal to human studies. In *Handbook of Famine, Starvation, and Nutrient Deprivation*; Preedy, V.R., Patel, V.B., Eds.; Springer: Cham, Switzerland, 2017; pp. 1–20. ISBN 987-3-319-40007-5.
14. Steinbrenner, H.; Al-Quraishy, S.; Dkhil, M.A.; Wunderlich, F.; Sies, H. Dietary selenium in adjuvant therapy of viral and bacterial infections. *Adv. Nutr.* **2015**, *6*, 73–82. [CrossRef] [PubMed]
15. Kieliszek, M.; Lipinski, B. Pathophysiological significance of protein hydrophobic interactions: An emerging hypothesis. *Med. Hypotheses* **2018**, *110*, 15–22. [CrossRef] [PubMed]
16. Lipinski, B. Sodium selenite as an anticancer agent. *Anticancer Agents Med. Chem.* **2017**, *17*, 658–661. [CrossRef] [PubMed]

17. Kieliszek, M.; Lipinski, B.; Blazejak, S. Application of sodium selenite in the prevention and treatment of cancers. *Cells* **2017**, *6*, 39. [CrossRef] [PubMed]
18. Lipinski, B. Rationale for the treatment of cancer with sodium selenite. *Med. Hypotheses* **2005**, *64*, 806–810. [CrossRef] [PubMed]
19. Kiremidjian-Schumacher, L.; Roy, M.; Cohen, M.W.; Stotzky, G. Supplementation with selenium augments the functions of natural killer and lymphokine-activated killer cells. *Biol. Trace Elem.* **1996**, *52*, 227–239. [CrossRef] [PubMed]
20. Brodin, O.; Eksborg, S.; Wallenberg, M.; Asker-Hagelberg, C.; Larsen, E.H.; Mohlkert, D.; Lenneby-Helleday, C.; Jacobsson, H.; Linder, S.; Misra, S.; et al. Pharmacokinetics and toxicity of sodium selenite in the treatment of patients with carcinoma in a phase I clinical trial: The SECAR study. *Nutrients* **2015**, *7*, 4978–4994. [CrossRef] [PubMed]
21. Jiang, X.R.; Macey, M.G.; Lin, H.X.; Newland, A.C. The anti-leukaemic effects and the mechanism of sodium selenite. *Leuk. Res.* **1992**, *16*, 347–352. [CrossRef]
22. Olm, E.; Jönsson-Videsäter, K.; Ribera-Cortada, I.; Fernandes, A.P.; Eriksson, L.C.; Lehmann, S.; Björnstedt, M. Selenite is a potent cytotoxic agent for human primary AML cells. *Cancer Lett.* **2009**, *282*, 116–123. [CrossRef] [PubMed]
23. Hawkes, W.C.; Tappel, A.L. In vitro synthesis of glutathione peroxidase from selenite. Translational incorporation of selenocysteine. *Biochim. Biophys. Acta* **1983**, *739*, 225–234. [CrossRef]
24. Kryukov, G.V.; Castellano, S.; Novoselov, S.V.; Lobanov, A.V.; Zehtab, O.; Guigo, R.; Gladyshev, V.N. Characterization of mammalian selenoproteomes. *Science* **2003**, *300*, 1439–1443. [CrossRef] [PubMed]
25. Zoidis, E.; Seremelis, I.; Kontopoulos, N.; Danezis, G.P. Selenium-dependent antioxidant enzymes: Actions and properties of selenoproteins. *Antioxidants* **2016**, *7*, 66. [CrossRef] [PubMed]
26. Bösl, M.R.; Takaku, K.; Oshima, M.; Nishimura, S. Early embryonic lethality caused by targeted disruption of the mouse selenocysteine tRNA gene (*Trsp*). *Proc. Natl. Acad. Sci. USA* **1997**, *94*, 5532–5534. [CrossRef]
27. Kasaikina, M.V.; Hatfield, D.L.; Gladyshev, V.N. Understanding selenoprotein function and regulation through the use of rodent models. *Biochim. Biophys. Acta* **2012**, *1823*, 1633–1642. [CrossRef] [PubMed]
28. Reeves, M.A.; Hoffmann, P.R. The human selenoproteome: Recent insights into functions and regulation. *Cell. Mol. Life Sci.* **2009**, *66*, 2427–2478. [CrossRef] [PubMed]
29. Hatfield, D.L.; Gladyshev, V.N. How selenium has altered our understanding of the genetic code. *Mol. Cell. Biol.* **2002**, *22*, 3565–3576. [CrossRef] [PubMed]
30. Hubert, N.; Walczak, R.; Sturchler, C.; Myslinski, E.; Schuster, C.; Westhof, E.; Carbon, P.; Krol, A. RNAs mediating cotranslational insertion of selenocysteine in eukaryotic selenoproteins. *Biochimie* **1996**, *78*, 590–596. [CrossRef]
31. Burk, R.F.; Hill, K.E. Orphan selenoproteins. *BioEssays* **1999**, *21*, 231–237. [CrossRef]
32. Kim, H.Y.; Gladyshev, V.N. Different catalytic mechanisms in mammalian selenocysteine- and cysteine-containing methionine-*R*-sulfoxide reductases. *PLoS Biol.* **2005**, *3*, e375. [CrossRef] [PubMed]
33. Schweizer, U.; Fradejas-Villar, N. Why 21? The significance of selenoproteins for human health revealed by inborn errors of metabolism. *FASEB J.* **2016**, *30*, 3669–3681. [CrossRef] [PubMed]
34. Moghadaszadeh, B.; Petit, N.; Jaillard, C.; Brockington, M.; Quijano Roy, S.; Merlini, L.; Romero, N.; Estournet, B.; Desguerre, I.; Chaigne, D.; et al. Mutations in SEPN1 cause congenital muscular dystrophy with spinal rigidity and restrictive respiratory syndrome. *Nat. Genet.* **2001**, *29*, 17–18. [CrossRef] [PubMed]
35. Schomburg, L.; Schweizer, U.; Holtmann, B.; Flöhe, L.; Sendtner, M.; Köhrle, J. Gene disruption discloses role of selenoprotein P in selenium delivery to target tissues. *Biochem. J.* **2003**, *370*, 397–402. [CrossRef] [PubMed]
36. Hill, K.E.; Zhou, J.; McMahan, W.J.; Motley, A.K.; Atkins, J.F.; Gesteland, R.F.; Burk, R.F. Deletion of selenoprotein P alters distribution of selenium in the mouse. *J. Biol. Chem.* **2003**, *278*, 13640–13646. [CrossRef] [PubMed]
37. Smith, A.C.; Mears, A.J.; Bunker, R.; Ahmed, A.; MacKenzie, M.; Swarzentruber, J.A.; Beaulieu, C.L.; Ferretti, E.; FORGE Canada Consortium; Majewski, J.; et al. Mutations in the enzyme glutathione peroxidase 4 cause Sedaghatian-type spondylometaphyseal dysplasia. *J. Med. Genet.* **2014**, *51*, 470–474. [CrossRef] [PubMed]
38. Conrad, M.; Jakupoglu, C.; Moreno, S.G.; Lippl, S.; Banjac, A.; Schneider, M.; Beck, H.; Hatzopoulos, A.K.; Just, U.; Sinowatz, F.; et al. Essential role for mitochondrial thioredoxin reductase in hematopoiesis, heart development, and heart function. *Mol. Cell. Biol.* **2004**, *24*, 9414–9423. [CrossRef] [PubMed]

39. Prasad, R; Chan, L.F.; Hughes, C.R.; Kaski, J.P.; Kowalczyk, J.C.; Savage, M.O.; Peters, C.J.; Nathwani, N.; Clark, A.J.L.; Storr, H.L.; et al. Thioredoxin reductase 2 (*TXNRD2*) mutation associated with familial glucocorticoid deficiency (FGD). *J. Clin. Endocrinol. Metab.* **2014**, *99*, E1556–E1563. [CrossRef] [PubMed]

40. Bassett, J.H.D.; Boyde, A.; Howell, P.G.T.; Bassett, R.H.; Galliford, T.M.; Archanco, M.; Evans, H.; Lawson, M.A.; Croucher, P.; St. Germain, D.L.; et al. Optimal bone strength and mineralization requires the type 2 iodothyronine deiodinase in osteoblasts. *Proc. Natl. Acad. Sci. USA* **2010**, *107*, 7604–7609. [CrossRef] [PubMed]

41. Dumitrescu, A.M.; Refetoff, S. The syndromes of reduced sensitivity to thyroid hormone. *Biochim. Biophys. Acta* **2013**, *1830*, 3987–4003. [CrossRef] [PubMed]

42. Schoenmakers, E.; Carlson, B.; Agostini, M.; Moran, C.; Rajanayagam, O.; Bochukova, E.; Tobe, R.; Peat, R.; Gevers, E.; Muntoni, F.; et al. Mutation in human selenocysteine transfer RNA selectively disrupts selenoprotein synthesis. *J. Clin. Investig.* **2016**, *126*, 992–996. [CrossRef] [PubMed]

43. Downey, C.M.; Horton, C.R.; Carlson, B.A.; Parsons, T.E.; Hatfield, D.L.; Hallgrimsson, B.; Jirik, F.R. Osteo-chondroprogenitor-specific deletion of the selenocysteine tRNA gene, *Trsp*, leads to chondronecrosis and abnormal skeletal development: A putative model for Kashin-Beck disease. *PLoS Genet.* **2009**, *5*, e1000616. [CrossRef] [PubMed]

44. Guo, X.; Ma, W.J.; Zhang, F.; Ren, F.L.; Qu, C.J.; Lammi, M.J. Recent advances in the research of an endemic osteochondropathy in China: Kashin-Beck disease. *Osteoarthritis Cartilage* **2014**, *22*, 1774–1783. [CrossRef] [PubMed]

45. Bulteau, A.L.; Chavatte, L. Update on selenoprotein biosynthesis. *Antioxid. Redox Signal.* **2015**, *23*, 775–794. [CrossRef] [PubMed]

46. Mäenpää, P.H.; Bernfield, M.R. A specific hepatic transfer RNA for phosphoserine. *Proc. Natl. Acad. Sci. USA* **1970**, *67*, 688–695. [CrossRef] [PubMed]

47. Hatfield, D.; Portugal, F.H. Seryl-tRNA in mammalian tissues: Chromatographic differences in brain and liver and a specific response to the codon, UGA. *Proc. Natl. Acad. Sci. USA* **1970**, *67*, 1200–1206. [CrossRef] [PubMed]

48. Low, S.C.; Harney, J.W.; Berry, M.J. Cloning and functional characterization of human selenophosphate synthetase, an essential component of selenoprotein synthesis. *J. Biol. Chem.* **1995**, *270*, 21659–21664. [CrossRef] [PubMed]

49. Xu, X.-M.; Carlson, B.A.; Irons, R.; Mix, H.; Zhong, N.; Gladyshev, V.N. Selenophosphate synthetase 2 is essential for selenoprotein biosynthesis. *Biochem. J.* **2007**, *404*, 115–120. [CrossRef] [PubMed]

50. Latreche, L.; Jean-Jean, O.; Driscoll, D.M.; Chavatte, L. Novel structural determinants in human SECIS elements modulate the translational recoding of UGA as selenocysteine. *Nucleic Acids Res.* **2009**, *37*, 5868–5880. [CrossRef] [PubMed]

51. Martin, G.W., 3rd; Harney, J.W.; Berry, M.J. Selenecysteine incorporation in eukaryotes: Insights into mechanism and efficiency from sequence, structure, and spacing proximity studies of the type I deiodinase SECIS element. *RNA* **1996**, *2*, 171–182. [PubMed]

52. Copeland, P.R.; Fletcher, J.E.; Carlson, B.A.; Hatfield, D.L.; Driscoll, D.M. A novel RNA binding protein, SBP2, is required for the translation of mammalian selenoprotein mRNAs. *EMBO J.* **2000**, *19*, 306–314. [CrossRef] [PubMed]

53. Low, S.L.; Grundner-Culemann, E.; Harney, J.W.; Berry, M.J. SECIS-SBP2 interactions dictate selenocysteine incorporation efficiency and selenoprotein hierarchy. *EMBO J.* **2000**, *19*, 6882–6890. [CrossRef] [PubMed]

54. Latreche, L.; Duhieu, S.; Touat-Hamici, Z.; Jean-Jean, O.; Chavatte, L. The differential expression of glutathione peroxidase 1 and 4 depends on the nature of SECIS element. *RNA Biol.* **2012**, *9*, 681–690. [CrossRef] [PubMed]

55. Squires, J.E.; Stoytchev, I.; Forry, E.P.; Berry, M.J. SBP2 binding affinity is a major determinant in differential selenoprotein mRNA translation and sensitivity to nonsense-mediated decay. *Mol. Cell. Biol.* **2007**, *27*, 7848–7855. [CrossRef] [PubMed]

56. Kervestin, S.; Jacobson, A. NMD: A multifaceted response to premature translational termination. *Nat. Rev. Mol. Cell. Biol.* **2012**, *13*, 700–712. [CrossRef] [PubMed]

57. Budiman, M.E.; Bubenik, J.L.; Miniard, A.C.; Middleton, L.M.; Gerber, C.A.; Cash, A.; Driscoll, D.M. Eukaryotic initiation factor 4a3 is a selenium-regulated RNA-binding protein that selectively inhibits selenocysteine incorporation. *Mol. Cell* **2009**, *35*, 479–489. [CrossRef] [PubMed]

58. Seyedali, A.; Berry, M.J. Nonsense-mediated decay factors are involved in the regulation of selenoprotein mRNA levels during selenium deficiency. *RNA* **2014**, *20*, 1248–1256. [CrossRef] [PubMed]

59. Chavatte, L.; Brown, B.A.; Driscoll, D.M. Ribosomal protein L30 is a component of the UGA-selenocysteine recoding machinery in eukaryotes. *Nat. Struct. Mol. Biol.* **2005**, *12*, 408–416. [CrossRef] [PubMed]

60. Wu, R.; Shen, Q.; Newburger, P.E. Recognition and binding of the human selenocysteine insertion sequence by nucleolin. *J. Cell. Biochem.* **2000**, *77*, 507–516. [CrossRef]

61. Lei, X.G.; Evenson, J.K.; Thompson, K.M.; Sunde, R.A. Glutathione peroxidase and phospholipid hydroperoxide glutathione peroxidase are differently regulated in rats by dietary selenium. *J. Nutr.* **1995**, *125*, 1438–1446. [CrossRef] [PubMed]

62. Sunde, R.A.; Raines, A.M.; Barnes, K.M.; Evenson, J.K. Selenium status highly regulates selenoprotein mRNA levels for only a subset of the selenoproteins in the selenoproteome. *Biosci. Rep.* **2009**, *29*, 329–338. [CrossRef] [PubMed]

63. Liu, C.P.; Fu, J.; Lin, S.L; Wang, X.S.; Li, S. Effect of dietary selenium deficiency on mRNA levels of twenty-one selenoprotein genes in the liver of layer chicken. *Biol. Trace Elem. Res.* **2014**, *159*, 192–198. [CrossRef] [PubMed]

64. Bermano, G.; Nicol, F.; Dyer, J.A.; Sunde, R.A.; Beckett, G.J.; Arthur, J.R.; Hesketh, J.E. Tissue-specific regulation of selenoenzyme gene expression in rats. *Biochem. J.* **1995**, *311*, 425–430. [CrossRef] [PubMed]

65. Couto, N.; Wood, J.; Barber, J. The role of glutathione reductase and related enzymes on cellular redox homeostasis network. *Free Radic. Biol. Med.* **2016**, *95*, 27–42. [CrossRef] [PubMed]

66. Brigelius-Flohe, R.; Maiorino, M. Glutathione peroxidases. *Biochim. Biophys. Acta* **2013**, *1830*, 3289–3303. [CrossRef] [PubMed]

67. Ottaviano, F.G.; Handy, D.E.; Loscalzo, J. Redox regulation in the extracellular environment. *Circ. J.* **2008**, *72*, 1–16. [CrossRef] [PubMed]

68. Meister, A. Glutathione metabolism. *Methods Enzymol.* **1995**, *251*, 3–7. [CrossRef] [PubMed]

69. Deponte, M. The incomplete glutathione puzzle: Just guessing at numbers and figures? *Antioxid. Redox Signal.* **2017**, *27*, 1130–1161. [CrossRef] [PubMed]

70. Go, Y.M.; Jones, D.P. Redox theory of aging: Implications for health and disease. *Clin. Sci.* **2017**, *131*, 1669–1688. [CrossRef] [PubMed]

71. Shimodaira, S.; Asano, Y.; Arai, K.; Iwaoka, M. Selenogluthione diselenide: Unique redox reactions in the GPx-like catalytic cycle and repairing of disulfide bonds in scrambled protein. *Biochemistry* **2017**, *56*, 5644–5653. [CrossRef] [PubMed]

72. Yoshida, S.; Kumakura, F.; Komatsu, I.; Arai, K.; Onuma, Y.; Hojo, H.; Singh, B.G.; Priyadarsini, K.I.; Iwaoka, M. Antioxidative glutathione peroxidase activity of selenoglutathione. *Angew. Chem. Int. Ed. Engl.* **2011**, *50*, 2125–2128. [CrossRef] [PubMed]

73. Bindoli, A.; Rigobello, M.P. Principles in redox signaling; from chemistry to functional significance. *Antioxid. Redox Signal.* **2013**, *18*, 1557–1593. [CrossRef] [PubMed]

74. Lu, J.; Holmgren, A. The thioredoxin system. *Free Radic. Biol. Med.* **2014**, *66*, 75–87. [CrossRef] [PubMed]

75. Mills, C.C. Hemoglobin catabolism. I. Glutathione peroxidase, an erythrocyte enzyme which protects hemoglobin from oxidative breakdown. *J. Biol. Chem.* **1957**, *229*, 189–197. [PubMed]

76. Chu, F.-F.; Esworthy, S.R. The expression of an intestinal form of glutathione peroxidase (GSHPx-GI) in rat intestinal epithelium. *Arch. Biochem. Biophys.* **1995**, *323*, 288–294. [CrossRef] [PubMed]

77. Chu, F.-F.; Doroshow, J.H.; Esworthy, R.S. Expression, characterization, and tissue distribution of a new cellular selenium-dependent glutathione peroxidase, GSHPx-GI. *J. Biol. Chem.* **1993**, *268*, 2571–2576. [PubMed]

78. Wingler, K.; Böcher, M.; Flohe, L.; Kollmus, H.; Brigeliu-Flohe, R. mRNA stability and selenocysteine insertion sequence efficiency rank gastrointestinal glutathione peroxidase high in the hierarchy of selenoproteins. *FEBS J.* **1999**, *259*, 149–157. [CrossRef]

79. Avissar, N.; Ornt, D.B.; Yagil, Y.; Horowitz, S.; Watkins, R.H.; Kerl, E.A.; Takahashi, K.; Palmer, I.S.; Cohen, H.J. Human kidney proximal tubules are the main source of plasma glutathione peroxidase. *Am. J. Physiol.* **1994**, *266*, C367–C375. [CrossRef] [PubMed]

80. Olson, G.E.; Whitin, J.C.; Hill, K.E.; Winfrey, V.P.; Motley, A.K.; Austin, L.M.; Deal, J.; Cohen, H.J.; Burk, R.F. Extracellular glutathione peroxidase (Gpx3) binds specifically to basement membranes of mouse renal cortex tubule cells. *Am. J. Physiol. Renal Physiol.* **2010**, *298*, F1244–F1253. [CrossRef] [PubMed]

81. Brigelius-Flohe, R.; Aumann, K.D.; Blocker, H.; Gross, G.; Kiess, M.; Kloppel, K.D.; Maiorino, M.; Roveri, A.; Schuckelt, R.; Ursini, F.; et al. Phospholipid-hydroperoxide glutathione peroxidase. Genomic DNA, cDNA, and deduced amino acid sequence. *J. Biol. Chem.* **1994**, *269*, 7342–7348. [PubMed]

82. McCann, J.C.; Ames, B.N. Adaptive dysfunction of selenoproteins from the perspective of the triage theory: Why modest selenium deficiency may increase risk of diseases of aging. *FASEB J.* **2011**, *25*, 1793–1814. [CrossRef] [PubMed]

83. Sun, Q.-A.; Wu, Y.; Zappacosta, F.; Jeang, K.-T.; Lee, B.J.; Hatfield, D.L.; Gladyshov, V.N. Redox regulation of cell signaling by selenocysteine in mammalian thioredoxin reductases. *J. Biol. Chem.* **1999**, *274*, 24522–24530. [CrossRef] [PubMed]

84. Lothrop, A.P.; Ruggles, E.L.; Hondal, R.J. No selenium required: Reactions catalyzed by mammalian thioredoxin reductase that are independent of selenocysteine residue. *Biochemistry* **2009**, *48*, 6213–6223. [CrossRef] [PubMed]

85. Lothrop, A.P.; Snider, G.W.; Ruggles, E.L; Hondal, R.J. Why is mammalian reductase 1 so dependent upon the use of selenium? *Biochemistry* **2014**, *53*, 554–565. [CrossRef] [PubMed]

86. Venardos, K.; Ashton, K.; Headrick, J.; Perkins, A. Effects of dietary selenium on post-ischemic expression of antioxidant mRNA. *Mol. Cell. Biochem.* **2005**, *270*, 131–138. [CrossRef] [PubMed]

87. Barrera, L.N.; Cassidy, A.; Wang, W.; Wei, T.; Belshaw, N.J; Johnson, I.T.; Brigelius-Flohe, R.; Bao, Y. TrxR1 and Gpx2 are potently induced by isothiocyanates and selenium, and mutually cooperate to protect Caco-2 cells against free radical-mediated cell death. *Biochim. Biophys. Acta* **2012**, *1823*, 1914–1924. [CrossRef] [PubMed]

88. Legrain, Y.; Touat-Hamici, Z.; Chavatte, L. Interplay between selenium levels, selenoprotein expression, and replicative senescence in WI-38 human fiubroblasts. *J. Biol. Chem.* **2014**, *289*, 6299–6310. [CrossRef] [PubMed]

89. Arnér, E.S.J. Targeting the selenoprotein thioredoxin reductase 1 for anticancer therapy. *Adv. Cancer Res.* **2017**, *136*, 139–151. [CrossRef] [PubMed]

90. Rundlöf, A.-K.; Arnér, E.S.J. Regulation of the mammalian selenoprotein thiredoxin reductase 1 in relation to cellular phenotype, growth, and signaling events. *Antioxid. Redox Signal.* **2004**, *6*, 41–52. [CrossRef] [PubMed]

91. Rundlöf, A.-K.; Janard, M.; Miranda-Vizuete, A.; Arnér, E.S.J. Evidence for intriguingly complex transcription of human thioredoxin reductase 1. *Free Radic. Biol. Med.* **2004**, *36*, 641–656. [CrossRef] [PubMed]

92. Björkhem-Bergman, L.; Jönsson, K.; Eriksson, L.C.; Olsson, J.M.; Lehmann, S.; Paul, C.; Björnstedt, M. Drug-resistant human lung cancer cells are more sensitive to selenium cytotoxicity. Effects on thioredoxin reductase and glutathione reductase. *Biochem. Pharmacol.* **2002**, *63*, 1875–1884. [CrossRef]

93. Hernandez, A.; St. Germain, D.L. Thyroid hormone deiodinases: Physiology and clinical disorders. *Curr. Opin. Pediatr.* **2003**, *15*, 416–420. [CrossRef] [PubMed]

94. Beckett, G.J; MacDougall, D.A.; Nicol, F.; Arthur, R. Inhibition of type I and type II iodothyronine deiodinase activity in rat liver, kidney and brain produced by selenium deficiency. *Biochem. J.* **1989**, *259*, 887–892. [CrossRef] [PubMed]

95. Behne, D.; Kyriakopoulos, A.; Meinhold, H.; Köhrle, J. Identification of type I iodothyronine 5'-deiodinase as a selenoenzyme. *Biochem. Biophys. Res. Commun.* **1990**, *173*, 1143–1149. [CrossRef]

96. Hill, K.E.; Lyons, P.R.; Burk, R.F. Differential regulation of rat liver selenoprotein mRNAs in selenium deficiency. *Biochem. Biophys. Res. Commun.* **1992**, *185*, 260–263. [CrossRef]

97. Koenig, R.J. Regulation of type I iodothyronine deiodinase in health and disease. *Thyroid* **2005**, *15*, 835–840. [CrossRef] [PubMed]

98. Foroughi, M.A.; Dehghani, H. Short communication: Quantitative comparison of iodothyronine deiodinase I and II mRNA expression in ovine tissues. *Res. Vet. Sci.* **2013**, *95*, 891–893. [CrossRef] [PubMed]

99. Burk, R.F.; Hill, K.E. Selenoprotein P: An extracellular protein with unique physical characteristics and a role in selenium homeostasis. *Annu. Rev. Nutr.* **2005**, *25*, 215–235. [CrossRef] [PubMed]

100. Guimaraes, M.J.; Peterson, D.; Vicari, A.; Cocks, B.G.; Copeland, N.G.; Gilbert, D.J.; Jenkins, N.A.; Ferrick, D.A.; Kastelein, R.A.; Bazan, J.F.; et al. Identification of a novel selD homolog from eukaryotes, bacteria, and archaea: Is there an autoregulatory mechanism in selenocysteine metabolism? *Proc. Natl. Acad. Sci. USA* **1996**, *93*, 15086–15091. [CrossRef] [PubMed]

101. Hoffmann, P.R.; Höge, S.C.; Hoffmann, F.W.; Hashimoto, A.C.; Barry, M.J. The selenoproteome exhibits widely varying, tissue-specific dependence on selenoprotein P for selenium supply. *Nucleic Acids Res.* **2007**, *35*, 3963–3973. [CrossRef] [PubMed]

102. Raines, A.M.; Sunde, R.A. Selenium toxicity but not deficient or supernutrional selenium status vastly alters the transcriptome in rodents. *BMC Genom.* **2011**, *12*, 26. [CrossRef] [PubMed]

103. Yan, J.; Zheng, Y.; Min, Z.; Ning, Q.; Lu, S. Selenium effect on selenoprotein trasncriptome in chondrocytes. *Biometals* **2013**, *26*, 285–296. [CrossRef] [PubMed]

104. Sunde, R.A.; Raines, A.M. Selenium regulation of the selenoprotein and nonselenoprotein transcriptomes in rodents. *Adv. Nutr.* **2011**, *2*, 138–150. [CrossRef] [PubMed]

105. Pohjanvirta, R.; Boutros, P.C.; Moffat, I.D.; Lindén, J.; Wendelin, D.; Okey, A.B. Genome-wide effects of acute progressive feed restriction in liver and white adipose tissue. *Toxicol. Appl. Pharmacol.* **2008**, *230*, 41–56. [CrossRef] [PubMed]

106. Rojo de la Vega, M.; Chapman, E.; Zhang, D.D. NRF2 and the hallmarks of cancer. *Cancer Cell* **2018**, *34*, 21–43. [CrossRef] [PubMed]

107. Maciel-Dominguez, A.; Swan, D.; Ford, D.; Hesketh, J. Selenium alters miRNA profile in an intestinal cell line. Evidence that miR-185 regulates expression of *GPX2* and *SEPSH2*. *Mol. Nutr. Food Res.* **2013**, *57*, 2195–2206. [CrossRef] [PubMed]

108. Matouskova, P.; Hanouskova, B.; Skalova, L. MicroRNAs as potential regulators of glutathione peroxidases expression and their role in obesity and related pathologies. *Int J. Mol. Sci.* **2018**, *19*, 1199. [CrossRef] [PubMed]

109. Potenza, N.; Castiello, F.; Panella, M.; Colonna, G.; Ciliberto, G.; Russo, A.; Costantini, S. Human miR.544a modulates SELK expression in hepatocarcinoma cell lines. *PLoS ONE* **2016**, *11*, e0156908. [CrossRef] [PubMed]

110. Chen, Z.; Gu, D.; Zhou, M.; Shi, H.; Yan, S.; Cai, Y. Regulatory role of miR-125a/b in the suppression by selenium of cadmium-induced apoptosis via the mitochondrial pathway in LLC-PK1 cells. *Chem. Biol. Interact.* **2016**, *243*, 35–44. [CrossRef] [PubMed]

111. Xing, Y.; Liu, Z.; Yang, G.; Gao, D.; Niu, X. MicroRNA expression profiles in rats with selenium deficiency and the possible role of the Wnt/β-catenin signaling pathway in cardiac dysfunction. *Int. J. Mol. Med.* **2015**, *35*, 143–152. [CrossRef] [PubMed]

112. McCann, M.J.; Rotjanapun, K.; Hesketh, J.E.; Roy, N.C. Expression profiling indicating low selenium-sensitive microRNA levels linked to cell cycle and stress response pathways in the Caco-2 cell line. *Br. J. Nutr.* **2017**, *117*, 1212–1221. [CrossRef] [PubMed]

113. Cao, C.; Fan, R.; Zhao, J.; Zhao, X.; Yang, J.; Zhang, Z.; Xu, S. Impact of exudative diathesis induced by selenium deficiency on lncRNAs and their roles in the oxidative reduction process in broiler chick veins. *Oncotarget* **2017**, *8*, 20695–20705. [CrossRef] [PubMed]

114. Fan, R.; Cao, C.; Zhao, X.; Shi, Q.; Zhao, J.; Xu, S. Downregulated long noncoding RNA ALDBGAL0000005049 induces inflammation in chicken muscle suffered from selenium deficiency by regulating stearoyl-CoA desaturase. *Oncotarget* **2017**, *8*, 52761–52774. [CrossRef] [PubMed]

115. Wang, S.; Lv, Y.; Wang, Y.; Du, P.; Tan, W.; Lammi, M.J.; Guo, X. Network analysis of Se- and Zn-related proteins in the serum proteomics expression profile of the endemic dilated cardiomyopathy Keshan disease. *Biol. Trace Elem. Res.* **2018**, *183*, 40–48. [CrossRef] [PubMed]

116. Wang, S.; Yan, R.; Wang, B.; Du, P.; Tan, W.; Lammi, M.J.; Guo, X. Prediction of co-expression genes and integrative analysis of gene microarray proteomics profile of Keshan disease. *Sci. Rep.* **2018**, *8*, 231. [CrossRef] [PubMed]

117. Wang, W.Z.; Guo, X.; Duan, C.; Ma, W.J.; Zhang, Y.G.; Xu, P.; Gao, Z.Q.; Wang, Z.F.; Yan, H.; Zhang, Y.F.; et al. Comparative analysis of gene expression profiles between the normal human cartilage and the one with endemic osteoarthritis. *Osteoarthritis Cartilage* **2009**, *17*, 83–90. [CrossRef] [PubMed]

118. Duan, C.; Guo, X.; Zhang, X.D.; Yu, H.J.; Yan, H.; Gao, Y.; Ma, W.J.; Gao, Z.Q.; Xu, P.; Lammi, M. Comparative analysis of gene expression profiles between primary knee osteoarthritis and an osteoarthritis endemic to Northwestern China, Kashin-Beck disease. *Arthritis Rheum.* **2010**, *62*, 771–780. [CrossRef] [PubMed]

119. Zhang, F.; Guo, X.; Duan, C.; Wu, S.; Yu, H.; Lammi, M. Identification of differentially expressed genes and pathways between primary osteoarthritis and endemic osteoarthritis (Kashin-Beck disease). *Scand. J. Rheumatol.* **2013**, *42*, 71–79. [CrossRef] [PubMed]

120. Li, C.; Wang, W.; Guo, X.; Zhang, F.; Ma, W.; Zhang, Y.; Li, Y.; Bai, Y.; Lammi, M.J. Pathways related to mitochondrial dysfunction in cartilage of endemic osteoarthritis patients in China. *Sci. China Life Sci.* **2012**, *55*, 1057–1063. [CrossRef] [PubMed]

121. Wu, C.; Liu, H.; Zhang, F.; Shao, W.; Yang, L.; Ning, Y.; Wang, S.; Zhao, G.; Lee, B.J.; Lammi, M.; et al. Long noncoding RNA expression profile reveals lncRNAs signature associated with extracellular matrix degradation in kashin-beck disease. *Sci. Rep.* **2017**, *7*, 17553. [CrossRef] [PubMed]

122. Du, B.; Zhou, J.; Zhou, J. Selenium status of children in Kashin-Beck disease endemic areas in Shaanxi, China: Assessment with mercury. *Environ. Geochem. Health* **2018**, *40*, 903–913. [CrossRef] [PubMed]

123. Ning, Y.; Wang, X.; Zhang, P.; Anatoly, S.V.; Prakash, N.T.; Li, C.; Zhou, R.; Lammi, M.; Zhang, F.; Guo, X. Imbalance of dietary nutrients and the associated differentially expressed genes and pathways may play important roles in juvenile Kashin-Beck disease. *J. Trace Elem. Med. Biol.* **2018**, in press. [CrossRef] [PubMed]

© 2018 by the authors. Licensee MDPI, Basel, Switzerland. This article is an open access article distributed under the terms and conditions of the Creative Commons Attribution (CC BY) license (http://creativecommons.org/licenses/by/4.0/).

International Journal of
Molecular Sciences

Article

A G-Quadruplex Structure in the Promoter Region of *CLIC4* Functions as a Regulatory Element for Gene Expression

Mu-Ching Huang [1], I-Te Chu [2], Zi-Fu Wang [2], Steven Lin [3], Ta-Chau Chang [2] and Chin-Tin Chen [1,*]

[1] Department of Biochemical Science and Technology, National Taiwan University, Taipei 106, Taiwan; d02b22005@ntu.edu.tw
[2] Institute of Atomic and Molecular Sciences, Academia Sinica, Taipei 115, Taiwan; r04223204@ntu.edu.tw (I.-T.C.); f96223151@gmail.com (Z.-F.W.); tcchang@po.iams.sinica.edu.tw (T.-C.C.)
[3] Institute of Biological Chemistry, Academia Sinica, Taipei 115, Taiwan; stevenlin@gate.sinica.edu.tw
* Correspondence: chintin@ntu.edu.tw; Tel.: +886-2-3366-9487; Fax: +886-2-3366-2271

Received: 17 July 2018; Accepted: 6 September 2018; Published: 10 September 2018

Abstract: The differential transcriptional expression of *CLIC4* between tumor cells and the surrounding stroma during cancer progression has been suggested to have a tumor-promoting effect. However, little is known about the transcriptional regulation of *CLIC4*. To better understand how this gene is regulated, the promoter region of *CLIC4* was analyzed. We found that a high GC content near the transcriptional start site (TSS) might form an alternative G-quadruplex (G4) structure. Nuclear magnetic resonance spectroscopy (NMR) confirmed their formation in vitro. The reporter assay showed that one of the G4 structures exerted a regulatory role in gene transcription. When the G4-forming sequence was mutated to disrupt the G4 structure, the transcription activity dropped. To examine whether this G4 structure actually has an influence on gene transcription in the chromosome, we utilized the CRISPR/Cas9 system to edit the G4-forming sequence within the *CLIC4* promoter in the cell genome. The pop-in/pop-out strategy was adopted to isolate the precisely-edited A375 cell clone. In CRISPR-modified A375 cell clones whose G4 was disrupted, there was a decrease in the endogenous *CLIC4* messenger RNA (mRNA) expression level. In conclusion, we found that the G4 structure in the *CLIC4* promoter might play an important role in regulating the level of transcription.

Keywords: G-quadruplex; transcriptional regulation; promoter; CRISPR/Cas9

---

1. Introduction

Chloride intracellular channel 4, CLIC4, is a multifunctional protein. In addition to its diverse physiological functions [1–4], the differential expression of CLIC4 between cancer cells and the surrounding stroma has been reported in various tumor types [5]. During cancer development, CLIC4 is downregulated in cancer cells, and it is recognized as a suppressor for tumor growth [6]. In the tumor stroma, on the other hand, CLIC4 is often upregulated and it plays a critical role in myofibroblast transition [7]. The opposite was found as early as at the transcription level [5], and both mechanisms could promote tumor progression. However, how CLIC4 is regulated remains unknown.

The regulation of gene expression includes multiple mechanisms such as transcription factor binding [8] and epigenetic modification, which involves DNA methylation, histone modification, and microRNA interaction [9]. Moreover, the DNA and messenger RNA (mRNA) sequences that would adopt alternative secondary structures have significant impacts on the transcription and translation of gene expression. Although CLIC4 has been identified for nearly 30 years [10,11], only NANOG, p53, c-Myc, and SP-1 transcription factors have been reported to regulate CLIC4 expression. Genome-scale location analysis revealed that NANOG and SOX2 were bound to CLIC4 early in embryonic stem cells [12]. CLIC4 was elevated in p53- or c-Myc-overexpressing cells that induced apoptosis with direct binding to the *CLIC4* promoter [13,14] and SP-1 was involved in Ca^{2+}-induced keratinocyte differentiation [15]. Nevertheless, transcriptional regulation of the *CLIC4* promoter is still unclear.

Guanine-rich (G-rich) sequences containing four runs of G-tracts can adopt four-stranded structures, named the G-quadruplex (G4). Four guanine bases can assemble into a square planar structure via Hoogsteen-type hydrogen bonds, and this is called a G-quartet. A stack of G-quartets is stabilized by monovalent cations, such as potassium or sodium, and this forms a tetrahelical G4 structure. The loop sequences connecting G-tracts determine the types of G4 topology [16]. G4 architecture was first discovered in vitro more than 50 years ago [17]. The biological relevance of G4s then emerged since the late 1980s [18]. In the past few years, the existence of G4s in the human genome was directly visualized by using an antibody against the G4 structure [19], and fluorescent G4 probes [20].

G4s were found predominately in telomeres, and in the gene regulatory region of the human genome [21]. Particularly, a number of G4s have been recently found in eukaryotic promoter regions, including c-Myc, KRAS, PDGF, BCL-2, etc. [22–26]. According to current knowledge, G4 structures in promoter regions may influence transcription in both positive and negative ways. This also depends on the strand that G4 locates in, and the function of the proteins that are bound to the G4 structures. In addition, G4 formation could also affect protein binding. This is based on the function of the affected proteins; for example, whether the protein is a transcriptional activator or a repressor, and the proteins that stabilize or resolve the G4 structure, which could also result in different outcome [27]. Upon analyzing the *CLIC4* promoter, we found that high GC contents could form G4 structures. We further evaluated the effect of the G4 structure in regulating *CLIC4* gene expression.

2. Results

2.1. CLIC4 Promoter Analysis

To examine the regulatory region of the *CLIC4* promoter, a series of truncated *CLIC4* promoter sequences from −1700 to +285, relative to the transcription start site (TSS, +1) were cloned into the secreted embryonic alkaline phosphatase (SEAP) reporter vector pSEAP2-basic, and transfected into A375 cells to estimate the promoter activity of each region. As shown in Figure 1A, the longest construct p(−1700, +285) displayed the highest SEAP activity. The deletion of −1700 to −1340 at the promoter region resulted in a significant decrease in the SEAP activity. However, in p(−518, +285), the SEAP activity reversed to as high as p(−1700, +285), but further deletions of −518 to −125 resulted in a 50% drop in promoter activity. These results indicated that there are positive regulatory sites that are located in −1700 to −1340, and in −518 to −125, and negative regulatory sites that are located in the region between −1340 to −518.

During the reporter plasmid constructions with different promoter regions, we found that the PCR and sequencing reactions were hampered. Since high GC contents were found in the *CLIC4* promoter region, we argued that DNA secondary structure might exist to interfere with the PCR and the sequencing process. Therefore, this prompted us to analyze its promoter sequence with DNA secondary structure. The QGRS program [28] predicted that there was high G4 density in the region between −518 and +285. The promoter sequence of *CLIC4* p(−518, +285) is shown in Figure 1B.

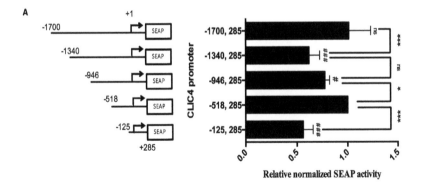

A

B

-518GATCGAAACGATCAGGAGCGTCAAAGGGAAAAACGAAGTCAGAGGCAGAAGAAAAAGGTGTTCCCCAGGCGAT
GAAAACTTTCCAGTAAGTTGTCTTGGAAAAGGCAACTCCGCGTAATAGAGGGACTTGGGGAGGTAGCCGCTGCGGGAA
AGGGCGATGAGTCCTGGGGTGCTGCAGCGGGGGAGTGGGTAGGAGGCGCCTCCTAGCTTCCCCAGGGCGGCCGAGGC
ATGAGGTCACCTGGCACAGCAACAGCAGCAGCTGCGGAGGACGCGCAGGCGCTGCGGCGGGAGGCAGTGTCAGGCTC
GGGCCCCTGGGGAATTCCTCCCCTCCTTTCCCCCCCGCACCAGTCCCCTAGAGAGGCGTACCCGGGTCCCGCCCGCCAGA
GCCTGAGCTTTGTACCGCCCACCAAGGCCACGCCCCCTACCAGCTGTGGGGCGCCGCAGAGGTCCTTTCTTCGGCTCCG
GAGCGGTCCCTCCGCCTTCCCCGCTCGAGGCCACGCCCCCGTGCTCCTGCCGCC+1TTATTTTCCCCGGAGAGTCCCGAGG
CGCCGCGCCTTGGCCCTGCCTACAGCCCGAGGCCCCGCCCCCGGCGCCCCTCCCAGCCGTTTGAAGCGGCTCGGGCTGC
GGCTGGCTCAGAGTGGCGCGGGGGGCGTGGGGCGGTGCTGAGGAGCTGAAGCCGTGGCCAGCTCGACGCCGGACAG
TCCAGCGAGCAGCACGGCGGGAACCGGCAGCCGGAGCAGTCCCGGAGCAGAAGCAGCAGCAGCAGCAGCCCTC
GCCGTTCGCGGAGCGCAGCCGAGCCGGC+285CATG

Figure 1. (**A**) A series of truncated *CLIC4* promoter regions were constructed in the pSEAP2-Basic vector, and the relative activities of the *CLIC4* promoter were compared to p(−518, +285). After transfection in A375 cells for 48 h, media were collected and a secreted embryonic alkaline phosphatase (SEAP) assay was performed. +1 indicates the transcription start site (TSS). Data are expressed as the means ± SD of three replicates. # $p < 0.05$, ### $p < 0.001$, as compared to p(−518, +285). * $p < 0.05$, *** $p < 0.001$ as compared to the adjacent promoter region. ns: non-significant difference; (**B**) Sequence of the *CLIC4* promoter (−518, +285). The underlined sequences are potential G-quadruplexes (PG4s).

2.2. Putative G-Quadruplexes in the CLIC4 Promoter

To further elucidate the formation of the G4 structure in the *CLIC4* promoter region, we examined three G-rich sequences with a higher potential to form G4 structures, and named them as putative G-quadruplex-1, -2, and -3 (PG4-1, -2 and -3), as shown in Figure 2A. PG4-1 and PG4-2 are located on −396 to −364, and −352 to −322 of the sense strand, respectively; PG4-3 is on +50 to +79 of the antisense strand. The location of each PG4 within the *CLIC4* promoter is presented in Figure 1B. Circular dichroism (CD) spectra and one-dimensional (1D) imino-proton nuclear magnetic resonance spectroscopy (NMR) experiments were conducted to examine whether the three PG4s were involved in DNA secondary structure formation. CD bands at ~265 and ~290 nm characterize the signature of G4 DNA. Different spectra represent different conformation of G4. As shown in Figure 2B, both absorption bands appeared in PG4-1 and PG4-2, and the only band near 295 nm was increased in PG4-3 by treatment of K$^+$ to induce G4 formation, implying the formation of G4 structure in all of these three PG4s. Furthermore, in NMR spectra (Figure 2C), multiple imino proton signals at 10.5~12 ppm were also found in all of these three PG4s in the presence of K$^+$, confirming the formation of Hoogsteen hydrogen bonding for the quadruplex structure. In addition, there were also signals shown near 13 ppm representing typical Watson-Crick hydrogen bonding, suggesting the existence of the hairpin structure in PG4-1 and PG4-3. These results indicated the existence of DNA secondary structures in the *CLIC4* promoter region.

To examine whether PG4s in the promoter region play an important role in regulating *CLIC4* transcription, we constructed SEAP reporter plasmids containing different lengths of the *CLIC4* promoter that excluded PG4 at a point from upstream of the promoter, to elucidate the biological significance of each PG4. As shown in Figure 2D, when the first and second PG4 (PG4-1 and PG4-2) were removed, there was no significant change in the SEAP activity, suggesting that the sequences or the structures of PG4-1 and PG4-2 were not critical for *CLIC4* transcription. The deletion of −321 to −125 resulted in a decreased of SEAP activity implying the positive regulatory site lies in −518 to −125 observed in Figure 1A had been narrowed down to this region. Since the significance of PG4-1 and PG4-2 has been excluded, the following studies were focused on verifying the role of PG4-3 in regulating *CLIC4* transcription.

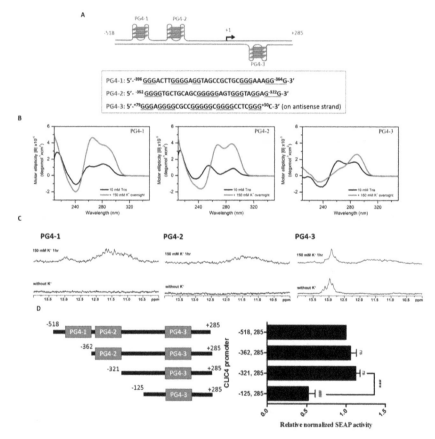

Figure 2. Putative G-quadruplexes (PG4s) in the *CLIC4* promoter (−518, +285) (**A**) Schematic representation of three PG4s and their sequences in *CLIC4* promoter. G-tracts that might participate in G4 formation were underlined; (**B**) The circular dichroism (CD) spectra of three PG4s in Tris-HCl buffer without (black line) and with (red line) 150 mM KCl for overnight; (**C**) The imino proton nuclear magnetic resonance spectroscopy (NMR) spectra of three PG4s in Tris-HCl buffer without (bottom panel) and with (top panel) 150 mM KCl for 1 h; (**D**) Progressive deletions of PG4s from the 5′-flanking regions in the *CLIC4* promoter were generated in the pSEAP2-Basic reporter plasmid. A375 cells were transfected with each reporter plasmid for 48 h. Media were collected and subjected to SEAP activity measurements. Data were expressed as means ± SD of three replicates. ### $p < 0.001$, as compared to p(−518, +285). *** $p < 0.001$ as compared to the adjacent promoter region. ns: non-significant difference.

2.3. PG4-3 Is Involved in Regulating CLIC4 Transcription

It has been shown that G4s in the promoter region could play a regulatory role for transcription in both positive and negative ways [27]. To elucidate the functional role of PG4-3 in regulating *CLIC4* transcription, we substituted G with T to disrupt the quadruplex structure. In the PG4-3 region, there are 29 bases (+79 to +50) comprising five G-tracts in the antisense strand of the *CLIC4* promoter. Three mutants were designed to disrupt the quadruplex formation. Mutant No. 1 was generated by changing 78G and 52G in the first and last G-tracts of the PG4-3 sequence to 78T and 52T, respectively. Mutant No. 1 was unable to form G4 as revealed by CD spectra with the absence of ~265 nm or ~290 nm bands in 150 mM K$^+$ solution (Figure 3A). The NMR signals of PG4-3 at 10.5–12 ppm were diminished in Mutant No. 1 (Figure 3B). Meanwhile, the transcription activity of the reporter construct containing Mutant No. 1 was mostly decreased (Figure 3C), indicating that PG4-3 did play an important role in regulating CLIC4 expression. To further verify the G4 structure of the PG4-3 in gene transcription, we further constructed another two PG4-3 mutants. Except for the replacement of 78G with 78T in the first G-tract, the second G-tract (+75 to +72) was replaced with T in Mutant No. 2. On the contrary, Mutant No. 3 was designed by replace G with T in the penultimate G-tracts (+61 to +58), and 52G in the last G-tract. As shown in Figure 3AB, CD and NMR signals of G4 were also negligible in these two mutants. In the reporter assay (Figure 3C), the transcription activity in Mutant No. 3 was decreased; however, there was no significant change of transcription activity in Mutant No. 2.

Despite the disappearance of the G4 NMR signal in the synthetic Mutant No. 2 oligonucleotide, the reporter construct containing Mutant No. 2 sequence still exhibited normal transcription activity. We therefore further examined this contradictory result between the reporter construct and the synthetic nucleotide. Except for the five G-tracts in PG4-3, we found another G-tract at +39 to +41 immediately adjacent to PG4-3. We argued that this additional G-tract might be incorporated with the remaining three G-tracts, and it may have participated in forming another G4 structure in the reporter construct when the first two G-tracts were mutated in Mutant No. 2. To address the possibility that the Mutant No. 2 reporter construct still contained a G4-forming sequence, the sequence from +67 to +39, which included three remaining G-tracts and the additional G-tract, was synthesized and analyzed by NMR. As shown in Figure 4A, there were appreciable signals of a quadruplex structure at the 10.5~12 ppm region, suggesting that G4 formation occurs in the sequence from +67 to +39. Therefore, we further constructed different reporter plasmids containing other mutation sites (Figure 4B). Mutant No. 4 harbors extra mutation sites: 58G, 59G to 58T, 59T respectively, compared to Mutant No. 2. Meanwhile, Mutant No. 5 was created by only mutating 58G, 59G to 58T, 59T, respectively. As shown in Figure 4C, the reporter activity was again decreased once the remaining G4-forming sequences were mutated. Furthermore, in order to understand whether different promoter lengths would affect G4 formation in the plasmid, we also compared the transcription activity in a longer promoter region. A mutation site for Mutant No. 5 was inserted in a reporter plasmid with the longest promoter *CLIC4* p(−1700, +285) that we had constructed. As shown in Figure 4D, a significant decrease of transcription activity was found in Mutant No. 5.

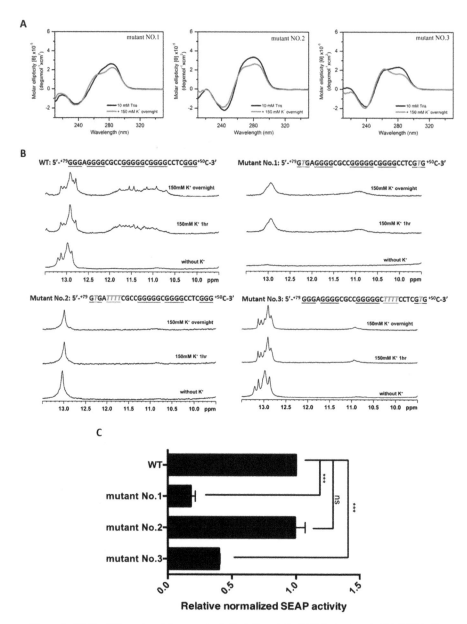

Figure 3. (**A**) The CD spectra of three mutants that disrupt the PG4-3 structure in Tris-HCl buffer without (black line) and with (red line) 150 mM KCl for overnight; (**B**) The imino proton NMR spectra of PG4-3 wild type (WT) and three mutants in Tris-HCl buffer without (**bottom** panel) or with 150 mM KCl for 1 h (**middle** panel) and overnight (**top** panel). G-tracts are underlined, and the mutation sites are marked in red italicized letters; (**C**) The effects of the mutations that disrupt the G4-3 structure. Each mutant of *CLIC4* p(−125, +285) was transfected in A375 cells. After 72 h, the media were collected for a SEAP assay. Data were expressed as the means ± SD of three replicates. *** $p < 0.001$, ns: non-significant difference as compared to WT control.

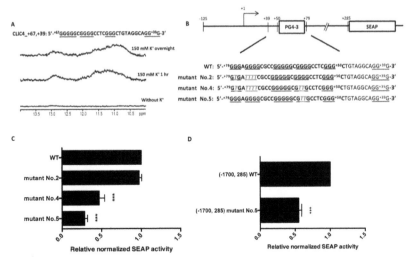

Figure 4. Further disruption of the remaining possible G4 structure formed in the Mutant No. 2 reporter plasmid. (**A**) NMR analysis of *CLIC4* +67 to +39, the possible G4-forming sequence in Mutant No. 2 in Tris-HCl buffer and 150 mM KCl, for 1 h and overnight; (**B**) Sequence of WT and mutants derived from Mutant No. 2—Mutants No. 4 and No. 5 contain the following G-tract at the 3′ end. The original PG4-3-forming region is shown in bold letters, G-tracts are underlined, and the mutated sequences in *CLIC4* p($-$125, +285) plasmids are marked in red italicized letters; (**C**) SEAP activity of *CLIC4* p($-$125, +285) mutants further disrupting the G4 forming motif in Mutant No. 2 were determined in A375 cells after transfection for 72 h. Data were expressed as the means $\pm$ SD of three replicates. *** $p < 0.001$ as compared to the WT. (**D**) SEAP activity of *CLIC4* p($-$1700, +285) mutant No. 5 in A375 cells after transfection for 72 h. Data were expressed as means $\pm$ SD of three replicates. *** $p < 0.001$ as compared to *CLIC4* p($-$1700, +285) WT.

To further examine the importance of the PG4-3 G4 structure in regulating *CLIC4* transcription, we designed another mutant that we named loop-3T, in which only the non-G-tract sites: 76A, 70G, and 56C in the loops were mutated, respectively, to 76T, 70T, and 55T, which would not destroy G4-forming elements. The NMR result in Figure 5A showed that the G4 signals were retained in the mutant loop-3T. The reporter activity was also similar to the PG4-3 wild-type (WT) control (Figure 5B). These results indicate that the PG4-3 G4 structure within the promoter region plays an important role in regulating *CLIC4* transcription.

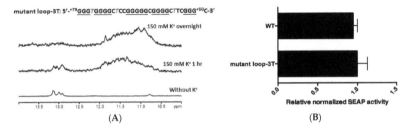

(A)   (B)

Figure 5. Strengthening of the PG4-3 structure. (**A**) The imino proton NMR spectra of the mutant *CLIC4* loop-3T that did not affect the G4-forming motif. G-tracts are underlined and the mutation sites are marked in red italicized letters; (**B**) The SEAP activity in A375 was determined after 72 h of transfection of *CLIC4* p($-$125, +285) WT and the mutant loop-3T reporter plasmid. Data are expressed as the means $\pm$ SD of three replicates.

To elucidate the role of PG4-3 in regulating *CLIC4* expression, we used software to predict the possible candidate which could binds to the nucleotide sequences of PG4-3. SP1 and MAZ are the two candidate genes predicted to bind on this sequence and also have been reported to bind on G4 structure [29,30]. However, we found that knockdown SP1 or MAZ did not affect *CLIC4* mRNA expression nor transcription activity (Figure S1), indicating these TFs were not the key proteins affecting *CLIC4* transcription.

2.4. PG4-3 Acts as a Regulatory Element in the CLIC4 Promoter Region of the Cell Chromosome

In order to verify whether this G4 structure actually has a regulatory function in the cell chromosome, we managed to disrupt the PG4-3 sequence in the *CLIC4* gene of the cell genome by using the CRISPR/Cas9 system. *CLIC4* is located on chromosome 1. The A375 cell line that was used in this study is hypotriploid, with three copies of chromosome 1 where the *CLIC4* gene is located. With regard to this fact, the two steps of editing: the pop-in/pop-out strategy developed by Cech and colleagues [31] was adopted with some modifications to assure precise editing by replacing the wild-type sequence (+79 to +50) with the Mutant No. 3 sequence in each chromosome. The workflow is illustrated in Figure 6A. In the pop-in step, the fluorescence markers were integrated into the cell genome by homology-directed repair (HDR). One of cell clones that co-expressed the three fluorescence markers with the expected genome size was then subjected to the pop-out step, in which the fluorescence markers were excised out and specifically repaired into the Mutant No. 3 sequence in cells that underwent HDR; therefore, triple-negative cells were isolated. In the end, two cell clones, HDR2 #90 and HDR2 #101, carrying PG4-3 Mutant No. 3 in the endogenous *CLIC4* promoter region, were generated. The representative sequencing results for *CLIC4* promoter region covering PG4-3 site of each cell clone could be found in supplementary folder (HDR2 single cell clones sequencing data). A lower level of endogenous mRNA expression was found in genome-edited cell clones carrying Mutant No. 3 that disrupted the PG4-3 structure (Figure 6B), indicated that PG4-3 in the promoter region of cell chromosome is important for *CLIC4* transcription.

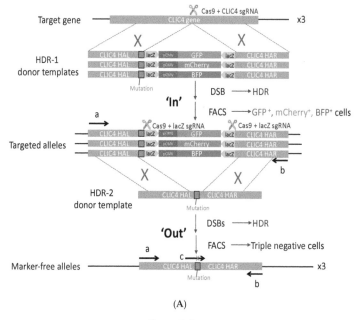

(A)

Figure 6. *Cont.*

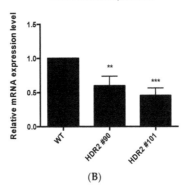

(B)

Figure 6. Mutation of the PG4-3 sequence at the endogenous *CLIC4* promoter. (**A**) Schematic diagram of the pop-in ('In') and pop-out ('Out') steps taken to modify PG4-3 to Mutant No. 3 in the *CLIC4* promoter. Briefly, in the pop-in step, CRISPR-Cas9 targeting by *CLIC4* sgRNA occurred next to the mutated site, and the HDR-1 donor templates consisting of a 1 kb sequence of *CLIC4* homology arms left and right (HAL and HAR, respectively), was interrupted by a fluorescence cassette: green fluorescent protein (*GFP*), *mCherry*, or blue fluorescent protein (*BFP*) gene driven by a CMV promoter and tagged with a lacZ sequence at both ends were introduced into the A375 cells. The lacZ sequence does not exist in the human genome and was later used in the pop-out step. Fluorescence-activated cell sorting (FACS) was used to isolate cells co-expressing the three fluorescence markers; primers 'a' and 'b' were used to check the genomic DNA size. In the pop-out step, the fluorescence markers were excised by two double strand breaks (DSBs) at the lacZ sites. The DSBs were repaired using the HDR-2 donor template. Loss of fluorescence expression in the cells (triple negative cells) were isolated. A TaqMan probe 'c', specifically purposed for recognizing the Mutant No. 3 sequence, was used to select clones for further sequencing of the PCR products that were generated with primers 'a' and 'b'. Cleavage points are indicated by scissors; grey X represents homologous recombination; mutation site of Mutant No. 3 is shown as red box; arrow of a, b and c indicates the direction of synthesis; (**B**) *CLIC4* messenger RNA (mRNA) expression level in A375 with the genome edited to the Mutant No. 3 sequence. Total RNA was extracted for RT-PCR and real-time qPCR to analyze the *CLIC4* mRNA expression level. Data are expressed as the means ± SD of three replicates. ** $p < 0.01$, *** $p < 0.001$ as compared to A375 WT cells.

3. Discussion

Several transcriptional factors in regulating *CLIC4* expression have been reported. Comparing the binding sites relative to the putative TSS in previous studies [13–15] with the known *CLIC4* mRNA TSS in the NCBI Reference Sequence, NM_013943.2, we found about 500 bp in differences. The promoter region from −500 to the TSS, and even to the translation start site of *CLIC4* have, in fact, not been well studied. This region is full of GC-rich sequences, implying the possible existence of DNA secondary structures. In this study, we demonstrated that the putative G-quadruplexes could be found in this high-GC-content region, and we further showed that this G4 structure did play an important role in the *CLIC4* promoter. Accordingly, we speculated that this secondary structure might cause past difficulties in amplifying and sequencing this promoter region, which might have impeded recent progress with studying the transcription mechanisms of *CLIC4*.

To investigate the effect of the G4 structure on the promoter with regard to transcription, studies have most commonly performed plasmid-based reporter assays. In this study, we first used a reporter assay to determine the significance of the PG4-3 structure in the *CLIC4* promoter, for its transcription. In addition, when the mutation sites that disrupted the G4 structure were placed in longer *CLIC4* promoter sequences within the reporter plasmid, this revealed similar results. Furthermore, it is known that chromatin status affects G4 formation [32]. Although the plasmid that was transfected into eukaryotic cells could be packed by histones and other proteins to become a nucleosome-like structure [33], it is not known whether it retained a similar status as a cellular chromosome, which may have had an impact on G4 formation. To this end, we then directly edited the promoter sequence in the cell chromosome by the CRISPR/Cas9 system, and we provide in vivo evidence to support the positive regulatory effect of PG4-3 on *CLIC4* transcription.

PG4-3 was on the anti-sense strand, which served as the template for RNA polymerase (RNAP)-mediated transcription. With regard to this, the formation of G4 could be the obstacle for RNAP, and the disruption of the G4 structure would result in the upregulation of transcriptional activity. However, as shown in Figure 3, we found the opposite phenomenon, with disruption of the G4-forming sequence leading to a decrease or no significant change in transcriptional activity. Therefore, we hypothesized that this PG4 might provide a binding site for protein(s) that favor *CLIC4* transcription or that prevent the binding of the repressor. SP1 and MAZ were two candidates that have been tested. However, we found that these two factors did not play the substantial role in regulating *CLIC4* expression. Meanwhile, in our preliminary EMSA studies, some unidentified proteins had increased binding onto the probes of Mutants 1 and 3 than that of the wild-type, suggesting the possible binding of repressors at the PG4-3 region.

There are five G-tracts in the originally identified PG4-3 sequence. In Mutant No. 2, we noticed that the additional G-tract might incorporate in the G4 formation when the first two G-tracts were mutated. The G4-forming sequence with a fifth G-tract has been observed in many oncogene promoters, and the 'spare tires' hypothesis was proposed [34]. For instance, guanine in G4 is susceptible to oxidation and has been shown to affect G4 stability [35]. The G4 in the vascular endothelial growth factor (*VEGF*) and endonuclease III-like protein 1 (*NTHL1*) promoters were modifiable by 8-oxo-7,8-dihydroguanine (OG) [34,36], and this caused the instability of the original G4 structure that was formed by the first four 4 G-tracts. After that, the fifth G-tract was able to act as a spare tire to maintain the G4 fold, and to allow the repair of DNA damage. Coincidentally, *CLIC4* is also a gene that can be regulated by reactive oxygen species (ROS)-induced oxidative stress. PDT-induced ROS downregulated *CLIC4* transcription [37], while TGF-β-induced ROS upregulated *CLIC4* transcription [38]. Whether OG modification take place on PG4-3 and affects G4 stability such that that the other two additional G-tracts can take part in G4 formation, deserves further investigation.

4. Materials and Methods

4.1. Cell Culturing

Human melanoma A375 cells (American Type Culture Collection, Manassas, VA, USA) were cultured in Dulbecco's modified Eagle's medium (DMEM) supplemented with 10% fetal bovine serum (FBS), and grown at 37 °C under 5% CO_2.

4.2. Circular Dichroism (CD) Spectroscopy

CD experiments were conducted using a spectropolarimeter (J-815, Jasco, Tokyo, Japan) with a bandwidth of 2 nm at a scan speed of 50 nm/min and a step resolution of 0.2 nm over the spectral range of 210–350 nm. The DNA sample concentrations were 4 µM in 10 mM Tris-HCl (pH 7.5), and a stock solution of 3 M KCl (Sigma-Aldrich, St. Louis, MO, USA) was added to the DNA samples to reach a final K^+ concentration of 150 mM. The observed signals were baseline subtracted.

4.3. NMR

The unlabeled oligonuleotides synthesized by Bio Basic (Markham, ON, Canada) were prepared to 100 µM in 10 mM Tris-HCl (pH 7.5) with or without 150 mM KCl, followed by denaturing at 95 °C for 5 min and slowly annealed to 25 °C. The strand concentrations of the NMR samples were 100 µM containing 10% D_2O in 10 mM Tris-HCl (pH 7.5) or 150 mM K^+ conditions with an internal reference of 0.01 mM DSS (4,4-dimethyl-4-silapentane-1-sulfonic acid), and they were analyzed by Bruker AVIII (Billerica, MA, USA) 500 MHz spectrometers equipped with a prodigy probehead, and on a Bruker AVIII 800 MHz NMR spectrometer equipped with a cryoprobe at 25 °C. 1D imino proton NMR spectra were recorded using a WATERGATE for water suppression.

4.4. Reporter Assay

The *CLIC4* sequence from -1700 to 285 relative to the transcription start site (+1) was generated from genomic DNA of A375 cells, and cloned into the pSEAP2-Basic vector (Clontech, Mountain View, CA, USA). After transfection in A375 cells with TurboFect (Thermo Scientific™, Waltham, MA, USA) for 48 to 72 h, the culture medium was collected and analyzed for SEAP activity by measurement of the hydrolysis of p-nitrophenyl phosphate (pNpp) with a spectrophometer at OD_{405}. MTT assay was used for the normalization of cell numbers.

4.5. CRISPR/Cas9

CLIC4 sgRNA targeting near *CLIC4* PG4-3 was selected and synthesized by in vitro transcription. HDR-1 donor templates containing different fluorescence cassettes: *GFP*, mCherry, and *BFP*, driven by a CMV promoter with a lacZ sequence on both ends (kindly provided by Dr. Steve Lin) flanking a 1 kb sequence of *CLIC4* homology arms upstream and downstream of the PG4-3 Mutant No. 3 mutation site were generated by Gibson assembly. Cas9, *CLIC4* sgRNA, and three fluorescence donor templates were introduced to A375 cells by nucleofection with the 4D-Nucleofector™ system and SF kit (Lonza, Basel, Switzerland) under the FF-120 program. Single cells co-expressing three fluorescence markers were isolated by FACS for clonal expansion, and the genomic sizes of the sequences containing the fluorescence cassettes were confirmed by PCR. In the pop-out step, Cas9, sgRNA targeting the lacZ sites, and the HDR-2 donor template only containing the *CLIC4* homology arms with the mutation sites, were again nucleofected in the pop-in cell clone. Single cells without fluorescence were sorted by FACS. The genomic DNA edited into the Mutant No. 3 sequence was analyzed by PCR and Custom TaqMan® Gene Expression Assays, SM ID: APFVMGD, which was designed to specifically anneal to the *CLIC4* mutation site. Cell clones harboring the mutant sequence were further confirmed by sequencing.

4.6. Real-Time PCR Analysis

Total RNA was extracted using TRIzol reagent (Invitrogen, Carlsbad, CA, USA) following the manufacturer's instructions accordingly. A total of 1 µg RNA was used to synthesize complementary DNA (cDNA) by reverse transcription. The cDNA product was used as a template for real-time PCR analysis using the ABI Fast SYBR® Green Master Mix Kit (Thermo Fisher Scientific, Waltham, MA, USA) with the ABI StepOne system (Thermo Fisher Scientific). The primer sequences were as follows: CLIC4 (sense), 5'-GCAGTGATGGTGAAAGCATAG-3'; CLIC4 (anti-sense), 5'-TATAAATGGTGGGTGGGTCC-3'; GAPDH (sense), 5'-GACCACAGTCCATGCCATCA-3'; GAPDH (anti-sense), 5'-GTCCACCACCCTGTTGCTGTA-3'.

4.7. Statistical Analysis

All results were obtained from three independent experiments, and each value was expressed as the mean $\pm$ SD. The two-tailed Student's *t*-test was used to compare the differences between pairs of means. $p < 0.05$ was considered significant.

5. Conclusions

The G4 structure formed by PG4-3 in the *CLIC4* promoter region may act as a regulatory element in regulating *CLIC4* gene transcription, as shown in the reporter assay, as well as in the CRISPR-modified A375 cell clone with mutated PG4-3.

Supplementary Materials: Supplementary materials can be found at http://www.mdpi.com/1422-0067/19/9/2678/s1. Figure S1: Regulation of MAZ and SP1. Supplementary Folder: HDR2 single cell clones sequencing data. The representative sequencing results of HDR2 #90 and HDR2 #101 for CLIC4 promoter region covering PG4-3 site.

Author Contributions: M.-C.H. Participated in the design of the experiment and carried out the work. Z.-F.W. and I.-T.C. participated in analyzing the DNA secondary structure. S.L. participated in the design of the CRISPR experiment. T.-C.C. participated in the study of the G4 structure. All of the authors read and approved the final manuscript. C.-T.C. conceived the study, participated in its design, and coordination and finalized the draft of the manuscript.

Funding: Financial support in the authors' laboratories were mainly given by the Ministry of Science and Technology, Taiwan (MOST-104-2320-B-002-041 and MOST 105-2320-B-002-056).

Conflicts of Interest: The authors declare no conflict of interest.

References

1. Berryman, M.A.; Goldenring, J.R. CLIC4 is enriched at cell-cell junctions and colocalizes with AKAP350 at the centrosome and midbody of cultured mammalian cells. *Cell Motil. Cytoskel.* **2003**, *56*, 159–172. [CrossRef] [PubMed]

2. Suh, K.S.; Mutoh, M.; Nagashima, K.; Fernandez-Salas, E.; Edwards, L.E.; Hayes, D.D.; Crutchley, J.M.; Marin, K.G.; Dumont, R.A.; Levy, J.M.; et al. The organellular chloride channel protein CLIC4/mtCLIC translocates to the nucleus in response to cellular stress and accelerates apoptosis. *J. Biol. Chem.* **2004**, *279*, 4632–4641. [CrossRef] [PubMed]

3. Littler, D.R.; Harrop, S.J.; Goodchild, S.C.; Phang, J.M.; Mynott, A.V.; Jiang, L.; Valenzuela, S.M.; Mazzanti, M.; Brown, L.J.; Breit, S.N.; et al. The enigma of the CLIC proteins: Ion channels, redox proteins, enzymes, scaffolding proteins? *FEBS Lett.* **2010**, *584*, 2093–2101. [CrossRef] [PubMed]

4. Shukla, A.; Malik, M.; Cataisson, C.; Ho, Y.; Friesen, T.; Suh, K.S.; Yuspa, S.H. TGF-beta signalling is regulated by Schnurri-2-dependent nuclear translocation of CLIC4 and consequent stabilization of phospho-Smad2 and 3. *Nat. Cell Biol.* **2009**, *11*, 777–784. [CrossRef] [PubMed]

5. Suh, K.S.; Crutchley, J.M.; Koochek, A.; Ryscavage, A.; Bhat, K.; Tanaka, T.; Oshima, A.; Fitzgerald, P.; Yuspa, S.H. Reciprocal modifications of CLIC4 in tumor epithelium and stroma mark malignant progression of multiple human cancers. *Clin. Cancer Res.* **2007**, *13*, 121–131. [CrossRef] [PubMed]

6. Suh, K.S.; Malik, M.; Shukla, A.; Ryscavage, A.; Wright, L.; Jividen, K.; Crutchley, J.M.; Dumont, R.A.; Fernandez-Salas, E.; Webster, J.D.; et al. CLIC4 is a tumor suppressor for cutaneous squamous cell cancer. *Carcinogenesis* **2012**, *33*, 986–995. [CrossRef] [PubMed]

7. Ronnov-Jessen, L.; Villadsen, R.; Edwards, J.C.; Petersen, O.W. Differential expression of a chloride intracellular channel gene, CLIC4, in transforming growth factor-beta1-mediated conversion of fibroblasts to myofibroblasts. *Am. J. Pathol.* **2002**, *161*, 471–480. [CrossRef]

8. Todeschini, A.L.; Georges, A.; Veitia, R.A. Transcription factors: Specific DNA binding and specific gene regulation. *Trends Genet.* **2014**, *30*, 211–219. [CrossRef] [PubMed]

9. Gibney, E.R.; Nolan, C.M. Epigenetics and gene expression. *Heredity* **2010**, *105*, 4–13. [CrossRef] [PubMed]

10. Edwards, J.C. A novel p64-related Cl⁻ channel: Subcellular distribution and nephron segment-specific expression. *Am. J. Physiol.* **1999**, *276*, F398–F408. [CrossRef] [PubMed]

11. Chuang, J.Z.; Milner, T.A.; Zhu, M.; Sung, C.H. A 29 kDa intracellular chloride channel p64H1 is associated with large dense-core vesicles in rat hippocampal neurons. *J. Neurosci.* **1999**, *19*, 2919–2928. [CrossRef] [PubMed]

12. Boyer, L.A.; Lee, T.I.; Cole, M.F.; Johnstone, S.E.; Levine, S.S.; Zucker, J.P.; Guenther, M.G.; Kumar, R.M.; Murray, H.L.; Jenner, R.G.; et al. Core transcriptional regulatory circuitry in human embryonic stem cells. *Cell* **2005**, *122*, 947–956. [CrossRef] [PubMed]

13. Fernandez-Salas, E.; Suh, K.S.; Speransky, V.V.; Bowers, W.L.; Levy, J.M.; Adams, T.; Pathak, K.R.; Edwards, L.E.; Hayes, D.D.; Cheng, C.; et al. mtCLIC/CLIC4, an organellular chloride channel protein, is increased by DNA damage and participates in the apoptotic response to p53. *Mol. Cell. Biol.* **2002**, *22*, 3610–3620. [CrossRef] [PubMed]

14. Shiio, Y.; Suh, K.S.; Lee, H.; Yuspa, S.H.; Eisenman, R.N.; Aebersold, R. Quantitative proteomic analysis of myc-induced apoptosis: A direct role for Myc induction of the mitochondrial chloride ion channel, mtCLIC/CLIC4. *J. Biol. Chem.* **2006**, *281*, 2750–2756. [CrossRef] [PubMed]

15. Suh, K.S.; Mutoh, M.; Mutoh, T.; Li, L.; Ryscavage, A.; Crutchley, J.M.; Dumont, R.A.; Cheng, C.; Yuspa, S.H. CLIC4 mediates and is required for Ca^{2+}-induced keratinocyte differentiation. *J. Cell Sci.* **2007**, *120*, 2631–2640. [CrossRef] [PubMed]

16. Burge, S.; Parkinson, G.N.; Hazel, P.; Todd, A.K.; Neidle, S. Quadruplex DNA: Sequence, topology and structure. *Nucleic Acids Res.* **2006**, *34*, 5402–5415. [CrossRef] [PubMed]

17. Gellert, M.; Lipsett, M.N.; Davies, D.R. Helix formation by guanylic acid. *Proc. Natl. Acad. Sci. USA* **1962**, *48*, 2013–2018. [CrossRef] [PubMed]

18. Henderson, E.; Hardin, C.C.; Walk, S.K.; Tinoco, I.; Blackburn, E.H. Telomeric DNA oligonucleotides form novel intramolecular structures containing guanine·guanine base pairs. *Cell* **1987**, *51*, 899–908. [CrossRef]

19. Biffi, G.; Tannahill, D.; McCafferty, J.; Balasubramanian, S. Quantitative visualization of DNA G-quadruplex structures in human cells. *Nat. Chem.* **2013**, *5*, 182–186. [CrossRef] [PubMed]

20. Tseng, T.Y.; Chien, C.H.; Chu, J.F.; Huang, W.C.; Lin, M.Y.; Chang, C.C.; Chang, T.C. Fluorescent probe for visualizing guanine-quadruplex DNA by fluorescence lifetime imaging microscopy. *J. Biomed. Opt.* **2013**, *18*, 101309. [CrossRef] [PubMed]

21. Du, Z.; Zhao, Y.; Li, N. Genome-wide colonization of gene regulatory elements by G4 DNA motifs. *Nucleic Acids Res.* **2009**, *37*, 6784–6798. [CrossRef] [PubMed]

22. Qin, Y.; Hurley, L.H. Structures, folding patterns, and functions of intramolecular DNA G-quadruplexes found in eukaryotic promoter regions. *Biochimie* **2008**, *90*, 1149–1171. [CrossRef] [PubMed]

23. Yang, D.; Hurley, L.H. Structure of the biologically relevant G-quadruplex in the c-MYC promoter. *Nucleosides Nucleotides Nucleic Acids* **2006**, *25*, 951–968. [CrossRef] [PubMed]

24. Paramasivam, M.; Membrino, A.; Cogoi, S.; Fukuda, H.; Nakagama, H.; Xodo, L.E. Protein hnRNP A1 and its derivative Up1 unfold quadruplex DNA in the human KRAS promoter: Implications for transcription. *Nucleic Acids Res.* **2009**, *37*, 2841–2853. [CrossRef] [PubMed]

25. Qin, Y.; Rezler, E.M.; Gokhale, V.; Sun, D.; Hurley, L.H. Characterization of the G-quadruplexes in the duplex nuclease hypersensitive element of the PDGF-A promoter and modulation of PDGF-A promoter activity by TMPyP4. *Nucleic Acids Res.* **2007**, *35*, 7698–7713. [CrossRef] [PubMed]

26. Dai, J.; Dexheimer, T.S.; Chen, D.; Carver, M.; Ambrus, A.; Jones, R.A.; Yang, D. An intramolecular G-quadruplex structure with mixed parallel/antiparallel G-strands formed in the human BCL-2 promoter region in solution. *J. Am. Chem. Soc.* **2006**, *128*, 1096–1098. [CrossRef] [PubMed]

27. Bochman, M.L.; Paeschke, K.; Zakian, V.A. DNA secondary structures: Stability and function of G-quadruplex structures. *Nat. Rev. Genet.* **2012**, *13*, 770–780. [CrossRef] [PubMed]

28. Kikin, O.; D'Antonio, L.; Bagga, P.S. QGRS Mapper: A web-based server for predicting G-quadruplexes in nucleotide sequences. *Nucleic Acids Res.* **2006**, *34*, W676–W682. [CrossRef] [PubMed]

29. Cogoi, S.; Shchekotikhin, A.E.; Xodo, L.E. HRAS is silenced by two neighboring G-quadruplexes and activated by MAZ, a zinc-finger transcription factor with DNA unfolding property. *Nucleic Acids Res.* **2014**, *42*, 8379–8388. [CrossRef] [PubMed]

30. Raiber, E.A.; Kranaster, R.; Lam, E.; Nikan, M.; Balasubramanian, S. A non-canonical DNA structure is a binding motif for the transcription factor SP1 in vitro. *Nucleic Acids Res.* **2012**, *40*, 1499–1508. [CrossRef] [PubMed]

31. Xi, L.; Schmidt, J.C.; Zaug, A.J.; Ascarrunz, D.R.; Cech, T.R. A novel two-step genome editing strategy with CRISPR-Cas9 provides new insights into telomerase action and TERT gene expression. *Genome Biol.* **2015**, *16*, 231. [CrossRef] [PubMed]

32. Hansel-Hertsch, R.; Beraldi, D.; Lensing, S.V.; Marsico, G.; Zyner, K.; Parry, A.; Di Antonio, M.; Pike, J.; Kimura, H.; Narita, M.; et al. G-quadruplex structures mark human regulatory chromatin. *Nat. Genet.* **2016**, *48*, 1267–1272. [CrossRef] [PubMed]

33. Mladenova, V.; Mladenov, E.; Russev, G. Organization of Plasmid DNA into Nucleosome-Like Structures after Transfection in Eukaryotic Cells. *Biotechnol. Biotechnol. Equip.* **2009**, *23*, 1044–1047. [CrossRef]

34. Fleming, A.M.; Zhou, J.; Wallace, S.S.; Burrows, C.J. A Role for the Fifth G-Track in G-Quadruplex Forming Oncogene Promoter Sequences during Oxidative Stress: Do These "Spare Tires" Have an Evolved Function? *ACS Cent. Sci.* **2015**, *1*, 226–233. [CrossRef] [PubMed]

35. Zhou, J.; Fleming, A.M.; Averill, A.M.; Burrows, C.J.; Wallace, S.S. The NEIL glycosylases remove oxidized guanine lesions from telomeric and promoter quadruplex DNA structures. *Nucleic Acids Res.* **2015**, *43*, 4039–4054. [CrossRef] [PubMed]

36. Fleming, A.M.; Ding, Y.; Burrows, C.J. Oxidative DNA damage is epigenetic by regulating gene transcription via base excision repair. *Proc. Natl. Acad. Sci. USA* **2017**, *114*, 2604–2609. [CrossRef] [PubMed]

37. Chiang, P.C.; Chou, R.H.; Chien, H.F.; Tsai, T.; Chen, C.T. Chloride intracellular channel 4 involves in the reduced invasiveness of cancer cells treated by photodynamic therapy. *Lasers Surg. Med.* **2013**, *45*, 38–47. [CrossRef] [PubMed]

38. Yao, Q.; Qu, X.; Yang, Q.; Wei, M.; Kong, B. CLIC4 mediates TGF-beta1-induced fibroblast-to-myofibroblast transdifferentiation in ovarian cancer. *Oncol. Rep.* **2009**, *22*, 541–548. [PubMed]

 © 2018 by the authors. Licensee MDPI, Basel, Switzerland. This article is an open access article distributed under the terms and conditions of the Creative Commons Attribution (CC BY) license (http://creativecommons.org/licenses/by/4.0/).

Article

PR/SET Domain Family and Cancer: Novel Insights from The Cancer Genome Atlas

Anna Sorrentino [1,2,†], **Antonio Federico** [2,3,†], **Monica Rienzo** [4], **Patrizia Gazzerro** [5],
Maurizio Bifulco [6], **Alfredo Ciccodicola** [2,3], **Amelia Casamassimi** [1,*] **and Ciro Abbondanza** [1,*]

[1] Department of Precision Medicine, University of Campania "Luigi Vanvitelli", Via L. De Crecchio,
     80138 Naples, Italy; anna.sorrentino@unicampania.it
[2] Department of Science and Technology, University of Naples "Parthenope", 80143 Naples, Italy;
     antonio.federico@igb.cnr.it (A.F.); alfredo.ciccodicola@igb.cnr.it (A.C.)
[3] Institute of Genetics and Biophysics "Adriano Buzzati Traverso", CNR, 80131 Naples, Italy
[4] Department of Environmental, Biological, and Pharmaceutical Sciences and Technologies,
     University of Campania "Luigi Vanvitelli", 81100 Caserta, Italy; monica.rienzo@unicampania.it
[5] Department of Pharmacy, University of Salerno, 84084 Salerno, Italy; pgazzerro@unisa.it
[6] Department of Molecular Medicine and Medical Biotechnologies, University of Naples "Federico II",
     80131 Naples, Italy; maubiful@unina.it
*   Correspondence: amelia.casamassimi@unicampania.it (A.C.); ciro.abbondanza@unicampania.it (C.A.);
    Tel.: +39-081-566-7579 (A.C.); +39-081-566-7568 (C.A.)
†   These two authors contributed equally to this work.

Received: 30 September 2018; Accepted: 17 October 2018; Published: 19 October 2018

Abstract: The PR/SET domain gene family (PRDM) encodes 19 different transcription factors that share a subtype of the SET domain [Su(var)3-9, enhancer-of-zeste and trithorax] known as the PRDF1-RIZ (PR) homology domain. This domain, with its potential methyltransferase activity, is followed by a variable number of zinc-finger motifs, which likely mediate protein–protein, protein–RNA, or protein–DNA interactions. Intriguingly, almost all PRDM family members express different isoforms, which likely play opposite roles in oncogenesis. Remarkably, several studies have described alterations in most of the family members in malignancies. Here, to obtain a pan-cancer overview of the genomic and transcriptomic alterations of *PRDM* genes, we reanalyzed the Exome- and RNA-Seq public datasets available at The Cancer Genome Atlas portal. Overall, *PRDM2*, *PRDM3/MECOM*, *PRDM9*, *PRDM16* and *ZFPM2/FOG2* were the most mutated genes with pan-cancer frequencies of protein-affecting mutations higher than 1%. Moreover, we observed heterogeneity in the mutation frequencies of these genes across tumors, with cancer types also reaching a value of about 20% of mutated samples for a specific *PRDM* gene. Of note, *ZFPM1/FOG1* mutations occurred in 50% of adrenocortical carcinoma patients and were localized in a hotspot region. These findings, together with OncodriveCLUST results, suggest it could be putatively considered a cancer driver gene in this malignancy. Finally, transcriptome analysis from RNA-Seq data of paired samples revealed that transcription of *PRDMs* was significantly altered in several tumors. Specifically, *PRDM12* and *PRDM13* were largely overexpressed in many cancers whereas *PRDM16* and *ZFPM2/FOG2* were often downregulated. Some of these findings were also confirmed by real-time-PCR on primary tumors.

Keywords: PRDM gene family; TCGA data analysis; somatic mutations; transcriptome profiling; human malignancies

1. Introduction

The positive regulatory domain (PRDM) gene family, a subfamily of Kruppel-like zinc finger gene products, currently includes 19 members in humans [1–4]. The protein products of this family share a conserved N-terminal PR (PRDI-BF1-RIZ1 homologous) domain, which is structurally and functionally similar to the catalytic SET domain that defines a large group of histone methyltransferases (HMTs) [5–8]. So far, enzymatic activity has been experimentally demonstrated for only a few family members; otherwise, PRDM proteins (PRDMs) lacking intrinsic enzymatic activity are able to recruit histone-modifying enzymes to mediate their function.

The PR domain is generally positioned at the protein N-terminal region and, with the exception of PRDM11, it is followed by repeated zinc fingers toward the C-terminus, potentially mediating sequence-specific DNA or RNA binding and protein–protein interactions [5–9] (Figure S1). Importantly, PRDMs have been established to tether transcription factors to target gene promoters by recognition of a specific DNA consensus sequence [10] or acting as non-DNA binding cofactors [11,12]. These features give PRDMs the ability to drive cell differentiation and to specify cell fate choice and, thus, contribute to many developmental processes [5,7,12–14].

A common characteristic of most *PRDM* genes is to express two main molecular variants, one lacking the PR domain (PR-minus isoform) but otherwise identical to the other PR-containing product (PR-plus isoform). These two isoforms, generated either by alternative splicing or alternative use of different promoters [7,8,15], play opposite roles, particularly in cancer. The full-length product PR-plus usually acts as a tumor suppressor, whereas the short isoform functions as an oncogene. This bivalent behavior has been tagged as 'yin-yang'. The imbalance in favor of the PR-minus is found in many human malignancies and it can be due to inactivating mutations or silencing of the complete form and/or to increased expression of the PR-minus form [8].

PRDM1 and *PRDM2* use alternative promoters to generate short isoforms lacking the PR domain, which show oncogenic properties. Increased levels of the short isoforms were reported in various cancer cell lines. A similar PR-less product was also described for *PRDM3/MECOM* (MDS1 and EVI1 complex locus), *PRDM16*, and *PRDM6*, thus suggesting that this 'yin-yang' expression pattern and its functional implications could be a hallmark of most, if not all, PRDMs [8].

Furthermore, several studies have described alterations (both mutations and/or gene expression changes) of most PRDMs in solid tumors and/or hematological malignancies [8]. For instance, frameshift mutations of microsatellite repeats within the *PRDM2* coding region are frequent events in various cancers. A recent study has described a frameshift mutation in the C-terminal region of PRDM2, affecting the (A)9 repeat within exon 8, as a microsatellite indel driver hotspot and as a driver mutation in microsatellite instability (MSI) colorectal cancer [15,16]. Notably, a similar frameshift mutation was found to occur in a mononucleotide repeat (A7) of *PRDM3/MECOM* gene in this cancer type [17]. Intriguingly, recent findings also indicate that PRDM2 methyltransferase is required for BRCA1-mediated genome maintenance [15,18]. Moreover, a significant reduction of *PRDM2* expression was observed in high-grade gliomas [19], and forced expression of *PRDM2* in glioma cell lines inhibits cell proliferation and increases apoptosis. This evidence strongly suggests a possible tumor suppressive role for PRDM2 [19]. Interestingly, PRDM9 HMTase activity is essential for meiotic DNA double-strand break formation at its binding sites [20,21]. Moreover, both PRDM1 and PRDM5 negatively modulate WNT/β-catenin signaling, a pathway involved in the occurrence of several cancers, including glioma and colorectal cancer [22,23].

This evidence suggests that PRDMs are involved in human cancer through modulation of several processes, such as epigenetic modifications, genetic reprogramming, inflammation, and metabolic homeostasis.

To date, both mutations and altered expression have been reported for some *PRDMs* in specific cancer entities. However, our understanding of the role played by different PRDM family members in cancer is still limited by the lack of a systematic and comprehensive approach in deciphering the mutational status and the complete transcriptional profile of all the *PRDMs* across a large number of different cancer types.

Here, The Cancer Genome Atlas (TCGA) deposited exome and RNA-Seq data [24] were used to obtain a complete pan-cancer overview of the genomic and transcriptomic alterations for all *PRDM* genes across 31 distinct human cancer types.

2. Results

2.1. Mutational Profiling of PRDM Genes Across Human Cancers

To systematically identify somatic mutations within genes encoding PRDMs, we started with a mutational profiling of these genes across human cancers. We downloaded Exome-sequencing datasets from the TCGA web portal for 31 cancer types and about 11,000 patients. The number of samples for each cancer type is illustrated in Table S1 [25].

Overall, we identified 3995 point mutations, 180 deletions (39 in-frame and 141 frameshift), and 22 insertions (16 in-frame and 6 frameshift) affecting PRDM genes. Silent or synonymous mutations were 1531 (26.7% of total mutations) and ranged between 11% (*PRDM6*) and 41% (*PRDM8*) of the total mutations for each gene (Figure 1).

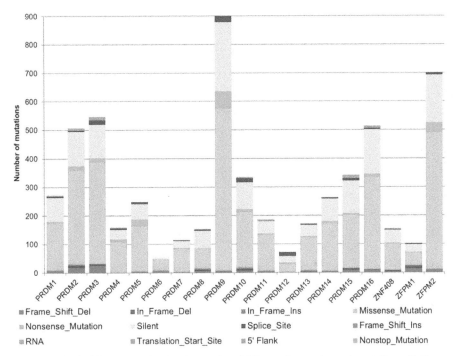

Figure 1. Stacked histograms showing the number of different classes of somatic mutations affecting *PRDM* genes as reported in the Mutation Annotation Files across all analyzed cancer entities.

According to our reanalysis, the most mutated genes were *PRDM2* (507 mutations; with 24% of silent mutations), *PRDM3/MECOM* (547 mutations; 22% silent), *PRDM9* (899 mutations; 27% silent), *PRDM16* (514 mutations; 31% silent), and *ZFPM2/FOG2* (700 mutations; 24% silent). Non-sense mutations were more recurrent in *PRDM5* (23), *PRDM9* (60), and *ZFPM2/FOG2* (34), whereas splice sites disrupting mutations were more frequently detected in *PRDM3/MECOM* (13), *PRDM9* (19), *PRDM10* (14) and *PRDM12* (13) (Figure 1).

To measure the frequencies of somatic mutations for each *PRDM* gene across all tumor types, only non-synonymous mutations were considered. We observed heterogeneity in the mutation frequencies of these genes in the different tumor types. A global low mutation rate (from 0 to 8.2%) was found, except for *PRDM3/MECOM, PRDM8, PRDM9, PRDM15, ZFPM1/FOG1*, and *ZFPM2/FOG2* (Table 1). In detail, *PRDM8* and *PRDM15* were mutated at low rates in most of the analyzed cancer types except PAAD where they were both frequently mutated (16.0% and 11.2%, respectively). *PRDM3/MECOM* was recurrently mutated in various cancer types, also reaching a value of 20.1% of mutated samples in SKCM. Similarly, *PRDM9* was mutated with a high mutation rate in many cancer types, achieving values of 10.0% in UCEC, 14.2% in LUAD, and 15.4% in SKCM. Otherwise, *ZFPM1/FOG1* was mutated at a low rate in a few cancer types, except in UCS (5.2%), COAD (6.6%), READ (9.4%), and ACC (50.5%). Finally, *ZFPM2/FOG2* was frequently mutated at a high rate in various cancer types, reaching a value of 11.1% in LUAD and 16.5% in SKCM.

We visualized the mutation data in each tumor type by Oncostrip function (Supplementary file 1). Through this approach, we evaluated the percentage of samples with at least one mutated *PRDM* gene in each tumor type ranging from 1.02% (2/196) in LAML samples to 55.43% (51/92) in ACC samples. Furthermore, this function allowed us to visualize the mutation type affecting *PRDM* genes in each sample. Interestingly, *ZFPM1/FOG1* revealed a high number of samples, especially in ACC, with "multi_hit" mutations (more than one mutation affecting the same gene in the same cancer sample). Specifically, we found 11/18 (61%) multi-hit mutations in COAD, 10/11 (90%) in READ, and 23/47 (48%) in ACC (see Supplementary file S1).

Int. J. Mol. Sci. **2018**, *19*, 3250

Table 1. Frequency of patients carrying mutations in the *PRDMs* across the 31 analyzed tumors.

Cancers	Genes																		
	PRDM1	PRDM2	MECOM	PRDM4	PRDM5	PRDM6	PRDM7	PRDM8	PRDM9	PRDM10	PRDM11	PRDM12	PRDM13	PRDM14	PRDM15	PRDM16	ZNF408	ZFPM1	ZFPM2
ACC	0	1.1	0	1.1	1.1	2.2	0	0	3.2	1.1	0	0	1.1	1.7	0	1.1	0	50.5	4.3
BLCA	1.7	4.1	5.3	3.4	3.4	0.2	1.9	1	3.9	2.2	1.7	0	1.4	1.7	2.7	4.1	0.7	1	3.6
BRCA	0.8	0.9	1	0.5	0.3	0.2	0.2	0.5	0.7	0.8	0.2	0.2	0.3	0.5	0.6	0.5	0.1	0.2	1.3
CESC	0.5	4	2	1.5	2	0.5	1	0.5	4.5	1	1.5	0	0	0	3	3	1	0.5	1
CHOL	2.7	0	0	5.4	0	2.7	0	0	5.4	0	0	0	0	0	0	0	2.7	0	0
COAD	2.2	6.6	5.5	1.5	2.6	2.9	0.7	1.1	5.5	4.4	1.1	0.4	1.5	1.8	2.2	4.4	1.5	6.6	2.6
DLBC	8.2	2	2	0	2	0	0	2	8.2	0	0	0	0	0	2	6.1	0	0	4.1
ESCA	1.1	1.6	3.2	2.1	1.6	2.7	1.1	1.6	7	4.8	1.6	0	2.1	2.7	4.8	2.7	1.6	0.5	8.6
GBM	0.6	0.6	2.8	0.3	1.1	0	0.3	0	3.3	1.9	0.6	0.8	0.3	0.3	1.9	0.6	0.6	0	0.3
HNSC	1.3	2.1	2.5	0.8	1.3	0	0.6	0.6	7.2	1.1	1.1	0.8	0.8	1.9	0.9	2.3	0.6	0.9	0.8
KICH	0	1.5	1.5	0	0	0	0	1.5	6	0	0	1.5	1.5	0	0	1.5	0	0	0
KIRC	0.2	0.8	1.2	1.5	0.3	0.2	0.8	0.3	1	1	0.2	0.3	0.7	0.5	1.2	0.5	0.5	0.2	1.2
KIRP	0.3	2.4	0	0.3	0.3	0.7	0.3	0.3	0.7	1	0.7	0.3	1.4	1	1.7	1.4	1.4	0.3	0.3
LAML	0	0	0	0	0	0	0	0	0.5	0	0	0	0	0	0	0.5	0	0	0
LIHC	1.6	3.2	2.9	1.8	1.1	0.8	0.5	0.8	2.4	1.6	1.8	0	0.8	1.3	2.4	2.1	0.8	0	3.9
LUAD	1.9	3.7	2.1	1.6	1.9	0	1.9	1.2	14.2	3	1.6	1.1	0.9	4.4	2.3	4.2	0.9	0.2	11.1
LUSC	2.8	3.4	5	0	2.8	0	0.6	0	7.3	3.4	2.2	1.1	3.4	0	2.2	3.4	1.7	0	5
OV	0.2	0.4	0	0	0	0	0.4	0	1.3	0	0.2	0	0	0.2	0.2	0	0	0	1.3
PAAD	0.5	4.3	2.7	1.1	1.6	0	2.1	16	1.6	3.2	2.7	2.7	1.6	0.5	11.2	2.1	1.1	0	2.7
PCPG	0	0.5	0	0	0	0	0	0.6	1.1	0.4	0.2	0.4	0	0	0	0	0.5	0	0.5
PRAD	0	0.8	1.4	0	0.6	0	0	0	0.8	1.7	0.9	0.9	0	0.4	0.4	0.8	0.6	0.6	1.4
READ	0	2.6	1.7	2.6	3.4	0.9	0.6	0	2.6	1.5	1.1	0.4	0.8	0	1.7	0.9	0	9.4	5.1
SARC	1.1	1.9	1.9	1.1	0.8	2.7	0.8	0.4	3	3.6	3	1.3	0.8	1.5	3.4	3	0.8	0.8	2.3
SKCM	5.3	4.2	20.1	1.3	3.6	0	2.3	1.5	15.4	4.3	1.5	0.3	2.5	3.6	3.6	7.8	1.9	0.8	16.5
STAD	2.3	7.8	3.8	1	3	0	1	2.3	5.8	2.5	0.6	0	2.8	2.5	3	4.8	1.8	1	5.8
TGCT	0	2.5	0.6	0	0	1.3	0.6	1.9	2.5	0.8	0.2	0.2	1.3	0	0	1.3	0.4	1.3	0.4
THCA	0.4	0.6	0	0.4	0.2	0	0.8	0.4	1.2	2.5	0.6	0	0.2	0	0.4	0.8	0.4	0.4	0.4
THYM	2.4	3.2	3.2	2.4	0	2.4	3.2	0.8	3.2	0.8	0.8	0.8	0.8	2.4	2.4	3.2	0.8	0.8	1.6
UCEC	4	7.2	5.6	2.4	4.4	2	0.8	1.6	10	5.2	2.4	1.2	2.8	4.8	3.2	5.6	4	5.2	4.8
UCS	5.2	1.7	0	0	3.4	0	1.7	1.7	1.7	1.7	1.7	0	1.7	0	3.4	0	1.7	0	5.2
UVM	0	1.2	0	1.2	1.2	0	0	1.2	4.9	0	1.2	0	0	0	1.2	0	0	0	0

To distinguish between damaging and tolerated missense mutations, we carried out a variant effect predictor (VEP) analysis (Table S2). Missense mutations with a SIFT score ranging in the interval 0.0–0.05 and/or with a PolyPhen score in the interval 0.5–1 were considered as deleterious or probably damaging, respectively. As shown in Table 2, adding all the other deleterious somatic mutations (frameshift, in-frame deletions, stop gained and start lost mutations, splice site, UTR, and intron variants) to the deleterious missense mutations classified with the VEP analysis, we obtained the total number of deleterious mutations affecting each *PRDM* gene. Thus, we obtained the percentage of deleterious somatic mutations across the tumor samples. This number was $\geq$50% for *MECOM/PRDM3* (52.7%), *PRDM4* (55%), *PRDM5* (54.8%), *PRDM6* (58.5%), *PRDM10* (55.2%), *PRDM11* (54%), *PRDM13* (51.7%), and *PRDM16* (50%).

Table 2. Percentage of deleterious and tolerated mutations in the *PRDMs* across the analyzed tumor samples.

Genes	Deleterious Mutations	Total Mutations	% Deleterious Mutations
PRDM1	120	272	44.1
PRDM2	223	507	44.0
MECOM/PRDM3	288	547	52.7
PRDM4	88	160	55.0
PRDM5	136	248	54.8
PRDM6	31	53	58.5
PRDM7	31	114	27.2
PRDM8	54	154	35.1
PRDM9	403	899	44.8
PRDM10	185	335	55.2
PRDM11	101	187	54.0
PRDM12	30	72	41.7
PRDM13	89	172	51.7
PRDM14	125	263	47.5
PRDM15	162	341	47.5
PRDM16	257	514	50.0
ZNF408/PRDM17	55	155	35.5
ZFPM1/FOG1	43	102	42.2
ZFPM2/FOG2	234	700	33.4

Additionally, to predict the potential functional effect of the identified *PRDM* somatic mutations on the affected proteins and to detect a possible mutation enrichment in some domains, we localized the deleterious missense mutations on the canonical protein isoform of each *PRDM* (Figure 2). Interestingly, a random sampling weighted on the size of the annotated protein domains demonstrated that somatic deleterious mutations were significantly enriched in the PR domain of *PRDM1, PRDM5, PRDM6, PRDM8, PRDM9, PRDM12* and *PRDM13* ($p < 0.005$).

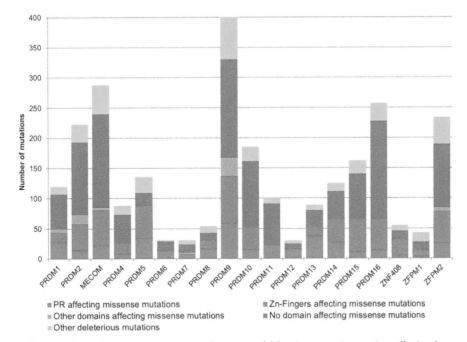

Figure 2. The stacked histogram represents the amount of deleterious somatic mutations affecting the different known domains of PRDM proteins. In detail, the missense mutations affecting the PR domain are reported in blue, the missense mutations affecting the Zn fingers are in red, and the missense mutations affecting other known domains (where present) are illustrated in green. The deleterious missense mutations not affecting known domains are shown in violet and the other classes of deleterious mutations (frameshift, in-frame deletions, stop gained and start lost mutations, splice site, UTR, and intron variants) are in orange.

Another important aspect of cancer genetic studies is the presence of possible recurrent and hotspot mutations. Figure 3 illustrates mutations in *PRDM* genes recurring in more than three tumor types. Interestingly, the frameshift mutation T/-→K678X, despite affecting *PRDM3/MECOM* in a region not containing known domains, was recurrent in different tumor types; similarly, also the missense mutation G/A→S237L occurred in a region without known domains but in many tumors. Otherwise, missense mutations affecting a Zn-finger domain and occurring in different tumors were observed for *PRDM9*, *PRDM14*, and *PRDM16*. Likewise, *PRDM12* was frequently mutated in a splice donor site in a region coding for the PR domain whereas in different tumor types, *ZFPM2/FOG2* was affected by the missense mutation C/T→R734C in a region without known domains. In addition, *PRDM2* and *PRDM15* revealed an in-frame deletion in various cancers and *PRDM11* a frameshift mutation. Finally, *ZFPM1/FOG1* showed several recurrent mutations; they all (frameshift mutations and in-frame deletions) hit a region without known domains (Figure 3).

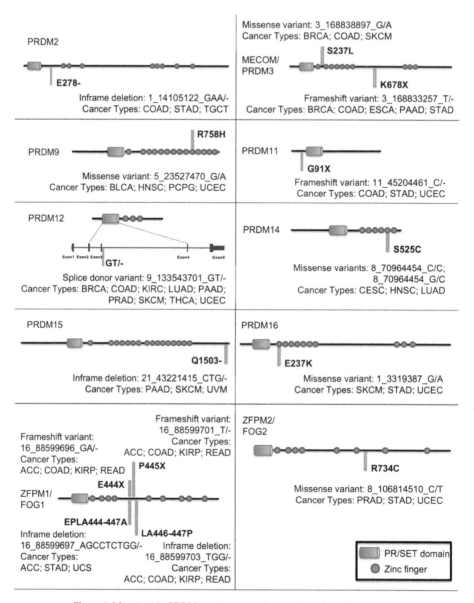

Figure 3. Mutations in *PRDM* genes recurrent in more than three tumor types.

Interestingly, all these mutations were particularly recurrent in ACC patients. In this cohort, *ZFPM1/FOG1* also displayed five hotspot mutations, all localized in the same region outside the known domains (Figure 4a). To establish whether these hotspot mutations could have an impact on the ZFPM1/FOG1 structure, we utilized the I-TASSER web-tool to predict the tertiary structure of the annotated ZFPM1/FOG1 protein (Figure 4b) and proteins carrying the missense mutations and the in-frame deletions (Figure 4c–e). As illustrated, these mutations completely altered the structure of the canonical protein. Otherwise, the frameshift mutations E444X and P445X led to premature stop codons at the residues 669 and 796, respectively; both of the mutated proteins shared only the first

443 residues with the canonical protein whereas they changed in the 444–669 and 444–796 regions and missed respectively 337 and 210 residues at the C-terminal, which contains the last five zinc fingers of ZFPM1/FOG1.

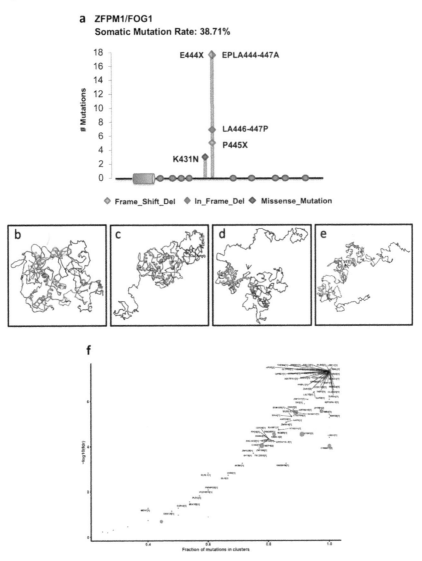

Figure 4. (**a**) *ZFPM1/FOG1* hotspot mutations in ACC obtained by Lollipop plot visualization function. These hotspot mutations are all localized outside the PR domain (green cylinder) and Zinc fingers (blue dots) (**b**–**e**) I-TASSER predicted tertiary structures of the annotated ZFPM1/FOG1 canonical protein (**b**) and ZFPM1/FOG1 proteins carrying the mutations ELPA444-447A (**c**), LA446-447P (**d**), and K431N (**e**); the arrows show the mutated protein regions. (**f**) The scatter plot shows the results of the OncodriveCLUST algorithm analysis for ACC. The dimension of the dots is proportional to the number of clusters found in a certain gene, also indicated in the squared bracket. Specifically, in the *ZFPM1/FOG1* locus, two mutation clusters were found (fdr < 2.87 × 10^{-6}).

Finally, to assess whether members of the PRDM family may be driver genes in a given cancer type, we used the OncodriveCLUST tool, which aims to identify genes whose mutations are biased towards a large spatial clustering. This method is based on the feature that cancer gene mutations frequently cluster in particular positions of the protein. Thus, mutations with a frequency higher than the background rate that tend to cluster in specific regions of protein-coding genes are likely to be driver genes. Based on the scores of this analysis, *ZFPM1/FOG1* can be considered as a cancer driver for ACC (Figure 4f) and *PRDM8* for PAAD (Figure S2).

2.2. Differentially Expressed PRDM Genes across Human Cancers

To evaluate whether the expression of *PRDM* genes is affected in human cancers, we took advantage of RNA-Seq datasets from paired samples (cancer vs. benign counterpart) available at the TCGA web portal. Globally, 585 patients across 21 cancer types were analyzed (Table S1). The gene expression profiles differed considerably between normal and tumor specimens, depending on the cancer type, as shown by the principal component analysis [26]. The results of gene expression profiling are summarized in Table S3 and Figure 5.

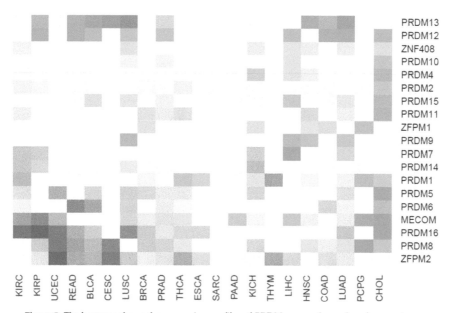

Figure 5. The heatmap shows the expression profiles of *PRDMs* across the analyzed cancer types.

Data indicate that a large subset of *PRDM* genes is consistently deregulated across several cancer types. Particularly, a significant overexpression was measured for *PRDM12* and *PRDM13*. On the other side, *ZFPM2/FOG2*, *PRDM8*, and *PRDM16* were more often downregulated across tumors (Figure 5). Strong upregulation of almost all *PRDM* genes was measured in CHOL where 13/19 PRDM genes were overexpressed in tumor versus healthy counterparts. Among them, the most upregulated were *MECOM/PRDM3* (FC = 12.64), *PRDM5* (FC = 10.7) and *PRDM16* (FC = 13.74). Conversely, the cancer types with the smallest number of deregulated *PRDM* genes were SARC with no *PRDM* genes deregulated, PAAD with 1 gene upregulated (*MECOM/PRDM3*, FC = 2.71), THYM with two genes strongly downregulated (*PRDM1*, FC = 0.13 and *ZFPM2/FOG2*, FC = 0.16), CESC with two genes strongly downregulated (*PRDM8*, FC = 0.07 and *ZFPM2/FOG2*, FC = 0.04) and one gene strongly upregulated (*PRDM13* FC = 29.86), and, finally, ESCA with two genes strongly downregulated (*PRDM16*, FC = 0.37 and *ZFPM2/FOG2*, FC = 0.44) and one gene upregulated (*PRDM1*, FC = 2.28).

2.3. PRDM Expression in Human Primary Tumors

In an attempt to validate the findings obtained on RNA-Seq datasets, we assayed TissueScan cDNA panel arrays containing eight different tumors (breast, colon, kidney, liver, lung, ovary, prostate, and thyroid) [25].

Specifically, we analyzed the differential expression of *PRDM3/MECOM*, *PRDM10*, *PRDM12*, *PRDM16*, and *ZFPM2/FOG2* genes (Figure S3). As illustrated in Figure 6, a general differential expression, even though not significant, was observed for both *ZFPM2/FOG2* and *PRDM16* in several tumor tissues (Figure 6a,b). *PRDM3/MECOM* was found to be significantly overexpressed in breast ($p < 0.001$), ovary ($p < 0.001$), and colon ($p < 0.05$) cancer specimens (Figure 6c). Similarly, *PRDM10* was overexpressed at significant levels in breast ($p < 0.001$) and colon ($p < 0.05$) cancer specimens (Figure 6d). No statistical differences in the expression of all these genes were measured between tumor and healthy samples of the other cancer entities (Figure 6a–d). Otherwise, the *PRDM12* gene was difficult to analyze. This gene was confirmed as being highly overexpressed in all the analyzed cancer tissues, except in ovarian cancer, as indicated by reanalysis of the TCGA RNA-Seq dataset. However, this gene was not expressed or was expressed at very low levels in all the healthy tissues used as controls; it was expressed in tumor specimens. For this reason, its relative gene expression could not be calculated using the $2^{-\Delta\Delta Ct}$ method for all the analyzed tissues. Particularly, relative expression was measured only for thyroid, ovary, prostate, and liver (Figure 6e) whereas in breast, colon, kidney, and lung normal tissues, the amplification products were undetectable when observed by agarose gel electrophoresis analysis (Figure 6f). Of note, these cancer specimens showed measurable levels of *PRDM12* (Figure 6e–f).

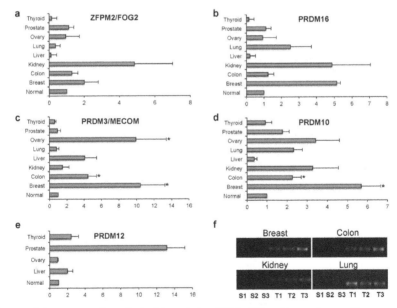

Figure 6. Relative expressions obtained by real-time PCR in different types of cancer tissues versus corresponding normal tissues (with arbitrary expression value equal to 1). The comparative threshold cycle (Ct) method was used with β-actin as the internal control. The results are expressed as the mean ± ES. The statistical significance of differences between experimental groups was calculated using the unpaired two-tailed Student's *t*-test. (*) Results with a *p*-value < 0.05 were considered significant. (**a**) *ZFPM2/FOG2*; (**b**) *PRDM16*; (**c**) *PRDM3/MECOM*; (**d**) *PRDM10*; (**e**) *PRDM12*. (**f**) Representative samples analyzed for *PRDM12* by agarose gel electrophoresis.

3. Discussion

In this study, we provide for the first time a systematic and comprehensive overview of both the mutational status and the expression profile of all the *PRDM* genes across a large number of different cancer types.

Recently, the availability of multi-omics datasets (such as TCGA) from human cancers, together with the development of advanced bioinformatics tools, represent a unique source to study human malignancies [24].

Our reanalysis of the TCGA Exome-sequencing datasets revealed *PRDM2, PRDM3/MECOM, PRDM9, PRDM16* and *ZFPM2/FOG2* as the most mutated genes. Heterogeneity in the mutation frequencies was observed in the different tumor types with higher mutation rates found for *PRDM3/MECOM, PRDM8, PRDM9, PRDM15, ZFPM1/FOG1* and *ZFPM2/FOG2* in specific cancers. Remarkably, VEP analysis indicated that the percentage of total deleterious mutations across the tumor samples was high for most genes. More interestingly, a random sampling weighted on the size of the annotated protein domains demonstrated that deleterious mutations were significantly enriched in the PR domain of *PRDM1, PRDM5, PRDM6, PRDM8, PRDM9, PRDM12* and *PRDM13*. Frequent mutations disrupting the PR domain in tumor samples would be a mechanism for removing the tumor suppressor function of the PR-plus isoform in favor of the oncogenic PR-minus form.

A big challenge in cancer biology studies is distinguishing between mutations conferring a selective growth advantage to cancer cells (drivers) and those randomly accumulating and without significant effects on the oncogenic process (passengers) [27,28]. Many algorithms employing different approaches are now utilized to recognize driver genes although, when compared for their performance, all display both strengths and weaknesses [25,29–31]. Moreover, recent studies have highlighted the existence of "mini-driver" genes with weaker tumor-promoting effects, thus expanding the previous driver–passenger dichotomy to a continuous model [25,30,32]. Besides, a sub-classification has been proposed to differentiate "mut-driver genes", usually altered by somatic gene mutations, from "epi-driver genes", which are deregulated through epigenetic modifications but are not frequently mutated [25,27].

In this study, OncodriveCLUST analysis revealed two putative cancer mut-driver genes: *PRDM8* in PAAD and *ZFPM1/FOG1* in ACC. In the latter case, we found that the involved gene was mutated in a very high percentage (about 55%) of tumor samples. Additionally, a mutational hotspot region localized between the amino acid positions 444 and 447, outside the known domains, was recognizable. All these findings agree with the key parameters commonly used to discern drivers from passengers [29]. Notably, these hotspot mutations recurred also in other malignancies, such as COAD, KIRP, READ, STAD, and UCS, supporting an important role of this gene in carcinogenesis. Moreover, the finding of "multi_hit" mutations in ACC, as well as in COAD and READ, advises that this gene could function as a tumor suppressor gene. This is conceivable with the current knowledge about the role of *ZFPM1/FOG1* in differentiation. Indeed, ZFPM1 is also known as a friend of GATA1 (FOG1) since it interacts with GATA1 and it is an essential cofactor for the transcription factor GATA-1 in erythroid and megakaryocytic differentiation. Reduced expression of ZFPM1/FOG1 was found in preleukemic progenitors of a mouse model of leukemia [33,34]. Besides, the high recurrence of these mutations together with results from 3D-modeling of the canonical and mutant *ZFPM1/FOG1* proteins suggest that, although without known domains, this is a critical region for the protein. Interestingly, this region (particularly K431 residue) is evolutionary highly conserved (Figure S4).

It is conceivable that other *PRDM* genes may play a key role in the initiation and progression of specific or multiple tumor types, as also reported by previous literature [8,15–17]. Indeed, it is accepted that cancer driver genes are mainly involved in three core cellular processes: cell fate, cell survival, and genome maintenance. For instance, *PRDM2* has a relevant role in all of them; specifically, it also participates to the formation of protein complexes involved in the DNA damage response and in genome maintenance [15]. Additionally, a recent study identified a driver *PRDM2* mutation in MSI colorectal cancer [16]. However, our analysis has not considered this gene as a driver. We cannot exclude, among the explanations, the limitation of the utilized bioinformatics tool [30].

Our pan-cancer study has also identified *PRDM12* as a possible epi-driver gene in multiple cancers. Of note, TCGA gene expression profiling of PRDM genes revealed significant overexpression of *PRDM12* and *PRDM13* in many tumor types whereas *ZFPM2/FOG2*, *PRDM8*, and *PRDM16* were more often downregulated in tumor tissues. Our qRT-PCR analysis was not able to confirm all the results obtained through TCGA analysis. The main reason could be the utilization of unpaired samples in the validation through TissueScan cDNA panel arrays; otherwise, our analysis on TCGA was carried out on paired samples. In addition, we have analyzed a small number of samples by qRT-PCR compared to the huge number of cases from TCGA. Noteworthy is that when we measured the differential expression of *PRDM12* in cDNA panel arrays, we found the expression of this gene only in cancer specimens but not in healthy samples of several tissues, suggesting that it could be putatively utilized as a biomarker in those malignancies. Our study represents the first analysis of all *PRDM* genes in pan-cancer; further studies using large cohorts are necessary to validate the most promising results, particularly for *PRDM12*. In addition, given the lack of literature data, we are aware that functional studies investigating the effect of altered expression both in vitro and in vivo are required to establish the possible impact of *PRDM12* in cancer.

Altogether, our results can be useful for identifying a subset of relevant *PRDMs* that are frequently mutated and/or transcriptionally deregulated in certain tumor types. Functional studies on specific *PRDM* gene mutations should be accomplished to definitely prove their potential oncogenic role. Moreover, it would be interesting to investigate whether these mutations contribute to cancer progression and metastasis, as well as whether they correlate with prognosis and/or with drug response and resistance. The epigenetic changes underlying the altered gene expression observed in tumor samples should also be explored. In this context, the availability of novel multi-omics data integration tools and methods also offer the opportunity to further integrate our analysis of *PRDM* gene expression by a systematic pan-cancer study of the epigenetic marks in these genes [25,35,36].

4. Materials and Methods

4.1. TCGA Data Source Selection and Processing for Mutation Analysis

In this manuscript, we analyzed both whole Exome- and RNA-Seq data retrieved from publicly accessible repositories. Specifically, we retrieved the whole exome sequencing data from the GitHub R data package for pre-compiled somatic mutations from TCGA cohorts "TCGA mutations" and analyzed it using the Bioconductor package "maftools" [37].

The selection and nomenclature of *PRDM* genes were based on the HUGO Gene Nomenclature Committee [38]. To estimate the mutation enrichment within the PR domain of each of the PRDM proteins, we performed a random sampling iterated 1000 times weighted on the size of the annotated domains.

To assess whether one or more PRDM proteins could be considered as cancer driver genes based on the positional clustering of the variants in the selected human cancers, we used a re-implementation of the software OncodriveCLUST within the maftools package [39].

Three-dimensional (3D) modeling of the human canonical and mutant ZFPM1/FOG1 proteins was carried out using I-TASSER [26,40,41].

4.2. TCGA Data Source Selection and Processing for Expression Analysis

The RNA-Seq gene expression data were downloaded from TCGA [42]. The analysis of gene expression and the identification of differentially expressed genes were performed comparing the expression profiles of cancer vs. matched normal samples in a paired analysis. Therefore, expression data taken from human primary cancers for which healthy samples were not available were discarded. According to this criterion, 22 tumor entities were analyzed. To have a more robust differential expression analysis in paired samples, we applied generalized linear models implemented in the EdgeR Bioconductor package version 3.17.10. *p*-values adjustment was performed through the application of the false discovery rate (FDR) method. We considered differentially expressed genes with a logFC ≤ -1 and logFC ≥ 1, and an FDR ≤ 0.01.

4.3. Real-Time RT-PCR Analysis

Quantitative real-time PCR (qRT-PCR) experiments were carried out on TissueScan Cancer Survey Panels, which contained cDNA samples from various normal and cancer tissues covering eight different tumors (breast, colon, kidney, liver, lung, ovary, prostate and thyroid) from independent patients diagnosed at various clinical disease stages and selected from mixed ages and genders. Tissue cDNAs of each array were synthesized from high-quality total RNAs of pathologist-verified tissues, normalized and validated with β-actin in two sequential qPCR analyses, and provided with clinical information and QC data [25].

To quantitatively determine the relative amount of *PRDM3/MECOM*, *PRDM8*, *PRDM10*, *PRDM12*, *PRDM16*, and *ZFPM2/FOG2* RNAs, qRT-PCR was performed [25]. Primers were designed using Primer3Plus [43] and specificity was verified with the BLAST program and through *in-silico* PCR analysis by the UCSC Genome Browser [44].

The selected sequences of oligonucleotides forward (F) and reverse (R) were: *PRDM3/MECOM*-F 5′-AGTGGCAGTGACCTGGAAAC-3′; *PRDM3/MECOM*-R 5′-ACCGCAGTCTGCTCCTCTAA-3′; *PRDM10*-F 5′-CAGCACATTCGAAAGAAGCA-3′; *PRDM10*-R 5′-GCGTTCGGTAGTCTGTCGTT-3′; *PRDM12*-F 5′-GGGAGTCCTTACGCAACCTT-3′; *PRDM12*-R 5′-TTCCATTGTGCCTCCACTCT-3′; *PRDM16*-F 5′-ATGATGGACAAGGCAAAACC-3′; *PRDM16*-R 5′-GATGTGGGAGGTAGCAGAGG-3′; *ZFPM2/FOG2*-F 5′-GACAGTGCCCATCAGATTTC-3′; *ZFPM2/FOG2*-R 5′-GGGCAGGAATTCTTC CATTTT-3′.

The amplification products were also analyzed by agarose gel electrophoresis [45]. Data were normalized with β-actin gene provided with arrays. The relative gene expression was calculated using the $2^{-\Delta\Delta Ct}$ method [25].

Supplementary Materials: Supplementary materials can be found at http://www.mdpi.com/1422-0067/19/10/3250/s1.

Author Contributions: Conceived and designed the analysis: C.A., M.R., A.C. (Alfredo Ciccodicola), and A.C. (Amelia Casamassimi) Analyzed the data and produced the report: A.F. and A.S. Contributed to the data analysis: A.F., A.S., M.R., and A.C. (Amelia Casamassimi) Interpreted and validated the results: C.A., M.R., and A.C. (Amelia Casamassimi) Wrote the paper: M.R., P.G., M.B., C.A., A.C. (Alfredo Ciccodicola), and A.C. (Amelia Casamassimi) All authors read and approved the manuscript.

Funding: This work was supported in part by ordinary funds from the University of Campania "Luigi Vanvitelli".

Conflicts of Interest: The authors declare no conflict of interest.

Abbreviations

FC	Fold change
FDR	False discovery rate
TCGA	The Cancer Genome Atlas
ACC	Adrenocortical carcinoma
BLCA	Bladder cancer
BRCA	Breast cancer
CESC	Cervical squamous cell carcinoma and endocervical adenocarcinoma
CHOL	Cholangiocarcinoma
COAD	Colon adenocarcinoma
DLBC	Lymphoid neoplasm diffuse large B-cell lymphoma
ESCA	Esophageal carcinoma
GBM	Glioblastoma
HNSC	Head and neck squamous cell carcinoma

KICH	Kidney chromophobe carcinoma
KIRC	Kidney renal clear cell carcinoma
KIRP	Kidney renal papillary cell carcinoma
LAML	Acute myeloid leukemia
LIHC	Liver hepatocarcinoma
LUAD	Lung adenocarcinoma
LUSC	Lung squamous cell carcinoma
OV	Ovarian cancer
PAAD	Pancreas adenocarcinoma
PCPG	Pheochromocytoma and paraganglioma
PRAD	Prostate adenocarcinoma
READ	Rectum adenocarcinoma
SARC	Sarcoma
SKCM	Skin cutaneous melanoma
STAD	Stomach adenocarcinoma
TGCT	Testicular germ cell tumors
THCA	Thyroid cancer
THYM	Thymoma
UCEC	Uterine corpus endometrial carcinoma
UCS	Uterine carcinosarcoma
UVM	Uveal melanoma

References

1. Fumasoni, I.; Meani, N.; Rambaldi, D.; Scafetta, G.; Alcalay, M.; Ciccarelli, F.D. Family expansion and gene rearrangements contributed to the functional specialization of PRDM genes in vertebrates. *BMC. Evol. Biol.* **2007**, *7*, 187. [CrossRef] [PubMed]

2. Sun, X.J.; Xu, P.F.; Zhou, T.; Hu, M.; Fu, C.T.; Zhang, Y.; Jin, Y.; Chen, Y.; Chen, S.J.; Huang, Q.H.; et al. Genome-wide survey and developmental expression mapping of zebrafish SET domain-containing genes. *PLoS ONE* **2008**, *3*, e1499. [CrossRef] [PubMed]

3. Clifton, M.K.; Westman, B.J.; Thong, S.Y.; O'Connell, M.R.; Webster, M.W.; Shepherd, N.E.; Quinlan, K.G.; Crossley, M.; Blobel, G.A.; Mackay, J.P. The identification and structure of an N-terminal PR domain show that FOG1 is a member of the PRDM family of proteins. *PLoS ONE* **2014**, *9*, e106011. [CrossRef] [PubMed]

4. Vervoort, M.; Meulemeester, D.; Béhague, J.; Kerner, P. Evolution of Prdm Genes in Animals: Insights from Comparative Genomics. *Mol. Biol. Evol.* **2016**, *33*, 679–696. [CrossRef] [PubMed]

5. Hohenauer, T.; Moore, A.W. The Prdm family: Expanding roles in stem cells and development. *Development* **2012**, *139*, 2267–2282. [CrossRef] [PubMed]

6. Fog, C.K.; Galli, G.G.; Lund, A.H. PRDM proteins: Important players in differentiation and disease. *BioEssays* **2012**, *34*, 50–60. [CrossRef] [PubMed]

7. Di Zazzo, E.; De Rosa, C.; Abbondanza, C.; Moncharmont, B. PRDM Proteins: Molecular Mechanisms in Signal Transduction and Transcriptional Regulation. *Biology* **2013**, *2*, 107–141. [CrossRef] [PubMed]

8. Mzoughi, S.; Tan, Y.X.; Low, D.; Guccione, E. The role of PRDMs in cancer: One family, two sides. *Curr. Opin. Genet. Dev.* **2016**, *36*, 83–91. [CrossRef] [PubMed]

9. Huang, S.; Shao, G.; Liu, L. The PR domain of the Rb-binding zinc finger protein RIZ1 is a protein binding interface and is related to the SET domain functioning in chromatin-mediated gene expression. *J. Biol. Chem.* **1998**, *273*, 15933–15939. [CrossRef] [PubMed]

10. Ren, B.; Chee, K.J.; Kim, T.H.; Maniatis, T. PRDI-BF1/Blimp-1 repression is mediated by corepressors of the Groucho family of proteins. *Genes Dev.* **1999**, *13*, 125–137. [CrossRef] [PubMed]

11. Kajimura, S.; Seale, P.; Kubota, K.; Lunsford, E.; Frangioni, J.V.; Gygi, S.P.; Spiegelman, B.M. Initiation of myoblast to brown fat switch by a PRDM16-C/EBP-beta transcriptional complex. *Nature* **2009**, *460*, 1154–1158. [CrossRef] [PubMed]

12. Seale, P.; Bjork, B.; Yang, W.; Kajimura, S.; Chin, S.; Kuang, S.; Scimè, A.; Devarakonda, S.; Conroe, H.M.; Erdjument-Bromage, H.; et al. PRDM16 controls a brown fat/skeletal muscle switch. *Nature* **2008**, *454*, 961–967. [CrossRef] [PubMed]

13. Okashita, N.; Suwa, Y.; Nishimura, O.; Sakashita, N.; Kadota, M.; Nagamatsu, G.; Kawaguchi, M.; Kashida, H.; Nakajima, A.; Tachibana, M.; et al. PRDM14 Drives OCT3/4 Recruitment via Active Demethylation in the Transition from Primed to Naive Pluripotency. *Stem Cell Rep.* **2016**, *7*, 1072–1086. [CrossRef] [PubMed]
14. Chi, J.; Cohen, P. The Multifaceted Roles of PRDM16: Adipose Biology and Beyond. *Trends Endocrinol. Metab.* **2016**, *27*, 11–23. [CrossRef] [PubMed]
15. Sorrentino, A.; Rienzo, M.; Ciccodicola, A.; Casamassimi, A.; Abbondanza, C. Human PRDM2: Structure, function and pathophysiology. *Biochim. Biophys. Acta* **2018**, *1861*, 657–671. [CrossRef] [PubMed]
16. Maruvka, Y.E.; Mouw, K.W.; Karlic, R.; Parasuraman, P.; Kamburov, A.; Polak, P.; Haradhvala, N.J.; Hess, J.M.; Rheinbay, E.; Brody, Y.; et al. Analysis of somatic microsatellite indels identifies driver events in human tumors. *Nat. Biotechnol.* **2017**, *35*, 951–959. [CrossRef] [PubMed]
17. Choi, E.J.; Kim, M.S.; Song, S.Y.; Yoo, N.J.; Lee, S.H. Intratumoral Heterogeneity of Frameshift Mutations in MECOM Gene is Frequent in Colorectal Cancers with High Microsatellite Instability. *Pathol. Oncol. Res.* **2017**, *23*, 145–149. [CrossRef] [PubMed]
18. Khurana, S.; Kruhlak, M.J.; Kim, J.; Tran, A.D.; Liu, J.; Nyswaner, K.; Shi, L.; Jailwala, P.; Sung, M.H.; Hakim, O.; et al. A macrohistone variant links dynamic chromatin compaction to BRCA1-dependent genome maintenance. *Cell Rep.* **2014**, *8*, 1049–1062. [CrossRef] [PubMed]
19. Zhang, C.; Zhu, Q.; He, H.; Jiang, L.; Qiang, Q.; Hu, L.; Hu, G.; Jiang, Y.; Ding, X.; Lu, Y. RIZ1: A potential tumor suppressor in glioma. *BMC Cancer* **2015**, *15*, 990. [CrossRef] [PubMed]
20. Kang, R.; Zelazowski, M.J.; Cole, F. Missing the Mark: PRDM9-Dependent Methylation Is Required for Meiotic DSB Targeting. *Mol. Cell* **2018**, *69*, 725–727. [CrossRef]
21. Diagouraga, B.; Clément, J.A.J.; Duret, L.; Kadlec, J.; de Massy, B.; Baudat, F. PRDM9 Methyltransferase Activity Is Essential for Meiotic DNA Double-Strand Break Formation at Its Binding Sites. *Mol. Cell* **2018**, *69*, 853–865. [CrossRef] [PubMed]
22. Shu, X.; Geng, H.; Li, L.; Ying, J.; Ma, C.; Wang, Y.; Poon, F.F.; Wang, X.; Ying, Y.; Yeo, W.; et al. The Epigenetic Modifier PRDM5 Functions as a Tumor Suppressor through Modulating WNT/β-Catenin Signaling and Is Frequently Silenced in Multiple Tumors. *PLoS ONE* **2011**, *6*, e27346. [CrossRef] [PubMed]
23. Meani, N.; Pezzimenti, F.; Deflorian, G.; Mione, M.; Alcalay, M. The tumor suppressor PRDM5 regulates Wnt signaling at early stages of zebrafish development. *PLoS ONE* **2009**, *4*, e4273. [CrossRef] [PubMed]
24. Weinstein, J.N.; Collisson, E.A.; Mills, G.B.; Shaw, K.R.; Ozenberger, B.A.; Ellrott, K.; Shmulevich, I.; Sander, C.; Stuart, J.M. The Cancer Genome Atlas Pan-Cancer analysis project. *Nat. Genet.* **2013**, *45*, 1113–1120. [CrossRef] [PubMed]
25. Federico, A.; Rienzo, M.; Abbondanza, C.; Costa, V.; Ciccodicola, A.; Casamassimi, A. Pan-Cancer Mutational and Transcriptional Analysis of the Integrator Complex. *Int. J. Mol. Sci.* **2017**, *18*, 936. [CrossRef] [PubMed]
26. Zhang, Y. I-TASSER server for protein 3D structure prediction. *BMC Bioinform.* **2008**, *9*, 40. [CrossRef] [PubMed]
27. Vogelstein, B.; Papadopoulos, N.; Velculescu, V.E.; Zhou, S.; Diaz, L.A., Jr.; Kinzler, K.W. Cancer genome landscapes. *Science* **2013**, *339*, 1546–1558. [CrossRef] [PubMed]
28. Garraway, L.A.; Lander, E.S. Lessons from the cancer genome. *Cell* **2013**, *153*, 17–37. [CrossRef] [PubMed]
29. Marx, V. Cancer genomes: Discerning drivers from passengers. *Nat. Methods* **2014**, *11*, 375–379. [CrossRef] [PubMed]
30. Porta-Pardo, E.; Kamburov, A.; Tamborero, D.; Pons, T.; Grases, D.; Valencia, A.; Lopez-Bigas, N.; Getz, G.; Godzik, A. Comparison of algorithms for the detection of cancer drivers at subgene resolution. *Nat. Methods* **2017**, *14*, 782–788. [CrossRef] [PubMed]
31. Tokheim, C.J.; Papadopoulos, N.; Kinzler, K.W.; Vogelstein, B.; Karchin, R. Evaluating the evaluation of cancer driver genes. *Proc. Natl. Acad. Sci. USA* **2016**, *113*, 14330–14335. [CrossRef] [PubMed]
32. Castro-Giner, F.; Ratcliffe, P.; Tomlinson, I. The mini-driver model of polygenic cancer evolution. *Nat. Rev. Cancer* **2015**, *15*, 680–685. [CrossRef] [PubMed]
33. Tsang, A.P.; Visvader, J.E.; Turner, C.A.; Fujiwara, Y.; Yu, C.; Weiss, M.J.; Crossley, M.; Orkin, S.H. FOG, a multitype zinc finger protein, acts as a cofactor for transcription factor GATA-1 in erythroid and megakaryocytic differentiation. *Cell* **1997**, *90*, 109–119. [CrossRef]
34. Cai, Q.; Jeannet, R.; Hua, W.K.; Cook, G.J.; Zhang, B.; Qi, J.; Liu, H.; Li, L.; Chen, C.C.; Marcucci, G.; et al. CBFβ-SMMHC creates aberrant megakaryocyte-erythroid progenitors prone to leukemia initiation in mice. *Blood* **2016**, *128*, 1503–1515. [CrossRef] [PubMed]

35. Ruffalo, M.; Koyutürk, M.; Sharan, R. Network-Based Integration of Disparate Omic Data to Identify "Silent Players" in Cancer. *PLoS Comput. Biol.* **2015**, *11*, e1004595. [CrossRef] [PubMed]

36. Huang, S.; Chaudhary, K.; Garmire, L.X. More Is Better: Recent Progress in Multi-Omics Data Integration Methods. *Front. Genet.* **2017**, *8*, 84. [CrossRef] [PubMed]

37. Mayakonda, A.; Koeffler, H.P. Maftools: Efficient analysis, visualization and summarization of MAF files from large-scale cohort based cancer studies. *bioRxiv* **2016**. [CrossRef]

38. HGNC. Available online: http://www.genenames.org (accessed on 19 October 2018).

39. Tamborero, D.; Gonzalez-Perez, A.; Lopez-Bigas, N. OncodriveCLUST: Exploiting the positional clustering of somatic mutations to identify cancer genes. *Bioinformatics* **2013**, *29*, 2238–2244. [CrossRef] [PubMed]

40. Roy, A.; Kucukural, A.; Zhang, Y. I-TASSER: A unified platform for automated protein structure and function prediction. *Nat. Protoc.* **2010**, *5*, 725–738. [CrossRef] [PubMed]

41. Yang, J.; Yan, R.; Roy, A.; Xu, D.; Poisson, J.; Zhang, Y. The I-TASSER Suite: Protein structure and function prediction. *Nat. Methods* **2015**, *12*, 7–8. [CrossRef] [PubMed]

42. TCGA. Available online: https://tcga-data.nci.nih.gov/tcga/ (accessed on 25 September 2017).

43. Primer3Plus. Available online: http://primer3plus.com/cgi-bin/dev/primer3plus.cgi (accessed on 12 October 2017).

44. UCSC-Genome Browser. Available online: https://genome.ucsc.edu (accessed on 8 January 2018).

45. Rienzo, M.; Schiano, C.; Casamassimi, A.; Grimaldi, V.; Infante, T.; Napoli, C. Identification of valid reference housekeeping genes for gene expression analysis in tumor neovascularization studies. *Clin. Transl. Oncol.* **2013**, *15*, 211–218. [CrossRef] [PubMed]

© 2018 by the authors. Licensee MDPI, Basel, Switzerland. This article is an open access article distributed under the terms and conditions of the Creative Commons Attribution (CC BY) license (http://creativecommons.org/licenses/by/4.0/).

International Journal of
Molecular Sciences

Review

Roles of Tristetraprolin in Tumorigenesis

Jeong-Min Park, Tae-Hee Lee and Tae-Hong Kang *

Department of Biological Science, Dong-A University, Busan 49315, Korea; zmpark@donga.ac.kr (J.-M.P.);
thlee@donga.ac.kr (T.-H.L.)
* Correspondence: thkang@dau.ac.kr; Tel.: +82-51-200-7261; Fax: +82-51-200-7269

Received: 5 October 2018; Accepted: 26 October 2018; Published: 29 October 2018

Abstract: Genetic loss or mutations in tumor suppressor genes promote tumorigenesis. The prospective tumor suppressor tristetraprolin (TTP) has been shown to negatively regulate tumorigenesis through destabilizing the messenger RNAs of critical genes implicated in both tumor onset and tumor progression. Regulation of TTP has therefore emerged as an important issue in tumorigenesis. Similar to other tumor suppressors, TTP expression is frequently downregualted in various human cancers, and its low expression is correlated with poor prognosis. Additionally, disruption in the regulation of TTP by various mechanisms results in the inactivation of TTP protein or altered TTP expression. A recent study showing alleviation of Myc-driven lymphomagenesis by the forced expression of TTP has shed light on new therapeutic avenues for cancer prevention and treatment through the restoration of TTP expression. In this review, we summarize key oncogenes subjected to the TTP-mediated mRNA degradation, and discuss how dysregulation of TTP can contribute to tumorigenesis. In addition, the control mechanism underlying TTP expression at the posttranscriptional and posttranslational levels will be discussed.

Keywords: tristetraprolin (TTP); tumorigenesis; posttranscriptional regulation; adenosine and uridine-rich elements (AREs)

1. Introduction

Posttranscriptional regulation of messenger RNA (mRNA) stability is essential for cells to rapidly respond to intracellular and extracellular stimuli [1,2]. The TPA-inducible sequence 11 (TIS11) family of RNA-binding proteins, composed of tristetraprolin (TTP) and butyrate response factors, modulates mRNA stability through direct binding to specific sequences located in the 3' untranslated region (3' UTR) of the target mRNA [3]. TTP, also known as TIS11A, G0/G1 switch regulatory protein 24 (GOS24), and growth factor-inducible nuclear protein NUP475, is encoded by the *ZFP36* gene. TTP contains a cysteine–cysteine-cysteine–histidine (CCCH) zinc finger motif for the recognition of cis-acting adenosine and uridine-rich elements (AREs) in the 3' UTR of target mRNA [4,5]. As illustrated in Figure 1, binding of TTP to AREs generally facilitates the decay of the mRNA by means of recruiting enzymes for the rapid shortening of the poly(A) tail [6]. For instance, TTP interacts with the carbon catabolite repressor protein 4 (Ccr4)-negative on TATA (Not1) deadenylase complex, the exosome components polymyositis/systemic sclerosis 75 (PM/Scl-75), and ribosomal RNA processing 4 (Rrp4) to hydrolyze the poly(A) tail in a processive manner [7]. Alternatively, TTP interacts with poly(A)-binding protein nuclear 1 (PABPN1) in the nucleus to inhibit 3'-polyadenylation of pre-mRNA [8]. The 5' to 3' degradation of mRNA is processed by a decapping complex, which includes mRNA-decapping enzyme 2 (Dcp2), enhancer of mRNA-decapping protein 3 (Edc3), and 5'–3' exoribonuclease (Xrn1), which interact with TTP [9,10].

The physiological importance of TTP in posttranscriptional coordination has been observed in TTP-deficient mice. These mice develop a complex syndrome of inflammatory arthritis, dermatitis, cachexia, autoimmunity, and myeloid hyperplasia [11]. These symptoms are recapitulated in the wild-type tumor necrosis factor alpha (TNF-α) transgenic and *TNF-α^{ΔARE}* mice [12,13]. Indeed, TTP has been shown to accelerate the degradation of *TNF-α* mRNA via direct binding to the ARE in the 3′ UTR of *TNF-α* mRNA [14].

It was revealed that approximately 16% of human protein-coding genes have at least one consensus motif of an ARE in their 3′ UTR [15]. Many of these genes are implicated in immune responses and tumorigenesis [16,17]. Importantly, TTP has been shown to negatively regulate tumorigenesis by destabilizing its target mRNA linked to tumor onset and progression [18,19]. Thus, dysregulation of TTP has been regarded as an important issue in tumorigenesis [20]. In this review, we describe the current understanding of TTP's roles in tumorigenesis, with a particular focus on the roles of TTP's target genes during tumorigenesis. We summarize key oncogenes and tumor-associated genes subjected to TTP-mediated mRNA decay, and discuss how dysregulation of this process potentially contributes to tumorigenesis.

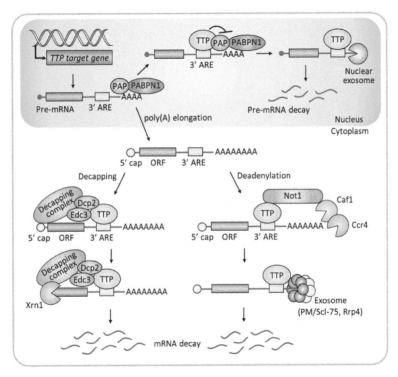

Figure 1. A schematic overview of posttranscriptional regulation of mRNA stability by TTP.

2. Oncogenes and Tumor Suppressor Genes Subjected to TTP-Mediated mRNA Decay

Tumorigenesis can be driven by the uncontrolled proliferation or the inappropriate survival of damaged cells due to the impairment of mRNA stability control of tumor-suppressor genes and oncogenes [21]. Table 1 shows the list of tumor-associated genes and their ARE sequences subjected to TTP-mediated mRNA degradation. The data indicate that TTP's target mRNAs during tumorigenesis are predominantly oncogenes as opposed to tumor suppressors; 21 oncogenes were targets, as compared to three tumor suppressor genes, such as cyclin-dependent kinase inhibitor 1 (*CDKN1A*), large tumor suppressor kinase 2 (*LATS2*), and aryl-hydrocarbon receptor repressor (*AHRR*) [22–24].

Table 1. List of oncogenes and tumor suppressor genes subjected to the ARE-mediated mRNA decay by TTP.

Gene Symbol	ARE Sequences	Regulation by TTP			
		ARE Binding [1]	3′ UTR Binding [2]	mRNA Decay [3]	Ref.
AHRR	TTCTGGCCTCTGGGCATTTATGGATTTAAGACCA GGATGGTATTTCAGAAGCTT	O	O	O	[23]
AKT-1	TTTTTTTACAACATTCAACTTTAGT	O	ND	O	[25,26]
BCL2	ATTTATTTATTTA	ND	O	O	[27]
BIRC3 (cIAP2)	TTTGGTTTCCTTAAAATTTTTATTTATTTACAACTC AAAAAACATTGTTTTG	O	O	O	[28,29]
CCNB1 (cyclin B1)	TTATTTACTTTTACCACTATTTAAG	O	O	O	[25,30]
CCND1 (cyclin D1)	TTATTATATTCCGTAGGTAGATGTG, ACATAATATATTCTATTTTTATACTCT	O	O	O	[25,31]
CDKN1A (p21)	TAGTCTCAGTTTGTGTGTCTTAATTATTATTTGTGT TTTAATTTAAACACCTCCT	O	O	O	[24]
CXCL1	TCTTCTATTTATTTATTTATTCATTAGTT	O	O	O	[25,32]
CXCL2 (MIP-2)	CACACTCTCCCATTATATTTATTG	O	ND	O	[25,33]
CXCR4	ACTTATTTATATAAATTTTTTTTG	O	O	O	[25,34]
E2F1	CTTTAATGGAGCGTTATTTATTTATCGAGGCC TCTTTG	O	O	O	[29,35]
FOS (c-Fos)	TAATTTATTTATT	ND	O	O	[36]
HMGA2	TGTAATTTAATGA	ND	O	O	[37]
IFN-γ	CTATTTATTAATATTTAA	O	O	O	[38]
JUN (c-Jun)	TTCTCTATTAGACTTTAGAAA, AGCACTCTGAGTTTACCATTTG	O	ND	ND	[25,39]
LATS2	TTCAAATTAGTATGATTCCTATTTAAAGTGATTTA TATTTGAGTAAAAAGTTCAA	O	O	O	[22]
Lin28A	TTTTATTTATTTG	O	O	O	[29,40]
MACC1	TATAATTTAATAT	ND	O	O	[41]
MYC (Myc)	AATTTCAATCCTAGTATATAGTACCTAGTATTAT AGGTACTATAAACCCTAATTTTTTTTATTTAA	O	O	O	[25,31]
PIM-1	CCTGGAGGTCAATGTTATGTATTTATTTATTTATT TATTTGGTTCCCTTCCTATTCC	O	O	O	[42]
PIM-3	TTTAATTTATTTG	ND	O	O	[43]
SNAI1 (Snail1)	GTTATATGTACAGTTTATTGATATTCAATAAAGC AGTTAATTTATATATTAAAAA	O	O	O	[44]
XIAP	CAAATTTATTTTATTTATTTAATT	O	O	O	[25,43]

[1] O: experimentally determined ARE sequences; ND: not determined experimentally, the predicted ARE sequences are from ARED-Plus web source; [2,3] O: experimentally confirmed; ND: not determined experimentally.

TTP has been shown to prevent malignant proliferation by suppressing the expression of genes for cell-cycle progression and cellular proliferation depicted in Figure 2. Among these, *CCNB1* (cyclin B1) is the key oncogenic driver whose overexpression itself leads to the chronic proliferation of cancer cells [45]. Previous studies reported that high CCNB1 expression levels were detected in various cancers such as breast, colon, and non-small cell lung cancer [46–48]. Ectopic overexpression of TTP suppressed CCNB1 expression but depletion of TTP promoted the accumulation of *CCNB1* mRNA in human lung cancer cells [30]. This is because the ARE motif in *CCNB1* 3′ UTR is subjected to TTP-mediated degradation [25]. High expression of *CCND1* (cyclin D1) also correlates with tumor onset and tumor progression [49]. A recent study showed that treatment with the mechanistic target of rapamycin kinase (mTOR) inhibitor, rapamycin, induced rapid *CCND1* mRNA decay due to the increased TTP expression in glioblastoma cells [31]. CCND1 binding with cyclin dependent kinase 4 (CDK4) or CDK6 is necessary for the G1/S transition [50]. The active CDK4/6 phosphorylates retinoblastoma 1 (RB), which results in the release of E2F1 [51]. Subsequently, E2F1 initiates the expression of genes required for the cell cycle transition [52]. E2F1 also contains three AREs in its 3′ UTR [35]. In the meantime, the *PIM-1* oncogene is also subjected to TTP-mediated mRNA decay. PIM-1 facilitates cell cycle progression via activating *CDC25a* and *CDC25c* oncogenes [53]. In pancreatic cancer, low TTP expression was correlated with high PIM-1 expression, and patients with such gene expression profiles showed unfavorable survival rates [54].

In addition, by suppressing the expression of lin-28 homolog A (Lin28A), TTP can increase the expression of the tumor suppressor microRNA (miRNA) let-7, whose expression is negatively regulated by Lin28A [29,55]. The miRNA let-7 has been implicated in the regulation of gene transcription including high mobility group A2 (HMGA2) [56]. HMGA2 is frequently upregulated in multiple cancers, and is associated with both malignant and benign tumor formation [57]. The bioinformatic analysis discovered that 3′ UTR of *HMGA2* mRNA contains the hairpin structure termed HMGA2-sh, which is further processed to a HMGA2-sh-3p20 fragment through the action of Drosha and Dicer [37]. Interestingly, HMGA2-sh-3p20 elevated HMGA2 expression in hepatoma cells by means of preventing TTP binding to the *HMGA* mRNA [37]. Thus, HMGA2-sh-3p20 facilitates hepatocarcinogenesis by antagonizing the TTP-mediated decay of *HMGA2* mRNA.

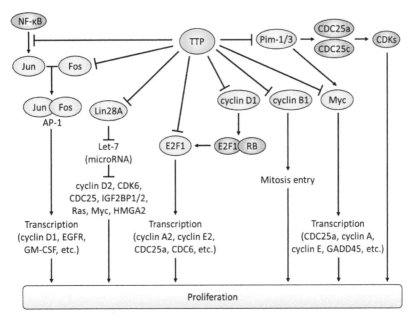

Figure 2. Attenuation of cellular proliferation by TTP-mediated suppression of oncogenic signalings.

The collapse of the homeostatic balance between cell death and cell proliferation is a hallmark of cancer [58]. Simultaneous overexpression of the anti-apoptotic protein BCL2 and the *Myc* oncogene induces lymphoma [59]. By downregulating the expression of both genes, TTP has been shown to alleviate Myc-driven lymphomagenesis [27,60]. IAP (inhibitors of apoptosis proteins) family anti-apoptotic protein BIRC3 and XIAP are also under control by TTP-mediated mRNA decay [28,43]. Therefore, loss of TTP function would confer resistance to cancer cells against apoptotic stimulus, and promotes cancer cell viability due to the impairment of the destabilizing of anti-apoptotic gene expression.

Aside from its canonical posttranscriptional role, TTP also has been implicated in the regulation of gene expression at the transcriptional level by participating in the nuclear factor kappa-light-chain-enhancer of activated B cells (NF-κB) pathway [39,61,62]. By blocking the nuclear import of NF-κB/p65, TTP suppresses the NF-κB-mediated transcription of oncogenes, including *c-Jun* [63]. c-Jun and c-Fos form the AP-1 early response transcription factor that promotes cell-cycle progression [64]. The stability of *c-Fos* mRNA is subjected to TTP-mediated posttranscriptional control [36]. Thereby, TTP regulates the activity of oncogenic AP-1 both at the transcriptional and posttranscriptional levels.

3. Roles of TTP in Tumor Progression

Recent studies have revealed novel TTP targets involved in the malignancy of tumors, such as epithelial-mesenchymal transition (EMT), invasion, and metastasis (Figure 3). Based on the gene expression profiles from 80 patient samples (23 normal colon mucosa, 30 primary colon carcinoma, and 27 liver metastases), lower TTP expression was detected in primary tumors as compared to normal mucosa [41]. Furthermore, TTP expression was remarkably downregulated in metastatic tumors as compared to primary tumors, suggesting the possibility that TTP is engaged in the EMT process. Consistent with this notion, recent studies have reported that TTP facilitates the mRNA decay of EMT marker genes including *Snail1* (zinc finger protein snail 1), *Twist1* (twist-related protein 1), *ZEB1* (zinc finger E-box binding homeobox 1), *MMP-2* (matrix metalloproteinase 2), and MMP-9 [41,44,65].

The molecular signature of low E-cadherin, high vimentin, and high N-cadherin is an indicator of cells undergoing EMT; this feature was also detected in circulating tumor cells [66]. NIH:OVCAR3 (ovarian adenocarcinoma) and HT29 (colorectal adenocarcinoma) cells with high TTP levels exhibited high E-cadherin, low N-cadherin, and low vimentin, whereas low TTP-expressing SKOV3 (ovarian adenocarcinoma) and H1299 (non-small lung carcinoma) cells displayed low E-cadherin expression [44]. Snail1, Twist1, and ZEB1 are transcription factors for the transcriptional repression of E-cadherin [67]. TTP binding to the ARE within the 3′ UTR of these three genes triggered their mRNA decay [44]. In the meantime, loss of TTP increased the growth rate and migration capability of colorectal cancer cells due to the upregulation of ZEB1, sex-determining region Y box 9 (SOX9), and metastasis-associated in colon cancer 1 (MACC1) [41]. Furthermore, one of the most significant alterations underlying colorectal cancer is the constitutive activation of the T-cell factor (Tcf)/β-catenin signaling, and the administration of Tcf/β-catenin inhibitor FH535 derepressed TTP expression [41]. Collectively, TTP downregulates Snail1, Twist1, ZEB1, SOX9, and MACC1 expression at the posttranscriptional level to inhibit EMT. Thus, the recovery of TTP expression seems to be promising to suppress EMT in some types of human cancers.

EMT facilitates the reorganization of the extracellular matrix (ECM), since many EMT-inducing factors activate the expression of MMPs [68]. MMPs induce ECM degradation and allow tumor cells to migrate out of the primary tumor to form metastases [69]. Specifically, MMP-1 is an interstitial collagenase that decomposes collagen types I, II, and III, and MMP-13 breaks down type II collagen more efficiently than types I and III [70]. MMP-2, along with MMP-9, cleaves type-IV collagen, which is the most abundant component of the basement membrane of which breakdown is a critical step in the invasion and metastatic progression of cancer cells [70]. An invasion experiment recapitulating the oral mucosa showed that the suppression of TTP activity gives rise to an accelerated invasion rate of head and neck cancer cells due to the secretion of MMP-2, MMP-9, and interleukin-6 (IL-6) [65]. Mechanistically, p38-mediated phosphorylation and the inactivation of TTP upregulated the stability of *MMP-2*, *MMP-9*, and *IL-6* transcripts [65]. Another study showed that ectopic re-expression of TTP in breast cancer cells attenuated the invasion rate because TTP suppressed MMP-1 and MMP-13 expression [71]. Urokinase-type plasminogen activator (uPA) and its specific receptor urokinase plasminogen activator receptor (uPAR) are also implicated in the degradation of the ECM [72]. Overexpression of uPA and uPAR has been observed in invasive glioblastomas [73], and the ectopic expression of TTP alleviated the invasiveness of these cancer cells by suppressing the expression of both uPA and uPAR [74,75]. In addition, a recent report showed that the upregulated TTP expression led to significant downregulation of uPA and MMP-9 protein expression in breast cancer [76]. Taken together, uPA and uPAR are physiological targets of TTP in various cancer types, and the concept of TTP-mediated downregulation of uPA and uPAR seems to be promising to attenuate the malignancy of tumors [75].

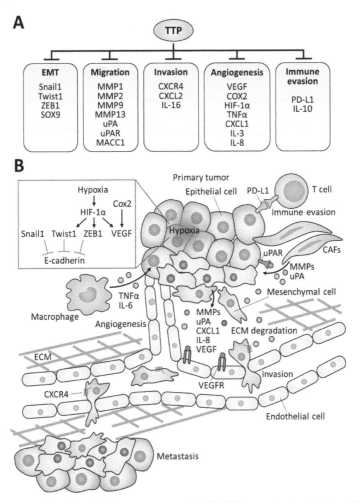

Figure 3. Suppressive roles of TTP in tumor progression. (**A**) TTP targets involved in the malignancy of tumors. EMT; epithelial-mesenchymal transition. (**B**) EMT and metastatic mechanisms driven by TTP target genes. CAFs; carcinoma-associated fibroblasts, ECM; extracellular matrix, VEGFR; vascular endothelial growth factor receptors.

Cancers develop in complex tissue environments known as tumor microenvironments, and these affect the growth and metastasis of tumor cells [77]. Tissues undergoing chronic inflammation due to the deregulation of the microenvironment generally exhibit a high incidence of cancer [78]. According to recent studies, programmed death-ligand 1 (PD-L1) is a novel target for TTP in gastric, lung, and colon cancer cells [79,80]. The expression of PD-L1 is essential for the development and functional maintenance of regulatory T cells [81], and its mRNA stability is negatively regulated by TTP [80]. Consequently, restoration of TTP expression enhanced anti-tumor immunity in a PD-L1-dependent manner [80]. Neoplastic cells are strongly influenced by the stroma, including surrounding and infiltrating cells. Immune cell infiltration into the tumor is an important determinant of tumor progression, and TTP depletion increases infiltration of monocytes/macrophages into the tumors [82]. IL-16 was identified as a critical TTP-regulated factor that contributes to the migration of immune cells [82]. IL-16 expression was increased in TTP-deficient 3D tumor spheroids, and elevated IL-16 levels enhanced the infiltration of monocytes into tumor spheroids [82]. Apparently, further studies are needed to determine the direct effect of TTP on IL-16 expression, but it seems clear that the loss of TTP allows immune cells within the microenvironment to promote tumor growth.

Tumor cells require a dedicated blood supply to obtain oxygen and nutrients for their maintenance and growth, and vascular endothelial growth factor (VEGF) is a crucial regulator of pathological angiogenesis [83]. TTP can bind to *VEGF* mRNA 3′ UTR and induce *VEGF* mRNA degradation [84]. Higher VEGF levels were detected in colorectal adenocarcinoma, as compared to normal tissues [85]. Cyclooxygenase 2 (COX-2) is also an important mediator of angiogenesis by facilitating the production of VEGF and BCL2 [86]. TTP binds between the nucleotides 3125 and 3232 in the 3′ UTR of *COX-2* mRNA and induces mRNA destabilization [87]. In colorectal cancer cells, low expression of TTP was responsible for the increased expression of COX-2 and VEGF, while overexpression of TTP in colon cancer cells markedly decreased the expression of both genes [88]. Moreover, cytokines related to tumor angiogenesis such as IL-3, IL-8, and TNF-α were reported as TTP targets which are suppressed by the way of ARE-mediated decay [32,89,90]. In contrast, TTP has been shown to increase human inducible nitric oxide synthase (iNOS) mRNA stability. TTP did not bind to human iNOS mRNA directly, but TTP destabilized the KH-type splicing regulatory protein (KSRP), which is responsible for iNOS mRNA decay, by facilitating recruitment of the exosome [91].

Although the precise mechanisms that determine the directional movement of tumor cells to distant sites are not well understood, this movement pattern seems similar to the chemokine-mediated leukocytes movement [92]. The expression of C-X-C motif chemokine receptor 4 (CXCR4) is low or absent in normal tissues, while it is highly expressed in various types of cancer, including colorectal cancer, ovarian cancer, and breast cancer [93,94], and the CXCR4 level was inversely correlated with TTP expression [34]. It has been revealed that CXCR4 is a TTP target containing a functional ARE in its 3′ UTR, and thus, induction of TTP results in the compromised CXCR4-mediated invasion and migration [34]. Furthermore, TTP depletion increased the production of several chemokines, such as C-X-C motif chemokine ligand 1 (CXCL1), CXCL2, and CXCL8 (also known as IL-8), which are involved in melanoma pathogenesis and angiogenesis [32,33,95]. Taken together, TTP has the ability to repress tumor metastasis by regulating chemokine-mediated migratory signaling.

As the tumor grows, consumption of nutrients and oxygen around it lead to a state of nutrient and oxygen deprivation [96]. Subsequently, hypoxia induces hypoxia-inducible factor 1α (HIF-1α), an important transcription factor involved in angiogenesis, leading to angiogenesis that allows nutrients to the microenvironment around tumor tissue [97]. Importantly, TTP expression was induced in hypoxic cells, and the overexpression of TTP repressed the hypoxic induction of HIF-1α protein in colorectal cancer cells [98]. Thus, it was proposed that cancer cells may benefit from the downregulation of TTP, which subsequently increases HIF-1α expression and assists with the adaptation of cancer cells to hypoxia.

4. Regulation of TTP Expression in Normal and Cancer Cells

Recent research has demonstrated that TTP is abnormally expressed in various human malignancies [60,85,99–105]. TTP was initially identified as a member of immediate early response genes that are rapidly induced by the stimulation of insulin [106], serum [107], or mitogen [108,109] in quiescent fibroblasts. Serum-stimulated *TTP* mRNA induction was dependent on consensus binding sites for several transcription factors, such as early growth response protein 1 (EGR1), specificity protein 1 (SP-1), and activator protein 1 (AP-2) in the 5′-proximal region of the TTP promoter [110]. A few studies have shown that transcription of *TTP* was induced by growth-inhibitory cytokines during an inflammatory response. For instance, transforming growth factor beta 1 (TGF-β1)-induced TTP transcription was mediated by the binding of Smad3/4 transcription factors to the putative Smad-binding elements of the TTP promoter in human T cells [111]. Parallel to this, TTP expression was necessary for TGF-β1-dependent growth inhibition in normal intestinal epithelium [112]. In addition, the TTP promoter contains putative binding sites for signal transducer and activator of transcription (STAT) proteins. Indeed, STAT1, STAT3, and STAT6 were recruited to these sites, and induced *TTP* gene transcription under stimulation by different cytokines. Interferon gamma (IFN-γ)-induced STAT1 phosphorylation promoted *TTP* gene transcription [113]. IL-10-activated STAT3 or IL-4-activated STAT6 induced TTP expression through the janus kinase 1 (JAK1) pathway [114,115]. Interestingly, *TTP* mRNA is highly unstable, and the rapid turnover of *TTP* mRNA is due to an auto-regulatory negative feedback loop [116,117].

TTP expression is often deficient in several cancer types (Figure 4). Rounbehler and colleagues [60] found that TTP was expressed at low levels in Myc-expressing cancers including breast, colorectal, and metastatic prostate cancer. The 5′-proximal region of the *TTP* gene includes a putative initiator element (Inr) near the TATA box [110]. Myc directly inhibits the transcription of *TTP* through direct binding to the Inr. In contrast, the tumor suppressor p53 activates *TTP* mRNA expression in human cancer cells [55]. Wild-type p53 stimulated by the DNA-damaging agent doxorubicin was recruited to the TTP promoter to activate *TTP* transcription, whereas mutant p53 failed to induce *TTP* transcription [55]. The epigenetic gene silencing of the TTP promoter has been shown as an alternative way to regulate TTP expression [105,118,119]. For instance, TGF-β1-dependent Smad-binding region located in the TTP promoter has a specific single CpG site. In hepatocellular carcinoma cell lines, TTP expression was attenuated frequently by methylation of the CpG site [105]. MicroRNA-29a (miR-29a) targets the 3′ UTR of *TTP* mRNA, leading to the degradation of *TTP* mRNA in cancer cells [101,120]. In pancreatic and breast cancer, miR-29a-mediated *TTP* mRNA degradation was associated with EMT, and promoted tumor growth, invasion, and metastasis [101,120].

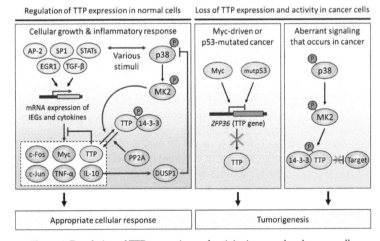

Figure 4. Regulation of TTP expression and activity in normal and cancer cells.

In addition to the loss of TTP expression, cells can become TTP deficient through a loss in TTP activity. Inactivation of TTP has been predominantly associated with its phosphorylation status. TTP can be phosphorylated by several kinases, including extracellular signal-regulated kinases (ERK) [121], p38 [121,122], c-Jun N-terminal kinases (JNK) [121], and v-akt murine thymoma viral oncogene (AKT) [31]. Among these, the p38 pathway is a major determinant for TTP activity but does not affect the protein level of TTP. p38 kinase phosphorylates and activates MAPK-activated protein kinase 2 (MK2); subsequently, MK2 phosphorylates TTP at serine 60 and 186 [123]. Subsequently, 14-3-3 proteins interact with phosphorylated TTP and inactivate it [31,124–126]. The TTP-14-3-3 complex cannot recruit the Ccr4-Not1 deadenylase complex, but has no impact on the binding affinity of ARE [124,125]. In addition, the interaction with 14-3-3 is required for cytoplasmic accumulation of TTP [126], which results in the inhibition of TTP's role in the nucleus. Cytoplasmic TTP promotes the decay of mRNA containing AREs [18], whereas nuclear TTP functions as a transcriptional corepressor of NF-κB [39,127] and several nuclear receptors [128,129]. In breast cancer cells, ectopic overexpression of TTP was capable of repressing the transactivation activity of nuclear receptors, including estrogen receptor alpha (ERα), progesterone receptor (PR), glucocorticoid receptor (GR), and androgen receptor (AR), via physically interacting with these factors [128]. Mechanistically, nuclear TTP attenuates ERα transactivation by disrupting its interaction with steroid receptor coactivator 1 (SRC-1) [129].

The phosphorylation-induced TTP inactivation is reversed by two phosphatases. Protein phosphatase 2A (PP2A) directly dephosphorylates and reactivates TTP, but this results in a decrease in TTP protein stability [130–134]. Another phosphatase for TTP reactivation is dual specificity phosphatase 1 (DUSP1), that indirectly regulates TTP activity through the dephosphorylation of p38, which results in the inactivation of p38 kinase activity [135]. Several reports have indicated that a high level of the phosphorylated, inactive form of TTP was found in head and neck [65] and brain cancer cells [99,136]. Thus, the pharmacological application of a p38 inhibitor against these cancers may provide new ways to treat cancers containing hyperphosphorylated TTP.

The first p38 inhibitors were identified in a screen for compounds that inhibited expression of TNF-α in human monocytic leukemia cell line [137]. In multiple myeloma and head and neck squamous cell carcinoma, p38 inhibitors were successfully used to limit tumor growth and angiogenesis, due indirectly to TTP-mediated inhibition of cytokine secretion [138,139]. p38 inhibitor also attenuates progression of malignant gliomas by inhibition of TTP phosphorylation [99]. Dufies and colleagues reported that p38 inhibitors may be a promising adjuvant therapy in cancer. Sunitinib, known as a first-line treatment for metastatic renal cell carcinoma, leads to patient relapse by p38 activation. While sunitinib mainly targets the host blood vessels via the inhibition of VEGF receptors, the mechanism of patient relapse is associated with increased lymphangiogenesis and lymph node metastasis via induction of *VEGFC* mRNA expression through p38-mediated inactivation of TTP. In renal cancer cells, the p38 inhibitor reduces the sunitinib-dependent increase in the *VEGFC* mRNA [140]. Several independent groups have identified effective drug candidates targeting TTP for anticancer therapies. For instance, Sorafenib targeting v-Raf murine sarcoma viral oncogene homolog B (B-Raf) kinase triggers re-expression of TTP in melanoma cells via the inhibition of B-Raf-dependent ERK activity [95]. Gambogic acid, the main active compound of *Gamboge hanburyi*, also induces upregulation of TTP expression through ERK inactivation, and efficiently inhibits the progression of colorectal cancer cells [141]. Histone deacetylase (HDAC) inhibitors in colorectal cancer cells induce TTP expression through activation of EGR1, which promote its binding affinity to ARE, and thus, reduce cell growth and angiogenesis [118,119]. Resveratrol, a natural polyphenolic compound present in many plant species, including grapes, peanuts, and berries, inhibits cell growth through TTP upregulation in several cancers, including breast [142], colorectal [29] and brain cancer [74]. Molecular activators of PP2A enhance the anti-inflammatory function of TTP in lung cancer cells, and thus, provide pharmacotherapeutic strategies to chronic inflammation-mediated cancer [133]. In addition, treatment with MK2 inhibitor triggers apoptosis in hepatocellular carcinoma. TTP knockdown rescued these cells from apoptosis in the presence of MK2 inhibitor, suggesting that the MK2-mediated TTP inactivation plays a role in cell survival of hepatocellular carcinoma [26]. These studies increase the understanding of the anti-cancer effects of various compounds and the molecular basis for further applications of therapeutic agents targeting TTP in clinical cancer therapy.

5. Conclusions

The role of TTP as a key factor in posttranscriptional gene regulation has been established, especially with regard to its function in promoting mRNA decay of ARE-containing genes, including oncogenes and cancer-related cytokines. What has become more obvious is that TTP participates extensively in gene regulatory networks for tumor suppression. Cumulative evidence was provided that the loss of TTP expression or function was closely related with tumor onset and tumor progression, and presented poor outcomes of cancer patients. Based on current knowledge, many factors and signal pathways have been identified to regulate TTP at the transcriptional, posttranscriptional, and posttranslational level. The abnormal expression or activity of these factors consequently affected TTP's expression or function. Therefore, endeavors for searching molecular pathways or chemical compounds upregulating TTP expression or activity will pave the way for potentially attractive therapeutics for cancer treatment.

Funding: This research was supported by Basic Science Research Program through the National Research Foundation of Korea (NRF) funded by the Ministry of Education (NRF-2018R1D1A3B07043817 and NRF-2015R1D1A1A01056994).

Conflicts of Interest: The authors declare no conflict of interest.

Abbreviations

3′ UTR	3′ untranslated regions
AHRR	Aryl-hydrocarbon receptor repressor
AKT-1	AKT serine/threonine kinase 1
AP-1	Activator protein 1
AR	Androgen receptor
AREs	Adenosine and uridine-rich elements
BCL2	B-cell CLL/Lymphoma 2
BIRC3	Baculoviral IAP repeat containing 3
B-Raf	v-Raf murine sarcoma viral oncogene homolog B
CAFs	Carcinoma-associated fibroblasts
CCCH	Cysteine–cysteine-cysteine–histidine
CCNB1	Cyclin B1
CCND1	Cyclin D1
Ccr4	Carbon catabolite repressor protein 4
CDC25A	Cell division cycle 25 homolog A
CDK4	Cyclin dependent kinase 4
CDK6	Cyclin dependent kinase 6
CDKN1A	Cyclin dependent kinase inhibitor 1A
COX-2	Cyclooxygenase 2
CXCL1	C-X-C motif chemokine ligand 1
CXCL2	C-X-C motif chemokine ligand 2
CXCR4	C-X-C motif chemokine receptor 4
Dcp2	mRNA-decapping enzyme 2
DUSP1	Dual specificity phosphatase 1
E2F1	E2F transcription factor 1
ECM	Extracellular matrix
EGR1	Early growth response protein 1
Edc3	Enhancer of mRNA-decapping protein 3
EMT	Epithelial-mesenchymal transition
ERK	Extracellular signal-regulated kinases
ERα	Estrogen receptor alpha
FOS	FBJ murine osteosarcoma viral oncogene homolog
GOS24	G0/G1 switch regulatory protein 24
GR	Glucocorticoid receptor
HDAC	Histone deacetylase

HIF-1α	Hypoxia-inducible factor 1α
HMGA2	High mobility group A2
IAP	Inhibitors of apoptosis proteins
IFN-γ	Interferon gamma
IL-6	Interleukin-6
iNOS	inducible nitric oxide synthase
Inr	Initiator element
JAK1	Janus kinase 1
JNK	c-Jun N-terminal kinases
JUN	v-jun avian sarcoma virus 17 oncogene homolog
KSRP	KH-type splicing regulatory protein
LATS2	Large tumor suppressor kinase 2
Lin28A	Lin-28 homolog A
MACC1	Metastasis associated in colon cancer 1
MAPK	Mitogen-activated protein kinase
miR-29a	MicroRNA-29a
MK2	MAPK-activated protein kinase 2
MMP-13	Matrix metalloproteinase 13
MMP-2	Matrix metalloproteinase 2
MMP-9	Matrix metalloproteinase 9
mRNA	Messenger RNAs
Myc	v-myc avian myelocytomatosis viral oncogene homolog
NF-κB	Nuclear factor kappa-light-chain-enhancer of activated B cells
Not1	Negative on TATA 1
NUP475	Growth factor-inducible nuclear protein NUP475
PABPN1	Poly-A-binding protein nuclear 1
PD-L1	Programmed death-ligand 1
PIM-1	Proto-oncogene serine/threonine-protein kinase pim 1
PM/Scl-75	Polymyositis/systemic sclerosis 75
PP2A	Protein phosphatase 2A
PR	Progesterone receptor
RB	Retinoblastoma 1
Rrp4	Ribosomal RNA processing 4
SNAI1	Zinc finger protein snail 1
SOX9	Sex-determining region Y box 9
SP-1	Specificity protein 1
SRC-1	Steroid receptor coactivator 1
STAT	Signal transducer and activator of transcription
Tcf	T-cell factor
TGF-β1	Transforming growth factor beta 1
TIS	12-*O*-tetradecanoylphorbol-13-acetate (TPA)-induced sequence
TNF-α	Tumor necrosis factor alpha
TTP	Tristetraprolin
Twist1	Twist-related protein 1
uPA	Urokinase-type plasminogen activator
uPAR	Urokinase plasminogen activator receptor
VEGF	Vascular endothelial growth factor
VEGFR	Vascular endothelial growth factor receptors
XIAP	X-linked inhibitor of apoptosis
Xrn1	5′-3′ exoribonuclease
ZEB1	Zinc finger E-box binding homeobox 1
ZFP36	Zinc finger protein 36

References

1. Guhaniyogi, J.; Brewer, G. Regulation of mRNA stability in mammalian cells. *Gene* **2001**, *265*, 11–23. [CrossRef]
2. Park, J.M.; Kang, T.H. Transcriptional and posttranslational regulation of nucleotide excision repair: The guardian of the genome against ultraviolet radiation. *Int. J. Mol. Sci.* **2016**, *17*, 1840. [CrossRef] [PubMed]
3. Sanduja, S.; Blanco, F.F.; Dixon, D.A. The roles of TTP and BRF proteins in regulated mRNA decay. *Wiley Interdisc. Rev. RNA* **2011**, *2*, 42–57. [CrossRef] [PubMed]
4. Fu, M.G.; Blackshear, P.J. RNA-binding proteins in immune regulation: A focus on ccch zinc finger proteins. *Nat. Rev. Immunol.* **2017**, *17*, 130–143. [CrossRef] [PubMed]
5. Hudson, B.P.; Martinez-Yamout, M.A.; Dyson, H.J.; Wright, P.E. Recognition of the mRNA AU-rich element by the zinc finger domain of tis11d. *Nat. Struct. Mol. Biol.* **2004**, *11*, 257–264. [CrossRef] [PubMed]
6. Lai, W.S.; Carballo, E.; Thorn, J.M.; Kennington, E.A.; Blackshear, P.J. Interactions of CCCH zinc finger proteins with mRNA—Binding of tristetraprolin-related zinc finger proteins to au-rich elements and destabilization of mRNA. *J. Biol. Chem.* **2000**, *275*, 17827–17837. [CrossRef] [PubMed]
7. Sandler, H.; Kreth, J.; Timmers, H.T.M.; Stoecklin, G. Not1 mediates recruitment of the deadenylase CAF1 to mRNAs targeted for degradation by tristetraprolin. *Nucleic Acids Res.* **2011**, *39*, 4373–4386. [CrossRef] [PubMed]
8. Su, Y.L.; Wang, S.C.; Chiang, P.Y.; Lin, N.Y.; Shen, Y.F.; Chang, G.D.; Chang, C.J. Tristetraprolin inhibits poly(a)-tail synthesis in nuclear mRNA that contains AU-rich elements by interacting with poly(a)-binding protein nuclear 1. *PLoS ONE* **2012**, *7*, e41313. [CrossRef] [PubMed]
9. Eulalio, A.; Behm-Ansmant, I.; Izaurralde, E. P bodies: At the crossroads of post-transcriptional pathways. *Nat. Rev. Mol. Cell Biol.* **2007**, *8*, 9–22. [CrossRef] [PubMed]
10. Fenger-Gron, M.; Fillman, C.; Norrild, B.; Lykke-Andersen, J. Multiple processing body factors and the are binding protein ttp activate mRNA decapping. *Mol. Cell* **2005**, *20*, 905–915. [CrossRef] [PubMed]
11. Taylor, G.A.; Carballo, E.; Lee, D.M.; Lai, W.S.; Thompson, M.J.; Patel, D.D.; Schenkman, D.I.; Gilkeson, G.S.; Broxmeyer, H.E.; Haynes, B.F.; et al. A pathogenetic role for tnf alpha in the syndrome of cachexia, arthritis, and autoimmunity resulting from tristetraprolin (TTP) deficiency. *Immunity* **1996**, *4*, 445–454. [CrossRef]
12. Probert, L.; Akassoglou, K.; Alexopoulou, L.; Douni, E.; Haralambous, S.; Hill, S.; Kassiotis, G.; Kontoyiannis, D.; Pasparakis, M.; Plows, D.; et al. Dissection of the pathologies induced by transmembrane and wild-type tumor necrosis factor in transgenic mice. *J. Leukoc. Biol.* **1996**, *59*, 518–525. [CrossRef] [PubMed]
13. Kontoyiannis, D.; Pasparakis, M.; Pizarro, T.T.; Cominelli, F.; Kollias, G. Impaired on/off regulation of tnf biosynthesis in mice lacking tnf AU-rich elements: Implications for joint and gut-associated immunopathologies. *Immunity* **1999**, *10*, 387–398. [CrossRef]
14. Lai, W.S.; Carballo, E.; Strum, J.R.; Kennington, E.A.; Phillips, R.S.; Blackshear, P.J. Evidence that tristetraprolin binds to au-rich elements and promotes the deadenylation and destabilization of tumor necrosis factor alpha mRNA. *Mol. Cell. Biol.* **1999**, *19*, 4311–4323. [CrossRef] [PubMed]
15. Gruber, A.R.; Fallmann, J.; Kratochvill, F.; Kovarik, P.; Hofacker, I.L. Aresite: A database for the comprehensive investigation of au-rich elements. *Nucleic Acids Res.* **2011**, *39*, D66–D69. [CrossRef] [PubMed]
16. Khabar, K.S.A. Hallmarks of cancer and AU-rich elements. *Wiley Interdiscip. Rev. RNA* **2017**, *8*, 1368. [CrossRef] [PubMed]
17. Bisogno, L.S.; Keene, J.D. RNA regulons in cancer and inflammation. *Curr. Opin. Genet. Dev.* **2018**, *48*, 97–103. [CrossRef] [PubMed]
18. Brooks, S.A.; Blackshear, P.J. Tristetraprolin (TTP): Interactions with mRNA and proteins, and current thoughts on mechanisms of action. *Biochim. Biophys. Acta Gene Regul. Mech.* **2013**, *1829*, 666–679. [CrossRef] [PubMed]
19. Guo, J.; Qu, H.H.; Chen, Y.; Xia, J.Z. The role of RNA-binding protein tristetraprolin in cancer and immunity. *Med. Oncol.* **2017**, *34*. [CrossRef] [PubMed]
20. Wang, H.; Ding, N.N.; Guo, J.; Xia, J.Z.; Ruan, Y.L. Dysregulation of TTP and hur plays an important role in cancers. *Tumor Biol.* **2016**, *37*, 14451–14461. [CrossRef] [PubMed]
21. Esquela-Kerscher, A.; Slack, F.J. Oncomirs—Micrornas with a role in cancer. *Nat. Rev. Cancer* **2006**, *6*, 259–269. [CrossRef] [PubMed]

22. Lee, H.H.; Vo, M.T.; Kim, H.J.; Lee, U.H.; Kim, C.W.; Kim, H.K.; Ko, M.S.; Lee, W.H.; Cha, S.J.; Min, Y.J.; et al. Stability of the lats2 tumor suppressor gene is regulated by tristetraprolin. *J. Biol. Chem.* **2010**, *285*, 17329–17337. [CrossRef] [PubMed]
23. Lee, H.H.; Kim, W.T.; Kim, D.H.; Park, J.W.; Kang, T.H.; Chung, J.W.; Leem, S.H. Tristetraprolin suppresses ahrr expression through mRNA destabilization. *FEBS Lett.* **2013**, *587*, 1518–1523. [CrossRef] [PubMed]
24. Al-Haj, L.; Blackshear, P.J.; Khabar, K.S.A. Regulation of P21/CIP1/WAF-1 mediated cell-cycle arrest by rnase l and tristetraprolin, and involvement of au-rich elements. *Nucleic Acids Res.* **2012**, *40*, 7739–7752. [CrossRef] [PubMed]
25. Mukherjee, N.; Jacobs, N.C.; Hafner, M.; Kennington, E.A.; Nusbaum, J.D.; Tuschl, T.; Blackshear, P.J.; Ohler, U. Global target mRNA specification and regulation by the RNA-binding protein ZFP36. *Genome Biol.* **2014**, *15*. [CrossRef] [PubMed]
26. Tran, D.D.H.; Koch, A.; Allister, A.; Saran, S.; Ewald, F.; Koch, M.; Nashan, B.; Tamura, T. Treatment with mapkap2 (MK2) inhibitor and DNA methylation inhibitor, 5-aza dc, synergistically triggers apoptosis in hepatocellular carcinoma (HCC) via tristetraprolin (TTP). *Cell. Signal.* **2016**, *28*, 1872–1880. [CrossRef] [PubMed]
27. Park, S.B.; Lee, J.H.; Jeong, W.W.; Kim, Y.H.; Cha, H.J.; Joe, Y.; Chung, H.T.; Cho, W.J.; Do, J.W.; Lee, B.J.; et al. TTP mediates cisplatin-induced apoptosis of head and neck cancer cells by down-regulating the expression of Bcl-2. *J. Chemother.* **2015**, *27*, 174–180. [CrossRef] [PubMed]
28. Kim, C.W.; Kim, H.K.; Vo, M.T.; Lee, H.H.; Kim, H.J.; Min, Y.J.; Cho, W.J.; Park, J.W. Tristetraprolin controls the stability of CIAP2 mRNA through binding to the 3' UTR of CIAP2 mRNA. *Biochem. Biophys. Res. Commun.* **2010**, *400*, 46–52. [CrossRef] [PubMed]
29. Lee, S.R.; Jin, H.; Kim, W.T.; Kim, W.J.; Kim, S.Z.; Leem, S.H.; Kim, S.M. Tristetraprolin activation by resveratrol inhibits the proliferation and metastasis of colorectal cancer cells. *Int. J. Oncol.* **2018**, *53*, 1269–1278. [CrossRef] [PubMed]
30. Zheng, X.T.; Xiao, X.Q. Sodium butyrate down-regulates tristetraprolin-mediated cyclin b1 expression independent of the formation of processing bodies. *Int. J. Biochem. Cell Biol.* **2015**, *69*, 241–248. [CrossRef] [PubMed]
31. Marderosian, M.; Sharma, A.; Funk, A.P.; Vartanian, R.; Masri, J.; Jo, O.D.; Gera, J.F. Tristetraprolin regulates Cyclin D1 and c-Myc mRNA stability in response to rapamycin in an AKT-dependent manner via p38 MAPK signaling. *Oncogene* **2006**, *25*, 6277–6290. [CrossRef] [PubMed]
32. Datta, S.; Biswas, R.; Novotny, M.; Pavicic, P.G.; Herjan, T.; Mandal, P.; Hamilton, T.A. Tristetraprolin regulates cxcl1 (KC) mRNA stability. *J. Immunol.* **2008**, *180*, 2545–2552. [CrossRef] [PubMed]
33. Jalonen, U.; Nieminen, R.; Vuolteenaho, K.; Kankaanranta, H.; Moilanen, E. Down-regulation of tristetraprolin expression results in enhanced IL-12 and MIP-2 production and reduced MIP-3alpha synthesis in activated macrophages. *Mediat. Inflamm.* **2006**, *2006*, 40691. [CrossRef] [PubMed]
34. Al-Souhibani, N.; Al-Ghamdi, M.; Al-Ahmadi, W.; Khabar, K.S.A. Posttranscriptional control of the chemokine receptor CXCR4 expression in cancer cells. *Carcinogenesis* **2014**, *35*, 1983–1992. [CrossRef] [PubMed]
35. Lee, H.H.; Lee, S.R.; Leem, S.H. Tristetraprolin regulates prostate cancer cell growth through suppression of E2F1. *J. Microbiol. Biotechnol.* **2014**, *24*, 287–294. [CrossRef] [PubMed]
36. Amit, I.; Citri, A.; Shay, T.; Lu, Y.L.; Katz, M.; Zhang, F.; Tarcic, G.; Siwak, D.; Lahad, J.; Jacob-Hirsch, J.; et al. A module of negative feedback regulators defines growth factor signaling. *Nat. Genet.* **2007**, *39*, 503–512. [CrossRef] [PubMed]
37. Wang, Y.; Chen, F.Q.; Yang, Z.; Zhao, M.; Zhang, S.Q.; Gao, Y.; Feng, J.Y.; Yang, G.; Zhang, W.Y.; Ye, L.H.; et al. The fragment HMGA2-sh-3p20 from HMGA2 mRNA 3'UTR promotes the growth of hepatoma cells by upregulating HMGA2. *Sci. Rep.* **2017**, *7*, 2070. [CrossRef] [PubMed]
38. Ogilvie, R.L.; SternJohn, J.R.; Rattenbacher, B.; Vlasova, I.A.; Williams, D.A.; Hau, H.H.; Blackshear, P.J.; Bohjanen, P.R. Tristetraprolin mediates interferon-gamma mRNA decay. *J. Biol. Chem.* **2009**, *284*, 11216–11223. [CrossRef] [PubMed]
39. Xu, L.; Ning, H.; Gu, L.; Wang, Q.; Lu, W.; Peng, H.; Cui, W.; Ying, B.; Ross, C.R.; Wilson, G.M.; et al. Tristetraprolin induces cell cycle arrest in breast tumor cells through targeting AP-1/c-jun and nf-kappab pathway. *Oncotarget* **2015**, *6*, 41679–41691. [CrossRef] [PubMed]

40. Kim, C.W.; Vo, M.T.; Kim, H.K.; Lee, H.H.; Yoon, N.A.; Lee, B.J.; Min, Y.J.; Joo, W.D.; Cha, H.J.; Park, J.W.; et al. Ectopic over-expression of tristetraprolin in human cancer cells promotes biogenesis of let-7 by down-regulation of Lin28. *Nucleic Acids Res.* **2012**, *40*, 3856–3869. [CrossRef] [PubMed]

41. Montorsi, L.; Guizzetti, F.; Alecci, C.; Caporali, A.; Martello, A.; Atene, C.G.; Parenti, S.; Pizzini, S.; Zanovello, P.; Bortoluzzi, S.; et al. Loss of zfp36 expression in colorectal cancer correlates to wnt/beta-catenin activity and enhances epithelial-to-mesenchymal transition through upregulation of zeb1, sox9 and macc1. *Oncotarget* **2016**, *7*, 59144–59157. [CrossRef] [PubMed]

42. Kim, H.K.; Kim, C.W.; Vo, M.T.; Lee, H.H.; Lee, J.Y.; Yoon, N.A.; Lee, C.Y.; Moon, C.H.; Min, Y.J.; Park, J.W.; et al. Expression of proviral integration site for moloney murine leukemia virus 1 (pim-1) is post-transcriptionally regulated by tristetraprolin in cancer cells. *J. Biol. Chem.* **2012**, *287*, 28770–28778. [CrossRef] [PubMed]

43. Selmi, T.; Martello, A.; Vignudelli, T.; Ferrari, E.; Grande, A.; Gemelli, C.; Salomoni, P.; Ferrari, S.; Zanocco-Marani, T. zfp36 Expression impairs glioblastoma cell lines viability and invasiveness by targeting multiple signal transduction pathways. *Cell Cycle* **2012**, *11*, 1977–1987. [CrossRef] [PubMed]

44. Yoon, N.A.; Jo, H.G.; Lee, U.H.; Park, J.H.; Yoon, J.E.; Ryu, J.; Kang, S.S.; Min, Y.J.; Ju, S.A.; Seo, E.H.; et al. Tristetraprolin suppresses the EMT through the down-regulation of twist1 and snail1 in cancer cells. *Oncotarget* **2016**, *7*, 8931–8943. [CrossRef] [PubMed]

45. Yuan, J.; Kramer, A.; Matthess, Y.; Yan, R.; Spankuch, B.; Gatje, R.; Knecht, R.; Kaufmann, M.; Strebhardt, K. Stable gene silencing of Cyclin b1 in tumor cells increases susceptibility to taxol and leads to growth arrest in vivo. *Oncogene* **2006**, *25*, 1753–1762. [CrossRef] [PubMed]

46. Kawamoto, H.; Koizumi, H.; Uchikoshi, T. Expression of the G2-M checkpoint regulators Cyclin b1 and cdc2 in nonmalignant and malignant human breast lesions: Immunocytochemical and quantitative image analyses. *Am. J. Pathol.* **1997**, *150*, 15–23. [PubMed]

47. Wang, A.; Yoshimi, N.; Ino, N.; Tanaka, T.; Mori, H. Overexpression of Cyclin b1 in human colorectal cancers. *J. Cancer Res. Clin. Oncol.* **1997**, *123*, 124–127. [CrossRef] [PubMed]

48. Soria, J.C.; Jang, S.J.; Khuri, F.R.; Hassan, K.; Liu, D.; Hong, W.K.; Mao, L. Overexpression of Cyclin b1 in early-stage non-small cell lung cancer and its clinical implication. *Cancer Res.* **2000**, *60*, 4000–4004. [PubMed]

49. Diehl, J.A. Cycling to cancer with Cyclin d1. *Cancer Biol. Ther.* **2002**, *1*, 226–231. [CrossRef] [PubMed]

50. Musgrove, E.A.; Caldon, C.E.; Barraclough, J.; Stone, A.; Sutherland, R.L. Cyclin d as a therapeutic target in cancer. *Nat. Rev. Cancer* **2011**, *11*, 558–572. [CrossRef] [PubMed]

51. Wang, S.; Nath, N.; Fusaro, G.; Chellappan, S. Rb and prohibitin target distinct regions of e2f1 for repression and respond to different upstream signals. *Mol. Cell. Biol.* **1999**, *19*, 7447–7460. [CrossRef] [PubMed]

52. Lee, S.R.; Roh, Y.G.; Kim, S.K.; Lee, J.S.; Seol, S.Y.; Lee, H.H.; Kim, W.T.; Kim, W.J.; Heo, J.; Cha, H.J.; et al. Activation of ezh2 and suz12 regulated by E2F1 predicts the disease progression and aggressive characteristics of bladder cancer. *Clin. Cancer Res.* **2015**, *21*, 5391–5403. [CrossRef] [PubMed]

53. Zhang, Y.; Wang, Z.; Magnuson, N.S. Pim-1 kinase-dependent phosphorylation of p21Cip1/WAF1 regulates its stability and cellular localization in H1299 cells. *Mol. Cancer Res.* **2007**, *5*, 909–922. [CrossRef] [PubMed]

54. Wei, Z.R.; Liang, C.; Feng, D.; Cheng, Y.J.; Wang, W.M.; Yang, D.J.; Wang, Y.X.; Cai, Q.P. Low tristetraprolin expression promotes cell proliferation and predicts poor patients outcome in pancreatic cancer. *Oncotarget* **2016**, *7*, 17737–17750. [CrossRef] [PubMed]

55. Lee, J.Y.; Kim, H.J.; Yoon, N.A.; Lee, W.H.; Min, Y.J.; Ko, B.K.; Lee, B.J.; Lee, A.; Cha, H.J.; Cho, W.J.; et al. Tumor suppressor p53 plays a key role in induction of both tristetraprolin and let-7 in human cancer cells. *Nucleic Acids Res.* **2013**, *41*, 5614–5625. [CrossRef] [PubMed]

56. Lujambio, A.; Lowe, S.W. The microcosmos of cancer. *Nature* **2012**, *482*, 347–355. [CrossRef] [PubMed]

57. Wu, J.J.; Zhang, S.Z.; Shan, J.L.; Hu, Z.J.; Liu, X.Y.; Chen, L.R.; Ren, X.C.; Yao, L.F.; Sheng, H.Q.; Li, L.; et al. Elevated HMGA2 expression is associated with cancer aggressiveness and predicts poor outcome in breast cancer. *Cancer Lett.* **2016**, *376*, 284–292. [CrossRef] [PubMed]

58. Cooper, J.P.; Youle, R.J. Balancing cell growth and death. *Curr. Opin. Cell Biol.* **2012**, *24*, 802–803. [CrossRef] [PubMed]

59. Otake, Y.; Soundararajan, S.; Sengupta, T.K.; Kio, E.A.; Smith, J.C.; Pineda-Roman, M.; Stuart, R.K.; Spicer, E.K.; Fernandes, D.J. Overexpression of nucleolin in chronic lymphocytic leukemia cells induces stabilization of bcl2 mRNA. *Blood* **2007**, *109*, 3069–3075. [PubMed]

60. Rounbehler, R.J.; Fallahi, M.; Yang, C.; Steeves, M.A.; Li, W.; Doherty, J.R.; Schaub, F.X.; Sanduja, S.; Dixon, D.A.; Blackshear, P.J.; et al. Tristetraprolin impairs Myc-induced lymphoma and abolishes the malignant state. *Cell* **2012**, *150*, 563–574. [CrossRef] [PubMed]

61. Gu, L.; Ning, H.; Qian, X.; Huang, Q.; Hou, R.; Almourani, R.; Fu, M.; Blackshear, P.J.; Liu, J. Suppression of il-12 production by tristetraprolin through blocking nf-kcyb nuclear translocation. *J. Immunol.* **2013**, *191*, 3922–3930. [CrossRef] [PubMed]

62. Chen, Y.L.; Jiang, Y.W.; Su, Y.L.; Lee, S.C.; Chang, M.S.; Chang, C.J. Transcriptional regulation of tristetraprolin by NF-kappaB signaling in lps-stimulated macrophages. *Mol. Biol. Rep.* **2013**, *40*, 2867–2877. [CrossRef] [PubMed]

63. Schichl, Y.M.; Resch, U.; Hofer-Warbinek, R.; de Martin, R. Tristetraprolin impairs NF-kappaB/p65 nuclear translocation. *J. Biol. Chem.* **2009**, *284*, 29571–29581. [CrossRef] [PubMed]

64. Schreiber, M.; Kolbus, A.; Piu, F.; Szabowski, A.; Mohle-Steinlein, U.; Tian, J.; Karin, M.; Angel, P.; Wagner, E.F. Control of cell cycle progression by c-jun is p53 dependent. *Genes Dev.* **1999**, *13*, 607–619. [CrossRef] [PubMed]

65. Van Tubergen, E.A.; Banerjee, R.; Liu, M.; Broek, R.V.; Light, E.; Kuo, S.; Feinberg, S.E.; Willis, A.L.; Wolf, G.; Carey, T.; et al. Inactivation or loss of TTP promotes invasion in head and neck cancer via transcript stabilization and secretion of MMP9, MMP2, and IL-6. *Clin. Cancer Res.* **2013**, *19*, 1169–1179. [CrossRef] [PubMed]

66. Roussos, E.T.; Keckesova, Z.; Haley, J.D.; Epstein, D.M.; Weinberg, R.A.; Condeelis, J.S. AACR special conference on epithelial-mesenchymal transition and cancer progression and treatment. *Cancer Res.* **2010**, *70*, 7360–7364. [CrossRef] [PubMed]

67. Nieto, M.A.; Huang, R.Y.J.; Jackson, R.A.; Thiery, J.P. EMT: 2016. *Cell* **2016**, *166*, 21–45. [CrossRef] [PubMed]

68. Radisky, E.S.; Radisky, D.C. Matrix metalloproteinase-induced epithelial-mesenchymal transition in breast cancer. *J. Mammary Gland Biol. Neoplasia* **2010**, *15*, 201–212. [CrossRef] [PubMed]

69. Egeblad, M.; Werb, Z. New functions for the matrix metalloproteinases in cancer progression. *Nat. Rev. Cancer* **2002**, *2*, 161–174. [CrossRef] [PubMed]

70. Kessenbrock, K.; Plaks, V.; Werb, Z. Matrix metalloproteinases: Regulators of the tumor microenvironment. *Cell* **2010**, *141*, 52–67. [CrossRef] [PubMed]

71. Al-Ahmadi, W.; Al-Ghamdi, M.; Al-Souhibani, N.; Khabar, K.S.A. miR-29a inhibition normalizes hur over-expression and aberrant AU-rich mRNA stability in invasive cancer. *J. Pathol.* **2013**, *230*, 28–38. [CrossRef] [PubMed]

72. Rao, J.S. Molecular mechanisms of glioma invasiveness: The role of proteases. *Nat. Rev. Cancer* **2003**, *3*, 489–501. [CrossRef] [PubMed]

73. Yamamoto, M.; Ueno, Y.; Hayashi, S.; Fukushima, T. The role of proteolysis in tumor invasiveness in glioblastoma and metastatic brain tumors. *Anticancer Res.* **2002**, *22*, 4265–4268. [PubMed]

74. Ryu, J.; Yoon, N.A.; Seong, H.; Jeong, J.Y.; Kang, S.; Park, N.; Choi, J.; Lee, D.H.; Roh, G.S.; Kim, H.J.; et al. Resveratrol induces glioma cell apoptosis through activation of tristetraprolin. *Mol. Cells* **2015**, *38*, 991–997. [PubMed]

75. Ryu, J.; Yoon, N.A.; Lee, Y.K.; Jeong, J.Y.; Kang, S.; Seong, H.; Choi, J.; Park, N.; Kim, N.; Cho, W.J.; et al. Tristetraprolin inhibits the growth of human glioma cells through downregulation of urokinase plasminogen activator/urokinase plasminogen activator receptor mRNAs. *Mol. Cells* **2015**, *38*, 156–162. [PubMed]

76. Xiong, T.; Liu, X.W.; Huang, X.L.; Xu, X.F.; Xie, W.Q.; Zhang, S.J.; Tu, J. Tristetraprolin: A novel target of diallyl disulfide that inhibits the progression of breast cancer. *Oncol. Lett.* **2018**, *15*, 7817–7827. [CrossRef] [PubMed]

77. Quail, D.F.; Joyce, J.A. Microenvironmental regulation of tumor progression and metastasis. *Nat. Med.* **2013**, *19*, 1423–1437. [CrossRef] [PubMed]

78. Grivennikov, S.I.; Greten, F.R.; Karin, M. Immunity, inflammation, and cancer. *Cell* **2010**, *140*, 883–899. [CrossRef] [PubMed]

79. Guo, J.; Qu, H.H.; Shan, T.; Chen, Y.G.; Chen, Y.; Xia, J.Z. Tristetraprolin overexpression in gastric cancer cells suppresses PD-L1 expression and inhibits tumor progression by enhancing antitumor immunity. *Mol. Cells* **2018**, *41*, 653–664. [PubMed]

80. Coelho, M.A.; Trecesson, S.D.; Rana, S.; Zecchin, D.; Moore, C.; Molina-Arcas, M.; East, P.; Spencer-Dene, B.; Nye, E.; Barnouin, K.; et al. Oncogenic Ras signaling promotes tumor immunoresistance by stabilizing PD-L1 mRNA. *Immunity* **2017**, *47*, 1083–1099. [CrossRef] [PubMed]

81. Francisco, L.M.; Salinas, V.H.; Brown, K.E.; Vanguri, V.K.; Freeman, G.J.; Kuchroo, V.K.; Sharpe, A.H. PD-L1 regulates the development, maintenance, and function of induced regulatory T cells. *J. Exp. Med.* **2009**, *206*, 3015–3029. [CrossRef] [PubMed]

82. Milke, L.; Schulz, K.; Weigert, A.; Sha, W.X.; Schmid, T.; Brune, B. Depletion of tristetraprolin in breast cancer cells increases interleukin-16 expression and promotes tumor infiltration with monocytes/macrophages. *Carcinogenesis* **2013**, *34*, 850–857. [CrossRef] [PubMed]

83. Neufeld, G.; Cohen, T.; Gengrinovitch, S.; Poltorak, Z. Vascular endothelial growth factor (VEGF) and its receptors. *FASEB J.* **1999**, *13*, 9–22. [CrossRef] [PubMed]

84. Essafi-Benkhadir, K.; Onesto, C.; Stebe, E.; Moroni, C.; Pages, G. Tristetraprolin inhibits ras-dependent tumor vascularization by inducing vascular endothelial growth factor mRNA degradation. *Mol. Biol. Cell* **2007**, *18*, 4648–4658. [CrossRef] [PubMed]

85. Lee, H.H.; Son, Y.J.; Lee, W.H.; Park, Y.W.; Chae, S.W.; Cho, W.J.; Kim, Y.M.; Choi, H.J.; Choi, D.H.; Jung, S.W.; et al. Tristetraprolin regulates expression of VEGF and tumorigenesis in human colon cancer. *Int. J. Cancer* **2010**, *126*, 1817–1827. [CrossRef] [PubMed]

86. Gately, S.; Li, W.W. Multiple roles of cox-2 in tumor angiogenesis: A target for antiangiogenic therapy. *Semin. Oncol.* **2004**, *31*, 2–11. [CrossRef] [PubMed]

87. Sawaoka, H.; Dixon, D.A.; Oates, J.A.; Boutaud, O. Tristetraprolin binds to the 3′-untranslated region of cyclooxygenase-2 mRNA. A polyadenylation variant in a cancer cell line lacks the binding site. *J. Biol. Chem.* **2003**, *278*, 13928–13935. [CrossRef] [PubMed]

88. Cha, H.J.; Lee, H.H.; Chae, S.W.; Cho, W.J.; Kim, Y.M.; Choi, H.J.; Choi, D.H.; Jung, S.W.; Min, Y.J.; Lee, B.J.; et al. Tristetraprolin downregulates the expression of both VEGF and COX-2 in human colon cancer. *Hepatogastroenterology* **2011**, *58*, 790–795. [PubMed]

89. Hau, H.H.; Walsh, R.J.; Ogilvie, R.L.; Williams, D.A.; Reilly, C.S.; Bohjanen, P.R. Tristetraprolin recruits functional mRNA decay complexes to are sequences. *J. Cell. Biochem.* **2007**, *100*, 1477–1492. [CrossRef] [PubMed]

90. Winzen, R.; Thakur, B.K.; Dittrich-Breiholz, O.; Shah, M.; Redich, N.; Dhamija, S.; Kracht, M.; Holtmann, H. Functional analysis of ksrp interaction with the au-rich element of interleukin-8 and identification of inflammatory mRNA targets. *Mol. Cell. Biol.* **2007**, *27*, 8388–8400. [CrossRef] [PubMed]

91. Fechir, M.; Linker, K.; Pautz, A.; Hubrich, T.; Forstermann, U.; Rodriguez-Pascual, F.; Kleinert, H. Tristetraprolin regulates the expression of the human inducible nitric-oxide synthase gene. *Mol. Pharmacol.* **2005**, *67*, 2148–2161. [CrossRef] [PubMed]

92. Zhu, Q.C.; Han, X.D.; Peng, J.Y.; Qin, H.L.; Wang, Y. The role of CXC chemokines and their receptors in the progression and treatment of tumors. *J. Mol. Histol.* **2012**, *43*, 699–713. [CrossRef] [PubMed]

93. Balkwill, F. Cancer and the chemokine network. *Nat. Rev. Cancer* **2004**, *4*, 540–550. [CrossRef] [PubMed]

94. Zeelenberg, I.S.; Ruuls-Van Stalle, L.; Roos, E. The chemokine receptor CXCR4 is required for outgrowth of colon carcinoma micrometastases. *Cancer Res.* **2003**, *63*, 3833–3839. [PubMed]

95. Bourcier, C.; Griseri, P.; Grepin, R.; Bertolotto, C.; Mazure, N.; Pages, G. Constitutive erk activity induces downregulation of tristetraprolin, a major protein controlling interleukin8/CXCL8 mRNA stability in melanoma cells. *Am. J. Physiol. Cell Physiol.* **2011**, *301*, C609–C618. [CrossRef] [PubMed]

96. Harris, A.L. Hypoxia—A key regulatory factor in tumour growth. *Nat. Rev. Cancer* **2002**, *2*, 38–47. [CrossRef] [PubMed]

97. Giatromanolaki, A.; Harris, A.L. Tumour hypoxia, hypoxia signaling pathways and hypoxia inducible factor expression in human cancer. *Anticancer Res.* **2001**, *21*, 4317–4324. [PubMed]

98. Kim, T.W.; Yim, S.; Choi, B.J.; Jang, Y.; Lee, J.J.; Sohn, B.H.; Yoo, H.S.; Il Yeom, Y.; Park, K.C. Tristetraprolin regulates the stability of hif-1 alpha mRNA during prolonged hypoxia. *Biochem. Biophys. Res. Commun.* **2010**, *391*, 963–968. [CrossRef] [PubMed]

99. Suswam, E.; Li, Y.; Zhang, X.; Gillespie, G.Y.; Li, X.; Shacka, J.J.; Lu, L.; Zheng, L.; King, P.H. Tristetraprolin down-regulates interleukin-8 and vascular endothelial growth factor in malignant glioma cells. *Cancer Res* **2008**, *68*, 674–682. [CrossRef] [PubMed]

100. Brennan, S.E.; Kuwano, Y.; Alkharouf, N.; Blackshear, P.J.; Gorospe, M.; Wilson, G.M. The mRNA-destabilizing protein tristetraprolin is suppressed in many cancers, altering tumorigenic phenotypes and patient prognosis. *Cancer Res.* **2009**, *69*, 5168–5176. [CrossRef] [PubMed]

101. Gebeshuber, C.A.; Zatloukal, K.; Martinez, J. Mir-29a suppresses tristetraprolin, which is a regulator of epithelial polarity and metastasis. *EMBO Rep.* **2009**, *10*, 400–405. [CrossRef] [PubMed]

102. Carrick, D.M.; Blackshear, P.J. Comparative expression of tristetraprolin (TTP) family member transcripts in normal human tissues and cancer cell lines. *Arch. Biochem. Biophys.* **2007**, *462*, 278–285. [CrossRef] [PubMed]

103. Sanduja, S.; Kaza, V.; Dixon, D.A. The mRNA decay factor tristetraprolin (ttp) induces senescence in human papillomavirus-transformed cervical cancer cells by targeting E6-AP ubiquitin ligase. *Aging* **2009**, *1*, 803–817. [CrossRef] [PubMed]

104. Young, L.E.; Sanduja, S.; Bemis-Standoli, K.; Pena, E.A.; Price, R.L.; Dixon, D.A. The mRNA binding proteins hur and tristetraprolin regulate cyclooxygenase 2 expression during colon carcinogenesis. *Gastroenterology* **2009**, *136*, 1669–1679. [CrossRef] [PubMed]

105. Sohn, B.H.; Park, I.Y.; Lee, J.J.; Yang, S.J.; Jang, Y.J.; Park, K.C.; Kim, D.J.; Lee, D.C.; Sohn, H.A.; Kim, T.W.; et al. Functional switching of TGF-beta1 signaling in liver cancer via epigenetic modulation of a single CPG site in ttp promoter. *Gastroenterology* **2010**, *138*, 1898–1908. [CrossRef] [PubMed]

106. Lai, W.S.; Stumpo, D.J.; Blackshear, P.J. Rapid insulin-stimulated accumulation of an mRNA encoding a proline-rich protein. *J. Biol. Chem.* **1990**, *265*, 16556–16563. [PubMed]

107. DuBois, R.N.; McLane, M.W.; Ryder, K.; Lau, L.F.; Nathans, D. A growth factor-inducible nuclear protein with a novel cysteine/histidine repetitive sequence. *J. Biol. Chem.* **1990**, *265*, 19185–19191. [PubMed]

108. Lim, R.W.; Varnum, B.C.; Herschman, H.R. Cloning of tetradecanoyl phorbol ester-induced 'primary response' sequences and their expression in density-arrested swiss 3T3 cells and a TPA non-proliferative variant. *Oncogene* **1987**, *1*, 263–270. [PubMed]

109. Gomperts, M.; Pascall, J.C.; Brown, K.D. The nucleotide sequence of a cdna encoding an EGF-inducible gene indicates the existence of a new family of mitogen-induced genes. *Oncogene* **1990**, *5*, 1081–1083. [PubMed]

110. Lai, W.S.; Thompson, M.J.; Taylor, G.A.; Liu, Y.; Blackshear, P.J. Promoter analysis of zfp-36, the mitogen-inducible gene encoding the zinc finger protein tristetraprolin. *J. Biol. Chem.* **1995**, *270*, 25266–25272. [CrossRef] [PubMed]

111. Ogawa, K.; Chen, F.; Kim, Y.J.; Chen, Y. Transcriptional regulation of tristetraprolin by transforming growth factor-beta in human T cells. *J. Biol. Chem.* **2003**, *278*, 30373–30381. [CrossRef] [PubMed]

112. Blanco, F.F.; Sanduja, S.; Deane, N.G.; Blackshear, P.J.; Dixon, D.A. Transforming growth factor beta regulates p-body formation through induction of the mRNA decay factor tristetraprolin. *Mol. Cell. Biol.* **2014**, *34*, 180–195. [CrossRef] [PubMed]

113. Sauer, I.; Schaljo, B.; Vogl, C.; Gattermeier, I.; Kolbe, T.; Muller, M.; Blackshear, P.J.; Kovarik, P. Interferons limit inflammatory responses by induction of tristetraprolin. *Blood* **2006**, *107*, 4790–4797. [CrossRef] [PubMed]

114. Gaba, A.; Grivennikov, S.I.; Do, M.V.; Stumpo, D.J.; Blackshear, P.J.; Karin, M. Cutting edge: IL-10-mediated tristetraprolin induction is part of a feedback loop that controls macrophage STAT3 activation and cytokine production. *J. Immunol.* **2012**, *189*, 2089–2093. [CrossRef] [PubMed]

115. Suzuki, K.; Nakajima, H.; Ikeda, K.; Maezawa, Y.; Suto, A.; Takatori, H.; Saito, Y.; Iwamoto, I. IL-4-STAT6 signaling induces tristetraprolin expression and inhibits TNF-alpha production in mast cells. *J. Exp. Med.* **2003**, *198*, 1717–1727. [CrossRef] [PubMed]

116. Tchen, C.R.; Brook, M.; Saklatvala, J.; Clark, A.R. The stability of tristetraprolin mRNA is regulated by mitogen-activated protein kinase p38 and by tristetraprolin itself. *J. Biol. Chem.* **2004**, *279*, 32393–32400. [CrossRef] [PubMed]

117. Brooks, S.A.; Connolly, J.E.; Rigby, W.F. The role of mRNA turnover in the regulation of tristetraprolin expression: Evidence for an extracellular signal-regulated kinase-specific, AU-rich element-dependent, autoregulatory pathway. *J. Immunol.* **2004**, *172*, 7263–7271. [CrossRef] [PubMed]

118. Sharma, A.; Bhat, A.A.; Krishnan, M.; Singh, A.B.; Dhawan, P. Trichostatin-a modulates claudin-1 mRNA stability through the modulation of hu antigen r and tristetraprolin in colon cancer cells. *Carcinogenesis* **2013**, *34*, 2610–2621. [CrossRef] [PubMed]

119. Sobolewski, C.; Sanduja, S.; Blanco, F.F.; Hu, L.; Dixon, D.A. Histone deacetylase inhibitors activate tristetraprolin expression through induction of early growth response protein 1 (EGR1) in colorectal cancer cells. *Biomolecules* **2015**, *5*, 2035–2055. [CrossRef] [PubMed]

120. Sun, X.J.; Liu, B.Y.; Yan, S.; Jiang, T.H.; Cheng, H.Q.; Jiang, H.S.; Cao, Y.; Mao, A.W. MicroRNA-29a promotes pancreatic cancer growth by inhibiting tristetraprolin. *Cell. Physiol. Biochem.* **2015**, *37*, 707–718. [CrossRef] [PubMed]

121. Cao, H.; Dzineku, F.; Blackshear, P.J. Expression and purification of recombinant tristetraprolin that can bind to tumor necrosis factor-alpha mRNA and serve as a substrate for mitogen-activated protein kinases. *Arch. Biochem. Biophys.* **2003**, *412*, 106–120. [CrossRef]

122. Carballo, E.; Cao, H.; Lai, W.S.; Kennington, E.A.; Campbell, D.; Blackshear, P.J. Decreased sensitivity of tristetraprolin-deficient cells to p38 inhibitors suggests the involvement of tristetraprolin in the p38 signaling pathway. *J. Biol. Chem.* **2001**, *276*, 42580–42587. [CrossRef] [PubMed]

123. Rigby, W.F.; Roy, K.; Collins, J.; Rigby, S.; Connolly, J.E.; Bloch, D.B.; Brooks, S.A. Structure/function analysis of tristetraprolin (TTP): P38 stress-activated protein kinase and lipopolysaccharide stimulation do not alter ttp function. *J. Immunol.* **2005**, *174*, 7883–7893. [CrossRef] [PubMed]

124. Marchese, F.P.; Aubareda, A.; Tudor, C.; Saklatvala, J.; Clark, A.R.; Dean, J.L. Mapkap kinase 2 blocks tristetraprolin-directed mRNA decay by inhibiting CAF1 deadenylase recruitment. *J. Biol. Chem.* **2010**, *285*, 27590–27600. [CrossRef] [PubMed]

125. Clement, S.L.; Scheckel, C.; Stoecklin, G.; Lykke-Andersen, J. Phosphorylation of tristetraprolin by MK2 impairs au-rich element mRNA decay by preventing deadenylase recruitment. *Mol. Cell. Biol.* **2011**, *31*, 256–266. [CrossRef] [PubMed]

126. Johnson, B.A.; Stehn, J.R.; Yaffe, M.B.; Blackwell, T.K. Cytoplasmic localization of tristetraprolin involves 14-3-3-dependent and -independent mechanisms. *J. Biol. Chem.* **2002**, *277*, 18029–18036. [CrossRef] [PubMed]

127. Liang, J.; Lei, T.; Song, Y.; Yanes, N.; Qi, Y.; Fu, M. RNA-destabilizing factor tristetraprolin negatively regulates NF-kappaB signaling. *J. Biol. Chem.* **2009**, *284*, 29383–29390. [CrossRef] [PubMed]

128. Barrios-Garcia, T.; Gomez-Romero, V.; Tecalco-Cruz, A.; Valadez-Graham, V.; Leon-Del-Rio, A. Nuclear tristetraprolin acts as a corepressor of multiple steroid nuclear receptors in breast cancer cells. *Mol. Genet. Metab. Rep.* **2016**, *7*, 20–26. [CrossRef] [PubMed]

129. Barrios-Garcia, T.; Tecalco-Cruz, A.; Gomez-Romero, V.; Reyes-Carmona, S.; Meneses-Morales, I.; Leon-Del-Rio, A. Tristetraprolin represses estrogen receptor alpha transactivation in breast cancer cells. *J. Biol. Chem.* **2014**, *289*, 15554–15565. [CrossRef] [PubMed]

130. Hitti, E.; Iakovleva, T.; Brook, M.; Deppenmeier, S.; Gruber, A.D.; Radzioch, D.; Clark, A.R.; Blackshear, P.J.; Kotlyarov, A.; Gaestel, M. Mitogen-activated protein kinase-activated protein kinase 2 regulates tumor necrosis factor mRNA stability and translation mainly by altering tristetraprolin expression, stability, and binding to adenine/uridine-rich element. *Mol. Cell. Biol.* **2006**, *26*, 2399–2407. [CrossRef] [PubMed]

131. Cristobal, I.; Torrejon, B.; Madoz-Gurpide, J.; Rojo, F.; Garcia-Foncillas, J. PP2A plays a key role in inflammation and cancer through tristetraprolin activation. *Ann. Rheum. Dis.* **2017**, *76*. [CrossRef] [PubMed]

132. O'Neil, J.D.; Ammit, A.J.; Clark, A.R. Mapk p38 regulates inflammatory gene expression via tristetraprolin: Doing good by stealth. *Int. J. Biochem. Cell. Biol.* **2018**, *94*, 6–9. [CrossRef] [PubMed]

133. Rahman, M.M.; Rumzhum, N.N.; Hansbro, P.M.; Morris, J.C.; Clark, A.R.; Verrills, N.M.; Ammit, A.J. Activating protein phosphatase 2a (PP2A) enhances tristetraprolin (TTP) anti-inflammatory function in a549 lung epithelial cells. *Cell Signal.* **2016**, *28*, 325–334. [CrossRef] [PubMed]

134. Sun, L.; Stoecklin, G.; Van Way, S.; Hinkovska-Galcheva, V.; Guo, R.F.; Anderson, P.; Shanley, T.P. Tristetraprolin (TTP)-14-3-3 complex formation protects ttp from dephosphorylation by protein phosphatase 2a and stabilizes tumor necrosis factor-alpha mRNA. *J. Biol. Chem.* **2007**, *282*, 3766–3777. [CrossRef] [PubMed]

135. Clark, A.R.; Dean, J.L. The control of inflammation via the phosphorylation and dephosphorylation of tristetraprolin: A tale of two phosphatases. *Biochem. Soc. Trans.* **2016**, *44*, 1321–1337. [CrossRef] [PubMed]

136. Suswam, E.A.; Shacka, J.J.; Walker, K.; Lu, L.; Li, X.; Si, Y.; Zhang, X.; Zheng, L.; Nabors, L.B.; Cao, H.; et al. Mutant tristetraprolin: A potent inhibitor of malignant glioma cell growth. *J. Neurooncol.* **2013**, *113*, 195–205. [CrossRef] [PubMed]

137. Lee, J.C.; Laydon, J.T.; McDonnell, P.C.; Gallagher, T.F.; Kumar, S.; Green, D.; McNulty, D.; Blumenthal, M.J.; Heys, J.R.; Landvatter, S.W.; et al. A protein kinase involved in the regulation of inflammatory cytokine biosynthesis. *Nature* **1994**, *372*, 739–746. [CrossRef] [PubMed]

138. Medicherla, S.; Reddy, M.; Ying, J.; Navas, T.A.; Li, L.; Nguyen, A.N.; Kerr, I.; Hanjarappa, N.; Protter, A.A.; Higgins, L.S. P38alpha-selective map kinase inhibitor reduces tumor growth in mouse xenograft models of multiple myeloma. *Anticancer Res.* **2008**, *28*, 3827–3833. [PubMed]

139. Banerjee, R.; Van Tubergen, E.A.; Scanlon, C.S.; Vander Broek, R.; Lints, J.P.; Liu, M.; Russo, N.; Inglehart, R.C.; Wang, Y.; Polverini, P.J.; et al. The G protein-coupled receptor GALR2 promotes angiogenesis in head and neck cancer. *Mol. Cancer Ther.* **2014**, *13*, 1323–1333. [CrossRef] [PubMed]

140. Dufies, M.; Giuliano, S.; Ambrosetti, D.; Claren, A.; Ndiaye, P.D.; Mastri, M.; Moghrabi, W.; Cooley, L.S.; Ettaiche, M.; Chamorey, E.; et al. Sunitinib stimulates expression of vegfc by tumor cells and promotes lymphangiogenesis in clear cell renal cell carcinomas. *Cancer Res.* **2017**, *77*, 1212–1226. [CrossRef] [PubMed]

141. Wei, F.; Zhang, T.; Yang, Z.; Wei, J.C.; Shen, H.F.; Xiao, D.; Wang, Q.; Yang, P.; Chen, H.C.; Hu, H.; et al. Gambogic acid efficiently kills stem-like colorectal cancer cells by upregulating zfp36 expression. *Cell. Physiol. Biochem.* **2018**, *46*, 829–846. [CrossRef] [PubMed]

142. Li, C.; Tang, C.; He, G. Tristetraprolin: A novel mediator of the anticancer properties of resveratrol. *Genet. Mol. Res.* **2016**, *15*. [CrossRef] [PubMed]

 © 2018 by the authors. Licensee MDPI, Basel, Switzerland. This article is an open access article distributed under the terms and conditions of the Creative Commons Attribution (CC BY) license (http://creativecommons.org/licenses/by/4.0/).

International Journal of
Molecular Sciences

Article

PWCDA: Path Weighted Method for Predicting circRNA-Disease Associations

Xiujuan Lei [1], Zengqiang Fang [1], Luonan Chen [2,3,4,*] and Fang-Xiang Wu [5,*]

[1]   School of Computer Science, Shaanxi Normal University, Xi'an 710119, China; xjlei@snnu.edu.cn (X.L.); fangzq@snnu.edu.cn (Z.F.)
[2]   Key Laboratory of Systems Biology, Center for Excellence in Molecular Cell Science, Institute of Biochemistry and Cell Biology, Shanghai Institutes for Biological Sciences, Chinese Academy of Sciences, Shanghai 200031, China
[3]   Center for Excellence in Animal Evolution and Genetics, Chinese Academy of Sciences, Kunming 650223, China
[4]   School of Life Science and Technology, Shanghai Tech University, Shanghai 201210, China
[5]   Department of Mechanical Engineering and Division of Biomedical Engineering, University of Saskatchewan, Saskatoon, SK S7N 5A9, Canada
*   Correspondence: lnchen@sibs.ac.cn (L.C.); faw341@mail.usask.ca (F.-X.W.); Tel.: +86-021-5492-0100 (L.C.); +1-(306)-966-5280 (F.-X.W.)

Received: 4 October 2018; Accepted: 26 October 2018; Published: 31 October 2018

Abstract: CircRNAs have particular biological structure and have proven to play important roles in diseases. It is time-consuming and costly to identify circRNA-disease associations by biological experiments. Therefore, it is appealing to develop computational methods for predicting circRNA-disease associations. In this study, we propose a new computational path weighted method for predicting circRNA-disease associations. Firstly, we calculate the functional similarity scores of diseases based on disease-related gene annotations and the semantic similarity scores of circRNAs based on circRNA-related gene ontology, respectively. To address missing similarity scores of diseases and circRNAs, we calculate the Gaussian Interaction Profile (GIP) kernel similarity scores for diseases and circRNAs, respectively, based on the circRNA-disease associations downloaded from circR2Disease database (http://bioinfo.snnu.edu.cn/CircR2Disease/). Then, we integrate disease functional similarity scores and circRNA semantic similarity scores with their related GIP kernel similarity scores to construct a heterogeneous network made up of three sub-networks: disease similarity network, circRNA similarity network and circRNA-disease association network. Finally, we compute an association score for each circRNA-disease pair based on paths connecting them in the heterogeneous network to determine whether this circRNA-disease pair is associated. We adopt leave one out cross validation (LOOCV) and five-fold cross validations to evaluate the performance of our proposed method. In addition, three common diseases, Breast Cancer, Gastric Cancer and Colorectal Cancer, are used for case studies. Experimental results illustrate the reliability and usefulness of our computational method in terms of different validation measures, which indicates PWCDA can effectively predict potential circRNA-disease associations.

Keywords: circRNA-disease associations; pathway; heterogeneous network

---

1. Introduction

In recent years, an increasing number of circRNAs [1] have been uncovered and have drawn more attention than before. CircRNA is a newly discovered category of non-coding RNAs. Non-coding RNAs also include a large number of different RNAs, such as miRNAs, lncRNAs, piRNAs [2]. The first discovery of circular RNA was in the Tetrahymena cell [3]. There is an obvious difference between circular RNAs and common linear RNAs. That is, circRNA has a circular closed loop RNA structure, yet have no free 5′ and 3′ compared with linear RNAs [4]. In addition, circRNAs can also be classified into 4 categories as follows: Exonic circRNAs, intronic circRNAs, exonintron circRNAs and intergenic circRNAs [4,5]. Because of such a closed loop structures, they are usually stable, abundant, conserved, and tissue-specifically expressed [5].

With the progress of high throughput sequencing technology [6], more and more circRNAs have been confirmed to play significant roles in different biological processes [7]. According to many experiments, a large amount of circRNAs functions have been found to work as a scaffold in the assembly of protein complexes [8], and local subcellular positions [9], and so on. They also regulate the expression of their ancestor genes [10] and acts as a microRNA (miRNA) sponge [11,12]. Especially, many studies have proved that circRNA can be biomarkers of tumors [13–15].

Recently, a sharply increasing number of circRNAs have been discovered and there are also some circRNA-disease databases being developed, such as circR2Disease [16], Circ2Traits [17] and Circ2Disease [18]. Simultaneously, circRNAs-related diseases also have been verified by classic biological experiments. However, they are both time-consuming and expensive. Therefore, it is appealing to develop computational methods that can produce reliable prediction results and reduce both time and cost. Although, some computational methods have been proposed for predicting miRNA-disease associations [19–21], lncRNA-disease associations [22,23] and drug-target associations [18,24,25], there is no computational method for predicting circRNA-disease associations yet.

In this study, we propose the first computational method, Path Weighed method for predicting CircRNA-Disease Associations (PWCDA). After building a heterogeneous network consisting of three sub-networks, the disease similarity network, the circRNA similarity network and circRNA-disease association network, we calculate an association score for each circRNA-disease pair based on the paths connecting them in the heterogeneous network to determine whether a circRNA-disease pair is associated. Our method is evaluated with leave one out cross validation (LOOCV) and five-fold cross validation. The average AUC (Area Under roc Curve) of LOOCV is 0.900, while the AUC value of five-fold cross validation is 0.890. For further investigating the performance of our proposed model, we conduct several case studies of some common cancers. What's more, we compare our method with some other computational prediction methods. The results show that our method outperforms other methods, which indicates that our proposed model has the better capability to predict potential circRNA-disease associations.

2. Results and Discussion

2.1. Effect of Parameter

Based on the previous study [26], we fix the maximum path length as 3. If the maximum path length is more than 3, not only do the running time of the method increases, but our method also takes some noisy information. In this study, we give a comprehensive analysis for the parameter α in our decaying function. After we calculate scores for each disease-circRNA pair, we can obtain a disease-circRNA association score matrix. Based on the scores matrix, we calculate the AUC. The results are represented in Table 1. It's obvious that the effect of different values of α on the final AUC value is quite small and it can take value from 1 to 3. Therefore, we adopt the best result setting the value of α as 1. In order to reduce the running time, we don't use any cross validation in this experiment. Furthermore, we also carry out an experiment to analyze another parameter, the threshold γ, which is represented in Table 2. For the sake of reducing the running time, any cross validation is not adopted. The result shows that the parameter γ might have tiny effect on the final AUC value. Thus, we set the γ value as 0.5, which gets the greatest AUC value.

Table 1. The Area Under roc Curve (AUC) value based on changing α and fixed pathway maximum length.

α	0.5	1	1.5	2	3	3.5	4	4.5	5
AUC	0.97100	0.97209	0.97206	0.97208	0.97202	0.97010	0.97010	0.97010	0.96879

Table 2. The AUC value based on changing γ and fixed pathway maximum length.

γ	0.1	0.2	0.3	0.4	0.5	0.6
AUC	0.96483	0.96483	0.96483	0.96500	0.97209	0.97205

2.2. LOOCV

For a given particular disease i, there are some associations between disease i and a number of circRNAs. In LOOCV, during each computational iteration, we leave one association out as a test data and use the remaining associations as a training dataset. If there is just one association between disease i and circRNAs in our dataset, we do not adopt LOOCV for this kind of disease. In LOOCV, we obtain an association score for each circRNA-disease pair and then rank all the prediction association scores. If a score value is greater than the pre-set threshold, we determine that the corresponding disease-circRNA is associated. With the change of the threshold, we can get a variety of true positive rates (TPRs) and false positive rates (FPRs), which can be used to draw the Receiver Operating Characteristic Curve (ROC) curve. In the end, we have compared our prediction method with other computational prediction methods [27,28]. The results can be found in Figure 1 and show that our proposed method outperforms the existing prediction methods.

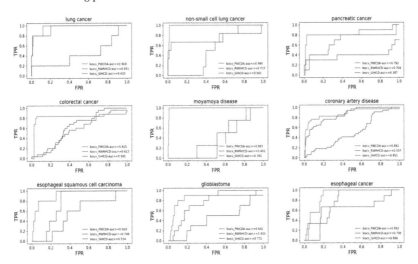

Figure 1. Comparison of Path Weighed method for predicting CircRNA-Disease Associations (PWCDA) with other models by leave one out cross validation (LOOCV). FPR, false positive rate.

2.3. Five-Fold Cross Validation

In order to further illustrate the performance of our proposed method, we have adopted five-fold cross validation verification method as well for investigating the prediction performance. In our study, we divide all disease-circRNA associations into 5 parts. Each time we pick up one part as the test dataset and the remaining four parts consist of the training set. Then we can obtain the scores of all circRNA-disease associations. Similarly, we follow the same procedure as LOOCV to draw the AUC curve based on five-fold cross validation. What's more, we have compared our proposed computational method with other prediction methods [27,28]. Our method gets more outstanding result than other methods, which is shown in the Figure 2.

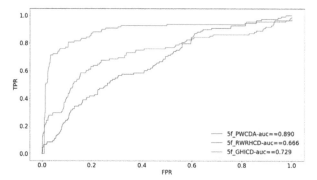

Figure 2. Comparison of PWCDA with other computational methods via five-fold cross validation.

2.4. Case Studies

Here, we also have conducted some case studies, which can help us further understand the associations between circRNAs and diseases. In this study, we choose three common diseases as prediction targets of our case studies, which are Breast Cancer [29], Gastric Cancer [30] and Colorectal Cancer [31]. In order to prove the prediction accuracy of our proposed method, we have used circRNA-disease database, and associations between circRNAs and diseases—which have been experimentally verified in the published articles [32].

Breast cancer is one the common cancers all over the world now [33], and breast cancer causes thousands of deaths every year. With the development of deep sequencing technology, circRNAs are confirmed to be biomarkers for diagnosing breast cancer. Based on our computational method, we have succeeded in predicting 29 of top 30 candidate circRNAs. For example, circpvt1 (top1) can be worked as miRNA spouse to regulate miRNA by moderating let-7 activity selected [30], and circRNA hsa_circ_104689 wasn't predicted by our method and the predicting result have been presented in Table 3.

Table 3. The top 30 breast cancer related candidates circRNAs.

		Breast Cancer			
Rank	circRNA Name/id	Evidences	Rank	circRNA Name/id	Evidences
1	circpvt1/hsa_circ_0001821	PMID:279280058	16	hsa_circ_0001667	circRNAdisease
2	circ-foxo3	circRNAdisease	17	hsa_circ_0085495	circRNAdisease
3	hsa_circ_0001313/circccdc66	PMID:28249903	18	hsa_circ_0086241	circRNAdisease
4	hsa_circ_0007534	PMID:29593432	19	hsa_circ_0092276	circRNAdisease
5	hsa_circ_0000284/circhipk3	PMID:27050392	20	hsa_circ_0003838	circRNAdisease
6	hsa_circ_0011946	PMID:29593432	21	circvrk1	PMID:29221160
7	hsa_circ_0093869	PMID: 29593432	22	circbrip	PMID: 29221160
8	hsa_circ_0001982	circRNAdisease	23	circola	PMID: 29221160
9	hsa_circ_0001785	circRNAdisease	24	circetfa	PMID: 29221160
10	hsa_circ_0108942	circRNAdisease	25	circmed13	PMID: 29221160
11	hsa_circ_0068033	circRNAdisease	26	circbc111b	PMID:28739726
12	circamot11/hsa_circ_0004214	circRNAdisease	27	circdennd4c	circRNAdisease
13	hsa_circ_0006528	circRNAdisease	28	hsa_circ_103110/hsa_circ_0004771	circRNAdisease
14	hsa_circ_0002113	circRNAdisease	29	hsa_circ_104689/hsa_circ_0001824	unconfirmed
15	hsa_circ_0002874	circRNAdisease	30	hsa_circ_104821/hsa_circ_0001875	circRNAdisease

Gastric cancer [34] causes a high mortality rate in human. It can be produced in any tissue of the human stomach. These tumors in the stomach are usually malignant tumors, and they can also destroy the surrounding nervous tissue. With our computational method, there are 25 of top 30 candidate circRNAs that have been confirmed by another database, circRNA disease. For example, hsa_circ_0076304 (top1) and hsa_circ_0076305 (top2) are identified to downregulate in a group of gastric cancer [35]. circpvt1 (top3) can be regarded as the sponge of the miR-125 family [13], which can upregulate in the gastric cells. The more details of results are shown in Table 4.

Table 4. The top 30 gastric cancer related candidates circRNAs.

Gastric Cancer					
Rank	circRNA Name/id	Evidences	Rank	circRNA Name/id	Evidences
1	hsa_circ_0076305	circRNAdisease	16	circma0138960/hsa-circma7690-15	circRNAdisease
2	hsa_circ_0076304	circRNAdisease	17	hsa_circ_0000181	circRNAdisease
3	circpvt1/hsa_circ_0001821	circRNAdisease	18	hsa_circ_0000745	circRNAdisease
4	hsa_circ_0001649	unconfirmed	19	hsa_circ_0085616	circRNAdisease
5	hsa_circ_0000284/circhipk3	unconfirmed	20	hsa_circ_0006127	circRNAdisease
6	hsa_circ_0014717	circRNAdisease	21	hsa_circ_0000026	circRNAdisease
7	cdr1as/cirs-7/hsa_circ_0001946	unconfirmed	22	hsa_circ_0000144	circRNAdisease
8	hsa_circ_0003195	circRNAdisease	23	hsa_circ_0032821	circRNAdisease
9	hsa_circ_0000520	circRNAdisease	24	hsa_circ_0005529	circRNAdisease
10	hsa_circ_0074362	circRNAdisease	25	hsa_circ_0061274	circRNAdisease
11	hsa_circ_0001017	circRNAdisease	26	hsa_circ_0005927	circRNAdisease
12	hsa_circ_0061276	circRNAdisease	27	hsa_circ_0092341	circRNAdisease
13	circ-zfr	unconfirmed	28	hsa_circ_0001561	unconfirmed
14	circma0047905/hsa_circ_0047905	circRNAdisease	29	circlarp4	circRNAdisease
15	circma0138960/hsa_circ_0138960	circRNAdisease	30	hsa_circ_0035431	circRNAdisease

Colorectal cancer [36] is one of the three most frequent cancers for women. Even though the incidence of colorectal cancer has been declined for a long time, a large proportion of patients die each year from colorectal cancer. In this study, we have succeeded in predicting 24 of top 30 candidate circRNAs. For example, hsa_circ_0001649 (top1) [31] has been identified to downregulate in colorectal cancer tissue. hsa_circ_0007534 (top2) [37] can upregulate in the different colorectal cancer cells. The more details of results are presented in Table 5.

Table 5. The top 30 colorectal cancer related candidates circRNAs.

Colorectal Cancer					
Rank	circRNA Name/id	Evidences	Rank	circRNA Name/id	Evidences
1	hsa_circ_0001649	PMID:29421663	16	has-circ_0006174	circRNAdisease
2	hsa_circ_0007534	PMID:29364478	17	hsa_circ_0008509	circRNAdisease
3	cdr1as/cirs-7/ hsa_circ_0001946	circRNAdisease	18	hsa_circ_0084021	circRNAdisease
4	hsa_circ_0000284/ circhipk3	PMID:27050392	19	circ_banp	circRNAdisease
5	hsa_circ_0001313/ circccdc66	circRNAdisease	20	hsa_circrna_103809	circRNAdisease
6	ciritch/hsa_circ_0001141/ hsa_circ_001763	unconfirmed	21	hsa_circrna_104700	circRNAdisease
7	hsa_circ_0014717	PMID:29571246	22	hsa_circ_0000069	circRNAdisease
8	hsa_circ_0000567	PMID:29333615	23	hsa_circ_001988/ hsa_circ_0001451	circRNAdisease
9	hsa_circ_000984/ hsa_circ_0001724	circRNAdisease	24	hsa_circ_0000677/ hsa_circ_001569/circabcc	circRNAdisease
10	hsa_circ_0020397	circRNAdisease	25	circ_kldhc10/ hsa_circ_0082333	PMID:26138677
11	hsa_circ_0007031	circRNAdisease	26	circ_stxbp51	unconfirmed
12	hsa_circ_0000504	circRNAdisease	27	circ-shkbp1	unconfirmed
13	hsa_circ_0007006	circRNAdisease	28	circ-fbxw7	unconfirmed
14	hsa_circ_0074930	circRNAdisease	29	hsa_circ_0046701	unconfirmed
15	hsa_circ_0048232	circRNAdisease	30	circttbk2/hsa_circ_0000594	unconfirmed

3. Materials and Methods

3.1. Human circRNA-Disease Associations Network

All the circRNA-disease associations are downloaded from the website of circR2Disease database [16] (http://bioinfo.snnu.edu.cn/CircR2Disease/). This initial dataset contains 739 associations between 661 circRNA entities and 100 disease entities that are found based on three main species—human, mouse and rat. In this study, we select 541 circRNA entities and 83 human disease entities from our initial dataset, which includes Gastric cancer, Breast cancer, Colorectal cancer, etc. Finally, we obtain 592 circRNA-disease associations, which have experimentally been verified. These make up our circRNA-disease association network with adjacency matrix M. If there is a verified association between disease i and circRNA j, the entry $M(i, j)$ is equal to 1, otherwise it is equal to 0.

3.2. CircRNA Semantic Similarity

For calculating circRNA semantic similarity, we download circRNA and its related gene targets dataset from circR2Disease. To measure circRNA semantic similarities, we also need to obtain gene related annotation terms that can be downloaded from Human Protein Reference Database (HPRD) database [38] (http://www.hprd.org/). Reviewing previous literature [39–41], there are some methods that can be referred to calculate the circRNA-related gene GO terms semantic similarities, including path-length-based methods, information-content-based methods, common-term-based methods and hybrid methods. In this study, we utilize a common-term-based method to measure circRNA similarity scores based on JACCARD index. In the previous studies [21,42], genes have been widely adopted to infer RNA similarity. Thus, the more gene related terms were shared by two circRNA C_i and C_j, the higher the similarity score they get. Denote CS as the circRNA semantic similarity matrix, and its entry $CS(i, j)$ can be calculated by the following formula:

$$CS(i,j) = \frac{|G_i \cap G_j|}{|G_i \cap G_j|} \tag{1}$$

where G_i/G_j denotes the GO terms that circRNA C_i/C_j target genes related.

3.3. Disease Functional Similarity

We adopt disease related gene annotations to measure disease functional similarities. These gene annotations are being extracted from two online databases. The first one is DisGeNET [43] (http://www.disgenet.org/web/DisGeNET/menu), which collects 381,056 gene-disease associations (GDAs) between 16,666 genes and 13,172 diseases. In addition, we also download disease phenotype data from OMIM [44]—Online Mendelian Inheritance in Man. OMIM is a biological database that is updated daily. We use the OMIM_2018_04_24 version. Then we integrate multiple annotation resources of diseases related genes, which help us get a more reliable performance.

There are also some methods for calculating disease similarities from previous studies[45]. The common methods include annotation-based measurements, function-based measurements and topology-based measurements [46–49]. We have adopted annotation-based methods to obtain disease similarities. We apply the JACCARD index, which is a standard method for computing similarities based on two collections of finite numbers of elements so as to estimate the similarity scores between diseases. Let g_{di} be a collection of annotations of a gene associated with disease d_i. We calculate the functional similarity score of two diseases d_i and d_j based on the JACCARD similarity coefficient score of g_{di} and g_{di}. Denote DS as the disease functional similarity matrix, then its entry $DS(i, j)$ can be calculated by the following formula:

$$DS(i,j) = \frac{|g_{d_i} \cap g_{d_j}|}{|g_{d_i} \cup g_{d_j}|} \tag{2}$$

We have constructed circRNA semantic similarity matrix based on their related GO terms and disease functional similarity based on its related annotating genes. However, one essential weakness that cannot be ignored is that the aforementioned similarity matrices are sparse, which indicates similarity of many pairs of diseases (or circRNAs) are unable to be calculated in their functional (or semantic) similarity matrices. To alleviate this weakness, the Gaussian interaction profile (GIP) kernel similarity [50,51] is adopted in this study to get additional information about the similarity of diseases and circRNAs.

3.4. CircRNA GIP Kernel Similarity

There is an assumption that the more similar the circRNA is, the more likely similar patterns of association and non-association with diseases. The GIP kernel similarity is adopted to calculate similarity based on the topological features of the known associations network widely, such miRNA-disease associations network [52], lncRNA-disease associations networks [53] and drug-target association network [54]. Accordingly, GIP kernel similarity is also used in this study to calculate the similarity of circRNA and disease. According to previous literature [54], we use a binary vector $C(i)$ to indicate whether circRNA i is associated with diseases. The GIP kernel similarity between circRNA $C(i)$ and $C(j)$ can be computed by the following formula:

$$KC(i,j) = exp(-\gamma_c \|C(i) - C(j)\|^2) \tag{3}$$

To overcome the shortcomings that the disease functional similarity matrix and circRNA semantic matrix are sparse matrices, the parameter γ_c is to adjust the kernel bandwidth, which can be calculated by the following formula:

$$\gamma_c = \gamma'_c \left/ \left(\frac{1}{n_c}\sum_i^{n_c}\|C(i)\|^2\right)\right. \tag{4}$$

where n_c is the number of circRNAs in our finial dataset. The parameter γ'_c is set as 1 based on the previous study [54], which has obtained a better performance.

3.5. Disease GIP Kernel Similarity

We also calculate the GIP kernel similarity score between disease i and j as follows:

$$KD(i,j) = exp(-\gamma_d\|d(i) - d(j)\|^2), \tag{5}$$

$$\gamma_d = \gamma'_d \left/ \left(\frac{1}{n_d}\sum_i^{n_d}\|d(i)\|^2\right)\right., \tag{6}$$

where $d(i)$ and $d(j)$ are the association profiles of diseases i and j, respectively, n_d is the number of diseases in our finial dataset, γ'_d is also set to 1 based on previous studies.

3.6. Combine Multiple Similarity (circRNA and Disease)

We integrate the GIP kernel similarity for circRNAs with the semantic similarity of circRNAs to construct the circRNA similarity network. Specifically, the elements of the adjacency matrix of this network is calculated as follows:

$$ICS(i,j) = \begin{cases} CS(i,j), & if \ CS(i,j) \neq 0 \\ KC(i,j), otherwise \end{cases} . \tag{7}$$

We also integrate the GIP kernel similarity for diseases with the functional similarity diseases to construct the diseases similarity network. Specifically, the elements of the adjacency matrix of this network is calculated as follows:

$$IDS(i,j) = \begin{cases} DS(i,j), & if \ DS(i,j) \neq 0 \\ KD(i,j), & otherwise \end{cases} \tag{8}$$

3.7. Constructing Heterogeneous Network

After we obtain the final disease similarity scores and circRNA similarity scores. We can construct an initial heterogeneous network, which is composed of disease similarity network, circRNA network and disease-circRNA associations network.

In this initial heterogeneous network, there are some small weighted edges, which may represent noises. Therefore, to weaken the effect of those unimportant or noisy edges, we set a threshold γ (γ is equal to 0.5 based on previous studies [26] and our experiment) to remove them. Specifically, let P_{final} and $P_{initial}$ be the adjacency matrices of the final and heterogeneous network, respectively, then we have:

$$P_{final}(i,j) = \begin{cases} P_{initial}(i,j) & P_{initial}(i,j) \geq \gamma \\ 0 & otherwise \end{cases} . \tag{9}$$

3.8. Perfomance Metrics

In this study, we adopt the AUC value to measure the prediction results. The AUC is the area under the ROC curve, which depicts the true positive rate (*TPR*) verse the false positive rate (*FPR*). The following equations are adopted to calculate the *TPR* and *FPR*:

$$TPR = \frac{TP}{TP + FN} \tag{10}$$

$$FPR = \frac{FP}{TN + FP} \tag{11}$$

where *TP* are positive samples (known associations), which are identified correctly, and *TN* are negative samples (unknown associations), which are identified correctly. *FP* are positive samples which are identified incorrectly while *FN* are negative samples, which are identified incorrectly.

3.9. PWCDA

In this study, we proposed a novel computational model called PWCDA (a Path-Weighted CircRNA-Disease Associations method) to predict potential associations between circRNAs and diseases. The framework of our method is depicted in Figure 3. The computational method PWCDA traverses each node in each pathway without repeating based on heterogeneous network. To avoid traversing the same node repeatedly, we adopt the depth-first search (DFS) algorithm and mark the traversed nodes during each turn. Depth first search is implemented as a recursive function traversing the graph moving along the edge. We modify it to mark nodes, because they are accessed in recursion, and then delete tags before returning from recursive calls. In this study, we set the maximum searching length η as 3 steps according to previous studies [26], i.e., for circRNA i and disease j, there are several pathways, such as circRNA i connecting disease j directly, circRNA i's neighbor circRNA connecting with disease j or circRNA i connecting with disease j's neighbor diseases, circRNA i's neighbor circRNAs connecting with disease j's neighbor diseases directly. The choice of these paths is based on a hypothesis that the larger similarity score is between two circRNAs, the higher probability that they have the same associations is. Thus, after the weight of each circRNA-disease pair within all three paths are summed up. We can obtain the final scores between each circRNA-disease pair.

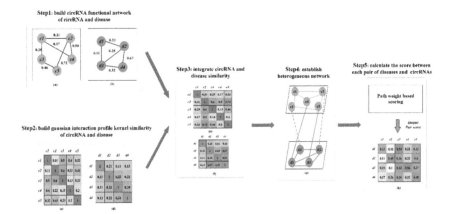

Figure 3. The flowchart of PWCDA is illustrated by five main steps. Step 1: Calculate circRNA semantic similarity and disease similarity scores, respectively. Step 2: Calculate GIP Kernel similarity scores for circRNAs and diseases. Step 3: Integrate circRNA (disease) semantic (functional) similarity with circRNA/disease GIP Kernel similarity, respectively. Step 4: Construct the heterogeneous network. Step 5: Calculate an association score for each circRNA-disease pair.

The more the number of paths between circRNA j and disease i exists, the greater the predictive score they obtain. Accordingly, the path set that connects circRNA C_j to disease di can be represented as $\{p1, p2, \dots, pm\}$, where m is the number of the paths that connect disease d_i and circRNA C_j with the length less than η. The final predictive scores of C_j and d_i can be calculated as follows:

$$score(d_i, C_j) = \sum_{k=1}^{m} (S_{path}(p_k))^{f_{weak}(len(p_k))} \tag{12}$$

where $S_{path}(P_k)$ is the score of the path $p_k = \{e_1, e_2, \dots, e_n\}$ [42] can be calculated as follows:

$$S_{path}(p_k) = \prod_{t=1}^{n} W_{e_t} \ (n \le \eta) \tag{13}$$

The longer the path is, the smaller the contribution it is made, which means that the longer path would have less effect on predicting potential circRNA-disease associations than the shorter one. Therefore, the decaying function is an exponential function to reduce the influence of long path on final prediction scores, which can be represented as Equation (14):

$$f_{weak}(len(p_k)) = \alpha \times exp(len(p_k)) \tag{14}$$

where α is a constraint factor and $len(p_k)$ is the length of path p_k.

An example for calculating the score between circRNA c_1 and disease d_2 is shown in Figure 4. In the Figure 4, three paths $\{c_1-c_4-d_2\}$, $\{c_1-c_3-d_1-d_2\}$ and $\{c_1-c_5-d_3-d_2\}$, which are marked as red, are used to calculate the score between c_1 and d_2. Therefore, the score of c_1 and d_2 can be calculated as follows: Score $(c_1, d_2) = \{c_1-c_4-d_2\} (w_2 \times w_5)^{3*exp(2)} + \{c_1-c_3-d_1-d_2\} (w_1 \times w_4 \times w_7)^{3*exp(3)} + \{c_1-c_5-d_3-d_2\} (w_3 \times w_6 \times w_8)^{3*exp(3)}$. There are also some other paths that can connect c_1 with d_2. Because the length of those paths, such as $\{c_1-c_2-c_5-d_3-d_2\}$, are more than 3, we don't consider this path.

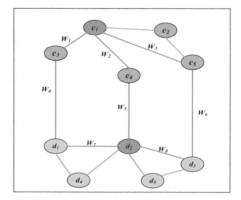

Figure 4. The path between c_1 and d_2 is within the maximum path length.

4. Conclusions

With the increasing number of diseases related to circRNAs being discovered, more and more researchers have been paying attention to investigate diseases-related circRNAs. Although, experimental methods can find potential circRNA-disease associations with a high precision, the process is not only time-consuming, but also expensive. Here, we have proposed an effective computational method called PWCDA, which can predict potential circRNA-disease associations. Firstly, we calculate disease/circRNA similarities by combining their functional/semantic similarity and GIP kernel similarity. Secondly, we build a heterogeneous network, including the circRNA-disease association sub-network, the disease similarity sub-network and the circRNA similarity sub-network. PWCDA searches all the paths within three steps to compute an association score for each circRNA-disease pair to determine if a circRNA-disease pair is associated.

To thoroughly investigate the performance of our proposed method, we adopt LOOCV and five-fold cross validation. Furthermore, we have also compared our method with two state-of-the-art prediction methods. The comparison results illustrate that our methods work much better than other methods. The AUC value of five-fold cross validation is 0.884. Moreover, we apply our method to three diseases: Breast Cancer, Gastric Cancer, Colorectal Cancer for case studies.

There are several significant factors, which may explain why our proposed method can get a better performance than other computational models. Firstly, we have taken into account the sparsity of disease/circRNA similarity sub-networks. Thus, we have integrated disease functional similarity scores and circRNA semantic similarity scores with their corresponding GIP kernel similarity scores. Secondly, according to previous studies, we just use the paths within three steps, which can reduce the noisy information. Although we have combined different similarity scores, there is still some information unavailable. Therefore, we set a threshold to remove those edges whose weights are less than the predefined threshold.

Although we get a much better performance than other computational models, we can't ignore the limitation. The prediction of associations between circRNAs and diseases is a relatively new research field, and the amount of data that we can use is limited. The ratio of positive samples to negative samples of circRNA-disease association is seriously unbalanced. To solve this problem, we may have two main solutions. One is that we can update the circRNA-disease database to obtain new data. The other is that we can extract the same number of positive samples as that of negative samples. Furthermore, our computational method tends to predict those circRNA-disease associations that are covered in the known associations' dataset, and it just predicts fewer novel circRNA-disease associations. Thus, we will adopt more biological data to overcome this weakness. As a future topic, we can apply this work to the disease diagnosis based on network biomarkers [55–57] and disease prediction based on dynamic network biomarkers [58–60] in an accurate and reliable manner.

Author Contributions: X.L. conceptualized the draft and proposed the methodology; Z.F. designed the software; L.C. and F.W. gave the validation and formal analysis; Z.F. performed the investigation and analyzed the resources and data curation; X.L. and Z.F. drafted the manuscript; L.C. and F.W. reviewed and edited the manuscript; F.W. polished the English expression; Z.F. visualized and X.L. supervised; X.L. is the project administrator and has acquired the funding. All the authors read and approved the manuscript.

Acknowledgments: This work was supported by the funding from National Natural Science Foundation of China (61672334, 61502290, 61401263, 31771476, 91529303), and the Strategic Priority Research Program of the Chinese Academy of Sciences (No. XDB13040700).

Conflicts of Interest: The authors declare no conflict of interest.

Abbreviations

PWCDA	Path Weighed method, for predicting CircRNA-Disease Associations
LOOCV	leave one out cross validation
GIP	Gaussian Interaction Profile
GO	Gene Ontology
AUC	Area Under roc Curve
ROC	Receiver Operating Characteristic Curve
TPR	True Positive Rate
FPR	False Positive Rate

References

1. Zhang, Y.; Zhang, X.-O.; Chen, T.; Xiang, J.-F.; Yin, Q.-F.; Xing, Y.-H.; Zhu, S.; Yang, L.; Chen, L.-L. Circular Intronic Long Noncoding RNAs. *Mol. Cell* **2013**, *51*, 792–806. [CrossRef] [PubMed]

2. Lasda, E.; Parker, R. Circular RNAs: Diversity of form and function. *RNA* **2014**, *20*, 1829–1842. [CrossRef] [PubMed]

3. Grabowski, P.J.; Zaug, A.J.; Cech, T.R. The intervening sequence of the ribosomal RNA precursor is converted to a circular RNA in isolated nuclei of Tetrahymena. *Cell* **1981**, *23*, 467–476. [CrossRef]

4. Meng, S.; Zhou, H.; Feng, Z.; Xu, Z.; Tang, Y.; Li, P.; Wu, M. CircRNA: Functions and properties of a novel potential biomarker for cancer. *Mol. Cancer* **2017**, *16*, 94. [CrossRef] [PubMed]

5. Wang, F.; Nazarali, A.J.; Ji, S. Circular RNAs as potential biomarkers for cancer diagnosis and therapy. *Am. J. Cancer Res.* **2016**, *6*, 1167–1176. [PubMed]

6. Jeck, W.R.; Sorrentino, J.A.; Wang, K.; Slevin, M.K.; Burd, C.E.; Liu, J.; Marzluff, W.F.; Sharpless, N.E. Circular RNAs are abundant, conserved, and associated with ALU repeats. *RNA* **2013**, *19*, 426. [CrossRef] [PubMed]

7. Wang, P.L.; Bao, Y.; Yee, M.-C.; Barrett, S.P.; Hogan, G.J.; Olsen, M.N.; Dinneny, J.R.; Brown, P.O.; Salzman, J. Circular RNA is expressed across the eukaryotic tree of life. *PLoS ONE* **2014**, *9*, e90859. [CrossRef] [PubMed]

8. Du, W.W.; Fang, L.; Yang, W.; Wu, N.; Awan, F.M.; Yang, Z.; Yang, B.B. Induction of tumor apoptosis through a circular RNA enhancing Foxo3 activity. *Cell Death Differ.* **2017**, *24*, 357–370. [CrossRef] [PubMed]

9. Armakola, M.; Higgins, M.J.; Figley, M.D.; Barmada, S.J.; Scarborough, E.A.; Diaz, Z.; Fang, X.; Shorter, J.; Krogan, N.J.; Finkbeiner, S.; et al. Inhibition of RNA lariat debranching enzyme suppresses TDP-43 toxicity in ALS disease models. *Nat. Genet.* **2012**, *44*, 1302–1309. [CrossRef] [PubMed]

10. Du, W.W.; Yang, W.; Liu, E.; Yang, Z.; Dhaliwal, P.; Yang, B.B. Foxo3 circular RNA retards cell cycle progression via forming ternary complexes with p21 and CDK2. *Nucleic Acids Res.* **2016**, *44*, 2846–2858. [CrossRef] [PubMed]

11. Du Toit, A. Circular RNAs as miRNA sponges. *Nat. Rev. Mol. Cell Boil.* **2013**, *14*, 195. [CrossRef]

12. Hansen, T.B.; Jensen, T.I.; Clausen, B.H.; Bramsen, J.B.; Finsen, B.; Damgaard, C.K.; Kjems, J. Natural RNA circles function as efficient microRNA sponges. *Nature* **2013**, *495*, 384–388. [CrossRef] [PubMed]

13. Chen, J.; Li, Y.; Zheng, Q.; Bao, C.; He, J.; Chen, B.; Lyu, D.; Zheng, B.; Xu, Y.; Long, Z.; et al. Circular RNA profile identifies circPVT1 as a proliferative factor and prognostic marker in gastric cancer. *Cancer Lett.* **2017**, *388*, 208–219. [CrossRef] [PubMed]

14. Sève, P.; Reiman, T.; Dumontet, C. The role of betaIII tubulin in predicting chemoresistance in non-small cell lung cancer. *Lung Cancer* **2010**, *67*, 136–143. [CrossRef] [PubMed]

15. Guo, S.; Xu, X.; Ouyang, Y.; Wang, Y.; Yang, J.; Yin, L.; Ge, J.; Wang, H. Microarray expression profile analysis of circular RNAs in pancreatic cancer. *Mol. Med. Rep.* **2018**, *17*, 7661–7671. [CrossRef] [PubMed]

16. Fan, C.; Lei, X.; Fang, Z.; Jiang, Q.; Wu, F.-X. CircR2Disease: A manually curated database for experimentally supported circular RNAs associated with various diseases. *Database* **2018**, *2018*. [CrossRef] [PubMed]

17. Ghosal, S.; Das, S.; Sen, R.; Basak, P.; Chakrabarti, J. Circ2Traits: A comprehensive database for circular RNA potentially associated with disease and traits. *Front. Genet.* **2013**, *4*, 283. [CrossRef] [PubMed]

18. Yao, D.; Zhang, L.; Zheng, M.; Sun, X.; Lu, Y.; Liu, P. Circ2Disease: A manually curated database of experimentally validated circRNAs in human disease. *Sci. Rep.* **2018**, *8*, 11018. [CrossRef] [PubMed]

19. Shao, B.; Liu, B.; Yan, C. SACMDA: MiRNA-Disease Association Prediction with Short Acyclic Connections in Heterogeneous Graph. *Neuroinformatics* **2018**, *16*, 373–382. [CrossRef] [PubMed]

20. Chen, X.; Wang, L.-Y.; Huang, L. NDAMDA: Network distance analysis for MiRNA-disease association prediction. *J. Cell. Mol. Med.* **2018**, *22*, 2884–2895. [CrossRef] [PubMed]

21. Liu, Y.; Zeng, X.; He, Z.; Zou, Q. Inferring microRNA-disease associations by random walk on a heterogeneous network with multiple data sources. *IEEE/ACM Trans. Comput. Biol. Bioinform.* **2017**, *14*, 905–915. [CrossRef] [PubMed]

22. Zhang, J.; Zhang, Z.; Chen, Z.; Deng, L. Integrating Multiple Heterogeneous Networks for Novel LncRNA-disease Association Inference. *IEEE/ACM Trans. Comput. Biol. Bioinform.* **2017**. [CrossRef] [PubMed]

23. Fu, G.; Wang, J.; Domeniconi, C.; Yu, G. Matrix factorization-based data fusion for the prediction of lncRNA-disease associations. *Bioinformatics* **2018**, *34*, 1529–1537. [CrossRef] [PubMed]

24. Jiang, J.; Wang, N.; Chen, P.; Zhang, J.; Wang, B. DrugECs: An Ensemble System with Feature Subspaces for Accurate Drug-Target Interaction Prediction. *Biomed. Res. Int.* **2017**, *2017*, 6340316. [CrossRef] [PubMed]

25. Zhang, W.; Chen, Y.; Li, D. Drug-Target Interaction Prediction through Label Propagation with Linear Neighborhood Information. *Molecules* **2017**, *22*, 2056. [CrossRef] [PubMed]

26. Ba-Alawi, W.; Soufan, O.; Essack, M.; Kalnis, P.; Bajic, V.B. DASPfind: New efficient method to predict drug-target interactions. *J. Cheminform.* **2016**, *8*, 15. [CrossRef] [PubMed]

27. Li, Y.; Patra, J.C. Genome-wide inferring gene-phenotype relationship by walking on the heterogeneous network. *Bioinformatics (Oxf. Engl.)* **2010**, *26*, 1219–1224. [CrossRef] [PubMed]

28. Chen, X.; Yan, C.C.; Zhang, X.; You, Z.-H.; Huang, Y.-A.; Yan, G.-Y. HGIMDA: Heterogeneous graph inference for miRNA-disease association prediction. *Oncotarget* **2016**, *7*, 65257–65269. [CrossRef] [PubMed]

29. Wang, M.; Yang, Y.; Xu, J.; Bai, W.; Ren, X.; Wu, H. CircRNAs as biomarkers of cancer: A meta-analysis. *BMC Cancer* **2018**, *18*, 303. [CrossRef] [PubMed]

30. Panda, A.C.; Grammatikakis, I.; Kim, K.M.; De, S.; Martindale, J.L.; Munk, R.; Yang, X.; Abdelmohsen, K.; Gorospe, M. Identification of senescence-associated circular RNAs (SAC-RNAs) reveals senescence suppressor CircPVT1. *Nucleic Acids Res.* **2017**, *45*, 4021–4035. [CrossRef] [PubMed]

31. Chen, S.; Zhang, L.; Su, Y.; Zhang, X. Screening potential biomarkers for colorectal cancer based on circular RNA chips. *Oncol. Rep.* **2018**, *39*, 2499–2512. [CrossRef] [PubMed]

32. Zhao, Z.; Wang, K.; Wu, F.; Wang, W.; Zhang, K.; Hu, H.; Liu, Y.; Jiang, T. circRNA disease: A manually curated database of experimentally supported circRNA-disease associations. *Cell Death Dis.* **2018**, *9*, 475. [CrossRef] [PubMed]

33. Rakha, E.A.; Reis-Filho, J.S.; Baehner, F.; Dabbs, D.J.; Decker, T.; Eusebi, V.; Fox, S.B.; Ichihara, S.; Jacquemier, J.; Lakhani, S.R.; et al. Breast cancer prognostic classification in the molecular era: The role of histological grade. *Breast Cancer Res.* **2010**, *12*, 207. [CrossRef] [PubMed]

34. Dang, Y.; Lan, F.; Ouyang, X.; Wang, K.; Lin, Y.; Yu, Y.; Wang, L.; Wang, Y.; Huang, Q. Expression and clinical significance of long non-coding RNA HNF1A-AS1 in human gastric cancer. *World J. Surg. Oncol.* **2015**, *13*, 302. [CrossRef] [PubMed]

35. Dang, Y.; Ouyang, X.; Zhang, F.; Wang, K.; Lin, Y.; Sun, B.; Wang, Y.; Wang, L.; Huang, Q. Circular RNAs expression profiles in human gastric cancer. *Sci. Rep.* **2017**, *7*, 9060. [CrossRef] [PubMed]

36. Siegel, R.L.; Miller, K.D.; Jemal, A. Cancer statistics, 2016. *CA Cancer J. Clin.* **2016**, *66*, 7–30. [CrossRef] [PubMed]

37. Zhang, R.; Xu, J.; Zhao, J.; Wang, X. Silencing of hsa_circ_0007534 suppresses proliferation and induces apoptosis in colorectal cancer cells. *Eur. Rev. Med. Pharmacol. Sci.* **2018**, *22*, 118–126. [PubMed]

38. Keshava Prasad, T.S.; Goel, R.; Kandasamy, K.; Keerthikumar, S.; Kumar, S.; Mathivanan, S.; Telikicherla, D.; Raju, R.; Shafreen, B.; Venugopal, A.; et al. Human Protein Reference Database—2009 update. *Nucleic Acids Res.* **2009**, *37*, D767–D772. [CrossRef] [PubMed]

39. Price, T.; Peña, F.I.; Cho, Y.-R. Survey: Enhancing protein complex prediction in PPI networks with GO similarity weighting. *Interdiscip. Sci.* **2013**, *5*, 196–210. [CrossRef] [PubMed]

40. Pedersen, T.; Pakhomov, S.V.S.; Patwardhan, S.; Chute, C.G. Measures of semantic similarity and relatedness in the biomedical domain. *J. Biomed. Inf.* **2007**, *40*, 288–299. [CrossRef] [PubMed]

41. Guzzi, P.H.; Mina, M.; Guerra, C.; Cannataro, M. Semantic similarity analysis of protein data: Assessment with biological features and issues. *Brief. Bioinform.* **2012**, *13*, 569–585. [CrossRef] [PubMed]

42. Huang, Y.-A.; Chan, K.C.C.; You, Z.-H. Constructing prediction models from expression profiles for large scale lncRNA-miRNA interaction profiling. *Bioinformatics* **2018**, *34*, 812–819. [CrossRef] [PubMed]

43. Pinero, J.; Queralt-Rosinach, N.; Bravo, A.; Deu-Pons, J.; Bauer-Mehren, A.; Baron, M.; Sanz, F.; Furlong, L.I. DisGeNET: A discovery platform for the dynamical exploration of human diseases and their genes. *Database* **2015**, *2015*. [CrossRef] [PubMed]

44. Oyston, J. Online Mendelian Inheritance in Man. *Anesthesiology* **1998**, *89*, 811–812. [CrossRef] [PubMed]

45. Lu, C.; Yang, M.; Luo, F.; Wu, F.-X.; Li, M.; Pan, Y.; Li, Y.; Wang, J. Prediction of lncRNA-disease associations based on inductive matrix completion. *Bioinformatics* **2018**, *34*, 3357–3364. [CrossRef] [PubMed]

46. Sun, K.; Gonçalves, J.P.; Larminie, C.; Przulj, N. Predicting disease associations via biological network analysis. *BMC Bioinform.* **2014**, *15*, 304. [CrossRef] [PubMed]

47. Hu, Y.; Zhou, M.; Shi, H.; Ju, H.; Jiang, Q.; Cheng, L. Measuring disease similarity and predicting disease-related ncRNAs by a novel method. *BMC Med. Genom.* **2017**, *10*, 71. [CrossRef] [PubMed]

48. Cheng, L.; Jiang, Y.; Wang, Z.; Shi, H.; Sun, J.; Yang, H.; Zhang, S.; Hu, Y.; Zhou, M. DisSim: An online system for exploring significant similar diseases and exhibiting potential therapeutic drugs. *Sci. Rep.* **2016**, *6*, 30024. [CrossRef] [PubMed]

49. Hu, Y.; Zhao, L.; Liu, Z.; Ju, H.; Shi, H.; Xu, P.; Wang, Y.; Cheng, L. DisSetSim: An online system for calculating similarity between disease sets. *J. Biomed. Semant.* **2017**, *8*, 28. [CrossRef] [PubMed]

50. Chen, X.; Liu, M.-X.; Yan, G.-Y. RWRMDA: Predicting novel human microRNA-disease associations. *Mol. Biosyst.* **2012**, *8*, 2792–2798. [CrossRef] [PubMed]

51. Chen, X.; Yan, G.-Y. Novel human lncRNA–disease association inference based on lncRNA expression profiles. *Bioinformatics* **2013**, *29*, 2617–2624. [CrossRef] [PubMed]

52. Sun, D.; Li, A.; Feng, H.; Wang, M. NTSMDA: Prediction of miRNA-disease associations by integrating network topological similarity. *Mol. Biosyst.* **2016**, *12*, 2224–2232. [CrossRef] [PubMed]

53. Chen, X. KATZLDA: KATZ measure for the lncRNA-disease association prediction. *Sci. Rep.* **2015**, *5*, 16840. [CrossRef] [PubMed]

54. van Laarhoven, T.; Nabuurs, S.B.; Marchiori, E. Gaussian interaction profile kernels for predicting drug–target interaction. *Bioinformatics* **2011**, *27*, 3036–3043. [CrossRef] [PubMed]

55. Zhang, W.; Zeng, T.; Liu, X.; Chen, L. Diagnosing phenotypes of single-sample individuals by edge biomarkers. *J. Mol. Cell Biol.* **2015**, *7*, 231–241. [CrossRef] [PubMed]

56. Yu, X.; Zhang, J.; Sun, S.; Zhou, X.; Zeng, T.; Chen, L. Individual-specific edge-network analysis for disease prediction. *Nucleic Acids Res.* **2017**, *45*, e170. [CrossRef] [PubMed]

57. Zhao, J.; Zhou, Y.; Zhang, X.J.; Chen, L. Part mutual information for quantifying direct associations in networks. *Proc. Natl. Acad. Sci. USA* **2016**, *113*, 5130–5135. [CrossRef] [PubMed]

58. Chen, L.; Liu, R.; Liu, Z.P.; Li, M.; Aihara, K. Detecting early-warning signals for sudden deterioration of complex diseases by dynamical network biomarkers. *Sci. Rep.* **2012**, *2*, 342. [CrossRef] [PubMed]

59. Yang, B.; Li, M.; Tang, W.; Liu, W.; Zhang, S.; Chen, L.; Xia, J. Dynamic network biomarker indicates pulmonary metastasis at the tipping point of hepatocellular carcinoma. *Nat. Commun.* **2018**, *9*, 678. [CrossRef] [PubMed]

60. Li, M.; Li, C.; Liu, W.; Liu, C.; Cui, J.; Li, Q.; Ni, H.; Yang, Y.; Wu, C.; Chen, C.; et al. Dysfunction of PLA2G6 and CYP2C44 associated network signals imminent carcinogenesis from chronic inflammation to hepatocellular carcinoma. *J. Mol. Cell Biol.* **2018**, *9*, 489–503. [CrossRef] [PubMed]

 © 2018 by the authors. Licensee MDPI, Basel, Switzerland. This article is an open access article distributed under the terms and conditions of the Creative Commons Attribution (CC BY) license (http://creativecommons.org/licenses/by/4.0/).

International Journal of
Molecular Sciences

Article

Role of Overexpressed Transcription Factor FOXO1 in Fatal Cardiovascular Septal Defects in Patau Syndrome: Molecular and Therapeutic Strategies

Adel Abuzenadah [1,2], Saad Alsaedi [3], Sajjad Karim [1,*] and Mohammed Al-Qahtani [1]

[1] Center of Excellence in Genomic Medicine Research, Faculty of Applied Medical Sciences, King Abdulaziz University, P.O. Box 80216, Jeddah 21589, Saudi Arabia; aabuzenadah@kau.edu.sa (A.A.); mhalqahtani@kau.edu.sa (M.A.-Q.)

[2] King Fahd Medical Research Center, King Abdulaziz University, P.O. Box 80216, Jeddah 21589, Saudi Arabia

[3] Department of Pediatric, Faculty of Medicine, King Abdulaziz University Hospital, King Abdulaziz University, P.O. Box 80215, Jeddah 21589, Saudi Arabia; salsaedi@hotmail.com

* Correspondence: skarim1@kau.edu.sa; Tel.: +966-55-7581741

Received: 15 September 2018; Accepted: 5 November 2018; Published: 10 November 2018

Abstract: Patau Syndrome (PS), characterized as a lethal disease, allows less than 15% survival over the first year of life. Most deaths owe to brain and heart disorders, more so due to septal defects because of altered gene regulations. We ascertained the cytogenetic basis of PS first, followed by molecular analysis and docking studies. Thirty-seven PS cases were referred from the Department of Pediatrics, King Abdulaziz University Hospital to the Center of Excellence in Genomic Medicine Research, Jeddah during 2008 to 2018. Cytogenetic analyses were performed by standard G-band method and trisomy13 were found in all the PS cases. Studies have suggested that genes of chromosome 13 and other chromosomes are associated with PS. We, therefore, did molecular pathway analysis, gene interaction, and ontology studies to identify their associations. Genomic analysis revealed important chr13 genes such as FOXO1, Col4A1, HMGBB1, FLT1, EFNB2, EDNRB, GAS6, TNFSF1, STARD13, TRPC4, TUBA3C, and TUBA3D, and their regulatory partners on other chromosomes associated with cardiovascular disorders, atrial and ventricular septal defects. There is strong indication of involving FOXO1 (Forkhead Box O1) gene—a strong transcription factor present on chr13, interacting with many septal defects link genes. The study was extended using molecular docking to find a potential drug lead for overexpressed FOXO1 inhibition. The phenothiazine and trifluoperazine showed efficiency to inhibit overexpressed FOXO1 protein, and could be potential drugs for PS/trisomy13 after validation.

Keywords: Patau Syndrome; cytogenetics; FOXO1; transcription factor; molecular pathways; bioinformatics; molecular docking; and drug design

1. Introduction

Patau Syndrome (PS) is a rare congenital anomaly due to the presence of an extra chromosome 13 popularly called trisomy 13 [1]. In spite of being the least common, it is the severest of all autosomal trisomies indicated by a prevalence rate of 1:5000 to 1:20,000 [2,3]. The syndrome is associated with a host of congenital anomalies including central nervous system (CNS) defects, midline abnormalities, eye and ear anomalies, cardiac defects, apnea, orofacial flaws, gastrointestinal and genitourinary aberrations, limb deformations, and developmental retardation [4,5]. Life expectancy is severely limited; more than 80% of PS patients do not survive long, and according to some estimates have median survival of 2.5 days [2,6,7]. Nevertheless, only a few can survive beyond 10 years but not with serious intellectual and physical disabilities [8–11]. Early death of PS is assigned to frequent CNS and cardiopulmonary system aberrations [12].

There is no specific treatment recommended for PS. Intensive care unit level of treatment for a couple of weeks is requisite for infants. Surgery for heart defects and other abnormalities like gastrointestinal or urogenital might be needed for six-month survivors. However, CNS disorders are difficult to treat by surgery. Children surviving more than a year suffer from severe intellectual disabilities, physical abnormalities and also have a high risk of developing cancer. Most studies indicate that older women are at higher risks of delivering trisomy 13 offspring [13]. Despite the fact that there are a number of trisomy 13 cases in Saudi Arabia, no systematic study has yet been done on causative factors like maternal age, consanguinity, and parity.

PS is a multigenic complex and lethal disease of multiple congenital abnormalities associated with poor prognosis [14]. Along with CNS disorders, heart ailments, especially septal defects are leading cause of deaths [2,15]. Septal defects is a complex disorder involving hundreds of altered gene regulations and these genes are located on multiple chromosomes including chromosome 13 [16]. Chromosome 13 is 114,364,328 bp in size, representing nearly 4% of the total DNA, and encodes 308 proteins. This chromosome has 343 protein-coding genes, 622 non-coding RNA genes, and 481 pseudogenes [17].

Molecular pathway and gene ontology analysis of chromosome 13 revealed the presence of important genes like *FOXO1, Col4A1, HMGBB1, FLT1, EFNB2, EDNRB, GAS6, TNFSF1, STARD13, TRPC4, TUBA3C, TUBA3D*. These genes are linked with cardiovascular disorders, atrial and ventricular septal defects commonly reported in PS [18–31]. Among them, *FOXO1* is a strong transcription factor which interacts and regulates several other genes on different chromosomes, (*GATA4* (8p23.1), *GATA6* (18q11.2), *GJA1* (6q22.31), *JAG1* (20p12.2), *CITED2* (6q24.1), *RYR2* (1q43), *NKX2-5* (5q35.1), *RARA* (17q21.2), *CXCL12* (10q11.21), *SIRT1* (10q21.3), *TBX5* (12q24.21), *AKT1* (14q32.33), *CDKN2A* (9p21.3), *PCK1* (20q13.31), etc.) and are associated with septal defects in PS [32–45]. Thus, some genes like *NODAL, FPR1, AFP, AGO2, UROD, ZIC2* are not located on chromosome 13 but have strong association with PS.

Forkhead Box O1 (*FOXO1*) gene needs special mention. It is a member of the forkhead box O family of transcription factors located on 13q14.11. The *FOXO1* exhibits its functions by binding to promoter of downstream genes or interacting with other transcription factors [46]; both its up- or down-regulation can lead to serious consequences. It has noticeable expression in the cardiovascular system, specifically in vascular and endothelial cells, and plays a substantial role in the crucial embryonic stage [22,47]. The specific function of FOXO1 has to be determined. However, some studies strongly suggest its key role in regulation of numerous cellular functions comprising proliferation, survival, cell cycle, metabolism, muscle growth differentiation, and myoblast fusion [48–50]. Other observations relate it to muscle fiber-type specification highly expressed in fast twitch fiber-enriched muscles, in comparison to slow muscles. The *FOXO1* is also involved in a host of other functions: metabolism regulation, cell proliferation, oxidative stress response, immune homeostasis, pluripotency in embryonic stem cells, and apoptosis [51,52]. Besides, *FOXO1* deletion or downregulation helps to rescue heart from diabetic cardiomyopathy and increases apoptosis under stress conditions like ischemia or myocardial infarction [52–55]. The *FOXO1* is a major transcription factor in cardiac development. Thus, we see *FOXO1* null mice have underdeveloped blood vessels, whereas overexpression of the *FOXO1* gene results in reduced heart size, myocardium thickening, and eventual heart failure [18–21]. Since *FOXO1* protects cardiac tissue from a variety of stress stimuli by up-regulating anti-apoptotic, antioxidant, and autophagy genes [47,56,57], and restores metabolic equilibrium to minimize cardiac injury due to apoptosis, therefore, in PS, *FOXO1* might be a chief regulator of cardiac disorders [52]. The fact is reinforced by reports where survival is improved by suppression of upregulated *FOXO1* [18]. Given the wide range of functions of *FOXO1*, its expression rate may play a vital role in PS and we checked its inhibition via molecular docking with certain drugs.

Molecular Docking between Candidate Drugs and FOXO1

Molecular docking, a computational simulation to screen inhibitor (ligand) compounds against biomolecule of interest, has become a crucial aspect of drug discovery approaches. Recently, repositioning or repurposing of the existing drugs is gaining attention for the treatment of diseases other than their known primary indications [58,59]. This approach could save enormous time, efforts and costs owing to the proven safety and quality of the drugs already available on the market, rather than to discover and develop novel chemical leads [60]. Similar observation on FOXO1, already implicated in a variety of functions, can specially be very promising for docking studies.

The FOXO1 protein contains 4 functional domains; (i) Forkhead domain (FKH), (ii) nuclear localization signal domain (NLS), (iii) nuclear export signal (NES), and (iv) transactivation domain (TAD). The FKH domain consists of four helices (H1–H4), two winged-loops (W1–W2), and three β strands (S1–S3), which mainly exhibits its functions as a DNA recognition and binding site. The FOXO1 regulates transcription of genes by directly binding with either 5′-GTAAA(T/C)AA-3′, or 5′-(C/A)(A/C)AAA(C/T)AA-3′ consensus sequence of downstream DNA [61–63]. The FOXO1 protein has thus become an extremely useful therapeutic target in many diseases including PS. Its expression can be regulated by acetylation, phosphorylation, and ubiquitination. Many potential inhibitors including leptomycin B [64], phenothiazines/trifluoperazine [65,66], bromotyrosine/psammaplysene A [67] or D4476 [68] and ETP-45658 [69], have been identified via virtual screening. Some drug candidates directly targeting FOXO1 have been patented [66]. For the docking study, we picked the FDA-approved drugs phenothiazine and its derivatives, trifluoperazine, which binds directly to the DNA binding domain of FOXO1 [70,71]. A brief introduction of both will be befitting here.

Phenothiazine (PTZ) and its derivatives are organic antihelmintic compounds presently used for important diseases like schizophrenia and bipolar disorder. Dopamine receptors are their main target. Repurposing PTZ has been tried earlier for developing novel antitumor agents [72] and Hepatitis C virus [73]. Trifluoperazine (TFP), the other derivative chosen in our studies, is a phenothiazine derivative and a dopamine antagonist, with antipsychotic and antiemetic properties. Their scaffold derivatives have also been suggested as an antiglioblastoma agent [74] and chemotherapeutic anticancer agent with high efficacy and reduced toxicity especially for oral cancer [72]. Lately, they have been shown as calmodulin antagonist [75,76].

In view of the fact that the exact mechanism is unknown as to how trisomy 13 disrupts development, heart disorders were identified as one of the most common disorders causing early death of PS patients. The present study, therefore, aims to explore the molecular interactions of 308 genes on this chromosome. We describe here the distinctive function of chromosome 13 and its key genes, especially FOXO1. We further intended to design a potential drug against FOXO1, a strong transcription factor which interacts with other key genes associated with lethal heart disorders in PS. The potential drugs to inhibit/reduce the transcriptional factor properties of FOXO1 are further explored with an aim to restore metabolic balance and limit apoptosis-induced cardiac damage.

2. Results

2.1. Cytogenetic Analysis of PS Patient

The prime aim of the current work was conducting genetic analysis of PS cases in the Saudi society (n = 37). Cytogenetic analyses were performed using G-banding technique-based karyotyping and found "full trisomy 13" in all 37 PS cases (Figure 1). The majority of individuals were newborns or children (up to 2 years), all with multiple abnormalities including heart disorders. Male to female ratio was found as 1.2:1. Analysis showed that mothers of affected individuals were above 35 years. The key clinical findings of PS observed: congenital heart defects (CHD) (61%), dysmorphic features (56%), polydactyly of hands and/or feet (53%), cryptorchidism (51%), abnormal auricles/low-set ears (47%), microphthalmia (40%), neurological disorders/microcephaly (35%), micrognathia (33%), scalp defects (31%), oral clefts (17%), microphthalmia/anophthalmia (9%), and duplication of the hallux (3%). Out of 37 cases, 31 underwent echocardiography and/or ultrasound, 21 of them showed heart defect and asymmetry of cardiac chambers. The main anatomical defects observed were arterial or ventricular septal defect, patent ductus arteriosus, pulmonic stenosis, coarctation of the aorta, tricuspid valve regurgitation, and mixed defects.

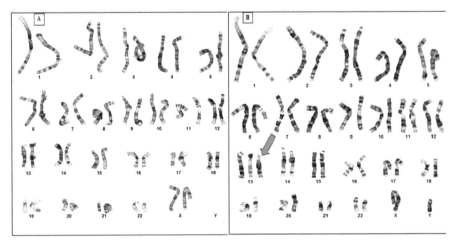

Figure 1. Karyotyping result; (**A**) Normal Karyotype of Healthy female and (**B**) Trisomy 13 in all cases (male = 20 and female = 17) of Patau Syndrome. Red arrow shows trisomy 13.

2.2. Molecular Pathway Analysis

Diploid status of chromosome 13 and normal expression of its genes are vital and a number of diseases are associated with its abnormalities (Table S1). However, molecular pathway and gene ontology analysis show as many as 308 protein coding genes on chr13; some of these pathogenic genes are *ATP7B, BRCA2, CAB39L, CKAP2, ESD, GJB2, GJB6, GPC5, HTR2A, MBNL2, RB1, SOX21, ZMYM2*, collectively noted for various disease associations. Other important genes such as *Col4A1, EFNB2, EDNRB, FLT1, FOXO1, GAS6, HMGB1, STARD13, TRPC4, TUBA3C, ZIC2* are specifically associated with cardiovascular disorders, atrial and ventricular septal defects—the key disorders of PS (Table 1). Ingenuity pathway analysis on 308 genes revealed canonical pathways like estrogen-mediated S-phase entry (Figure 2), gap junction signaling, cancer signaling, nitric oxide signaling in the cardiovascular system, adipogenesis pathway, VEGF signaling, cell cycle: G1/S checkpoint regulation, angiopoietin signaling, and 14-3-3-mediated signaling (Table 2). For a comprehensive idea, canonical pathways based on protein coding genes are summarized in Table 2. A cursory look shows FOXO1 to be involved in most of the canonical pathways. We focused our attention on it being strong transcription factor, interacting with and regulating many other genes on different chromosomes associated with septal defects in PS.

Table 1. Important pathogenic genes located on chromosome 13.

Gene Symbol	Gene Name	Cytoband	Associated Disease	Associated Pathways	Paralog
ATP7B	ATPase Copper Transporting Beta	13q14.3	Wilson Disease, Menkes Disease	Cardiac conduction; Ion channel transport; Transmembrane transport of small molecules	ATP7A
BRCA2	Breast cancer 2, early onset	13q13.1	Fanconi Anemia, and Breast Cancer	DNA Damage and Role of BRCA1 and BRCA2 in DNA repair	
CAB39L	Calcium-binding protein 39-like	13q14.2	Acute Monocytic Leukemia	RET signaling and mTOR signaling pathway	CAB39
COL4A1	Collagen Type IV Alpha 1 Chain	13q34	Coronary artery disease	Collagen chain trimerization, Integrin Pathway, ERK Signaling.	COL4A5
DZIP1	DAZ interacting zinc finger protein 1	13q32.1	Acrodermatitis Enteropathica, Zinc-Deficiency Type	Hedgehog signaling and GPCR signaling.	DZIP1L
EDNRB	Endothelin receptor type B	13q22.3	Waardenburg Syndrome	Calcium signaling pathway and Prostaglandin Synthesis and Regulation	EDNRA
ESD	S-formylglutathione hydrolase	13q14.2	Wilson Disease and Leukocoria	Glutathione metabolism	
FOXO1	Forkhead box O1	13q14.11	Rhabdomyosarcoma 2, Alveolar and Rhabdomyosarcoma	RET signaling; PI3K/AKT activation; Common Cytokine Receptor Gamma-Chain Family Signaling Pathways; AGE/RAGE pathway	FOXO3
FLT1	Fms-related tyrosine kinase 1	13q12.3	Anal Canal Squamous Cell Carcinoma and Eclampsia	p70S6K Signaling and Focal Adhesion	KDR
GAS6	Growth Arrest Specific 6	13q34	Sticky platelet Syndrome, Acute Maxillary Sinusitis, Mesangial Proliferative Glomerulonephritis	Apoptotic Pathways in Synovial Fibroblasts, GPCR Pathway, ERK Signaling	PROS1
GJB2	Gap junction protein, beta 2, 26 kDa (connexin 26)	13q12.11	Volwinkel Syndrome and Bart–Pumphrey Syndrome	Development Slit-Robo signaling and Gap junction trafficking.	GJB6.
GJB6	Gap junction protein, beta 6 (connexin 30)	13q12.11	Ectodermal Dysplasia 2, Clouston Type and Deafness, Autosomal Dominant 3B	Gap junction trafficking; Vesicle-mediated transport	GJB2
GPC5	Glypican-5	13q31.3	Simpson–Golabi–Behmel Syndrome and Tetralogy of Fallot	Glycosaminoglycan metabolism	GPC3
HMGB1	Box 5 Box 1	13q12.3	13q12.3 Microdeletion Syndrome, Adenosquamous Gallbladder Carcinoma	Activated TLR4 signaling; Cytosolic sensors of pathogen-associated DNA; Innate Immune System	HMGB2
HTR2A	5-HT2A receptor	13q14.2	Schizophrenia; Major Depressive Disorder	Calcium signaling pathway; Signaling by GPCR	HTR2C
MIPEP	Mitochondrial intermediate peptidase	13q12.12	Combined Oxidative Phosphorylation Deficiency 31		
PCCA	Propionyl Coenzyme A carboxylase, alpha polypeptide	13q32.3	Propionicacidemia and PCCA-Related Propionic Acidemia.	Metabolism and HIV Life Cycle.	MCCC1
RB1	Retinoblastoma 1	13q14.2	Retinoblastoma and Small-Cell Cancer of the Lung, Somatic.	Arrhythmogenic right ventricular cardiomyopathy (ARVC) and DNA Damage	RBL2
RCBTB1	RCC1 and BTB domain-containing protein 1	13q14.2	Retinal Dystrophy with Or Without Extraocular Anomalies.		RCBTB2
RGCC RNRI	Regulator of cell cycle RGCC Encoding RNA, ribosomal 45S cluster 1	13q14.11 13p12	Renal Fibrosis and Retinal Cancer Idiopathic Bilateral Vestibulopathy and Congenital Cytomegalovirus	TP53 Regulates Transcription of Cell Cycle Genes Viral mRNA Translation	
SLITRK6	SLIT and NTRK-like protein 6	13q31.1	Deafness and Yopia and Autosomal Recessive Non-Syndromic Sensorineural Deafness		SLITRK5
SOX21	Transcription factor SOX-21	13q32.1	Mesodermal Commitment Pathway and ERK Signaling.	Mesodermal Commitment Pathway; ERK Signaling	SOX14
STARD13	StAR-Related Lipid Transfer Domain Containing 13	13q13	Hepatocellular Carcinoma, Arteriovenous Malformations of the Brain, Fibrosarcoma of Bone	p75 NTR receptor-mediated signaling, Signaling by GPCR, Signaling by Rho GTPases	STARD8
TPT1	Translationally controlled tumor protein (TCTP)	13q14.13	Urticaria and Asthma	DNA Damage and Cytoskeletal Signaling	
TRPC4	Transient Receptor Potential Cation Channel Subfamily C Member 4	13q13.3	Photosensitive Epilepsy	Developmental Biology, Ion channel transport, Netrin-1 signaling	TRPC5

Table 1. *Cont.*

Gene Symbol	Gene Name	Cytoband	Associated Disease	Associated Pathways	Paralog
TSC22D1	TSC22 domain family protein 1	13q14.11	Salivary Gland Cancer and Brain Sarcoma	Development TGF-beta receptor signaling and Ectoderm Differentiation	*TSC22D2*
TUBA3C	Tubulin Alpha 3C	13q12.11	Clouston Syndrome, nonsyndromic Deafness, Kabuki Syndrome 1	Development Slit-Robo signaling, Cooperation of Prefoldin and TriC/CCT in actin and tubulin folding	*TUNA3D*
XPO4	Exportin-4	13q12.11	Conjunctival Degeneration and Pinguecula	eIF5A regulation in response to inhibition of the nuclear export system and Ran Pathway	
ZIC2	Zic Family Member 2	13q32.3	Holoprosencephaly 5 and Zic2-Related Holoprosencephaly	Mesodermal Commitment Pathway	*ZIC1*
ZMYM2	Zinc finger MYM-type protein 2	13q12.11	Lymphoblastic Lymphoma and 8P11 Myeloproliferative Syndrome	HIV Life Cycle and FGFR1 mutant receptor activation	*ZMYM3*

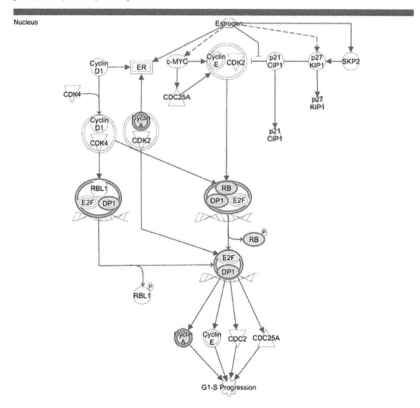

Figure 2. Estrogen-mediated S-phase Entry pathway derived from 308 protein coding genes of triosomy 13 (chromosome 13) using Ingenuity Pathway Analysis Tool.

Table 2. Top canonical pathways determined by Ingenuity pathway analysis tools based on protein coding genes located on chromosome 13.

Canonical Pathways	−log (*p* Value)	Ratio	Molecules
Estrogen-mediated S-phase Entry	2.06	0.115	RB1, CCNA1, TFDP1
Cancer Signaling	1.69	0.052	RB1, FOXO1, TFDP1, KL, IRS2, CDK8, SMAD9, TFDP1, ARHGEF7
Extrinsic Prothrombin Activation Pathway	1.56	0.125	F10, F7
Role of p14/p19ARF in Tumor Suppression	1.5	0.071	RB1, KL, IRS2
Gap Junction Signaling	1.41	0.036	GJB6, KL, GJA3, TUBA3C/TUBA3D, IRS2, GJB2, HTR2A
Docosahexaenoic Acid (DHA) Signaling	1.27	0.057	FOXO1, KL, IRS2
Aldosterone Signaling in Epithelial Cells	1.24	0.035	SACS, KL, HSPH1, DNAJC3, IRS2, DNAJC15
FGF Signaling	1.2	0.044	KL, FGF9, FGF14, IRS2
GP6 Signaling Pathway	1.18	0.038	COL4A1, KL, IRS2, COL4A2, KLF12
Adipogenesis pathway	1.17	0.037	RB1, SAP18, SMAD9, FOXO1, KLF5
VEGF Signaling	1.08	0.040	FOXO1, FLT1, KL, IRS2
Cell Cycle: G1/S Checkpoint Regulation	1.04	0.046	RB1, FOXO1, TFDP1
ErbB2-ErbB3 Signaling	0.994	0.044	FOXO1, KL, IRS2

Table 2. *Cont.*

Canonical Pathways	−log (*p* Value)	Ratio	Molecules
Nitric Oxide Signaling in the Cardiovascular System	0.988	0.037	FLT1, KL, SLC7A1, IRS2
Coagulation System	0.948	0.057	F10, F7
Angiopoietin Signaling	0.875	0.039	FOXO1, KL, IRS2
Role of NANOG in Mammalian Embryonic Stem Cell Pluripotency	0.866	0.0333	SMAD9, KL, CDX2, IRS2
IL-3 Signaling	0.805	0.036	FOXO1, KL, IRS2
Actin Cytoskeleton Signaling	0.801	0.027	KL, FGF9, DIAPH3, ARHGEF7, FGF14, IRS2
14-3-3-mediated Signaling	0.778	0.030	FOXO1, KL, TUBA3C/TUBA3D, IRS2
IL-7 Signaling Pathway	0.774	0.034	FOXO1, KL, IRS2
HMGB1 Signaling	0.77	0.030	HMGB1, KL, IL17D, IRS2
NF-κB Signaling	0.769	0.028	TNFSF11, FLT1, KL, IRS2, TNFSF13B

2.3. Genomic Analysis and Protein–Protein Interaction Study

The result of STRING displayed direct interaction and predicted functional relationship amid FOXO1 and its interacting proteins. The following proteins showed noticeable interactions with FOXO1: GATA4 (8p23.1), SIRT1 (10q21.3), CITED2 (6q24.1), NFATc1 (18q23), and TBX5 (12q24.21) (Figure 3). FOXO1 as transcription factor interacted with the following relevant target genes: FASLG (1q24.3), IGFBP1 (7p12.3), SOD2 (6q25.3), PPARGC1A (4p15.2), ADIPOQ (3q27.3), APOC3 (11q23.3), OSTN (3q28), BCL2L11 (2q13), CCND2 (12p13.32), and CDKN1B (12p13.1). This was predicted by text-mining application and UCSC genome browser. However, genomic analysis of PS had shown that many genes (NODAL on 10q22, FPR1 on 19q13.41, AFP on 4q13.3, AGO2 on 8q24.3, UROD on 1p34.1, ZIC2 on 13q32.3, etc.) are not directly regulated by FOXO1, rather strongly associated with PS (Table 3).

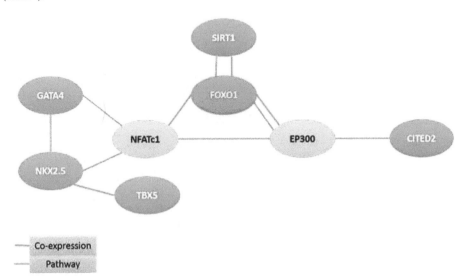

Figure 3. Protein–protein Interaction Partners (GATA4, NKX2-5, SIRT1, CITED, NFATc1, TBX5) of FOXO1.

Table 3. Key genes strongly associated with the survival of PS patient.

Gene Symbol	Gene Name	Cytoband	Associated Disease	Associated Pathways	Paralog
NODAL	Nodal Growth Differentiation Factor	10q22	Visceral Heterotaxy 5 (HTX5) and Nodal-Related Visceral Heterotaxy	Mesodermal Commitment Pathway and Signaling pathways regulating pluripotency of stem cells	GDF3
FPR1	Formyl Peptide Receptor 1	19q13.41	Susceptibility to Localized Juvenile Periodontitis and Periodontitis 1, Juvenile	Signaling by GPCR and Peptide ligand-binding receptors	FPR2
AFP	Alpha Fetoprotein	4q13.3	Alpha-Fetoprotein Deficiency and Hereditary Persistence of Alpha-Fetoprotein	Glucocorticoid receptor regulatory network and Embryonic and Induced Pluripotent Stem Cell Differentiation Pathways and Lineage-specific Markers	ALB
AGO2	Argonaute RISC Catalytic Component 2	8q24.3	Chromosome 18P Deletion Syndrome and Gum Cancer	RET signaling and Translational Control.	AGO1
UROD	Uroporphyrinogen Decarboxylase	1p34.1	Porphyria Cutanea Tarda and Urod-Related Porphyrias	Metabolism and Porphyrin and chlorophyll metabolism	
GATA4	GATA Binding Protein 4	8p23.1	Testicular Anomalies with or without Congenital Heart Disease and Atrial Septal Defect 2	Response to elevated platelet cytosolic Ca2+ and Human Embryonic Stem Cell Pluripotency	GATA6
GATA6	GATA Binding Protein 6	18q11.2	Pancreatic Agenesis and Congenital Heart Defects and Atrioventricular Septal Defect 5	Mesodermal Commitment Pathway and Response to elevated platelet cytosolic Ca2+	GATA4
GJA1	Gap Junction Protein Alpha 1	6q22.31	Oculodentodigital Dysplasia and Syndactyly, Type Iii	Development Slit-Robo signaling and Arrhythmogenic right ventricular cardiomyopathy	GJA3
JAG1	Jagged 1	20p12.2	Alagille Syndrome 1 and Tetralogy of Fallot	Signaling by NOTCH1 and NOTCH2 Activation and Transmission of Signal to the Nucleus	JAG2
CITED2	Cbp/P300 Interacting Transactivator with Glu/Asp Rich Carboxy-Terminal Domain 2	6q24.1	Atrial Septal Defect 8 and Ventricular Septal Defect 2	Cellular Senescence (REACTOME) and Transcriptional regulation by the AP-2 (TFAP2) family of transcription factors	CITED1
RYR2	Ryanodine Receptor 2	1q43	Ventricular Tachycardia, Catecholaminergic Polymorphic, 1 and Arrhythmogenic Right Ventricular Dysplasia 2	Calcium signaling pathway and Arrhythmogenic right ventricular cardiomyopathy	RYR3
NKX2-5	NK2 Homeobox 5	5q35.1	Atrial Septal Defect 7, With or Without Av Conduction Defects and Tetralogy of Fallot	Human Embryonic Stem Cell Pluripotency and NFAT and Cardiac Hypertrophy	NKX2-3
RARA	Retinoic Acid Receptor Alpha	17q21.2	Leukemia, Acute Promyelocytic, Somatic and Myeloid Leukemia	Nuclear Receptors in Lipid Metabolism and Toxicity and Activated PKN1 stimulates transcription of AR (androgen receptor) regulated genes KLK2 and KLK3.	RARB
CXCL12	C-X-C Motif Chemokine Ligand 12	10q11.21	HIV-1 and AIDS Dementia Complex	p70S6K Signaling and Akt Signaling	
SIRT1	Sirtuin 1	10q21.3	Xeroderma Pigmentosum, Group D and Ovarian Endodermal Sinus Tumor	Longevity regulating pathway and E2F transcription factor network	SIRT4
TBX5	T-Box 5	12q24.21	Holt-Oram Syndrome and Aortic Valve Disease 2	Human Embryonic Stem Cell Pluripotency and Cardiac conduction.	TBX4
AKT1	AKT Serine/Threonine Kinase 1	14q32.33	Cowden Syndrome 6 and Proteus Syndrome, Somatic	Transcription Androgen Receptor nuclear signaling and E-cadherin signaling in keratinocytes	AKT3
CDKN2A	Cyclin Dependent Kinase Inhibitor 2A	9p21.3	Pancreatic Cancer/Melanoma Syndrome and Melanoma and Neural System Tumor Syndrome	DNA Damage and Bladder cancer	CDKN2B
PCK1	Phosphoenolpyruvate Carboxykinase 1	20q13.31	Pepck 1 Deficiency and Phosphoenolpyruvate Carboxykinase-1, Cytosolic, Deficiency	Abacavir transport and metabolism and Citrate cycle (TCA cycle)	PCK2

Docking using the Lamarckian Genetic Algorithm approach was employed to elucidate the basis of structural binding of PTZ and TFP to FOXO1. The result demonstrated favored binding energies ∆G in the range of −4.17 kcal/mol to −1.87 kcal/mol, respectively, with 1 molecule of PTZ showing hydrogen bond with the active site residue Ser193. Other predominant interactions for PTZ were hydrophobic (Leu163, Leu168, Val194, and Pro195) and pi–pi ring stacking non-covalent interaction with Trp189. Estimated inhibition constant, Ki values were 879.98 µM (FOXO1:PTZ) and 42.27 mM (FOXO1:TFP).

It was further revealed that the PTZ hydrophobic binding pocket was lined mainly with residues Leu163, Leu168, Lys171, Trp189, Val194, and Pro195, and the hydroxylic Ser193 showed crucial interactions with the ligand (Figure 4). Similarly, TFP binding site was also hydrophobic with residues Leu183, Tyr187, Leu217, Arg225, Ser234, Ser235, and Trp237. Weak interactions with TFP were seen through Ser184, Ser218, Ser234, and Ser235. Besides, non-covalent hydrogen bonding was evident between TFP's electrophilic F1, F2, and F3 and nucleophilic O of Arg214 (Figure 5). Molecular docking analysis was done to understand the binding efficiency of the selected drugs; PTZ was found to be the better FOXO1 inhibitor as it displayed a higher negative binding energy as compared to TFP, hence, it promises to be a more effective inhibitor.

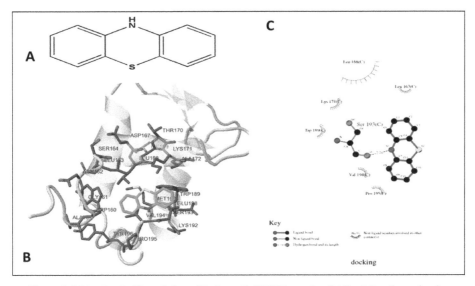

Figure 4. Molecular docking of phenothiazine with FOXO1 protein. (**A**) Depicting the molecular structure of phenothiazine; (**B**) Structure visualization of FOXO1 protein bound with ligand PTZ. The interacting residues are labeled in the binding site. (**C**) 2D plot of phenothiazine of FOXO1 showing ligand–protein interaction profiled by AutoDock software of Docking Server. Leu163, Leu168, Lys171, Trp189, Val194, Pro195, and Ser193 residues of FOXO1 showed crucial interactions with the phenothiazine.

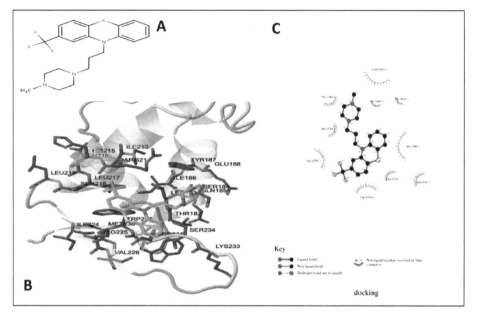

Figure 5. Molecular docking of trifluoperazine with FOXO1 protein. (**A**) Depicting the molecular structure of trifluoperazine; (**B**) Structure visualization of FOXO1 protein bound with ligand TFP. The binding site is shown and the interacting residues are labeled. (**C**) 2D plot of trifluoperazine of FOXO1 showing ligand–protein interaction profiled by AutoDock software of Docking Server. Leu183, Tyr187, Leu217, Arg225, Arg234, Ser184, Ser218, Ser234, Ser235 and Trp237 residues of FOXO1 showed crucial interactions with the trifluoperazine.

3. Discussion

Generally, normal development requires only two copies of autosomal chromosomes; the presence of a third copy of chromosome (trisomy) is mostly lethal to the embryo. However, trisomy 13, 18, and 21 are the only cases where development can proceed to live birth. In the present study, the age of PS patients ranged from 1 day to 2 years, the majority (*n* = 19) died within a week, 8 within a month, 9 passed a month barring 1 surviving 2 years as an exception. Studies also showed survival with trisomy 13 being miserably limited with median life expectancy of 2.5 days. The overall observation reinforces other studies where 85% of PS patients could hardly survive beyond a month [2] and rarely survive beyond 10 years [8–10].

It is typical to have different types of abnormalities related to chromosome 13 manifested in various disorders. Apart from PS others include 13q deletion syndrome, propionic academia, retinoblastoma, Waardenburg Syndrome, Wilson's Syndrome, Young–Madders Syndrome and also bladder and breast cancer. In the present case, all cases had confirmed trisomy 13. However, other researchers have reported full trisomy of chromosome 13 in 70–80% of cases, mosaicism in 10–20% and translocations involving chromosomes 13 in 5–10%, besides other types of chromosomal abnormalities in 5–10% of cases [15].

The frequency of CHD in patients was 61%, which falls in the range (56 to 86%) of frequency reported by other studies [3,15,77,78]. However, relatively low frequency has been reported by Rasmussen et al. (45.7%) and Pont et al. (34.8%) [4,79].

No plausible explanation is forthcoming as to how extra genetic material (trisomy 13) causes a plethora of abnormal features like abnormal cerebral functions, a small cranium, retardation, nonfunctional eyes, and heart imperfections. We made an attempt to identify important pathogenic genes such as *EDNRB, ZIC2, ATP7B, GJB2, HTR2A* located on chromosome 13 to associate these with diseases and pathways; however, none was alone capable of the symptoms of PS.

As for exploring the pathways, it is known that the *EDNRB* gene located at 13q22.3 codes for endothelin receptor type B protein, a GPCR located on the cell surface which functions via interaction with endothelins. It activates a phosphatidylinositol-calcium, the second messenger system transmitting information from outside the cell to inside. Its highest expression is in placental tissues. Mutations in this gene have been previously linked to the congenital genetic disorder, Hirschsprung disease, alternatively called Congenital Aganglionic Megacolon. It is a neural crest development disorder characterized by absence of enteric ganglia along a variable length of the intestine causing intestinal obstruction [80]. The Zic family member 2 (*ZIC2*) gene is present at 13q32.3 genomic region and encodes a type of zinc finger protein that functions as a transcriptional repressor and regulates both early and late stages of forebrain development. Mutations in *ZIC2* gene, involving expansion of an alanine repeat at C-terminus, cause holoprosencephaly-5, a structural anomaly of the human brain [81–83]. It appears that a polyhistidine tract gene polymorphism is probably associated with increased risk of Holoprosencephaly. The defect appears to be due to changes in the organizer region leading to defective anterior notochord, further resulting in degradation of the prechordal plate. As a result shh signal cannot reach to the developing forebrain, vital for the formation of the two hemispheres [84]. *ZIC2* has also been linked to neural tube defects [85] and heart defects [86].

The present endeavor extended search beyond chromosome 13 and identified genes such as *NODAL, FPR1, AFP, AGO2, UROD, GATA4, GJA1, JAG1, CETED2, RYR2, NKX2-5, RARA, SIRT1, TBX5, AKT1,* and *PCK1* across genome with a view to exploring their role in PS. In doing so, two facts emerged clearly; one, the majority of PS patients had CHD and, two, all patients showed trisomy13. It thus appears that there could be a strong link between genes located on trisomy13 and heart disorders. Ingenuity pathway analysis of chr13 genes explored indicated hundreds of canonical pathways and many of them had FOXO1 as key molecules of such pathways. We applied a bioinformatics approach and searched scientific literature and identified pathogenic genes involved in CHD located on chr13 beside other chromosomes. It appears that genes of chr13 and other chromosomes might work together, either as transcription factor regulator or interacting partners. Nevertheless, *FOXO1* is a strong transcription factor activating many genes, these being: *FASLG, IGFBP1, SOD2, PPARGC1A, ADIPOQ, APOC3, OSTN, BCL2L11, CCND2,* and *CDKN1B*. The protein–protein interaction study also showed key interacting partners like GATA4, NKX2-5, SIRT1, CITED2, NFATc1 and TBX5, which are actively implicated in heart disorders and thus partly responsible for PS.

It will not be out of place to mention GATA4, an interaction partner of FOXO1, a strong transcription factor regulating cardiac repair and remodeling. It plays an important role in cardiac development and differentiation as its abnormal expression leads to embryonic lethality [87–89]. Likewise, overexpression of NKX2-5 is reported as hypertrophic stimuli [90]. Interestingly, GATA4 and NKX2-5 act synergistically and regulate a myriad of cardiac genes [91,92]. Other studies showed that TBX5 is also an interaction partner of FOXO1, GATA4, and NKX2-5 and encodes transcription factors involved in the regulation of forelimb and heart development [93–95]. Thus, the role of GATA4, NKX2-5, and TBX5 is established in cardiogenesis; however, their role in regulating the heart septal formation is a matter of further investigation [45,96,97]. Sperling et al. are credited for reporting a direct role of *CITED2* gene mutation in CHD epigenetic factor like methylation in the promoter region of *CITED2* plays a vital role in heart disease [98,99]. The sirtuins, a family of enzymes, encoded by SIRT1–SIRT7, are highly expressed in the heart tissue and the vascular endothelium, and are pivotal regulators of lifespan and health. The SIRT1 executes its function by deacetylation of FOXO transcription factor and other key substrates; all closely linked to cardio vascular ailments. The SIRT1 inhibition is shown to be associated with septal and valvular heart defects, as well as vascular dysfunction [100–102].

One thing is evident clearly though multiple studies—there is a direct and indirect involvement of *FOXO1* in heart disease [52–55]. Activated *FOXO1* has a direct impact on cell survival via alteration

Int. J. Mol. Sci. **2018**, *19*, 3547

in metabolism and turning on the cell death signaling cascade [103,104]. Overexpression of FOXO1 also causes autophagy in heart, leading to death [56,57]. A latest study had shown that knock down of *FOXO1* and *FOXO3* in the heart of $Lmna^{-/-}$ mice results in attenuation of apoptosis with a twice increase in the survival rate [18].

A bit of inhibiting expression of FOXO1 protein will further classify its important role in regulation. It is a monomeric nuclear protein and functions primarily as transcription factors by binding to a consensus DNA sequence of promoter region of downstream genes with a DNA-binding domain, 158–248 amino acid region [63,105–107]. The nuclear localization of FOXO1 is tightly regulated by the post translational modifications like acetylation, methylation, phosphorylation, and ubiquitination, or simply by its interaction with proteins like 14-3-3 [105]. Studies on these lines have identified a long list of FOXO1 inhibitors to be classified into two groups: one, drugs targeting nuclear transport of FOXO1 including leptomycin B, curcumin, psammaplysene A, phenothiazines/triflouperazine, calmodulin inhibitor/calmidazolium, intracellular Ca^{2+} chelator- BAPTA-AM l; and two, drugs targeting FOXO1 signaling pathway including epigallocatechin- 3-gallate, theaflavins, hyaluronan oligosaccharides, resveratrol, apigenin, luteolin, and psammaplysene A [64,65,67].

Phenothiazines and its derivatives (trifluoperazine) are chosen from FDA-approved drugs and binds directly with DBD of FOXO1 [70,71]. The molecular docking approach was applied to determine inhibition constant, predicting binding modes and defining the specific binding sites. The results showed that both the drugs can potentially inhibit FOXO1 protein. The drug PTZ mainly interacts with hydrophobic amino acids of the DNA-binding region (158–235) of the protein target. The TFP also binds inside the DNA-binding domain but the interacting residues are different from those in the case of PTZ binding. The contact or interaction surface value of docked ligand and protein is 421.011 $Å^2$ for PTZ and 612.637 $Å^2$ for TFP.

The present study is mainly based on a bioinformatics approach, so it can be associated with few limitations. It is proposed to undertake in silico finding to resolve the issue, and predictions are advised to be validated before final conclusion. Our finding suggests genetic engineering potentials in future.

4. Materials and Methods

4.1. Patients

A total of 37 cases including PS, dysmorphic features, multiple congenital anomalies, CHD and cleft palate were registered from Western region of Saudi Arabia through the King Abdulaziz University Hospital, Jeddah. The majority of individuals were newborns with multiple abnormalities including heart and neurological disorders. Peripheral blood (5–10 mL) was obtained after informed consent and a complete clinical and case history was recorded. Ethical approval for the study (G/017/27) was obtained from the King Abdulaziz University clinical research ethics board dated 09-06-2009 and the study strictly followed the standard Helsinki ethical guidelines during this research work.

4.2. Cytogenetics Study

A standard 72 h lymphocyte culture and GTG banding (G banding by Trypsin and Giemsa) were applied to peripheral blood in all patients. Microscopic examinations were done using 50 cells for each patient. In cases of suspected mosaicism, the number expanded to one hundred cells. Chromosomes were analyzed by semi-automatic Applied Imaging Karyotyper and karyotypes were described as per the International System for Human Cytogenetic Nomenclature (ISCN, 2016) [108].

4.3. Molecular Pathway and Gene Ontology Analysis

Biological significance of protein coding genes of chromosome 13 was interpreted by the Ingenuity Pathways Analysis software version 338830M (Ingenuity Systems, Redwood City, CA, USA). Significance of relationships between genes and functional frameworks was indicated by Fisher's exact test *p*-values. The percentage and number of uploaded genes/molecules matching to genes of a canonical pathway were measured for significance, expressed as a score. The Molecule Activity Predictor was employed to predict the effects of a gene/molecule on other molecules of pathway.

4.4. Identification of Functionally Significant Interacting Proteins of FOXO1

Search Tool for the Retrieval of Interacting Genes/Proteins (STRING version 9.1, https://string-db.org/) was used to identify significant proteins interacting with FOXO1. The biological database and web resource of known and predicted protein interactions were utilized, derived from high-throughput experimental sources, text mining and co-expression [109–111].

4.5. Molecular Docking and Drug Design

A search was made for available three-dimensional structures of FOXO1 protein in the RCSB's PDB database and retrieved five entries: 3CO6, 3CO7, 3COA, 4LG0, and 5DUI. All these structures were DNA-bound protein complexes. We proceeded with PDB code 3CO7:C. It corresponds to UniProtKB (Q12778, https://www.uniprot.org/help/uniprotkb) and the residues 1–154 were missing from the protein chain C.

Information was collected for structure of two selected compounds; Phenothiazine and trifluoperazine from ZINC database (available online: http://zinc.docking.org). It is a database of commercially available compounds [112]. We downloaded the mol2 file for ZINC ID 00028150 and 19418959 respectively. Structural analogs of TFP (31350265 and 39546119) were not considered for the present study.

Docking calculations for predicting binding modes and energies of two ligands phenothiazine (PTZ) and trifluoperazine (TFP) to protein (FOXO1) employed DockingServer [113], and AutoDock software for gasteiger partial charges addition to the ligand atoms, combining non-polar hydrogen atoms and defining rotatable bonds. Affinity grids were generated using the Autogrid tool [114]. AutoDock parameter set- and distance-dependent dielectric functions were used in the calculation of the van der Waals and the electrostatic terms, respectively. Docking simulations were performed using Lamarckian genetic algorithm and the Solis & Wets local search algorithm (http://autodock.scripps.edu) [115]. Initial position, torsions, and orientation of the ligands were set randomly. All rotatable torsions were released during docking. All experimentation was resultant of 10 different runs set to finish after 250,000 energy evaluations. The population size was fixed to 150. Translational step of 0.2 Å, and quaternion and torsion steps of 5 were applied during the search.

4.6. Statistical Analysis

χ^2 analysis and Fisher's exact test were used to compare the clinical features and proportion of chromosomal abnormalities in PS patients. The statistical analysis was carried out using MATLAB ver R2007a (The MathWorks, Natick, MA, USA).

5. Conclusions

Cytogenetic analysis of 37 Saudi PS patients showed full trisomy 13 without exception. Molecular interactions study of 308 protein coding genes located on chromosome 13 led to identification of significant genes such as: *FOXO1*, *RB1*, *CCNA1*, *TFDP1*, *KL*, *IRS2*, *F10*, *F7 GJB6*, *GJA3*, *TUBA3C/TUBA3D*, *COL4A1*, *FLT1*, *KLF12*, and *ZIC2*. The pathways (Estrogen-mediated S-phase entry, Extrinsic prothrombin activation pathway, Gap junction signaling, Docosahexaenoic acid signaling, VEGF signaling, Cell cycle: G1/S checkpoint regulation, IL-3 Signaling) were explored to find an association with PS. Molecular network analysis and protein–protein interaction study indicated *FOXO1* as strong transcription factor which interacts with other key genes like *GATA4*, *CITED* and *TBX5* located on different chromosomes but associated with lethal heart disorders in PS. Lethal genetic disorders are toughest to treat and many PS newborns die within a couple of days with severe complications without proper treatment. However, patients with a less severe condition have some

Int. J. Mol. Sci. **2018**, *19*, 3547

chance of survival and could be diagnosed with an actual problem and treated (surgery or medicine) accordingly. The in silico molecular docking studies done separately indicated phenothiazine and trifluoperazine as efficient inhibitor for FOXO1 protein as potential drugs for septal defects patients and PS. Molecular docking indicated phenothiazine to be an efficient inhibitor for FOXO1 and a candidate for future drug target, especially in septal defects patients and PS cases. It is recommended to utilize the present outcome after validation in vitro and in vivo animal model approaches.

Supplementary Materials: Supplementary materials can be found at http://www.mdpi.com/1422-0067/19/11/3547/s1.

Author Contributions: Conceptualization, A.A., S.A., M.A.-Q., and S.K.; Methodology, S.K. and M.A.-Q.; Formal Analysis, S.K.; Investigation, M.A.-Q.; Writing—Original Draft Preparation, S.K. and M.A.-Q.; Writing—Review & Editing, A.A. and S.A.; Visualization, S.K.; Supervision, M.A.-Q. and A.A.; Project Administration, M.A.-Q. and A.A.; Funding Acquisition, S.A. and M.A.-Q. All authors read and approved the final manuscript.

Funding: This project was funded by the Deanship of Scientific Research (Project Award Number-G/017/27) King Abdulaziz University, Jeddah, Saudi Arabia.

Acknowledgments: The authors would like to thank Zeenat Mirza and Heba Abusamara for their significant contribution. Our special thanks to Waseem Ahmed for editing and improving the English language and Mohammad Amir Khan for help in typing. We would also like to thank lab-staffs for technical support and the Center of Excellence in Genomic Medicine Research for administrative support.

Conflicts of Interest: The authors declare no conflict of interests.

References

1. Patau, K.; Smith, D.W.; Therman, E.; Inhorn, S.L.; Wagner, H.P. Multiple congenital anomaly caused by an extra autosome. *Lancet (London, England)* **1960**, *1*, 790–793. [CrossRef]
2. Duarte, A.C.; Menezes, A.I.; Devens, E.S.; Roth, J.M.; Garcias, G.L.; Martino-Roth, M.G. Patau syndrome with a long survival. A case report. *Genet. Mol. Res.* **2004**, *3*, 288–292. [PubMed]
3. Wyllie, J.P.; Wright, M.J.; Burn, J.; Hunter, S. Natural history of trisomy 13. *Arch. Dis. Child.* **1994**, *71*, 343–345. [CrossRef] [PubMed]
4. Pont, S.J.; Robbins, J.M.; Bird, T.M.; Gibson, J.B.; Cleves, M.A.; Tilford, J.M.; Aitken, M.E. Congenital malformations among liveborn infants with trisomies 18 and 13. *Am. J. Med. Genet.* **2006**, *140*, 1749–1756. [CrossRef] [PubMed]
5. Torrelo, A.; Fernandez-Crehuet, P.; Del Prado, E.; Martes, P.; Hernandez-Martin, A.; De Diego, V.; Carapeto, F. Extensive comedonal and cystic acne in Patau syndrome. *Pediatr. Dermatol.* **2010**, *27*, 199–200. [CrossRef] [PubMed]
6. Locock, L.; Crawford, J.; Crawford, J. The parents' journey: Continuing a pregnancy after a diagnosis of Patau's syndrome. *BMJ* **2005**, *331*, 1186–1189. [CrossRef] [PubMed]
7. Hassold, T.J.; Jacobs, P.A. Trisomy in man. *Annu. Rev. Genet.* **1984**, *18*, 69–97. [CrossRef] [PubMed]
8. Hsu, H.F.; Hou, J.W. Variable expressivity in Patau syndrome is not all related to trisomy 13 mosaicism. *Am. J. Med. Genet.* **2007**, *143a*, 1739–1748. [CrossRef] [PubMed]
9. Iliopoulos, D.; Sekerli, E.; Vassiliou, G.; Sidiropoulou, V.; Topalidis, A.; Dimopoulou, D.; Voyiatzis, N. Patau syndrome with a long survival (146 months): A clinical report and review of literature. *Am. J. Med. Genet.* **2006**, *140*, 92–93. [CrossRef] [PubMed]
10. Tunca, Y.; Kadandale, J.S.; Pivnick, E.K. Long-term survival in Patau syndrome. *Clin. Dysmorphol.* **2001**, *10*, 149–150. [CrossRef] [PubMed]
11. Redheendran, R.; Neu, R.L.; Bannerman, R.M. Long survival in trisomy-13-syndrome: 21 cases including prolonged survival in two patients 11 and 19 years old. *Am. J. Med. Genet.* **1981**, *8*, 167–172. [CrossRef] [PubMed]
12. Goel, M.; Rathore, R. Trisomy 13 (Patau syndrome). *Indian Pediatr.* **2000**, *37*, 1140. [PubMed]
13. Hassold, T.; Hunt, P. Maternal age and chromosomally abnormal pregnancies: What we know and what we wish we knew. *Curr. Opin. Pediatr.* **2009**, *21*, 703–708. [CrossRef] [PubMed]
14. Carey, J.C. Trisomy 18 and trisomy 13 syndromes. *Manag. Genet. Syndr.* **2005**. [CrossRef]

15. Polli, J.B.; Groff Dde, P.; Petry, P.; Mattos, V.F.; Rosa, R.C.; Zen, P.R.; Graziadio, C.; Paskulin, G.A.; Rosa, R.F. Trisomy 13 (Patau syndrome) and congenital heart defects. *Am. J. Med. Genet.* **2014**, *164A*, 272–275. [CrossRef] [PubMed]

16. Yukifumi, M.; Hirohiko, S.; Fukiko, I.; Mariko, M. Trisomy 13 in a 9-year-old girl with left ventricular noncompaction. *Pediatr. Cardiol.* **2011**, *32*, 206–207. [CrossRef] [PubMed]

17. Gilbert, F. Chromosome 13. *Genet. Test.* **2000**, *4*, 85–94. [PubMed]

18. Auguste, G.; Gurha, P.; Lombardi, R.; Coarfa, C.; Willerson, J.T.; Marian, A.J. Suppression of Activated FOXO Transcription Factors in the Heart Prolongs Survival in a Mouse Model of Laminopathies. *Circ. Res.* **2018**, *122*, 678–692. [CrossRef] [PubMed]

19. Govindsamy, A.; Naidoo, S.; Cerf, M.E. Cardiac Development and Transcription Factors: Insulin Signalling, Insulin Resistance, and Intrauterine Nutritional Programming of Cardiovascular Disease. *J. Nutr. Metab.* **2018**, *2018*, 8547976. [CrossRef] [PubMed]

20. Potente, M.; Ghaeni, L.; Baldessari, D.; Mostoslavsky, R.; Rossig, L.; Dequiedt, F.; Haendeler, J.; Mione, M.; Dejana, E.; Alt, F.W.; et al. SIRT1 controls endothelial angiogenic functions during vascular growth. *Genes Dev.* **2007**, *21*, 2644–2658. [CrossRef] [PubMed]

21. Owens, G.K.; Kumar, M.S.; Wamhoff, B.R. Molecular regulation of vascular smooth muscle cell differentiation in development and disease. *Physiol. Rev.* **2004**, *84*, 767–801. [CrossRef] [PubMed]

22. Cai, B.; Wang, N.; Mao, W.; You, T.; Lu, Y.; Li, X.; Ye, B.; Li, F.; Xu, H. Deletion of FoxO1 leads to shortening of QRS by increasing Na+ channel activity through enhanced expression of both cardiac NaV1. 5 and β3 subunit. *J. Mol. Cell. Cardiol.* **2014**, *74*, 297–306. [CrossRef] [PubMed]

23. Yang, W.; Ng, F.L.; Chan, K.; Pu, X.; Poston, R.N.; Ren, M.; An, W.; Zhang, R.; Wu, J.; Yan, S.; et al. Coronary-Heart-Disease-Associated Genetic Variant at the COL4A1/COL4A2 Locus Affects COL4A1/COL4A2 Expression, Vascular Cell Survival, Atherosclerotic Plaque Stability and Risk of Myocardial Infarction. *PLoS Genet.* **2016**, *12*, e1006127. [CrossRef] [PubMed]

24. Di Marco, G.S.; Kentrup, D.; Reuter, S.; Mayer, A.B.; Golle, L.; Tiemann, K.; Fobker, M.; Engelbertz, C.; Breithardt, G.; Brand, E.; et al. Soluble Flt-1 links microvascular disease with heart failure in CKD. *Basic Res Cardiol.* **2015**, *110*, 30. [CrossRef] [PubMed]

25. Welten, S.M.; Goossens, E.A.; Quax, P.H.; Nossent, A.Y. The multifactorial nature of microRNAs in vascular remodelling. *Cardiovasc. Res.* **2016**, *110*, 6–22. [CrossRef] [PubMed]

26. Mazzuca, M.Q.; Khalil, R.A. Vascular endothelin receptor type B: Structure, function and dysregulation in vascular disease. *Biochem. Pharmacol.* **2012**, *84*, 147–162. [CrossRef] [PubMed]

27. Sunbul, M.; Cagman, Z.; Gerin, F.; Ozgen, Z.; Durmus, E.; Seckin, D.; Ahmad, S.; Uras, F.; Agirbasli, M. Growth arrest-specific 6 and cardiometabolic risk factors in patients with psoriasis. *Cardiovasc. Ther.* **2015**, *33*, 56–61. [CrossRef] [PubMed]

28. Hage, C.; Michaelsson, E.; Linde, C.; Donal, E.; Daubert, J.C.; Gan, L.M.; Lund, L.H. Inflammatory Biomarkers Predict Heart Failure Severity and Prognosis in Patients With Heart Failure With Preserved Ejection Fraction: A Holistic Proteomic Approach. *Circ. Cardiovasc. Genet.* **2017**, *10*. [CrossRef] [PubMed]

29. Lin, Y.; Chen, N.T.; Shih, Y.P.; Liao, Y.C.; Xue, L.; Lo, S.H. DLC2 modulates angiogenic responses in vascular endothelial cells by regulating cell attachment and migration. *Oncogene* **2010**, *29*, 3010–3016. [CrossRef] [PubMed]

30. Camacho Londono, J.E.; Tian, Q.; Hammer, K.; Schroder, L.; Camacho Londono, J.; Reil, J.C.; He, T.; Oberhofer, M.; Mannebach, S.; Mathar, I.; et al. A background Ca2+ entry pathway mediated by TRPC1/TRPC4 is critical for development of pathological cardiac remodelling. *Eur. Heart J.* **2015**, *36*, 2257–2266. [CrossRef] [PubMed]

31. Kurtenbach, S.; Kurtenbach, S.; Zoidl, G. Gap junction modulation and its implications for heart function. *Front. Physiol.* **2014**, *5*, 82. [CrossRef] [PubMed]

32. Furuyama, T.; Kitayama, K.; Shimoda, Y.; Ogawa, M.; Sone, K.; Yoshida-Araki, K.; Hisatsune, H.; Nishikawa, S.; Nakayama, K.; Nakayama, K.; et al. Abnormal angiogenesis in Foxo1 (Fkhr)-deficient mice. *J. Biol. Chem.* **2004**, *279*, 34741–34749. [CrossRef] [PubMed]

33. Liu, Z.; Ren, Y.A.; Pangas, S.A.; Adams, J.; Zhou, W.; Castrillon, D.H.; Wilhelm, D.; Richards, J.S. FOXO1/3 and PTEN Depletion in Granulosa Cells Promotes Ovarian Granulosa Cell Tumor Development. *Mol. Endocrinol.* **2015**, *29*, 1006–1024. [CrossRef] [PubMed]

34. Kobayashi, S.; Volden, P.; Timm, D.; Mao, K.; Xu, X.; Liang, Q. Transcription factor GATA4 inhibits doxorubicin-induced autophagy and cardiomyocyte death. *J. Biol. Chem.* **2010**, *285*, 793–804. [CrossRef] [PubMed]

35. Oh, M.H.; Collins, S.L.; Sun, I.H.; Tam, A.J.; Patel, C.H.; Arwood, M.L.; Chan-Li, Y.; Powell, J.D.; Horton, M.R. mTORC2 Signaling Selectively Regulates the Generation and Function of Tissue-Resident Peritoneal Macrophages. *Cell Rep.* **2017**, *20*, 2439–2454. [CrossRef] [PubMed]

36. Gomis, R.R.; Alarcon, C.; He, W.; Wang, Q.; Seoane, J.; Lash, A.; Massague, J. A FoxO-Smad synexpression group in human keratinocytes. *Proc. Natl. Acad. Sci. USA* **2006**, *103*, 12747–12752. [CrossRef] [PubMed]

37. Herndon, M.K.; Law, N.C.; Donaubauer, E.M.; Kyriss, B.; Hunzicker-Dunn, M. Forkhead box O member FOXO1 regulates the majority of follicle-stimulating hormone responsive genes in ovarian granulosa cells. *Mol. Cell Endocrinol.* **2016**, *434*, 116–126. [CrossRef] [PubMed]

38. Wang, X.; Lockhart, S.M.; Rathjen, T.; Albadawi, H.; Sorensen, D.; O'Neill, B.T.; Dwivedi, N.; Preil, S.R.; Beck, H.C.; Dunwoodie, S.L.; et al. Insulin Downregulates the Transcriptional Coregulator CITED2, an Inhibitor of Proangiogenic Function in Endothelial Cells. *Diabetes* **2016**, *65*, 3680–3690. [CrossRef] [PubMed]

39. Bround, M.J.; Wambolt, R.; Luciani, D.S.; Kulpa, J.E.; Rodrigues, B.; Brownsey, R.W.; Allard, M.F.; Johnson, J.D. Cardiomyocyte ATP production, metabolic flexibility, and survival require calcium flux through cardiac ryanodine receptors in vivo. *J. Biol. Chem.* **2013**, *288*, 18975–18986. [CrossRef] [PubMed]

40. Hariharan, N.; Maejima, Y.; Nakae, J.; Paik, J.; Depinho, R.A.; Sadoshima, J. Deacetylation of FoxO by Sirt1 Plays an Essential Role in Mediating Starvation-Induced Autophagy in Cardiac Myocytes. *Circ. Res.* **2010**, *107*, 1470–1482. [CrossRef] [PubMed]

41. Vasquez, Y.M.; Mazur, E.C.; Li, X.; Kommagani, R.; Jiang, L.; Chen, R.; Lanz, R.B.; Kovanci, E.; Gibbons, W.E.; DeMayo, F.J. FOXO1 is required for binding of PR on IRF4, novel transcriptional regulator of endometrial stromal decidualization. *Mol. Endocrinol.* **2015**, *29*, 421–433. [CrossRef] [PubMed]

42. Farhan, M.; Wang, H.; Gaur, U.; Little, P.J.; Xu, J.; Zheng, W. FOXO Signaling Pathways as Therapeutic Targets in Cancer. *Int. J. Biol. Sci.* **2017**, *13*, 815–827. [CrossRef] [PubMed]

43. Furukawa-Hibi, Y.; Kobayashi, Y.; Chen, C.; Motoyama, N. FOXO transcription factors in cell-cycle regulation and the response to oxidative stress. *Antioxid. Redox Signal.* **2005**, *7*, 752–760. [CrossRef] [PubMed]

44. Dharaneeswaran, H.; Abid, M.R.; Yuan, L.; Dupuis, D.; Beeler, D.; Spokes, K.C.; Janes, L.; Sciuto, T.; Kang, P.M.; Jaminet, S.S.; et al. FOXO1-mediated activation of Akt plays a critical role in vascular homeostasis. *Circ. Res.* **2014**, *115*, 238–251. [CrossRef] [PubMed]

45. Luna-Zurita, L.; Stirnimann, C.U.; Glatt, S.; Kaynak, B.L.; Thomas, S.; Baudin, F.; Samee, M.A.; He, D.; Small, E.M.; Mileikovsky, M.; et al. Complex Interdependence Regulates Heterotypic Transcription Factor Distribution and Coordinates Cardiogenesis. *Cell* **2016**, *164*, 999–1014. [CrossRef] [PubMed]

46. Zhao, H.H.; Herrera, R.E.; Coronado-Heinsohn, E.; Yang, M.C.; Ludes-Meyers, J.H.; Seybold-Tilson, K.J.; Nawaz, Z.; Yee, D.; Barr, F.G.; Diab, S.G.; et al. Forkhead homologue in rhabdomyosarcoma functions as a bifunctional nuclear receptor-interacting protein with both coactivator and corepressor functions. *J. Biol. Chem.* **2001**, *276*, 27907–27912. [CrossRef] [PubMed]

47. Sengupta, A.; Chakraborty, S.; Paik, J.; Yutzey, K.E.; Evans-Anderson, H.J. FoxO1 is required in endothelial but not myocardial cell lineages during cardiovascular development. *Dev. Dyn.* **2012**, *241*, 803–813. [CrossRef] [PubMed]

48. Xu, M.; Chen, X.; Chen, D.; Yu, B.; Huang, Z. FoxO1: A novel insight into its molecular mechanisms in the regulation of skeletal muscle differentiation and fiber type specification. *Oncotarget* **2017**, *8*, 10662–10674. [CrossRef] [PubMed]

49. Buckingham, M. Skeletal muscle formation in vertebrates. *Curr. Opin. Genet. Dev.* **2001**, *11*, 440–448. [CrossRef]

50. Gross, D.N.; van den Heuvel, A.P.; Birnbaum, M.J. The role of FoxO in the regulation of metabolism. *Oncogene* **2008**, *27*, 2320–2336. [CrossRef] [PubMed]

51. Maiese, K.; Hou, J.; Chong, Z.Z.; Shang, Y.C. A fork in the path: Developing therapeutic inroads with FoxO proteins. *Oxid. Med. Cell Longev.* **2009**, *2*, 119–129. [CrossRef] [PubMed]

52. Puthanveetil, P.; Wan, A.; Rodrigues, B. FoxO1 is crucial for sustaining cardiomyocyte metabolism and cell survival. *Cardiovasc. Res.* **2012**, *97*, 393–403. [CrossRef] [PubMed]

53. Kandula, V.; Kosuru, R.; Li, H.; Yan, D.; Zhu, Q.; Lian, Q.; Ge, R.S.; Xia, Z.; Irwin, M.G. Forkhead box transcription factor 1: Role in the pathogenesis of diabetic cardiomyopathy. *Cardiovasc. Diabetol.* **2016**, *15*, 44. [CrossRef] [PubMed]

54. Feben, C.; Kromberg, J.; Krause, A. An unusual case of Trisomy 13. *S. Afr. J. Child Health* **2015**, *9*, 61–62.

55. Kajimura, D.; Paone, R.; Mann, J.J.; Karsenty, G. Foxo1 regulates Dbh expression and the activity of the sympathetic nervous system in vivo. *Mol. Metab.* **2014**, *3*, 770–777. [CrossRef] [PubMed]

56. Ferdous, A.; Battiprolu, P.K.; Ni, Y.G.; Rothermel, B.A.; Hill, J.A. FoxO, autophagy, and cardiac remodeling. *J. Cardiovasc. Transl. Res.* **2010**, *3*, 355–364. [CrossRef] [PubMed]

57. Sengupta, A.; Molkentin, J.D.; Yutzey, K.E. FoxO transcription factors promote autophagy in cardiomyocytes. *J. Biol. Chem.* **2009**, *284*, 28319–28331. [CrossRef] [PubMed]

58. Ashburn, T.T.; Thor, K.B. Drug repositioning: Identifying and developing new uses for existing drugs. *Nat. Rev. Drug Discov.* **2004**, *3*, 673. [CrossRef] [PubMed]

59. Mirza, Z.; Beg, M.A. Possible Molecular Interactions of Bexarotene—A Retinoid Drug and Alzheimer's Abeta Peptide: A Docking Study. *Curr. Alzheimer Res.* **2017**, *14*, 327–334. [PubMed]

60. Tobinick, E.L. The value of drug repositioning in the current pharmaceutical market. *Drug News Perspect.* **2009**, *22*, 119–125. [CrossRef] [PubMed]

61. Zanella, F.; Dos Santos, N.R.; Link, W. Moving to the Core: Spatiotemporal Analysis of Forkhead Box O (FOXO) and Nuclear Factor-κB (NF-κB) Nuclear Translocation. *Traffic* **2013**, *14*, 247–258. [CrossRef] [PubMed]

62. Huang, H.; Tindall, D.J. Dynamic FoxO transcription factors. *J. Cell Sci.* **2007**, *120*, 2479–2487. [CrossRef] [PubMed]

63. Obsil, T.; Obsilova, V. Structural basis for DNA recognition by FOXO proteins. *Biochim. Biophys. Acta* **2011**, *1813*, 1946–1953. [CrossRef] [PubMed]

64. Mutka, S.C.; Yang, W.Q.; Dong, S.D.; Ward, S.L.; Craig, D.A.; Timmermans, P.B.; Murli, S. Identification of nuclear export inhibitors with potent anticancer activity in vivo. *Cancer Res.* **2009**, *69*, 510–517. [CrossRef] [PubMed]

65. Kau, T.R.; Schroeder, F.; Ramaswamy, S.; Wojciechowski, C.L.; Zhao, J.J.; Roberts, T.M.; Clardy, J.; Sellers, W.R.; Silver, P.A. A chemical genetic screen identifies inhibitors of regulated nuclear export of a Forkhead transcription factor in PTEN-deficient tumor cells. *Cancer Cell* **2003**, *4*, 463–476. [CrossRef]

66. Lu, H.; Huang, H. FOXO1: A potential target for human diseases. *Curr. Drug Targets* **2011**, *12*, 1235–1244. [CrossRef] [PubMed]

67. Schroeder, F.C.; Kau, T.R.; Silver, P.A.; Clardy, J. The psammaplysenes, specific inhibitors of FOXO1a nuclear export. *J. Nat. Prod.* **2005**, *68*, 574–576. [CrossRef] [PubMed]

68. Rena, G.; Bain, J.; Elliott, M.; Cohen, P. D4476, a cell-permeant inhibitor of CK1, suppresses the site-specific phosphorylation and nuclear exclusion of FOXO1a. *EMBO Rep.* **2004**, *5*, 60–65. [CrossRef] [PubMed]

69. Link, W.; Oyarzabal, J.; Serelde, B.G.; Albarran, M.I.; Rabal, O.; Cebriá, A.; Alfonso, P.; Fominaya, J.; Renner, O.; Peregrina, S. Chemical interrogation of FOXO3a nuclear translocation identifies potent and selective inhibitors of phosphoinositide 3-kinases. *J. Biol. Chem.* **2009**, *284*, 28392–28400. [CrossRef] [PubMed]

70. Jaszczyszyn, A.; Gąsiorowski, K.; Świątek, P.; Malinka, W.; Cieślik-Boczula, K.; Petrus, J.; Czarnik-Matusewicz, B. Chemical structure of phenothiazines and their biological activity. *Pharmacol. Rep.* **2012**, *64*, 16–23. [CrossRef]

71. Qi, L.; Ding, Y. Potential antitumor mechanisms of phenothiazine drugs. *Sci. China Life Sci.* **2013**, *56*, 1020–1027. [CrossRef] [PubMed]

72. Wu, C.-H.; Bai, L.-Y.; Tsai, M.-H.; Chu, P.-C.; Chiu, C.-F.; Chen, M.Y.; Chiu, S.-J.; Chiang, J.-H.; Weng, J.-R. Pharmacological exploitation of the phenothiazine antipsychotics to develop novel antitumor agents–A drug repurposing strategy. *Sci. Rep.* **2016**, *6*, 27540. [CrossRef] [PubMed]

73. Chen, Z.; Rice, C.M. Repurposing an old drug: A low-cost allergy medication provides new hope for hepatitis C patients. *Hepatology* **2015**, *62*, 1911–1913. [CrossRef] [PubMed]

74. Kang, S.; Lee, J.M.; Jeon, B.; Elkamhawy, A.; Paik, S.; Hong, J.; Oh, S.-J.; Paek, S.H.; Lee, C.J.; Hassan, A.H.E.; et al. Repositioning of the antipsychotic trifluoperazine: Synthesis, biological evaluation and in silico study of trifluoperazine analogs as anti-glioblastoma agents. *Eur. J. Med. Chem.* **2018**, *10*, 186–198. [CrossRef] [PubMed]

75. Abdelmonem, M.; Zarrin, B.; Asif, N. High Throughput Screening and Molecular Docking of Calmodulin with Antagonists of Trifluoperazine and Phenothiazine Chemical Class. *Lett. Drug Des. Discov.* **2018**, *15*, 136–142.

76. Pan, D.; Yan, Q.; Chen, Y.; McDonald, J.M.; Song, Y. Trifluoperazine Regulation of Calmodulin Binding to Fas: A Computational Study. *Proteins* **2011**, *79*, 2543–2556. [CrossRef] [PubMed]

77. Maeda, J.; Yamagishi, H.; Furutani, Y.; Kamisago, M.; Waragai, T.; Oana, S.; Kajino, H.; Matsuura, H.; Mori, K.; Matsuoka, R. The impact of cardiac surgery in patients with trisomy 18 and trisomy 13 in Japan. *Am. J. Med. Genet.* **2011**, *155*, 2641–2646. [CrossRef] [PubMed]

78. Sugayama, S.; Kim, C.; Albano, L.; Utagawa, C.; Bertola, D.; Koiffmann, C. Clinical and genetic study of 20 patients with trisomy 13 (Patau's syndrome). *Pediatria (São Paulo)* **1999**, *21*, 21–29.

79. Rasmussen, S.A.; Wong, L.-Y.C.; Yang, Q.; May, K.M.; Friedman, J. Population-based analyses of mortality in trisomy 13 and trisomy 18. *Pediatrics* **2003**, *111*, 777–784. [CrossRef] [PubMed]

80. Fuchs, S.; Amiel, J.; Claudel, S.; Lyonnet, S.; Corvol, P.; Pinet, F. Functional characterization of three mutations of the endothelin B receptor gene in patients with Hirschsprung's disease: Evidence for selective loss of Gi coupling. *Mol. Med.* **2001**, *7*, 115–124. [CrossRef] [PubMed]

81. Brown, S.A.; Warburton, D.; Brown, L.Y.; Yu, C.-y.; Roeder, E.R.; Stengel-Rutkowski, S.; Hennekam, R.C.; Muenke, M. Holoprosencephaly due to mutations in ZIC2, a homologue of Drosophila odd-paired. *Nat. Genet.* **1998**, *20*, 180–183. [CrossRef] [PubMed]

82. Brown, L.; Paraso, M.; Arkell, R.; Brown, S. In vitro analysis of partial loss-of-function ZIC2 mutations in holoprosencephaly: Alanine tract expansion modulates DNA binding and transactivation. *Hum. Mol. Genet.* **2004**, *14*, 411–420. [CrossRef] [PubMed]

83. Jobanputra, V.; Burke, A.; Kwame, A.Y.; Shanmugham, A.; Shirazi, M.; Brown, S.; Warburton, P.E.; Levy, B.; Warburton, D. Duplication of the ZIC2 gene is not associated with holoprosencephaly. *Am. J. Med. Genet.* **2012**, *158*, 103–108. [CrossRef] [PubMed]

84. Warr, N.; Powles-Glover, N.; Chappell, A.; Robson, J.; Norris, D.; Arkell, R.M. Zic2-associated holoprosencephaly is caused by a transient defect in the organizer region during gastrulation. *Hum. Mol. Genet.* **2008**, *17*, 2986–2996. [CrossRef] [PubMed]

85. Solomon, B.D.; Lacbawan, F.; Mercier, S.; Clegg, N.J.; Delgado, M.R.; Rosenbaum, K.; Dubourg, C.; David, V.; Olney, A.H.; Wehner, L.-E. Mutations in ZIC2 in human holoprosencephaly: Description of a novel ZIC2 specific phenotype and comprehensive analysis of 157 individuals. *J. Med. Genet.* **2010**, *47*, 513–524. [CrossRef] [PubMed]

86. Barratt, K.S.; Glanville-Jones, H.C.; Arkell, R.M. The Zic2 gene directs the formation and function of node cilia to control cardiac situs. *Genesis* **2014**, *52*, 626–635. [CrossRef] [PubMed]

87. Molkentin, J.D. The zinc finger-containing transcription factors GATA-4, -5, and -6. Ubiquitously expressed regulators of tissue-specific gene expression. *J. Biol. Chem.* **2000**, *275*, 38949–38952. [CrossRef] [PubMed]

88. Pikkarainen, S.; Tokola, H.; Kerkela, R.; Ruskoaho, H. GATA transcription factors in the developing and adult heart. *Cardiovasc. Res.* **2004**, *63*, 196–207. [CrossRef] [PubMed]

89. Charron, F.; Paradis, P.; Bronchain, O.; Nemer, G.; Nemer, M. Cooperative interaction between GATA-4 and GATA-6 regulates myocardial gene expression. *Mol. Cell. Biol.* **1999**, *19*, 4355–4365. [CrossRef] [PubMed]

90. Kohli, S.; Ahuja, S.; Rani, V. Transcription factors in heart: Promising therapeutic targets in cardiac hypertrophy. *Curr. Cardiol. Rev.* **2011**, *7*, 262–271. [CrossRef] [PubMed]

91. Valimaki, M.J.; Tolli, M.A.; Kinnunen, S.M.; Aro, J.; Serpi, R.; Pohjolainen, L.; Talman, V.; Poso, A.; Ruskoaho, H.J. Discovery of Small Molecules Targeting the Synergy of Cardiac Transcription Factors GATA4 and NKX2-5. *J. Med. Chem.* **2017**, *60*, 7781–7798. [CrossRef] [PubMed]

92. Kinnunen, S.M.; Tolli, M.; Valimaki, M.J.; Gao, E.; Szabo, Z.; Rysa, J.; Ferreira, M.P.A.; Ohukainen, P.; Serpi, R.; Correia, A.; et al. Cardiac Actions of a Small Molecule Inhibitor Targeting GATA4-NKX2-5 Interaction. *Sci. Rep.* **2018**, *8*, 4611. [CrossRef] [PubMed]

93. Takeuchi, J.K.; Ohgi, M.; Koshiba-Takeuchi, K.; Shiratori, H.; Sakaki, I.; Ogura, K.; Saijoh, Y.; Ogura, T. Tbx5 specifies the left/right ventricles and ventricular septum position during cardiogenesis. *Development* **2003**, *130*, 5953–5964. [CrossRef] [PubMed]

94. Boogerd, C.J.; Evans, S.M. TBX5 and NuRD Divide the Heart. *Dev. Cell* **2016**, *36*, 242–244. [CrossRef] [PubMed]

95. Steimle, J.D.; Moskowitz, I.P. TBX5: A Key Regulator of Heart Development. *Curr. Top. Dev. Biol.* **2017**, *122*, 195–221. [PubMed]

96. Stefanovic, S.; Barnett, P.; van Duijvenboden, K.; Weber, D.; Gessler, M.; Christoffels, V.M. GATA-dependent regulatory switches establish atrioventricular canal specificity during heart development. *Nat. Commun.* **2014**, *5*, 3680. [CrossRef] [PubMed]

97. Xie, L.; Hoffmann, A.D.; Burnicka-Turek, O.; Friedland-Little, J.M.; Zhang, K.; Moskowitz, I.P. Tbx5-hedgehog molecular networks are essential in the second heart field for atrial septation. *Dev. Cell* **2012**, *23*, 280–291. [CrossRef] [PubMed]

98. Xu, M.; Wu, X.; Li, Y.; Yang, X.; Hu, J.; Zheng, M.; Tian, J. CITED2 mutation and methylation in children with congenital heart disease. *J. Biomed. Sci.* **2014**, *21*, 7. [CrossRef] [PubMed]

99. Sperling, S.; Grimm, C.H.; Dunkel, I.; Mebus, S.; Sperling, H.P.; Ebner, A.; Galli, R.; Lehrach, H.; Fusch, C.; Berger, F.; et al. Identification and functional analysis of CITED2 mutations in patients with congenital heart defects. *Hum. Mutat.* **2005**, *26*, 575–582. [CrossRef] [PubMed]

100. Zeng, L.; Chen, R.; Liang, F.; Tsuchiya, H.; Murai, H.; Nakahashi, T.; Iwai, K.; Takahashi, T.; Kanda, T.; Morimoto, S. Silent information regulator, Sirtuin 1, and age-related diseases. *Geriatr. Gerontol. Int.* **2009**, *9*, 7–15. [CrossRef] [PubMed]

101. Borradaile, N.M.; Pickering, J.G. NAD(+), sirtuins, and cardiovascular disease. *Curr. Pharm. Des.* **2009**, *15*, 110–117. [CrossRef] [PubMed]

102. Corbi, G.; Bianco, A.; Turchiarelli, V.; Cellurale, M.; Fatica, F.; Daniele, A.; Mazzarella, G.; Ferrara, N. Potential mechanisms linking atherosclerosis and increased cardiovascular risk in COPD: Focus on Sirtuins. *Int. J. Mol. Sci.* **2013**, *14*, 12696–12713. [CrossRef] [PubMed]

103. Gross, D.N.; Wan, M.; Birnbaum, M.J. The role of FOXO in the regulation of metabolism. *Curr. Diab. Rep.* **2009**, *9*, 208–214. [CrossRef] [PubMed]

104. Peng, S.L. Immune regulation by Foxo transcription factors. *Autoimmunity* **2007**, *40*, 462–469. [CrossRef] [PubMed]

105. Pandey, A.; Kumar, G.S.; Kadakol, A.; Malek, V.; Bhanudas Gaikwad, A. FoxO1 inhibitors: The future medicine for metabolic disorders? *Curr. Diab. Rev.* **2016**, *12*, 223–230. [CrossRef]

106. Weigelt, J.; Climent, I.; Dahlman-Wright, K.; Wikström, M. Solution structure of the DNA binding domain of the human forkhead transcription factor AFX (FOXO4). *Biochemistry* **2001**, *40*, 5861–5869. [CrossRef] [PubMed]

107. Xuan, Z.; Zhang, M.Q. From worm to human: Bioinformatics approaches to identify FOXO target genes. *Mech. Ageing. Dev.* **2005**, *126*, 209–215. [CrossRef] [PubMed]

108. Nomenclature, I.S.C.o.H.C. *ISCN: An International System for Human Cytogenomic Nomenclature (2016)*; Jean, M.-J., Annet, S., Michael, S., Eds.; Karger: Basel, Switzerland, 2016.

109. Karim, S.; Merdad, A.; Schulten, H.J.; Jayapal, M.; Dallol, A.; Buhmeida, A.; Al-Thubaity, F.; Mirza, Z.; Gari, M.A.; Chaudhary, A.G.; et al. Low expression of leptin and its association with breast cancer: A transcriptomic study. *Oncol. Rep.* **2016**, *36*, 43–48. [CrossRef] [PubMed]

110. Jensen, L.J.; Kuhn, M.; Stark, M.; Chaffron, S.; Creevey, C.; Muller, J.; Doerks, T.; Julien, P.; Roth, A.; Simonovic, M.; et al. STRING 8—a global view on proteins and their functional interactions in 630 organisms. *Nucleic Acids Res.* **2009**, *37*, D412–416. [CrossRef] [PubMed]

111. Franceschini, A.; Szklarczyk, D.; Frankild, S.; Kuhn, M.; Simonovic, M.; Roth, A.; Lin, J.; Minguez, P.; Bork, P.; von Mering, C.; et al. STRING v9.1: Protein-protein interaction networks, with increased coverage and integration. *Nucleic Acids Res.* **2013**, *41*, D808–815. [CrossRef] [PubMed]

112. Irwin, J.J.; Shoichet, B.K. ZINC—A Free Database of Commercially Available Compounds for Virtual Screening. *J. Chem. Inf. Model.* **2005**, *45*, 177–182. [CrossRef] [PubMed]

113. Bikadi, Z.; Hazai, E. Application of the PM6 semi-empirical method to modeling proteins enhances docking accuracy of AutoDock. *J. Cheminform.* **2009**, *1*, 209–215. [CrossRef] [PubMed]

114. Morris, G.M.; Goodsell, D.S.; Halliday, R.S.; Huey, R.; Hart, W.E.; Belew, R.K.; Olson, A.J. Automated docking using a Lamarckian genetic algorithm and an empirical binding free energy function. *J. Comput. Chem.* **1998**, *19*, 1639–1662. [CrossRef]

115. Solis, F.J.; Wets, R.J.-B. Minimization by Random Search Techniques. *Math. Oper. Res.* **1981**, *6*, 19–30. [CrossRef]

© 2018 by the authors. Licensee MDPI, Basel, Switzerland. This article is an open access article distributed under the terms and conditions of the Creative Commons Attribution (CC BY) license (http://creativecommons.org/licenses/by/4.0/).

International Journal of
Molecular Sciences

Article

Causal Transcription Regulatory Network Inference Using Enhancer Activity as a Causal Anchor

Deepti Vipin [1], Lingfei Wang [2], Guillaume Devailly [1], Tom Michoel [2,3] and Anagha Joshi [1,4,*]

[1] Division of Developmental Biology, The Roslin Institute, The University of Edinburgh, Easter Bush, Midlothian, EH25 9RG Scotland, UK; Deepti.Vipin@roslin.ed.ac.uk (D.V.); Guillaume.Devailly@roslin.ed.ac.uk (G.D.)

[2] Division of Genetics and Genomics, The Roslin Institute, The University of Edinburgh, Easter Bush, Midlothian, EH25 9RG Scotland, UK; Lingfei.wang@roslin.ed.ac.uk (L.W.); Tom.Michoel@roslin.ed.ac.uk (T.M.)

[3] Computational Biology Unit, Department of Informatics, University of Bergen, DataBlokk, 5th Floor, Thormohlensgt 55, N-5008 Bergen, Norway

[4] Computational Biology Unit, Department of Clinical Science, University of Bergen, DataBlokk, 5th Floor, Thormohlensgt 55, N-5008 Bergen, Norway

* Correspondence: anagha.joshi@uib.no; Tel.: +47-55-58-54-35

Received: 18 September 2018; Accepted: 8 November 2018; Published: 15 November 2018

Abstract: Transcription control plays a crucial role in establishing a unique gene expression signature for each of the hundreds of mammalian cell types. Though gene expression data have been widely used to infer cellular regulatory networks, existing methods mainly infer correlations rather than causality. We developed statistical models and likelihood-ratio tests to infer causal gene regulatory networks using enhancer RNA (eRNA) expression information as a causal anchor and applied the framework to eRNA and transcript expression data from the FANTOM Consortium. Predicted causal targets of transcription factors (TFs) in mouse embryonic stem cells, macrophages and erythroblastic leukaemia overlapped significantly with experimentally-validated targets from ChIP-seq and perturbation data. We further improved the model by taking into account that some TFs might act in a quantitative, dosage-dependent manner, whereas others might act predominantly in a binary on/off fashion. We predicted TF targets from concerted variation of eRNA and TF and target promoter expression levels within a single cell type, as well as across multiple cell types. Importantly, TFs with high-confidence predictions were largely different between these two analyses, demonstrating that variability within a cell type is highly relevant for target prediction of cell type-specific factors. Finally, we generated a compendium of high-confidence TF targets across diverse human cell and tissue types.

Keywords: transcription regulation; gene expression; causal inference; enhancer activity

1. Introduction

Despite having the same DNA, gene expression is unique to each cell type in the human body. Cell type-specific gene expression is controlled by short DNA sequences called enhancers, located distal to the transcription start site of a gene. Collaborative efforts such as the FANTOM [1] and Roadmap Epigenomics [2] projects have now successfully built enhancer and promoter repertoires across hundreds of human cell types, with an estimated 1.4% of the human genome associated with putative promoters and about 13% with putative enhancers. Enhancers physically interact with promoters to activate gene expression. Although the general rules governing these interactions (if any) remain poorly understood, experimental techniques such as chromosome conformation capture (3C, 4C) combined with next generation sequencing (Hi-C) [3], as well as computational methods based on correlations between histone modifications or DNase I hypersensitivity at enhancers with the expression of nearby promoters [4,5] are continually improving in predicting enhancer-promoter interactions. In contrast, understanding how the activation of one gene leads to the activation or repression of other genes, i.e., uncovering the structure of cell type, specific transcriptional regulatory networks, remains a major challenge. It is known that promoter expression levels of transcription factors (TFs) co-express and cluster together with promoters of functionally-related genes [6], but without any additional information, such associations are merely correlative and do not indicate a causal regulation by the TF.

Statistical causal inference aims to predict causal models where the manipulation of one variable (e.g., expression of gene A) alters the distribution of the other (e.g., expression of gene B), but not necessarily vice versa [7]. A key role in causal inference is played by causal anchors, variables that are known *a priori* to be causally upstream of others and that can be used to orient the direction of causality between other, relevant variables. A major application of this principle has been found in genetical genomics or systems genetics: genetic variations between individuals alter molecular and organismal phenotypes, but not vice versa, so these quantitative trait loci (QTL) can be used as causal anchors to determine the direction of causality between correlated traits from population-based data [8–11]. Such pairwise causal associations can then be assembled into causal gene networks to model how genetic variation at multiple loci collectively affect the status of molecular networks of genes, proteins, metabolites and downstream phenotypes [12].

Interestingly, several experiments have recently shown that enhancer regions can be transcribed to form short (around 1000 bp), non-coding, often bi-directional transcripts, called enhancer RNAs or eRNAs [13]. eRNAs have other distinguishing features, including nonpolyadenylated and unspliced tails, and tend to remain nuclear, rather than reaching the cytoplasm for translation. Although the functional role of some eRNAs has been studied in great detail to demonstrate that they can enhance or suppress enhancer activity by enhancer-promoter looping [14], the full repertoire of functional mechanisms of eRNAs remains to be understood. Nevertheless, the presence of eRNAs from a regulatory region is an indicator of enhancer activity [15], and eRNA expression has been successfully used to predict transcription factor activity [16]. Moreover, eRNA expression is correlated with and, crucially, *temporally precedes* the expression of target genes [17]. We therefore hypothesized that eRNA expression as a readout of enhancer activity could act as a causal anchor, opening new avenues to reconstruct causal gene regulatory networks.

To test this hypothesis, we developed novel statistical models and likelihood-ratio tests for using (continuous) eRNA expression data in causal inference, based on existing methods for discrete eQTL data, and implemented these in the Findr software [18]. We applied this new method to Cap Analysis of Gene Expression (CAGE) data generated by the FANTOM5 Consortium. Unlike RNA sequencing, CAGE allows sequencing of only the five prime ends of mRNAs, providing genome-wide transcription start site quantification at much lower sequencing depth (and therefore, lower sequencing cost per sample) than RNA-seq. The FANTOM Consortium has generated a unique resource of enhancer and promoter expression across hundreds of human and mouse cell types [6,17] and validated predictions using ChIP-seq and perturbation data. Our analysis of the FANTOM data showed that continuous eRNA expression values increased target prediction performance for some factors, while for other factors, a binarized presence or absence of the enhancer signal performed better. Leveraging this observation, we found that a data-driven approach to classify enhancer expression as either binary or continuous was sufficient to select the best target prediction method automatically, allowing parameter-free application of the method to organisms and cell types where validation data are not currently available.

2. Results

2.1. Development of a Causal Transcription Network Inference Framework

We hypothesized that enhancer activity could be used as a causal anchor to predict causality between two co-expressed genes (Figure 1A). To this end, we used enhancer expression as a causal anchor to infer causal gene interactions, within the Findr framework [18]. Findr provides accurate and efficient inference of gene regulations using eQTLs as causal anchors by accounting for hidden confounding factors and weak regulations. This is achieved by performing and combining five likelihood ratio tests (Figure 1B), each of which consists of a null ($\mathcal{H}_{\text{null}}$) and an alternative ($\mathcal{H}_{\text{alt}}$) hypothesis, to support or reject the causal model $E \to A \to B$, where E is an enhancer in the regulatory region of gene A, and B is a putative target gene: primary linkage ($E \to A$), secondary linkage ($E \to B$), conditional independence ($E \to B$ only through A), B's relevance ($E \to B$ or correlation between A and B) and excluding pleiotropy (partial correlation between A and B after conditioning on E). The log-likelihood ratios (LLRs) are computed for all possible targets of each gene and then converted into p-values and posterior probabilities [19] of the alternative hypothesis being true; see [18] for details.

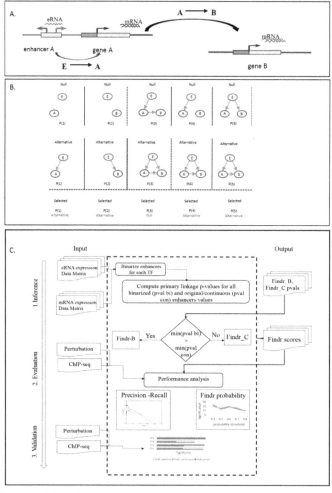

Figure 1. Overview of the Findr framework. (**A**) The schematic representation of causal gene regulatory network inference using enhancer activity as a causal anchor. (**B**) Five statistical tests used by Findr for causal inference. (**C**) Workflow of the Findr-A framework. eRNA, enhancer RNA; TF, transcription factor; B, binary; C, continuous.

We applied three treatments to enhancer expression data. First, we regarded enhancers as binary (on/off) variables, and after binarizing the data (see Methods), we used the existing Findr to predict TF targets directly. This approach will be referred to as Findr-B ("binary"). Second, we adapted all five tests in Findr to use continuous instead of discrete causal anchor data, and we used this method on (untransformed) eRNA data. This approach will be referred to as Findr-C ("continuous"). Third, to accommodate the co-existence of binary and continuous enhancers for different TFs within the same dataset, we developed an automatic adaptive method to treat each enhancer independently as binary or continuous, depending on the relative strength of the primary enhancer-TF linkage with either method. We call this approach Findr-A ("adaptive"). The workflow of the Findr framework is summarized in Figure 1C. We note that the implementation of the method is generic, i.e., it can be used by defining either eQTL genotypes or enhancer activity as causal anchors.

2.2. Causal Inference from Enhancer and Transcript/Gene Expression CAGE Data

To test our hypothesis that causal inference using enhancer expression as a causal anchor predicts true TF targets, we used CAGE data generated by the FANTOM Consortium across hundreds of cell types and tissues in human and mouse [6]. We used bi-directional expression in non-promoter regions as an indicator of likely enhancer activity in each CAGE sample [1]. Specifically, non-promoter regions with a similar number of sequence tags in both the forward and reverse direction were predicted as putative enhancer regions. Moreover, these regions also showed a high overlap (over 90 percent) with H3K4me1 modification. Importantly, CAGE data offer a powerful resource as each sample contains both enhancer and gene activity information. The ability to quantify enhancer expression (as a proxy for enhancer activity) and gene expression from the same sample is crucial for the ability to apply causal inference techniques. We first selected three mouse cell types (embryonic stem cells, macrophages and erythroblastic leukaemia) for systematic characterization, as these had more than 20 samples per cell type, with diverse treatments or time series. Furthermore, ChIP-seq and TF knock-out validation datasets were available for each of these cell types. In order to assign both promoter proximal, as well as promoter distal putative enhancers for each transcription factor, we selected predicted enhancers [1] within 50 kb of the transcription start site of each transcription factor in a cell type. This resulted in 109 enhancers for 48 transcription factors in ES cells, 55 enhancers for 8 transcription factors in macrophages and 5 enhancers for 4 transcription factors in erythroleukaemia, with an average of 3.8 enhancers per transcription factor across cell types.

We inferred causal transcription factor-target interactions for each transcription factor using the enhancer element most strongly linked to each TF in each cell type. We predicted targets for 48 transcription factors with two methods, one using continuous enhancer data ("Findr-C") and one using discretized, binary (on/off) enhancer data ("Findr-B") (see Methods). The Findr software outputs a score representing the putative probability of a causal interaction for each transcription factor-target pair (see Methods). For both methods, the targets with a predicted probability of a causal interaction greater than 0.8 (see Methods) were validated using a compendium of ChIP-seq data [20], containing 78 factors in ES cells, 12 factors in macrophages and 17 factors in erythroblasts. Of these factors, 18 in ES cells, 7 in macrophages and 4 in erythroblasts had enhancer expression in CAGE data. We noted that the suitability of Findr-B or Findr-C for causal inference was dependent on the factor, i.e., using continuous enhancer data performed better for some factors (Figure 2: Myc,Klf2, Figure S1: Fcgr3), while on/off data performed better for others (Figure 2: Gata1, Fli1, Figure S2: Junb, Jarid2). Because the number of putative ChIP-seq targets for each factor varied widely across factors and cell types, from only 420 gene targets for JunD in erythroblasts to over 12,000 gene targets for ESRRB in ES cells, the background precision levels differed highly between factors (Figure 2).

We further tested whether these results were sensitive to the target probability prediction threshold. The enrichments of true positives were stable over a wide range around this threshold for both methods (Figure 3, Figure S2).

Transcription factor binding inferred using ChIP-seq data is thought to be mostly opportunistic and therefore might not provide direct clues about the functional targets of the factor [21]. We therefore collected perturbation data from publications, specifically expression data after knock-out or knock-down (KO) of a factor. We generated differentially-expressed gene lists for 85 factors in ES cells and 11 factors in macrophages (see Methods). Using these gene lists as known targets, we evaluated the predictions of both methods. This confirmed the factor-specific suitability of either the Findr-B or Findr-C method (Figures 4, Figure S3).

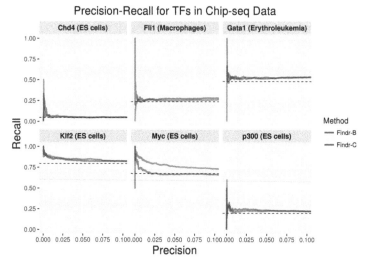

Figure 2. Recall-precision curves for target predictions by Findr-B and Findr-C using ChIP-seq. The dotted line represents the background or the the random classifier precision.

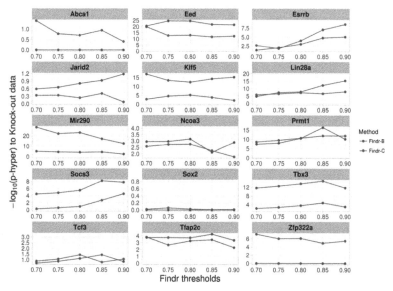

Figure 3. Robustness of Findr performance demonstrated by using different score thresholds.

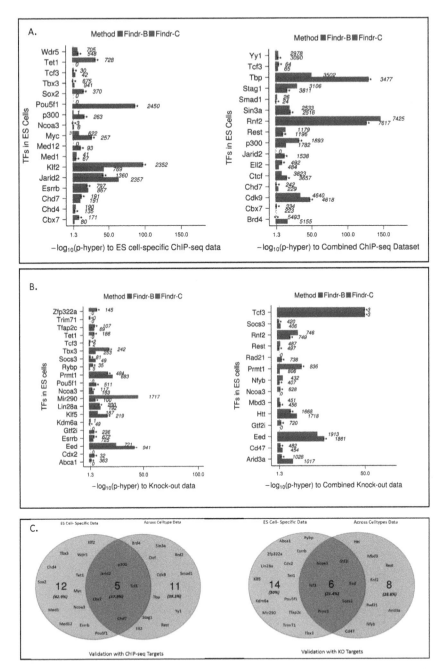

Figure 4. Comparison of Findr-adaptive (**A**) predictions using mouse ES cells and all cell types samples. (**A**) Bar plots representing enrichments for Findr-B, -C and -A predictions using ChIP-seq data as a validation dataset for ES cells (left) and all cell types (right). (**B**) Bar plots representing enrichments for Findr-B, -C and -A predictions using knock-out data as a validation dataset for ES cells (left) and all cell types (right). (**C**) Overlap of factors between ES and all cell types using ChIP-seq (left) and knock-out (right) as validation datasets.

The evaluation so far used the default combination of likelihood-ratio tests in Findr (Figure 1B, Tests 2, 4 and 5; see the Methods for details), which was previously shown to perform better than the traditional causal inference method, which relies on a conditional independence test (Figure 1B, Test 3), by accounting for hidden confounding due to common upstream regulatory factors between a TF and its targets [18]. To test whether this remains true for CAGE data, we implemented all five tests in both Findr-B and Findr-C. We found that the new test combination improved predictive power for the CAGE data in all three mouse cell types compared to the traditional test combination, similar to the previous results on eQTL data. We have therefore set it as the default choice for the Findr tool.

2.3. Development and Validation of an Adaptive Model-Selection Approach for Causal Inference Using Discretized or Continuous Data

As the optimal prediction performance depended on a factor-specific choice between discretizing enhancer expression data or not, we investigated if this decision could be made in a data-driven, adaptive approach (called "Findr-adaptive" or "Findr-A"; see Figure 1C), in the absence of validation data. In short, Findr-A selects for each TF among all its candidate enhancers, both continuous and binary, the one with the strongest primary linkage to the TF's expression and then uses that enhancer and its corresponding method (Findr-B or Findr-C) to predict downstream targets for that TF (see the Methods for details). This adaptive approach was indeed able to select the best performing method for most of the factors (Figure 4A,B, Findr-A selection marked by double stars).

We further performed functional enrichment analysis of gene target sets predicted by Findr-A. 1119 targets of JunB in macrophages were enriched for 'LPS signalling pathway' ($p < 10^{-8}$) and 'RNA binding' ($p < 10^{-8}$) (Tables S1 and S2). The macrophage CAGE samples indeed measured the response to LPS signalling and JunB is known to be a delayed response gene, attenuating transcriptional activity of immediate early genes and RNA binding proteins, specifically terminating translation of mRNAs induced by immediate early genes [22]. 131 targets for Cbx7 in ES cells (Figure 3) were enriched for 'regulation of transcription from RNA polymerase II promoter' ($p < 10^{-10}$), and included several developmental genes such as Hox family proteins (Tables S3 and S4). This enrichment was much stronger than for the ChIP bound targets of Cbx7 ($p < 10^{-5}$). Cbx7 is a part of the PRC complex, which binds predominantly at bivalent chromatin at promoters of transcription regulators in ES cells [23].

2.4. Perturbations within and across Cell Types Provide Causal Targets for a Distinct Set of Transcription Factors

We noted that most factors performed better using binary enhancer expression values and wondered if this might be due to the limited enhancer expression data available for each cell type. We therefore explored whether the enhancer activity across perturbations within cell type was comparable to variation of enhancer activity across different cell and tissue types. To test this, we used the FANTOM5 CAGE data containing over 1000 samples across 360 distinct mouse cell types and tissues, called "all-data". Findr-C indeed performed marginally better on all-data rather than cell type-specific data. Importantly, all-data and cell type (ES)-specific data resulted in causal targets for a distinct set of transcription factors. In particular, variation within a cell type was more informative for causal target predictions of cell type-specific factors. For example, the targets of key pluripotency factors Sox2 and Esrrb [24] were enriched in ES-data, but not all-data (Figure 4A,B).

There were only five common factors with causal targets predicted using both all-data and ES-specific data that were validated by ChIP-seq data and only six common factors validated by KO data (Figure 4C). Interestingly, the target genes predicted from ES-data and all-data for the same factor overlapped significantly. For example, 72% of Eed predicted targets using ES-data overlapped with Eed predicted targets using all-data.

2.5. Multiple Enhancers of the Same Factor Have a Highly Correlated Expression

Mammalian genes are controlled by multiple enhancers. We investigated the stability of inference outcome under different choices of enhancers as causal anchors. For all factors for which Findr-A predicted targets in ES cells that overlapped significantly with perturbation data (Figure 4), we predicted additional target sets using other available enhancers, resulting in target sets for 76 enhancer-transcription factor pairs for 46 unique transcription factors.

The hierarchical clustering of transcription factor-enhancer pairs based on these target sets clustered mostly by transcription factors, indicating that the expression of multiple enhancers contains highly redundant information about the activity of the associated transcription factor (Figure S5). Reassuringly, factors known to form regulatory complexes, including Max and Mxi1 or Runx1 and Smad1, also clustered together, i.e., shared predicted causal targets (Figure S5).

To investigate whether combining multiple enhancers was more informative for determining causal targets, we compared two integrative methods against the predictions of taking individual enhancers. Firstly, we used the median expression level of all the putative enhancers for each transcription factor as a 'meta-enhancer' in Findr-A (Figure S4A). Secondly, we calculated the first principal component of the binary target prediction matrix for all enhancer of a TF in order to 'average' predictions (Figure S4B). However, we did not observe any significant overall improvement in performance using either method.

2.6. Causal Inference Using CAGE Expression Data across Human Cell Types

Finally, we inferred causal interactions between transcription regulators and targets using CAGE enhancer and TSS expression data in humans. Specifically, we inferred causal interactions for 20 transcription factors (with eRNA expression) using Findr-A (see Methods). We firstly validated the predicted interactions using a database of experimentally-validated regulatory interactions in human [25] and noted a statistically-significant overlap between the predicted and experimentally-validated gene sets ($p < 10^{-5}$). Figure 5 represents the hierarchical clustering of the predicted top 200 interactions for each factor. We noted that multiple enhancers of the same factor predicted highly overlapping targets for that factor. Moreover, the factors involved in biologically-related processes shared predicted causal targets. For example, two members of the SMAD family, SMAD3 and SMAD6, as well as BCOR and SIN3A involved in histone deacetylase activity showed a high overlap of predicted targets.

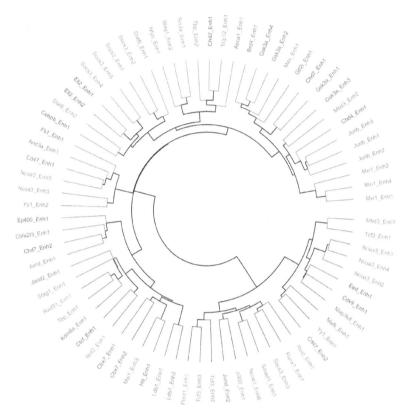

Figure 5. Hierarchical representation of similarities between transcription factor-target sets predicted using Findr-A causal inference on the FANTOM5 human dataset.

3. Discussion

We explored the utility of eRNA expression as a causal anchor to predict transcription regulatory networks, by leveraging the observation that eRNAs mark the activity of regulatory regions. Previous studies support this notion, as eRNA expression has been shown to precede the expression of its effector gene temporally [17] and to correlate strongly with active regulatory regions across cell types [15]. We therefore developed a novel statistical framework to infer causal gene networks (Findr-A), by extending the Findr software for causal inference using eQTL data [18].

We demonstrated the applicability of Findr-A by predicting causal interactions from CAGE data generated through the FANTOM Consortium and validating them with ChIP-seq and perturbation data for three mouse cell types, as well as on the entire FANTOM5 data. Notably, different factors were enriched for within cell type analysis as compared to across cell types. The causal regulatory network of cell type-specific factors (e.g., Sox2, Esrrb in ES cells) could be inferred only using expression variation within a cell type and not across cell types. Due to the limited availability of validation data, a more comprehensive assessment was not possible.

The current approach can be extended in several aspects in the future. Firstly, Findr assumes equal (or no) relations between all sample pairs [18], which hold for the majority of eQTL datasets. By accounting for heterogeneous sample relationships, such as biological and/or technical replicates, time series or population structure, we may be able to reconstruct more accurate networks. Secondly, the assumption that eRNAs act as causal anchors is only approximately true, because their activity ultimately is regulated by other regulatory factors, i.e., the assumption that they are a priori causally upstream of correlated TF-gene pairs will not hold for all genes. Because eRNAs are temporally expressed before their direct target genes, we hypothesize that explicit modelling of gene expression dynamics in the Findr framework will allow detecting and correcting for such feedback loops. Thirdly, eRNAs are expressed at relatively low levels, and therefore susceptible to noise, and a reliable eRNA signal was available for only a limited number of known transcription factors in mouse or human. Generating deep sequencing data for CAGE, utilizing GRO-seq or epigenetic or transcription factor ChIP-seq data to estimate enhancer activity could be possible ways to get around this.

In this work, we have used a very basic approach for associating enhancers with their likely promoters based on their mutual proximity, as associating enhancers with promoters was not our main focus. This could be refined by using existing data generated by methods such as chromatin confirmation capture or by integrating multiple genomic features into a statistical predictor for associating enhancers with promoters [5].

Finally, we used ChIP-seq and KO data for validation, because these datasets were available for the same cell types as the expression data in mouse. These data are far from ideal as ground-truth sets, as ChIP-seq data tend to be noisy and differential expression in transcription factor knock-outs could include indirect effects. Moreover, the targets predicted by ChIP-seq and KO experiments show a very poor overlap between them. For example, a large-scale study where 59 TFs were knocked down in a human lymphoblastoid cell line (GM12878) concluded that only a small subset of genes bound by a factor were differentially expressed following knock-down of that factor [21]. It is therefore important to note that there is no universal agreement on ground-truth TF-target regulatory networks for validation, including networks obtained from literature mining. For example, there are only 151 common TF-target interactions among four literature-curated databases, which together comprise more than 10,000 interactions [26]. As shown, our method predicts targets for many TFs w.r.t. to one ground-truth as well or better than the other ground-truth, reaffirming our hypothesis that eRNA data can be used to infer TF targets.

In conclusion, we have demonstrated that enhancer activity can be used to infer causal gene regulatory networks. We foresee this approach to be of high value in the context of human medicine, by combining genetic, epigenetic and transcriptomic information across individuals to unravel causal disease networks.

4. Materials and Methods

4.1. Datasets

- We used Cap Analysis of Gene Expression (CAGE) data (TPM expression values) from the FANTOM5 Consortium for enhancer and transcription start sites (TSS) in mouse embryonic stem (ES) cells (36 experiments), macrophages (224 experiments) and erythroblastic leukaemia (52 experiments) [6,17]. We also selected 1036 samples from all cell types and tissues in mouse.
- We use ChIP-seq data from the Codex Consortium [20]. Data available for 78 TFs in mouse ES cells, 12 TFs in macrophages and 17 TFs in erythroleukaemia cells were used.
- For validation using knock-out data, we have collected differentially-expressed gene lists after perturbation of factors from published studies in mouse and gene lists after overexpression of factors in mouse ES cells from [27].
- We obtained CAGE data (TPM expression values) from the FANTOM5 Consortium for enhancer and TSSs for human cell types and tissues [6,17]. The enhancer regions identified using bi-directional expression were obtained from [1]. The FANTOM5 Consortium has provided data for 1826 samples from which we selected 360 samples from all cell types and tissues, one sample for each cell and tissue type with the highest sequencing depth (removing technical and biological replicates).

4.2. Data Processing

- CAGE data were processed to clear unannotated and non-expressed genes, and expression levels were log-transformed. Genes with a TPM value of 1 or more in at least one sample were considered expressed. Only enhancers expressed in more than one third of experiments were retained.
- For each TF, we selected the promoter with the highest median expression level as the promoter for that TF.
- For each TF, all enhancers within 50 kb of the TF promoter region were detected using the GenomicRanges package in Bioconductor [28] and considered as candidate causal anchor enhancers for that TF. The Findr framework (described below in detail) includes a "primary linkage" step such that targets are only predicted for TFs with significantly correlated eRNAs.
- Enhancer data were binarized by setting all experiments with zero read count to zero and all others to one.
- For the ChIP-seq data, genes with a TF binding site within 1 kb of their TSS were defined as targets for that TF.
- For the knock-out and over-expression data, genes with differential expression q-value < 0.05 were defined as targets for the TF.

4.3. Likelihood Ratio Tests with Continuous Causal Anchor Data

Given a causal relation $E \rightarrow A \rightarrow B$ to test, where E is a (continuous) enhancer for TF A and B is a putative target gene, with their expression data samples $1, \ldots, n$ annotated in subscripts, we first convert each continuous variable into a standard normal distribution by rank. Each variable is modelled as a normal distribution with the mean linearly and additively dependent on its regulators in the five tests below (illustrated in Figure 1B).

1. **Primary linkage test**: The primary linkage test verifies that the enhancer E regulates the regulator gene A. Its null and alternative hypotheses are:

$$\mathcal{H}_{\text{null}}^{(1)} \equiv E \quad A, \qquad \mathcal{H}_{\text{alt}}^{(1)} \equiv E \rightarrow A.$$

   The log likelihood ratio (LLR) and its null distribution are identical to the correlation test in [18]. Therefore, the LLR is:

$$\text{LLR}^{(1)} = -\frac{n}{2} \ln(1 - \hat{\rho}_{EA}^2),$$

   where:

$$\hat{\rho}_{XY} \equiv \frac{1}{n} \sum_{i=1}^{n} X_i Y_i .$$

   Its null distribution is:

$$\text{LLR}_{\text{null}}^{(1)} / n \sim \mathcal{D}(1, n-2).$$

   The probability density function (PDF) for $z \sim \mathcal{D}(k_1, k_2)$ is defined as: for $z > 0$,

$$p(z \mid k_1, k_2) = \frac{2}{B(k_1/2, k_2/2)} \left(1 - e^{-2z}\right)^{(k_1/2 - 1)} e^{-k_2 z},$$

   and for $z \leq 0$, $p(z \mid k_1, k_2) = 0$, where $B(a, b)$ is the Beta function.

2. **Secondary linkage test**: The secondary linkage test verifies that the enhancer E regulates the target gene B. The LLR and its null distribution are identical to those of the primary linkage test, except by replacing A with B.

3. **Conditional independence test**: The conditional independence test verifies that E and B become independent after conditioning on A, with its null and alternative hypotheses as:

$$
\begin{aligned}
\mathcal{H}^{(3)}_{\text{null}} &\equiv E \to A \to B, \\
\mathcal{H}^{(3)}_{\text{alt}} &\equiv B \leftarrow E \to A \wedge (A \text{ correlates with } B).
\end{aligned}
$$

Correlated genes are modelled as having a multi-variant normal distribution, whose mean linearly depends on their regulator gene. Therefore,

$$
\begin{aligned}
\text{LLR}^{(3)} = {} & -\frac{n}{2}\ln\left((1 - \hat{\rho}^2_{EA})(1 - \hat{\rho}^2_{EB}) - (\hat{\rho}_{AB} - \hat{\rho}_{EA}\hat{\rho}_{EB})^2\right) \\
& + \frac{n}{2}\ln(1 - \hat{\rho}^2_{EA}) + \frac{n}{2}\ln(1 - \hat{\rho}^2_{AB}).
\end{aligned}
$$

Following the same definition of the null data, their null distribution is:

$$
\text{LLR}^{(3)}_{\text{null}}/n \sim \mathcal{D}(1, n - 3).
$$

4. **Relevance test**: The relevance test verifies that B is regulated by either E or A. Its hypotheses are:

$$
\begin{aligned}
\mathcal{H}^{(4)}_{\text{null}} &\equiv E \to A \quad B, \\
\mathcal{H}^{(4)}_{\text{alt}} &\equiv E \to A \wedge E \to B \leftarrow A.
\end{aligned}
$$

Similarly,

$$
\begin{aligned}
\text{LLR}^{(4)} = {} & -\frac{n}{2}\ln\left((1 - \hat{\rho}^2_{EA})(1 - \hat{\rho}^2_{EB}) - (\hat{\rho}_{AB} - \hat{\rho}_{EA}\hat{\rho}_{EB})^2\right) \\
& + \frac{n}{2}\ln(1 - \hat{\rho}^2_{EA}).
\end{aligned}
$$

$$
\text{LLR}^{(4)}_{\text{null}}/n \sim \mathcal{D}(2, n - 3).
$$

5. **Controlled test**: The controlled test verifies that E regulates B through A, partially or fully, with the hypotheses as:

$$
\begin{aligned}
\mathcal{H}^{(5)}_{\text{null}} &\equiv B \leftarrow E \to A, \\
\mathcal{H}^{(5)}_{\text{alt}} &\equiv B \leftarrow E \to A \wedge A \to B.
\end{aligned}
$$

Its LLR is

$$
\begin{aligned}
\text{LLR}^{(5)} = {} & -\frac{n}{2}\ln\left((1 - \hat{\rho}^2_{EA})(1 - \hat{\rho}^2_{EB}) - (\hat{\rho}_{AB} - \hat{\rho}_{EA}\hat{\rho}_{EB})^2\right) \\
& + \frac{n}{2}\ln(1 - \hat{\rho}^2_{EA}) + \frac{n}{2}\ln(1 - \hat{\rho}^2_{EB}),
\end{aligned}
$$

with the null distribution:

$$
\text{LLR}^{(5)}_{\text{null}}/n \sim \mathcal{D}(1, n - 3).
$$

The LLR and its null distribution then allow one to compute the p-values and the posterior probabilities of the null and alternative hypotheses separately for each subtest, as detailed in [18].

4.4. Findr-B and Findr-C

In [18], it was shown that a combined causal inference test performs best in terms of sensitivity and specificity for recovering true regulatory interactions, using both real and simulated test data. The combined test score is:

$$P = \frac{1}{2}(P_2 P_5 + P_4)$$

where P_i is the posterior probability for subtest i.

The Findr-B method returns this combined p-value using the original Findr on binarized enhancer data. Findr-C does the same using the new tests on continuous enhancer data.

4.5. Adaptive Method Findr-A

Given a set of TFs and for every TF, a set of candidate causal anchor enhancers, the adaptive Findr-A method performs the following, for each TF A (Figure 1C):

1. Compute the primary linkage test p-value for all candidate enhancers of A, both continuous and binarized.
2. Find the enhancer E with the lowest p-value overall.
3. If the lowest p-value occurred for a binarized enhancer, use Findr-B for TF A with E as its causal anchor, else use Findr-C.

Findr-A, -B and -C have been implemented in the Findr software, available at https://github.com/lingfeiwang/findr.

4.6. Validation Methods

For the purpose of evaluation, we calculated the Findr-B and Findr-C scores for all TF-gene combinations. Genes with scores exceeding a threshold of 0.8 of Findr score were considered as predicted targets for each TF. Precision-recall curves were calculated using the the the "PRROC" package. We used the FDR corrected (BH procedure) hyper-geometric test for enrichment analysis, where the overlap with respect to known targets from ChIP-seq and knock-out data in the same cell type was tested, and the resulting p-values were used to compare the performance of the two methods.

Supplementary Materials: The following are available online at http://www.mdpi.com/1422-0067/19/11/3609/s1.

Author Contributions: A.J. and T.M. conceived of the project and designed the analysis. L.W. developed the computational tool. D.V. and G.D. analysed the data. A.J. and T.M. wrote the paper.

Funding: This work was funded by grants from the Biotechnology and Biological Sciences Research Council (BBSRC) (BB/P013732/1, BB/M020053/1). A.J. is currently supported by Bergen Research Foundation Grant No. BFS2017TMT01.

Conflicts of Interest: The authors declare no conflict of interest.

References

1. Andersson, R.; Gebhard, C.; Miguel-Escalada, I.; Hoof, I.; Bornholdt, J.; Boyd, M.; Chen, Y.; Zhao, X.; Schmidl, C.; Suzuki, T.; et al. An atlas of active enhancers across human cell types and tissues. *Nature* **2014**, *507*, 455–461. [CrossRef] [PubMed]
2. Kundaje, A.; Meuleman, W.; Ernst, J.; Bilenky, M.; Yen, A.; Heravi-Moussavi, A.; Kheradpour, P.; Zhang, Z.; Wang, J.; Ziller, M.J.; et al. Integrative analysis of 111 reference human epigenomes. *Nature* **2015**, *518*, 317–330. [CrossRef] [PubMed]
3. Mifsud, B.; Tavares-Cadete, F.; Young, A.N.; Sugar, R.; Schoenfelder, S.; Ferreira, L.; Wingett, S.W.; Andrews, S.; Grey, W.; Ewels, P.A.; et al. Mapping long-range promoter contacts in human cells with high-resolution capture Hi-C. *Nat. Genet.* **2015**, *47*, 598–606. [CrossRef] [PubMed]
4. Thurman, R.E.; Rynes, E.; Humbert, R.; Vierstra, J.; Maurano, M.T.; Haugen, E.; Sheffield, N.C.; Stergachis, A.B.; Wang, H.; Vernot, B.; et al. The accessible chromatin landscape of the human genome. *Nature* **2012**, *489*, 75–82. [CrossRef] [PubMed]

5.   He, B.; Chen, C.; Teng, L.; Tan, K. Global view of enhancer-promoter interactome in human cells. *Proc. Natl. Acad. Sci. USA* **2014**, *111*, E2191–E2199. [CrossRef] [PubMed]

6.   Forrest, A.; Kawaji, H.; Rehli, M.; Baillie, J.; de Hoon, M.; Haberle, V.; Lassmann, T.; Kulakovskiy, I.; Lizio, M.; Itoh, M.; et al. A promoter-level mammalian expression atlas. *Nature* **2014**, *507*, 462–470. [CrossRef] [PubMed]

7.   Pearl, J. *Causality*; Cambridge University Press: Cambridge, UK, 2009.

8.   Schadt, E.E.; Lamb, J.; Yang, X.; Zhu, J.; Edwards, S.; GuhaThakurta, D.; Sieberts, S.K.; Monks, S.; Reitman, M.; Zhang, C.; et al. An integrative genomics approach to infer causal associations between gene expression and disease. *Nat. Genet.* **2005**, *37*, 710–717. [CrossRef] [PubMed]

9.   Chen, L.S.; Emmert-Streib, F.; Storey, J.D. Harnessing naturally randomized transcription to infer regulatory relationships among genes. *Genome Biol.* **2007**, *8*, R219. [CrossRef] [PubMed]

10. Rockman, M.V. Reverse engineering the genotype–phenotype map with natural genetic variation. *Nature* **2008**, *456*, 738–744. [CrossRef] [PubMed]

11. Li, Y.; Tesson, B.M.; Churchill, G.A.; Jansen, R.C. Critical reasoning on causal inference in genome-wide linkage and association studies. *Trends Genet.* **2010**, *26*, 493–498. [CrossRef] [PubMed]

12. Schadt, E.E. Molecular networks as sensors and drivers of common human diseases. *Nature* **2009**, *461*, 218. [CrossRef] [PubMed]

13. Natoli, G.; Andrau, J.C. Noncoding transcription at enhancers: general principles and functional models. *Annu. Rev. Genet.* **2012**, *46*, 1–19. [CrossRef] [PubMed]

14. Lam, M.T.; Cho, H.; Lesch, H.P.; Gosselin, D.; Heinz, S.; Tanaka-Oishi, Y.; Benner, C.; Kaikkonen, M.U.; Kim, A.S.; Kosaka, M.; et al. Rev-Erbs repress macrophage gene expression by inhibiting enhancer-directed transcription. *Nature* **2013**, *498*, 511–515. [CrossRef] [PubMed]

15. Danko, C.G.; Hyland, S.L.; Core, L.J.; Martins, A.L.; Waters, C.T.; Lee, H.W.; Cheung, V.G.; Kraus, W.L.; Lis, J.T.; Siepel, A. Identification of active transcriptional regulatory elements from GRO-seq data. *Nat. Methods* **2015**, *12*, 433–438. [CrossRef] [PubMed]

16. Azofeifa, J.G.; Allen, M.A.; Hendrix, J.R.; Read, T.; Rubin, J.D.; Dowell, R.D. Enhancer RNA profiling predicts transcription factor activity. *Genome Res.* **2018**. [CrossRef] [PubMed]

17. Arner, E.; Daub, C.O.; Vitting-Seerup, K.; Andersson, R.; Lilje, B.; Drablos, F.; Lennartsson, A.; Ronnerblad, M.; Hrydziuszko, O.; Vitezic, M.; et al. Transcribed enhancers lead waves of coordinated transcription in transitioning mammalian cells. *Science* **2015**, *347*, 1010–1014. [CrossRef] [PubMed]

18. Wang, L.; Michoel, T. Efficient and accurate causal inference with hidden confounders from genome-transcriptome variation data. *PLoS Comput. Biol.* **2017**, *13*, e1005703. [CrossRef] [PubMed]

19. Storey, J.D.; Tibshirani, R. Statistical significance for genomewide studies. *Proc. Natl. Acad. Sci. USA* **2003**, *100*, 9440–9445. [CrossRef] [PubMed]

20. Sánchez-Castillo, M.; Ruau, D.; Wilkinson, A.C.; Ng, F.S.; Hannah, R.; Diamanti, E.; Lombard, P.; Wilson, N.K.; Gottgens, B. CODEX: A next-generation sequencing experiment database for the haematopoietic and embryonic stem cell communities. *Nucleic Acids Res.* **2015**, *43*, D1117–D1123. [CrossRef] [PubMed]

21. Cusanovich, D.A.; Pavlovic, B.; Pritchard, J.K.; Gilad, Y. The functional consequences of variation in transcription factor binding. *PLoS Genet.* **2014**, *10*, e1004226. [CrossRef] [PubMed]

22. Healy, S.; Khan, P.; Davie, J.R. Immediate early response genes and cell transformation. *Pharmacol. Ther.* **2013**, *137*, 64–77. [CrossRef] [PubMed]

23. Mantsoki, A.; Devailly, G.; Joshi, A. CpG island erosion, polycomb occupancy and sequence motif enrichment at bivalent promoters in mammalian embryonic stem cells. *Sci. Rep.* **2015**, *5*, 16791. [CrossRef] [PubMed]

24. Dunn, S.J.; Martello, G.; Yordanov, B.; Emmott, S.; Smith, A.G. Defining an essential transcription factor program for naïve pluripotency. *Science* **2014**, *344*, 1156–1160. [CrossRef] [PubMed]

25. Han, H.; Cho, J.W.; Lee, S.; Yun, A.; Kim, H.; Bae, D.; Yang, S.; Kim, C.Y.; Lee, M.; Kim, E.; et al. TRRUST v2: An expanded reference database of human and mouse transcriptional regulatory interactions. *Nucleic Acids Res.* **2018**, *46*, D380–D386. [CrossRef] [PubMed]

26. Han, H.; Shim, H.; Shin, D.; Shim, J.E.; Ko, Y.; Shin, J.; Kim, H.; Cho, A.; Kim, E.; Lee, T.; et al. TRRUST: A reference database of human transcriptional regulatory interactions. *Sci. Rep.* **2015**, *5*, 11432. [CrossRef] [PubMed]

27. Xu, H.; Baroukh, C.; Dannenfelser, R.; Chen, E.Y.; Tan, C.M.; Kou, Y.; Kim, Y.E.; Lemischka, I.R.; Ma'ayan, A. ESCAPE: database for integrating high-content published data collected from human and mouse embryonic stem cells. *Database (Oxford)* **2013**, *2013*, bat045. [CrossRef] [PubMed]
28. Lawrence, M.; Huber, W.; Pages, H.; Aboyoun, P.; Carlson, M.; Gentleman, R.; Morgan, M.T.; Carey, V.J. Software for computing and annotating genomic ranges. *PLoS Comput. Biol.* **2013**, *9*, e1003118. [CrossRef] [PubMed]

 © 2018 by the authors. Licensee MDPI, Basel, Switzerland. This article is an open access article distributed under the terms and conditions of the Creative Commons Attribution (CC BY) license (http://creativecommons.org/licenses/by/4.0/).

International Journal of
Molecular Sciences

Review

Insect Transcription Factors: A Landscape of Their Structures and Biological Functions in *Drosophila* and beyond

Zhaojiang Guo [1,†], Jianying Qin [1,2,†], Xiaomao Zhou [2] and Youjun Zhang [1,*]

[1] Department of Plant Protection, Institute of Vegetables and Flowers, Chinese Academy of Agricultural Sciences, Beijing 100081, China; guozhaojiang@caas.cn (Z.G.); qinjianying0203@163.com (J.Q.)
[2] Longping Branch, Graduate School of Hunan University, Changsha 410125, China; zhouxm1972@126.com
* Correspondence: zhangyoujun@caas.cn; Tel.: +86-10-82109518
† These authors contributed equally to this work.

Received: 23 October 2018; Accepted: 16 November 2018; Published: 21 November 2018

Abstract: Transcription factors (TFs) play essential roles in the transcriptional regulation of functional genes, and are involved in diverse physiological processes in living organisms. The fruit fly *Drosophila melanogaster*, a simple and easily manipulated organismal model, has been extensively applied to study the biological functions of TFs and their related transcriptional regulation mechanisms. It is noteworthy that with the development of genetic tools such as CRISPR/Cas9 and the next-generation genome sequencing techniques in recent years, identification and dissection the complex genetic regulatory networks of TFs have also made great progress in other insects beyond *Drosophila*. However, unfortunately, there is no comprehensive review that systematically summarizes the structures and biological functions of TFs in both model and non-model insects. Here, we spend extensive effort in collecting vast related studies, and attempt to provide an impartial overview of the progress of the structure and biological functions of current documented TFs in insects, as well as the classical and emerging research methods for studying their regulatory functions. Consequently, considering the importance of versatile TFs in orchestrating diverse insect physiological processes, this review will assist a growing number of entomologists to interrogate this understudied field, and to propel the progress of their contributions to pest control and even human health.

Keywords: insect; transcription factors; structures and functions; research methods; progress and prospects

---

1. Introduction

Transcription factors (TFs) are a plethora of proteins that are present in all living organisms, and they can intricately control transcription of different functional genes in response to internal physiological processes, as well as external environmental stimuli [1,2]. Strikingly, it has been estimated that TFs make up approximately 3.5–7.0% of the total number of coding sequences in a reference eukaryotic genome [3,4]. TFs can be divided into diverse classes based on their structural characteristics, and they can influence the transcription of their targets in different mode of actions [5,6]. Most eukaryotic TFs with a DNA-binding domain (DBD) are thought to exert regulatory functions by binding to enhancers and recruiting transcriptional complexes or cofactors [7]. Nevertheless, recent studies have shown that some TFs do not perform the classical functions of binding enhancers and recruiting transcriptional complexes, but rather merely promote regulatory elements such as enhancers and promoters to form looped structures to facilitate transcriptional regulation [8]. TFs have a significant level of functional diversity, and they participate in various biological processes that are pivotal to ensure the proper expression levels of targets in the appropriate time and tissue in organisms.

To date, there are several reviews summarizing the role of certain TF families in certain biochemical processes in insects, especially in *Drosophila* [2,9–12]. In recent years, the vast range of available genome resources and novel genetic methods greatly promote the identification and investigation of regulatory mechanisms of TFs in diverse insects beyond *Drosophila*, especially lepidopteran insects (Figure 1). However, there is no comprehensive review covering TF structures and functions in different physiological and biochemical processes in *Drosophila* and beyond. Herein, this review describes the structures of the dominating TF types existing in insects, and presents a comprehensive summary of the roles of these TFs in various insect biological processes, which provides new insights into the studies of transcriptional regulation in both model and non-model insects, and elaborates on their application prospects as potential new strategies for pest control in the field, and even as potential targets for human disease treatment and health care.

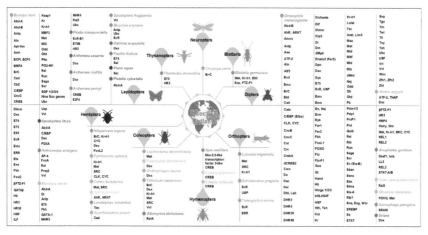

Figure 1. The insect transcription factor (TF) atlas, including the well-studied TFs and their related references collected in the review (See Table S1 for more detailed information about these TFs, and we apologize to researchers whose work could not be discussed and cited in the main text due to space limitations). As yet, diverse TFs have been documented in at least nine different insect orders including Diptera, Hemiptera, Lepidoptera, Coleoptera, Orthoptera, Thysanoptera, Blattaria, Neuroptera, and Hymenoptera. Different insect species are also denoted by colored circles on the vertical line.

2. Structures

Most TFs contain a DNA-binding domain (DBD) that specifically recognizes and binds to TF-binding sites in the enhancer or promoter of the target gene to read out information and modulate target transcription [13]. Therefore, eukaryotic TFs are generally classified based on their DBD. In humans and mice, TFs are divided into 10 superclasses, the first three of which are the helix-turn-helix (HTH) domain TFs (mostly homeodomain TFs), the basic DBD TFs, and the zinc finger (ZF) TFs [5,14]. These three superclasses occupy 90% and 86% of almost all TFs in humans and mice, respectively [14]. In insects, these three superclass TFs also account for the majority, and we summarize the structures of the three superclasses in insects as follows.

2.1. Homeodomain TFs

The HTH superclass TFs contain the HTH motif as the DBD. According to differences in HTH structure, the HTH superclass can be divided into several classes, including homeodomain, Forkhead (Fkh), and Heat shock factor (HSF). Homeodomain proteins play vital biological functions throughout the entire life of insects, especially during growth and development. The homeodomain was originally identified from the homeotic genes in *Drosophila melanogaster* [15,16], which typically contains a short N-terminal arm that contributes to the DNA binding affinity, as well as four α-helices with an HTH structure that is responsible for DNA binding and recognition [16,17]. Helices I and II are antiparallel to each other, helices II and III are separated by a β-turn to form a helix-turn-helix (HTH) structure, and helix III functions as the recognition helix for contacting and recognizing specific DNA sequences (Figure 2A). The homeodomain alone in some homeodomain protein is insufficient for specific DNA binding. Therefore, additional domains and/or cofactors are required to elevate the specificity of DNA binding. For example, Hox proteins generally form homodimers [18] or heterodimers through the conserved YPWM motif upstream homeodomain to increase their selectivity and affinity for DNA binding [19,20]. *Drosophila* contains a total of 103 homeodomain genes, and 54% of TFs are split among families such as paired-like, paired-domain, POU (Pit-Oct-Unc), and LIM proteins that contain other domains for higher binding affinity or protein dimerization, in addition to the homeodomain [21] (Figure 2A).

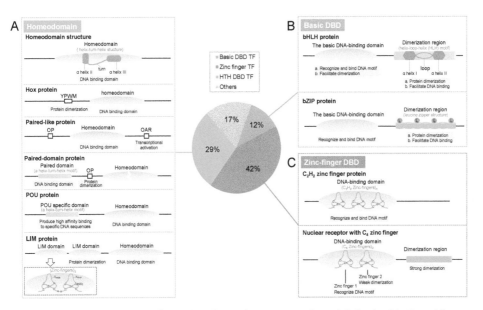

Figure 2. The main structures of insect TFs. The pie chart represents the statistical ratio of the *Drosophila* TFs in different superclasses based on Hammonds' study [22]. (**A**) Structures of homeodomain TFs. Homeodomain: a typical homeodomain contains a short N-terminal arm facilitating DNA binding and four α-helices with a helix II-turn-helix III structure that is responsible for DNA binding and recognition; Hox protein: in addition to the homeodomain, a Hox protein typically contains a YPWM motif in the N-terminus mediating protein dimerization; Paired-like protein: in addition to the homeodomain, some paired-like proteins include an N-terminal octapeptide (OP) motif, and some contain a C-terminal OAR motif that can be involved in transcriptional activation; Paired-domain protein: paired-domain proteins carry an N-terminal paired box with a helix-turn-helix (HTH) structure that mediates DNA binding, and some proteins have a full-length or truncated homeodomain in the C-terminus, as well as an OP motif between the paired box and the homeodomain to mediate protein dimerization; POU (Pit-Oct-Unc) protein: POU proteins have a POU-specific domain with an HTH structure in the N-terminus that contributes to the generation of high-affinity DNA binding, and the homeodomain in the C-terminus is responsible for DNA binding; LIM protein: LIM proteins possess two LIM domains with zinc finger (ZF) structures upstream of the homeodomain that mainly mediate protein-protein interactions. (**B**) Structures of basic DBD TFs. The basic leucine zipper (bZIP) protein: bZIP proteins harbor a basic

DNA-binding domain (DBD) for specific DNA recognition and binding, and a leucine zipper domain in the C-terminus for protein dimerization and DNA binding. The leucine zipper domain forms dextrorotatory α-helices, and a leucine appears in the seventh position of every seven amino acids; thus, an adjacent leucine appears every two turns on the same side of the helix. The basic helix-loop-helix (bHLH) protein: bHLH proteins are composed of a basic DBD in the N-terminus, followed by an HLH domain. The basic DBD accounts for DNA motif recognition and binding and facilitates protein dimerization. In the HLH domain, two hydrophilic and lipophilic α-helices are separated by a loop to form an HLH structure mediating protein dimerization and contributing to DNA binding. (**C**) Structures of the ZF TFs. C_2H_2 ZF protein: the C_2H_2 ZF protein has multiple connected ZF DBDs. In the root of every ZF, two cysteines and two histidines link Zn^{2+} to form a tetrahedron. Nuclear receptor (NR) with C_4 ZF: NR contains a C_4-type ZF region as a DBD that consists of eight conserved cysteine residues coordinated with two Zn^{2+} to form two ZFs with a tetrahedral coordination structure. The first ZF provides DNA-binding specificity, and the second ZF has a weak dimerization interface, allowing for dimerization of the receptor molecule. In addition, a ligand-binding domain is typically found in the C-terminus and functions as the main dimerization region.

2.2. Basic DBD TFs

2.2.1. bHLH Proteins

The basic helix-loop-helix (bHLH) TFs are widely distributed in insects, and they play crucial roles in the regulation of insect growth and development. The bHLH region contains approximately 60 amino acids, consisting of a basic domain and an HLH domain. The basic domain at the N-terminus is responsible for DNA binding, and it is followed by the HLH region (Figure 2B). In the HLH region, two α-helices that are both hydrophilic and lipophilic are separated by different lengths of a linking loop to form an HLH structure (Figure 2B). In general, bHLH TFs form homo- or heterodimers through α-helix interactions to regulate target transcription [23].

In 2002, Ledent et al. categorized *Drosophila* bHLH proteins and divided them into six groups (A–F) according to their DNA binding and structural characteristics [24] (Table S1). Group A proteins usually bind the 5′-CAC/GCTG-3′ motif, and they primarily participate in the regulation of mesodermal subdivisions and the development of the nervous system, muscle cells and glands. Group B proteins generally bind the 5′-CACGTG-3′ or 5′-CATGTTG-3′ motifs, and they are involved in the regulation of development, lipid metabolism and glucose tolerance. Group C proteins have a PAS domain followed a bHLH region to combine with a 5′-A/GCGTG-3′ motif [25], which participate in the development of the nervous system and trachea, as well as in the dictation of circadian rhythms and the hypoxia response. Group D proteins contain only an HLH region and they lack a basic domain; these proteins can inhibit the binding of other bHLH proteins to DNA by forming dimers with those bHLH proteins [26]. In *Drosophila*, only one protein Extramacrochaetae (Emc) exists in group D [27,28]. Group E proteins preferentially bind N-boxes (5′-CACGA/CG-3′) and have an additional "Orange" domain and WRPW peptide motif at the C-terminus [29]. Proteins in this group are mainly involved in nervous system development. Group F proteins contain a COE domain, which plays an important role in both dimerization and DNA binding [24,30]. *Drosophila* has only one COE family member, Collier (Col) [31].

2.2.2. bZIP proteins

The basic leucine zipper (bZIP) proteins contain a basic domain that specifically recognizes and binds DNA sequences, followed by a leucine zipper region that is responsible for protein dimerization, and thus promotes the binding of basic regions to DNA [17] (Figure 2B). In the leucine zipper domain, a leucine appears in the seventh position of every seven amino acids, and this region forms dextrorotatory α-helices, with an adjacent leucine appearing every two turns on the same side of the helix [32] (Figure 2B). Multiple bZIP proteins have been identified in *Drosophila*, and they play essential roles in *Drosophila* development and reproduction (Table S1).

2.3. Zinc-Finger TFs

ZF TFs contain ZF structure motifs as DBDs, and they can be mainly classified into C_2H_2, C_4, and C_6 classes, according to differences in the conserved ZF domain. TFIIIA is the first ZF protein that has been identified in *Xenopus laevis* in 1983 [33,34]. Subsequently, ZF TFs have been extensively discovered, and they have been found to be the most widely distributed proteins in eukaryotic genomes. In insects, ZF TFs primarily include C_2H_2 ZF proteins and nuclear receptors (NRs) with a C_4 structure (Figure 2C).

2.3.1. C_2H_2 Zinc Finger

C_2H_2 ZFs can be connected in series to recognize DNA sequences of different lengths [35]. For instance, *Drosophila* Krüppel (Kr) contains four tandemly repeated ZFs [17]. Each C_2H_2 ZF is an independent domain with a consensus amino acid motif: X_2-C-$X_{2,4}$-C-X_{12}-H-$X_{3,4,5}$-H (C represents cysteine, H represents histidine, D represents aspartic acid, and X represents any amino acid) [36]. These sequences are tightly folded to form $\beta\beta\alpha$ structures in the presence of Zn^{2+}, and Zn^{2+} is sandwiched between α-helices and two antiparallel β-sheets [36–38]. Two Cys and two His link Zn^{2+} to form a tetrahedron (Figure 2C), in which the two Cys are located on the β-sheet, and the two His are located at the C-terminus of the α-helix. The α-helix acts as a recognition helix to insert into the major groove of DNA and contact specific DNA sequences [17,35].

2.3.2. Nuclear Receptors

NRs in insects are receptors that are responsible for sensing and responding to ecdysone. The C_4-type ZF domain in NRs is the DBD, and has a consensus sequence: C-X_2-C-X_{13}-C-X_2-C-$X_{15,17}$-C-X_5-C-X_9-C-X_2-Cys-X_4-C. This motif consists of eight Cys coordinated with two Zn^{2+} to form two ZFs with a tetrahedral structure (Figure 2C). The first ZF provides DNA binding specificity, and the second ZF has a weak dimerization interface, allowing for the dimerization of the receptor molecule [9]. Besides the conserved DBD, NRs commonly contain a ligand-binding domain in the C-terminus, which is the main region for NR dimerization. NRs typically form dimers to bind DNA motifs, and each receptor molecule recognizes and binds half of the sequence (abbreviated to half-site). The distance between two half-sites and the sequence arrangement facilitate receptor binding to specific DNA motifs [9]. The second ZF plays a critical role in identifying the optimal distance between the two half-sites [17]. Some NRs such as *Drosophila* E75 and E78, can act as "orphan receptors" to bind a single response element.

3. Biological Functions

3.1. Internal Responses

Numerous TFs in insects precisely regulate spatiotemporal gene expression in response to their internal physiological needs, such as growth and development, metamorphosis, and reproduction. In this regard, the growth and development processes such as embryonic axis establishment, eye development, and gland formation coordinated by substantial numbers of TFs have been thoroughly studied in the model insect *Drosophila*. Additionally, some internal responses of non-model insects modulated by diverse TFs have also recently made great progress. In this section, we systematically review the TF-regulating internal physiological responses in these model and non-model insects.

3.1.1. Embryonic Axis Establishment

In *Drosophila*, the establishment of the embryonic axis has been well documented, and the TF-mediated transcriptional hierarchy has been comprehensively investigated [39,40]. Four morphogens are essential for the establishment of the anterior-posterior (A-P) axis: Bicoid (Bcd) and Hunchback (Hb) regulate the anterior region, and Nanos (Nos) and Caudal (Cad) control the posterior region. Morphogen gradients control the expression of gap genes. Different concentrations of gap proteins activate distinct pair-rule genes, and form seven stripes perpendicular to the A-P axis. Pair-rule proteins in different combinations then activate the transcription of segment polarity genes, further subdividing the embryos into 14 body segments. Finally, Gap, pair-rule, and segment polarity proteins together orchestrate the expression of Hox genes, which determine the developmental fate of each segment (Figure 3).

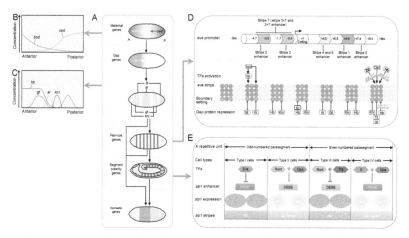

Figure 3. Establishment of the embryonic anterior–posterior (A-P) axis in *Drosophila*. The diagram is adapted from Gilbert's works [40,41]. (**A**) Regulatory hierarchy of the formation of *Drosophila* A-P axis patterning. The maternal effect proteins Bicoid (Bcd) and Caudal (Cad) form a concentration gradient along the A-P axis and generate specific positional information to activate the expression of the gap gene *hunchback* (*Hb*). Hh further initiates the proper expression of other gap genes along the A-P axis. Gap proteins subsequently activate the expression of pair-rule genes, which form seven stripes perpendicular to the A-P axis and divide those discontinuous regions defined by the gap gene into body segments. The pair-rule proteins then regulate the expression of segment polarity genes in specific cells of each somite, and their 14 expression stripes establish the boundaries of parasegments. Finally, each segment is characterized by specifically expressed Hox genes. (**B**) The concentration gradient of the maternal effect proteins Bcd and Cad along the A-P axis in the early cleavage embryo. (**C**) The concentration changes in gap genes along the A-P axis. (**D**) Regulation of expression of the pair-rule gene *even-skipped* (*eve*) in seven stripes. The above region represents a partial promoter of the *eve* gene that contains five different enhancers responsible for the distinct stripes. The lower part illustrates how TFs regulate *eve* expression in different stripes. The black box represents the TF, the green characters indicate the enhancer of *eve*, and the orange circles display the cells expressing *eve*. The vertical bars with the letters "A" and "P" denote the anterior and posterior boundaries of the eve stripe, respectively, and the numbers show the number of *eve* bands. (**E**) Regulation of a repetitive unit of *sloppy paired 1* (*slp1*) stripe [42]. The 14-stripe patterning constitutes seven repetitive units, each containing odd-numbered and even-numbered parasegments. The odd-numbered parasegment consists of two types of cells: two type I cells in the posterior half do not express *slp1* and two type II cells in the posterior half that express *slp1*. The even-numbered parasegment also contains two types of cells: type III cells that do not express *slp1* in the first half and the latter type IV cells expressing *slp1*. The expression of *slp1* in different cell types is regulated by different pair-rule proteins in a specific combination to regulate the proximal early stripe element (PESE) or the distal early stripe element (DESE) of *slp1*. The colored hexagons indicate the pair-rule proteins, the orange quadrants represent the enhancer, the gray ovals exhibit the cells that do not express *slp1*, the blue ovals denote the cells expressing *slp1*, the gray rectangles with "No" show no *slp1* strips, and the blue rectangles represent *slp1* strips. TF abbreviations: Hb: Hunchback; Gt: Giant; Kr: Krüppel; Kni: Knirps; Tll: Tailess; Zld: Zelda; Opa: Odd-paired; Ftz: Fushi tarazu; X: represents an as yet unidentified Factor X.

Dorsal (Dl) protein is critical for the formation of dorsal–ventral (D-V) patterning. The Dl protein is initially uniformly distributed in the egg, but after the ninth cell division, the ventral Dl protein begins to migrate into the nucleus, and the dorsal Dl protein remains in the cytoplasm, causing the formation of a gradient of the Dl protein along the D-V axis. This Dl gradient triggers the formation of mesoderm, neuroectoderm, dorsal ectoderm, and amnioserosa by specifically ordering the expression of different regulatory genes along the D-V axis in early *Drosophila* embryos [43]. A high level of Dl protein activates the expression of *twist* (*twi*) and *snail* (*sna*), and inhibits the expression of *decapentaplegic* (*dpp*) and *zerknüllt* (*zen*) in the mesoderm, and the intermediate levels of Dl activates the expression of *rhomboid* (*rho*), while a low level of Dl activates the expression of *tolloid* (*tld*) and *dpp* in dorsal ectoderm and *zen* in amnioserosa [40,43–47]. Dl-mediated transcriptional repression requires additional factors. Groucho (Gro) is a corepressor of Dl, and it is recruited by Dl to bind to Gro-binding sites closing to Dl-binding sites and other TF-binding sites [46,48,49].

3.1.2. Nervous System Development

Drosophila is an excellent model for studying the developmental mechanisms of the central nervous system (CNS). CNS development is a multistep and complex process, and it requires a multitude of TFs to precisely govern the expression of multiple neural development-related genes (Table S1).

The CNS initially forms in early embryos. In neuroblasts (NBs), TFs are sequentially expressed in a cascade of Hb–Kr–Pdm–Castor (Cas)–Grainyhead (Grh), whose temporal regulation is crucial for the generation of neuronal diversity [50]. Asymmetric cell division (ACD) of embryonic NBs produces two daughter cells: a larger NBs and smaller ganglion mother cells (GMCs). GMCs perform a differentiation function, and they differentiate to produce two neurons or glial cells after mitosis. Snail family proteins play redundant and essential functions in GMC formation by controlling NB ACD [51,52]. Additionally, Worniu (Wor) is continuously required in NBs to maintain NB self-renewal [53].

A much larger and more complicated CNS is established in the larval phase. A different TF cascade (Hth–Ey–Slp–D–Tll) is sequentially activated in the optic lobe NBs, and it regulates temporal expressed genes in type I NBs, resulting in the production of various types of nerve cells [50,54].

Approximately 100 NBs exist in the central brain, the majority of which are type I NBs, whereas only eight are type II NBs. Type I NBs produce terminal dividing GMCs, while ACD of type II NBs first produce immature INPs (imINPs), that need to differentiate into mature INPs before dividing [55]. The marker proteins Asense (Ase) and Prospero (Pros), which determine NB identity, are expressed in type I NBs and mature INPs, but are not expressed in type II NBs. At the end of ACD, Pros is asymmetrically localized to the budding GMC, promoting GMC differentiation.

The ETS family protein Pointed P1 (PntP1) is specifically expressed in type II NBs and imINPs. In type II NBs, Notch signaling inhibits *erm* activation by PntP1, allowing type II NBs to maintain self-renewal and identity [56]. Furthermore, PntP1 represses Ase in type II NBs and promotes the generation of INPs [57]. In imINPs, PntP1 prevents both the premature differentiation and dedifferentiation of imINPs. The Sp family protein Buttonhead (Btd) also functions together with PntP1 to prevent premature differentiation by inhibiting *pros* expression in newly generated imINPs [58].

The ZF protein Earmuff (Erm) is indispensable for INP maturation. In imINPs, *erm* is activated by PntP1 due to a lack of Notch signaling, and by rapid down-regulation of the activities of some self-renewal transcriptional repressors [58,59]. Erm restricts the developmental potential of imINPs, and exerts a negative-feedback effect on PntP1, allowing imINPs to express *ase* and mature [58,60].

3.1.3. Eye Development

The development and formation of *Drosophila* eyes depend on the retinal determination gene network (RDGN), which consists of highly conserved genes encoding TFs and related cofactors that are essential for eye formation [61]. As yet, some TFs are also found to be involved in the regulation of eye development. Orthodenticle (Otd) and ecdysone receptor (EcR) is crucial for the regulation of photoreceptor maturation [62]. Kr regulates the differentiation of photoreceptor neurons (PRs) in the

Drosophila larval eye [63]. Camta, Lola, Defective proventriculus (Dve) and Hazy are key regulators of PR differentiation in adult ocelli [63].

3.1.4. Trachea and Gland Formation

Gland formation is essential for insect development. The differentiation and morphogenesis of trachea, corpora allata (CA), prothoracic glands (PG), and corpora cardiaca (CC) requires a proper TF cascade during *Drosophila* embryogenesis, respectively [64]. The tracheal epithelial tubes develop from 10 trunk placodes, where Antennapedia (Antp) and STAT activate the expression of *ventral veins lacking* (*vvl*) and *tracheales* (*trh*) for trachea formation. The homologous ectodermal cells in the maxilla and labium form CA and PG, respectively. In the maxilla, Deformed (Dfd) and STAT induce *vvl* and *sna* expression, forming CA with specific-expressed mark *seven-up* (*svp*) after the epithelial-mesenchymal transition (EMT). In the labium, Sex combs reduced (Scr) and STAT regulate *vvl* and *sna* expression, forming PG with the mark gene *spalt* (*sal*) after MET. The CC cells are derived from anterior mesodermal cells, and they specifically express the marker *glass* (*gl*). Likewise, the formation of salivary glands (SGs) in *Drosophila* embryos is also regulated by a series of TFs [65] (Table S1). SG specification requires Scr and two cofactors, Extradenticle (Exd) and Homothorax (Hth), which work together to activate several early SG TFs, including Fkh, CrebA, Sage, and Huckebein (Hkb). Subsequently, Scr, Extradenticle (Exd), and Homothorax (Hth) disappear, and they are not involved in maintaining SG-specific gene expression. Hkb is transiently expressed in SGs, while Fkh, Sage, and CrebA are continuously expressed in SGs, accounting for the maintenance and implementation of the SG fate decision.

3.1.5. Sex Determination

The sex determination mechanisms exhibit high diversity within and between insects, which promotes the amazing diversity of insects on our planet [66]. The primary signal (e.g., X-chromosome dose, M factor, parental imprinting) that commences sex determination is processed by the master gene (e.g., *sexlethal* (*sxl*), *transformer* (*tra*) or *feminizer* (*fem*)) that carries out alternative splicing and differentially expresses in different genders. The master gene then transmits the sex determination signal to the conserved switch *doublesex* (*dsx*) to control sexual differentiation.

By contrast, the molecular basis of sex determination is well studied in *Drosophila*. Sxl is the master factor in *Drosophila* somatic sex determination, which contains two promoters, *Sxl-Pe* and *Sxl-Pm* [67]. In female embryos, the first *Sxl* establishment promoter, *Sxl-Pe*, is transiently activated by a double dose of X-linked signal elements (XSE) or molecular genes to produce the functional protein Sxl, which acts on pre-messenger RNAs (mRNAs) produced by the second constitutive promoter *Sxl-Pm*, to establish the splicing loop and to maintain *Sxl* in an active state [68]. In contrast, in male embryos, *Sxl-Pe* remains inactive, producing a nonfunctional Sxl, and thereby directing the male fate [69]. Several XSE that regulate *Sxl-Pe* have been identified in *Drosophila*, including *scute* (*sc*), *sisA*, *runt* (*run*) and *unpaired* that encode TFs that serve as the primary determinants of X dose [70]. *Sxl-Pm* and *Sxl-Pe* share a common regulatory element (a 1.4 kb region) that responds to the X chromosome dose [67]. Some, but not all, of the X-linked signal TFs that regulate *Sxl-Pe*, including Sc, Daughterless (Da), and Run, are also required for the earliest expression of *Sxl-Pm* [67].

The sex-determining initial signal of other insects is different from *Drosophila*'s X-chromosome dose [71]. In addition, most insects use *tra/fem* as the master factor to sense and transmit the primary signal instead of *sxl* [71,72]. Similar to *Drosophila sxl*, *tra/fem* autoregulates to produce the corresponding protein and perform gender differentiation.

The conserved TF *dsx* is downstream of master gene, and it is located at the bottom of the sex determination cascade. Most insects contain only one *dsx* gene in their genome, while the generation of multiple *dsx* splice variants (including Dsx^F and Dsx^M) occurs via sex-specific alternative splicing [73] (Table S1). Dsx isoforms are sex-specifically expressed in the male or female to regulate the expression of gender-related genes which then control sexual differentiation (Table S1).

The upstream regulation mechanism for the sex determination of the lepidopteran model insect *B. mori* has yet to be fully uncovered, and it is still a research hotspot in developmental biology. Studying and clarifying the sex determination cascade of the representative insect can contribute to our understanding of the insect sex determination molecular mechanism during adaptive evolution, and provide new strategies for pest management.

3.1.6. Wing Imaginal Disc Development

The molecular mechanism underlying insect wing development and differentiation has been a research hotspot for insect development. Decades of research have gradually revealed the molecular mechanism of wing development, especially in *Drosophila*. *Drosophila* wings and legs originate from a common pool of ectodermal cells that express the homeodomain gene *Distal-less* (*Dll*) [74,75]. The concentration of Dpp protein decreases from dorsal to ventral. Under high concentrations of Dpp, the dorsal-most cells expressing *Dll* migrate dorsally and induce the expression of *vestigial* (*vg*) to form the original wing primordium, and later, two ZF genes, *escargot* (*esg*) and *sna*, are expressed to maintain wing disc cell fate [75]. A low concentration of Dpp promotes leg primordium formation from cells expressing *Dll* in the middle and lower regions. After the division and proliferation of wing primordia, four compartments containing different cell populations are generated: cells in the posterior (P) compartment express the homeodomain gene *engrailed* (*en*) while cells in the anterior (A) compartment do not express *en*, and an A-P axis is formed in the first instar larvae; cells in the dorsal (D) compartment express the homeobox gene *apterous* (*ap*), while cells in the ventral (V) compartment do not express *ap*, and a D-V boundary is generated in the third instar larvae [76–78]. The proximal–distal (P-D) axis is also required for *Drosophila* wing development. Several TFs such as Stat92E and Zinc finger homeodomain 2 (Zfh2) participate in the establishment of the P-D axis [79–82]. The wing imaginal disc along the P-D axis is partitioned into the distal pouch, hinge, surrounding pleura, and notum [79]. Hth, Exd, Teashirt (Tsh), and three MADF-BESS family proteins, including Hinge1, Hinge2, and Hinge3, are essential for wing hinge formation [83–85].

3.1.7. Lipid Metabolism

Insect fat body is the central organization of insect growth, development, metamorphosis, and reproduction, and many TFs in this tissue play an important role in the regulation of insect lipid metabolism. Among them, FOXO is the major terminal TF for insulin/insulin-like growth factor signaling (IIS), and it modulates lipid metabolism in some insects, including *D. melanogaster*, *Bombyx mori*, and *Glossina morsitans* [86–89]. Har-Relish responds to 20E signaling, and it regulates fat body dissociation in *Helicoverpa armigera* [90]. In *Drosophila*, activating transcription factor-2 (ATF-2) and βFTZ-F1 participate in lipid metabolism [91,92]. In *Aedes aegypti*, hormone receptor 3 (HR3), Thanatos-associated protein (THAP) and ATF-2 regulate the transcription of *Sterol carrier protein 2* (*SCP2*), which is a critical factor for sterol absorption and transport [93,94]. Moreover, C/EBP may directly regulate *SlSCPx* expression in *Spodoptera litura* [95].

3.1.8. Circadian Clock Adjustment

The circadian clock system of most living organisms participates in the regulation of various rhythmic behaviors and physiological functions. The *Drosophila* circadian rhythm is mainly regulated by a transcriptional translation feedback loop (TTFL) that is centered on the master circadian transcription complex Clock-Cycle (CLK-CYC). The CLK-CYC heterodimer directly activates the transcription of the core circadian genes *period* (*per*) and *timeless* (*tim*) during the night [96]. Phosphorylated Per and Tim in turn repress the transcription of *clk* and *cyc* [97]. Later, Per and Tim are degraded in the presence of light, allowing CLK-CYC to initiate the next cycle of transcription [98]. In addition, CLK-CYC activates the expression of hundreds of clock-controlled output genes such as *vrille* (*vri*), *Par Domain Protein 1ε* (*Pdp1ε*), *clockwork orange* (*cwo*), and *Mef2* [99–101]. In turn, VRI, PDP1m and Cwo can also regulate the expression of *clk* and *cyc* [99], and Mef2 is involved in neuronal remodeling to facilitate locomotor activity rhythms [101]. Meanwhile, the NRs induced by ecdysone are also involved in the regulation of insect rhythms. E75 and Unfulfilled (UNF; DHR51) strengthen the CLK/CYC-mediated transcription of *per* by directly binding to the regulatory element [102]. In firebrat *Thermobia domestica*, the normal circadian expression of *E75* and *HR3* is necessary for the maintenance of locomotor rhythms [103].

3.1.9. Diapause Control

Diapause can facilitate insect survival from adverse environmental conditions, such as extreme weather or reduced food availability. In *H. armigera*, multiple TFs, including Har-AP-4, POU-M2 (Vvl), and FoxA control diapause by directly regulating the expression of *diapause hormone* and *pheromone biosynthesis-activating neuropeptide (DH-PBAN)* [104–107]. In *B. mori*, POU-M2 is also essential for the regulation of *DH-PBAN* [108], and BmILF is involved in the transcriptional regulation of *POUM2* [109]. Additionally, the diapause status of *Pyrrhocoris apterus* guts is triggered under a short photoperiod in winter. Low-level JH leads to *cry2* expression overriding $Pdp1_{iso1}$, thus initiating the diapause-specific program and activating the expression of the diapause downstream genes *superoxide dismutase* (*sod*) and *transferrin* (*tf*) in the gut [110].

3.1.10. Cuticular Protein Synthesis

Insect cuticular protein is one of the main components that constitute the insect stratum corneum. Spatiotemporal expression of insect cuticular protein genes (ICPGs) is regulated by multiple TFs, especially a series of 20E-response genes (Table S1). For example, the NR βFTZ-F1 is one of the early found TFs to regulate IGPG expression, it can regulate the expression of *EDG84A* and *EDG74E* in *Drosophila* [111], MSCP14.6 in *Manduca sexta* [112], and many ICPGs in *B. mori*. In addition, the homeoprotein BmPOUM2 interacts with BmAbd-A to regulate *BmWCP4* gene expression [113,114]. Although several TFs that regulate some ICPG expression have been identified, multiple uncharacterized TFs that control other ICPGs warrant further investigation.

3.1.11. Cuticle Coloration Dictation

Insect body color and markings pattern are of great significance for insect survival and reproduction. Abundant coloring patterns are displayed in butterfly adult wings and in the epidermis of silkworm larvae and different species of fruit fly. Pigment patterning in *Drosophila* adults has been intensively studied, and is regulated by pleiotropic regulatory TFs, including sex-determination genes (e.g., *Dsx*), HOX genes (e.g., *Abdominal-B (Abd-B)*), and selector genes (e.g., *Optomotor-blind (Omb)* and *Engrailed (En)* via the control of the expression of effector genes that encode the enzymes and co-factors required for pigment biosynthesis [115,116]. The butterfly wing pattern is also regulated by multiple patterning TFs such as Omb, Abd-B, Dsx, Sal, and En [116,117]. The TF Apt-like participates in *B. mori* larval epidermal pigmentation or the melanin biosynthesis pathway by regulating the single genetic *p* locus that contains at least 15 alleles, and produces a phenotypic diversity of pigment patterns [118].

Insect hyperpigmentation is a good model for studying insect adaptation, evolution and development. Pigmentation diversity in insects can be attributed to changes in the expression levels of transcriptional activators and changes in the *cis*-regulatory elements of the pigment synthesis gene for TF binding [115]. Deciphering the mechanism of insect coloration regulated by TFs provides an important reference for agricultural and forestry pest control, and ecological adaptability exploration. Nonetheless, the current accumulated knowledge is not enough to allow us to fully understand the complete regulation network of insect coloring pattern; thus, further studies are required to identify regulatory TFs and to expound the regulatory mechanism.

3.1.12. Silk Protein Production

Silk is mainly composed of fibroin and sericin. Fibroin consists of the fibroin heavy chain (fibH), fibroin light chain (fibL) and P25 proteins. These genes are specifically expressed in posterior silk gland (PSG) cells during the feeding stage of silkworm larval development, but they are suppressed during the molting stage. The *sericin-1* (*ser1*) gene is expressed in the posterior of the middle silk gland (MSG) before the fifth instar larvae, and its expressional region extends to the middle in the fifth instar larvae. A variety of TFs jointly regulate the spatiotemporal expression of these silk protein synthesis-related genes (Table S1). Many TFs have been reported to regulate *fibH* gene expression. Among them, the bHLH TFs Dimmed (Dimm) and Sage usually form heterodimers with other proteins to regulate *fibH* expression. For instance, Dimm directly activates *fibH* expression by interacting with Sage [119]. Dimm can also act as a repressor of *fibH* by interacting with repressor MBF2 [120]. Sage forms a complex with Fkh to enhance *fibH* expression [121]. Whether Dimm, Sage, and Fkh can form a triplet to activate *fibH* transcription merits further study [119]. Relatively few TFs are known to regulate *fibL* and *P25* genes. Fkh and SGF2 positively regulate the expression of the *fibL* gene [122,123]. Fkh, SGF2, PSGF, and BMFA are involved in the regulation of *P25* expression [124–126]. Some TFs that positively regulate the expression of the *sericin-1* gene have also been identified, including Fkh [127], POU-M1 [128], and Antp [129]. Additionally, POU-M1 participates in the restriction of the anterior boundary of the *ser1* expression region [130]. Nevertheless, although many transcriptional activators controlling the expression of silk protein synthesis-related genes have been identified, transcriptional repressors inhibiting the expression of these genes at the molting stage, thereby limiting their spatial expression, still remain largely unknown.

3.1.13. Molting and Metamorphosis Initiation

Insect larval molting and metamorphosis are coordinated by ecdysone and juvenile hormones (JHs). The 20-hydroxyecdysone (20E, the biologically active form of ecdysone) induces larval–larval molting in the presence of JHs, while 20E induces larval–pupal and pupal–adult metamorphosis upon the disappearance of JHs [131].

20E regulates various physiological and biochemical processes in insects, especially molting and metamorphosis [132]. TFs play an essential role in the regulation of ecdysone titers. Several TFs have been identified to specifically regulate Halloween genes encoding a series of ecdysone biosynthetic enzymes, to promote steroidogenesis (Figure 4A and Table S1). Among them, Séance (Séan), Ouija board (Ouib), and Molting defective (Mld) are only found in *Drosophila*, they are, therefore, thought to be evolved specifically to control the transcription of the two Halloween genes *neverland* (*nvd*) and *spookier* (*spok*) in *Drosophila* [133]. Reduction of ecdysone titers regulated by TFs occurs in two ways: inhibition of Halloween gene expression and the direct degradation of ecdysone (Figure 4A). Hence, changes in ecdysone titer in insects are regulated by TFs via manipulating the synthesis and degradation of ecdysone. Accordingly, the 20E regulatory cascades have been proposed [9]. In general, 20E signaling is transduced by NRs. Firstly, 20E binds to the EcR/Ultraspiracle (USP) complex, and then the 20E/EcR/USP complex directly induces the early 20E-response genes including *E74*, *E75*, and *Broad-Complex* (*Br-C*). Products of these early genes activate the later 20E-response genes, which encode TFs to regulate the spatiotemporal expression of downstream targets. Furthermore, the expression of some of the 20E-response genes is also controlled by both 20E/EcR/USP and early responsive products.

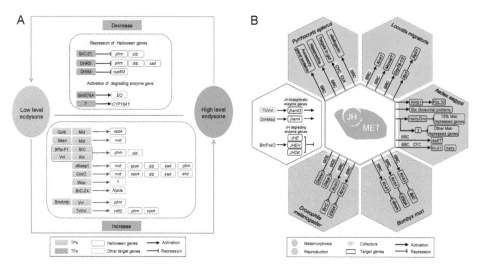

Figure 4. The 20-hydroxyecdysone (20E) and juvenile hormone (JH) signaling pathways regulating insect molting, metamorphosis, and reproduction. (**A**) Regulation of ecdysone titer by TFs. Changes in the ecdysone titers of insects are synergistically controlled by the synthesis and degradation of ecdysone. A low level of ecdysone promotes TFs to specifically regulate the Halloween genes in PG, thereby increasing steroidogenesis. A high level of ecdysone drives TFs to repress the expression of Halloween genes and activate the expression of degradative enzyme genes to directly degrade ecdysone, thus decreasing ecdysteroid titers. (**B**) Regulation of insect reproduction and metamorphosis by JH signaling pathway-related TFs. The white hexagon on the left shows TFs regulating the synthesis and degradation of JHs; the central hexagon represents JH binding to the JH receptor Met to regulate the transcription of downstream target genes; the orange hexagons display the JH/Methoprene-tolerant (Met) complex regulates downstream targets to control insect reproduction; the green hexagons indicate metamorphosis; and the purple gradient ellipses denote the cofactors of Met.

JHs are synthesized and secreted by the corpora allata (CA) in insects. The prominent role of JHs is to prevent the premature transition of immature larvae to pupae and adults [131]. There have been several studies on JH regulation by TFs, as summarized above (Figure 4B). In *Drosophila* CA cells, phosphorylated Mad shuttles into the nucleus, together with co-Smad, and triggers the expression of the JH biosynthetic enzyme, *jhamt* [134]. TcVvl is upstream of JH signaling, and it is important for the normal expression of the JH synthetic gene *jhamt3* [135]. In addition, BmFOXO regulates JH degradation by regulating the expression of JH-degrading enzyme genes *JHE*, *JHEH*, and *JHDK* [136]. In insects, JH signaling is primarily transduced by the JH receptor Met, which is a member of the bHLH-PAS family and was originally identified in *Drosophila* mutants [137]. Met typically forms dimers to directly regulate target gene transcription. SRC (also known as "FISC" or "Taiman", hereinafter referred to as SRC) is the most common coactivator of Met in multiple insects, including *Drosophila*, *A. aegypti*, *B. mori*, and *Tribolium castaneum* [138–143]. In *Drosophila*, Met also forms homodimers or forms heterodimers with another bHLH-PAS family member, Gce, to perform functions [144,145]. The JH/Met-SRC/Krüppel homolog 1 (Kr-h1) cascade is conserved in both holometabolous and hemimetabolous insects, and it mediates JH-repressed metamorphosis. The *Kr-h1* gene is a direct target of Met, and it encodes a C_2H_2 ZF protein that plays a central role in the JH-mediated inhibition of metamorphosis [142,143,146–148]. Kr-h1 inhibits the expression of the pupal specifier *Br-C* to prevent premature metamorphosis from larva to pupa in the larval stage [134,149]. The transient peak of *Kr-h1* at the end stage of the last-instar larvae upregulates the expression of *Br-C* to allow for the correct formation of the pupae, and inhibits the premature upregulation of *E93* to prevent larvae from bypassing the pupal stage and directly developing into adults [150,151].

3.1.14. Reproduction Manipulation

One of the main reasons of pest outbreaks is the high fecundity of insects based on oogenesis, which can be divided into three developmental stages: previtellogenesis, vitellogenesis, and choriogenesis. Insect oogenesis is a complicated biological process that is coordinatively controlled by various signaling pathways, especially the 20E and JH signaling pathways, as well as multifarious TFs.

JH-regulated insect reproduction is mediated through the JH-receptor complex Met–SRC. In the previtellogenesis of *A. aegypti*, Met–SRC regulates the expression of downstream genes, preparing for subsequent vitellogenesis and egg development [152]. Met is capable of directly activating target transcription, while the suppressing action of Met on targets is indirect and requires other mediators such as Hairy [153]. Studies have shown that Hairy and its corepressor Gro in female *A. aegypti* mediate the repression of 15% of Met-repressed genes [154]. Further studies are required to reveal other mechanisms mediated Met action in gene repression [155]. In the migratory locust, the Met–SRC complex directly regulates the expression of *Mcm4*, *Mcm7*, and *Cdc6* to promote DNA replication and polyploidy for vitellogenesis and oocyte maturation [156,157]. The complex also induces the expression of Grp78-2, which is required for insect fat body cell homeostasis and vitellogenesis [158]. In addition, Met-SRC directly activates the transcription of *Kr-h1* to promote vitellogenesis and oocyte maturation [159]. In *Cimex lectularius*, Met–SRC also regulates vitellogenesis and ovigenesis by the indirect regulation of *Vg* synthesis, but TFs downstream of Met that regulate the expression of *vitellogenin* (*Vg*) still remain mysterious [160]. In female *P. apterus*, the Met–SRC complex is required for JH-induced *Vg* expression during vitellogenesis [148]. Regulation of the reproductive status of the *P. apterus* gut requires Met, as well as its cofactors CLK and CYC, to activate the expression of $Pdp1_{iso1}$, which in turn upregulates the reproduction-associated genes *lipase* (*lip*), *esterase* (*est*) and *defensin* (*def*), and suppresses *Cryptochrome 2* (*Cry2*) and the diapause-related downstream genes *superoxide dismutase* (*sod*) and *transferrin* (*tf*) [110]. Met–SRC is also involved in the regulation of male reproduction by controlling the accessory gland proteins and hexamerins in fat bodies of male *P. apterus* [161].

Ecdysteroid-dependent regulation of insect oogenesis is induced by a series of NRs and 20E response genes [152]. In *A. aegypti*, 20E regulates the vitellogenesis of female mosquitoes by regulating the transcription of the *Vg* gene. In addition to regulation of midoogenesis, oocyte maturation and oviposition, ecdysone regulates the very early steps of oogenesis, including niche formation, germline stem cell (GSC) behavior, and cyst cell differentiation in *Drosophila* [152,162].

Most studies on the regulation of chorion gene expression by TFs mainly focus on the insect model *B. mori* during the choriogenesis period [2]. The CCAAT-enhancer-binding protein (C/EBP) is a major regulator of early/early-middle chorion gene expression in *B. mori* [95]. The relative concentration of C/EBP is correlated with its differential binding affinity to the response elements, leading to the activation or repression of targets [163]. Another two C/EBP-like proteins, the chorion bZIP factor (CbZ) and C/EBPγ, can form the CbZ-C/EBPγ heterodimer to repress chorion gene expression by antagonizing the binding of C/EBP homodimers to the promoter [164]. The expression of the late chorionic gene in silkworm is generally regulated by *Bombyx* Chorion Factors I (BCFI) and GATAβ. The Forkhead box transcription factor L2 (NlFoxL2) in *Nilaparvata lugens* directly activates *follicle cell protein 3C* (*NlFcp3C*) to regulate chorion formation [165].

3.2. External Responses

Insect TFs also play an important role in the external response, which can improve the tolerance of insects to adverse environments and protect them from diverse external stress.

3.2.1. Biotic Factor Responses

Insect responses to biotic factors primarily includes immune responses to pathogens and viruses. TFs play an essential role in humoral immunity, especially in the regulation of the production of antimicrobial peptides (AMPs).

In insects, the regulation of inducible AMP genes relies mainly on NF-κB factors that are activated by the intracellular Toll or Imd signaling pathway when infected by pathogens and parasites. *Drosophila* is used as a model to investigate the innate immune mechanism. Three NF-κB proteins have been identified in *Drosophila*: Dl, Dorsal-related immunity factor (Dif), and Relish. These proteins contain the N-terminal Rel-homology domain (RHD) that is used for DNA binding and dimerization [166]. However, only Relish contains an inhibitor of inhibitor κB (IκB) domain [167]. These NF-κB proteins are activated by two distinct pathways: the Toll and Imd signaling pathways. When *Drosophila* are infected with Gram-positive bacteria, fungi, and viruses, the Toll pathway is responsible for the activation of the NF-κB proteins Dl and Dif. It first causes the dissociation of Dif–Cactus and Dorsal–Cactus complexes in the cytoplasm, and then Dif and Dl translocate to the nucleus and activate the expression of specific AMP genes such as *Drosomycin* (*Drs*) [168–170]. The immune deficiency (Imd) pathway is activated upon infection by Gram-negative bacteria, leading to the endoproteolytic cleavage of the Relish protein in the cytoplasm. Subsequently, its N-terminal fragment containing RHD translocates to the nucleus and activates expression of AMP genes [171].

NF-κB factors can form homodimers or heterodimers to regulate AMPs expression. The *Drosophila* Dif-Relish heterodimer linked by a flexible peptide linker can activate *Diptericin* (*Dipt*) and *CecA1* [172]. However, it is still unclear whether Dif–Relish or Dl–Relish heterodimers actually form *in vivo* [172,173]. In addition, several cofactors interacting with NF-κB proteins have also been identified in *Drosophila*, including the coactivators Dip4/Ubc9 [174] and Dip3 [174,175], the three POU proteins Pdm1, Pdm2, and Dfr/Vv1 [176], as well as corepressors such as Cautus and Gro [46,48,168].

The homologs of *Drosophila Dl* and *Relish* have been found in other insects, which also regulate the expression of inducible AMP genes in responses to pathogens and parasites (Table S1). *A. gambiae* and *A. aegypti* have two NF-κB genes: *REL1* and *REL2*, which are the homologues of *DmDl* and *DmRelish*, respectively [177–182]. *AgREL2* gene encodes two *REL2* isoforms REL2-Full and REL2-Short, through alternative splicing of the *REL2* gene [179]. The IMD/REL2-Full cascade defends against Gram-positive *Staphylococcus aureus*, and regulates the intensity of mosquito infection with the malaria parasite *Plasmodium berghei*, whereas REL2-Short is resistant against the Gram-negative *Escherichia coli* [179]. The *AaREL1* gene encodes two isoforms, AaREL1-A and AaREL1-B, which are the key activators of the Toll-mediated antifungal immune pathway to activate the expression of *Dipt* and *Drs*, and elevate defense against the fungus *Beauveria bassiana* [180]. The *AaREL2* gene encodes three isoforms: REL2-Full, containing the RHD and the IkB-like domain, REL2-Short, comprising RHD, and REL-IkB, with only the IkB-like domain [183]. All three *Relish* transcripts are activated when *A. aegypti* is infected by bacteria [183]. REL2 prevents *Aedes* against infection by Gram-positive and Gram-negative bacteria and *Plasmodium gallinaceum* [184,185]. In *Culex quinquefasciatus*, the *DmRelish* homolog *Rel2* is activated by a TEAE-dependent pathway after WNV infection, and binds to the NF-κB site of the upstream promoter of the *Vago* gene to induce *Vago* expression, thereby triggering antiviral responses [186]. In *B. mori*, the *BmRel* gene, a homolog of *Dl*, encodes two isoforms: BmRelA (long) and BmRelB (short). These two isoforms act differentially to activate antibacterial peptide genes: BmRelB strongly activates the *Attacin* (*Att*) gene, while *BmRelA* strongly activates the *lebocin 4* (*Leb4*) gene and weakly activates the *Att* and *lebocin3* (*Leb3*) genes [187]. *BmRelish* (gene homologous to *Drosophila Relish*) also encodes two proteins: BmRelish1 and BmRe0lish2. BmRelish1 can activate the expression of *CecB1*, *Att* and *Leb4*. BmRelish2 is a dominant negative factor of the *BmRelish1* active form, and it inhibits the *CecB1* gene activated by BmRelish1 [188]. There are also two NF-κB genes in *M. sexta*: one is *MsDorsal* (the homologous gene of *DmDl*), and the other is *MsRel2* (the homologous gene of *DmRelish*), which produces two isoforms, MsRel2A and MsRel2B [173]. These three NF-κB factors can form homodimers and activate promoters of different AMP genes. Moreover, *MsDorsal* and *MsRel2* form heterodimers to repress the activation of AMP gene promoters and prevent their overactivation [173].

In addition to NF-κB proteins, other TFs are also involved in the regulation of inducible AMP expression (Table S1). However, the relationship between NF-κB proteins and Toll and Imd pathways remains to be clarified in other insects beyond *Drosophila*.

Constitutive AMP genes are expressed continuously in an NF-κB-independent manner in defined tissues, to function as a first line of defense against microbial infection during development and reproduction. For example, the *Drosophila* homeodomain protein Cad regulates the continuous expression of *Cec* and *Drs* in epithelia in a NF-κB-independent manner [189]. Cad also regulates commensal–gut mutualism by inhibiting NF-κB-dependent AMPs [190]. The POU protein Vvl synergizes with other proteins to regulate constitutive AMP gene expression in a range of immunocompetent tissues, including the male ejaculatory duct [191]. In *M. sexta*, Fkh activates a series of AMPs under non-infectious conditions to protect them from microbial infections during insect molting and metamorphosis [192]. This activation is possibly essential for the defense against microbial infection during insect molting and metamorphosis [192]. During *B. mori* metamorphosis, 20E activates Br-Z4 and Ets to regulate *Leb* expression in the midgut to protect the midgut from infection [193].

Insect immune response is highly homologous to mammals, and insects are relatively simple and easy to manipulate, compared with mammals. Therefore, the study of insect TF regulating immune response can not only enable us to understand the entire immune system in insects, but also inspire our understanding and exploration of the human immune regulatory mechanism. Thus far, the studies on AMP gene expression regulated by insect TFs have focused on holometabolous insects such as fruit fly, mosquito, silkworm, etc., while there are few studies on hemimetabolous insects. It is possible that the TF immunoregulatory mechanisms in hemimetamorphosis insects is different from holometabolous insects, because functional annotation of immune and defense-related genes in the aphid genome revealed that some of the AMPs commonly found in metamorphosis insects are not expressed in aphids [194]. Therefore, studies of TFs regulating immune responses in insects can be concentrated in hemimetamorphosis insects in the future.

3.2.2. Abiotic Factor Responses

Insect abiotic factor responses are mostly comprised of the resistance to xenobiotics (including chemical pesticides, biological pesticides, and secondary metabolites), and the response to high temperature and oxygen stress. Numerous studies have shown that diverse TFs are involved in xenobiotic resistance in insects (Table S1).

CncC and aryl hydrocarbon receptor (AhR) can regulate insect tolerance to plant secondary toxicants. CncC participates in the *Leptinotarsa decemlineata* adaptation to potato plant allelochemicals and *Aphis gossypii* tolerance to gossypol [195,196]. AhR heterodimerizes with aryl hydrocarbon receptor nuclear translocator (ARNT) to directly activate the expression of P450s after exposure to plant secondary metabolites [197,198]. In addition, two other potential TFs (FK506 binding protein (FKBP) and Prey2) were reported to regulate the expression of *CYP6B6* in *H. armigera* under 2-tridecanone stress [199].

Insect resistance to chemical insecticides is regulated by TFs such as CncC and DHR96 [200,201]. CncC extensively participates in insect resistance to different insecticides other than plant secondary toxicant [200]. Although CncC has a short half-life, its constitutive activation in some insects can confer a resistance phenotype [200]. Recently, a genome-wide analysis of TFs in *Plutella xylostella* found that the altered expression of multiple TFs may be involved in *P. xylostella* insecticide resistance, but their precise functions remain to be further validated [202]. Moreover, it has been reported that some unidentified TFs downstream of the G protein-coupled receptor (GPCR) signaling can be involved in regulation of P450-mediated permethrin resistance in *C. quinquefasciatus* [203,204].

Insect resistance to biopesticides such as *Bacillus thuringiensis* (Bt) can also be modulated by some TFs. Altered expression of midgut functional genes can lead to Bt resistance in many insects, but the potential TF-mediated regulation mechanisms of their expression alteration still remain to be unveiled [205]. For example, our previous studies have shown that high-level resistance to Bt Cry1Ac toxin in *P. xylostella* is associated with differential expression of a suite of midgut functional genes, including *ALP*, *ABCC1*, *ABCC2*, *ABCC3*, and *ABCG1*, which are *trans*-regulated by the MAPK signaling pathway [206], and we can speculate that the novel MAPK-mediated *trans*-regulatory mechanism may be further controlled by diverse downstream TFs such as FOXA [207]. Further study deserves to be conducted in order to characterize these downstream TFs, to comprehensively understand Bt resistance mechanisms in different insects.

Under heat shock or other stresses, HSF relocalizes within the nucleus to form a trimer to activate heat shock (HS) gene expression by binding to HS elements [208,209]. In addition to *HS*, some genes under non-stress conditions might be the targets of HSF, since that HSF is also required for oogenesis and early larval development under normal growth conditions [210].

The HIF family member Similar (Sima) is required for the hypoxia response and normal development in *Drosophila*. Under hypoxic conditions, Sima is upregulated and heterodimerizes with Tgo to activate the expression of related genes [28]. In addition, Sima activates Notch signaling to facilitate the survival of *Drosophila* blood cells under both normal hematopoiesis and hypoxic stress [211].

Altogether, countering the selection pressures of non-biological factors offers an evolutionary force for insects to adapt to the surrounding environment, and TFs play a pivotal role in this adaptive evolution process. Unfortunately, little is known about the TFs that are response to resistance-related signaling cascades and that regulate the expression of xenobiotic-resistant genes. Thus, this area will become a research hotspot, and it will facilitate insect resistance management in the near future.

4. Research Methods

4.1. TF-Binding Site (TFBS) Prediction

TFs regulate the transcription of target genes by specifically binding to their TFBS located in the regulatory region. Therefore, TFBS prediction in the target promoter is a critical step to studying transcriptional regulation. A number of databases specifically collecting TFBS-related information have been established, with the advancement of experimental techniques for TFBS identification, and multiple online software and websites have been developed with the rapid development of bioinformatics, which allows researchers to predict TFBS in target promoters *in silico*, which lays a critical foundation for further transcriptional regulation studies (Table 1).

Table 1. *In silico* TF-binding site (TFBS) prediction by utilizing different TF databases and TFBS searching tools.

Names	Organisms	Websites	Descriptions	References
TRANSFAC	Eukaryotes	http://gene-regulation.com/	Partially commercial. License required to access some restricted areas.	[212]
JASPAR	Eukaryotes	http://jaspar.genereg.net/	Contains a curated, non-redundant set of profiles, derived from published collections of experimentally defined eukaryotic TFBS.	[213]
DBD	Cellular organisms	http://www.transcriptionfactor.org/	Contains TF predictions of more than 1000 cellular organisms.	[212]
UniPROBE	Cellular organisms	http://the_brain.bwh.harvard.edu/uniprobe/	Contains DNA binding data for 638 non-redundant proteins and complexes from a diverse collection of organisms.	[214]
PlantTFDB	Plants	http://planttfdb.cbi.pku.edu.cn/	Contains 320,370 TFs from 165 plant species; enables regulation prediction and functional enrichment analyses.	[215]
LASAGNA-Search	Organisms	http://biogrid-lasagna.engr.uconn.edu/lasagna_search/	An integrated web tool for TFBS search and visualization.	[216]
PROMO	Organisms	http://alggen.lsi.upc.es/cgi-bin/promo_v3/promo/promoinit.cgi?dirDB=TF_8.3	A virtual laboratory for the identification of putative TFBS in DNA sequences from a species or groups of species of interest.	[217]
MatInspector	Organisms	http://www.genomatix.de/matinspector.html	A software tool that utilizes a large library of matrix descriptions for TFBS to locate matches in DNA sequences.	[218]
INSECT 2.0	Insects	http://bioinformatics.ibioba-mpsp-conicet.gov.ar/INSECT2/	A web-server for genome-wide *cis*-regulatory module prediction.	[219]

4.2. DNA–Protein Interaction Detection

TFs act mainly through binding directly to sequence-specific DNA motifs in the promoters or enhancers of target genes. Therefore, identifying the interaction between TFs and DNA is particularly crucial for TF functional studies. In this section, we comprehensively elaborate the basic principles, merits, faults, and applications of several current techniques that are extensively applied in investigating DNA–protein interactions, which will provide theoretical and technical guidelines for researchers to study TF functions.

4.2.1. Dual-Luciferase Reporter Assay System

The luciferase reporter gene assay is extensively used for detecting the interaction between TFs and DNA motifs, and it is characterized by a high level of sensitivity, good specificity, short detection time, and a wide linear range. Researchers often use one luciferase (such as firefly luciferase) to monitor gene expression, and another type of luciferase (such as *Renilla* luciferase) as an internal control to construct the dual-luciferase reporter assay system that reduces external interference and improves system detection sensitivity and reliability [220]. Since the introduction of the dual-luciferase assay system in the mid-1990s, this reporter assay has been widely applied in the study of TF–DNA interaction [221–223]. For example, the functional interaction between Kr-h1 and Kr-h1 binding site (KBS) in the E93 promoter was examined by the dual-luciferase assay system in *B. mori* [131].

4.2.2. Electrophoretic Mobility Shift Assay (EMSA)

EMSA is a classical technique for the rapid and sensitive investigation of protein–DNA interactions *in vitro*, it has the advantages of simple operation and high detection sensitivity. The currently widely used assays are based on methods originally described by Garner and Revzin [224] and Fried and Crothers [225,226]. The labeled nucleic acid probes bind protein to form nucleic acid–protein complexes that migrate more slowly in gel electrophoresis than do the corresponding free nucleic acids, whereupon the nucleic acid–protein complexes are separated out. Additionally, it can also use competitive experiments and supershift assays to evaluate the properties of protein–DNA binding. For instance, EMSA was succeeded to confirm the specific binding of BmE74A to the E74A binding site in the *ecdysone oxidase* (*EO*) promoter in *B. mori* [227]. However, this method *in vitro* does not truly reflect the interaction between proteins and nucleic acids in organisms [228]. Moreover, this method requires nucleic acids be labeled with radioisotopes, fluorophores, or biotin, which takes a long time and has a high cost. In recent years, emerging microfluidic-based EMSAs were not just limited to the investigation of protein–nucleic acid interactions; these assays include high-throughput and multiplexed analyses that could be applied for molecular conformational analysis, immunoassays, affinity analysis and genomics study [229].

4.2.3. Yeast One-Hybrid System (Y1H)

Y1H assay is an effective method for studying protein–DNA interaction in yeast cells. The Y1H assay consists essentially of two components: a reporter gene that contains the known specific DNA sequence, and a construct that contains the complementary DNA (cDNA) encoding the test TF that is fused to the activation domain (AD) of the yeast Gal4 (Gal4-AD), both of which are transferred to a suitable yeast strain [230]. If the TF can bind to the DNA sequence, Gal4-AD will activate reporter gene expression [231]. Of course, the Y1H assay also has some disadvantages. It usually takes a long time and has difficultly detecting interacting dimers or proteins that require posttranslational modifications to bind DNA [230]. In addition, it may lead to some uncharacterized TFs binding to the target DNA, owing to the incompleteness of the TF library in a species.

4.2.4. Chromatin Immunoprecipitation (ChIP)

ChIP is based on the principle of antigen–antibody binding, and it can reflect the interactions of proteins and DNA that occur in living cells. ChIP was originally used for the study of histone covalent modification, and was later widely used in the study of TF–DNA interaction. ChIP generally first fixes the protein–DNA complexes that occur in the cell with formaldehyde. The cells are then lysed, and the chromatin is randomly cleaved into small segments of a certain length. The protein–DNA complexes are then selectively immunoprecipitated using specific protein antibodies against the target protein. DNA fragments that bind to the target protein are then specifically enriched, purified, and identified [232,233].

There are three predominant methods for the identification of immunoprecipitated DNA: ChIP-qPCR, ChIP-chip and ChIP-seq. ChIP-qPCR is the earliest method to identify the specific binding of proteins to DNA and it is suitable for identifying the known sequence of precipitated DNA fragments and quantifying the binding of TFs to specific target DNA [233]. ChIP-chip has become a common method for studying protein–DNA interactions since Ren et al. applied the ChIP-chip method for the first time to identify the genome-wide binding sites of the transcriptional activators Gal4 and Stel2 in yeast [234]. The ChIP-seq method was first reported in 2007, which combines ChIP with massively parallel DNA sequencing, and can efficiently detect genome-wide DNA fragments that interact with TFs or histones [235]. ChIP-chip and ChIP-seq do not require knowledge of the target DNA sequences in advance, and it can identify whole-genome targets and quantify binding levels [233]. ChIP-seq can quickly decode a large number of DNA fragments at a higher efficiency and at a relatively low cost, compared with ChIP-chip [232]. In addition, the data provided by ChIP-seq are of higher resolution, and the information obtained is more accurate and quantitative than that from ChIP-chip [232]. Thus, ChIP-seq is currently one of the most frequently used methods for studying protein–DNA interactions. Recently, the genome-wide ChIP-seq analysis in *B. mori* has identified a consensus KBS in the *E93* promoter [131], which provides a paradigm to use this technique in insects.

ChIP has a broad application prospect and can capture the interaction between TFs and binding sites *in vivo*, and identify the distinct regulatory mechanisms of differentially expressed genes [232]. However, this approach has its limitations as well: it requires highly specific antibodies. Acquisition of highly abundant binding fragments requires a high level of simulation of an intracellular environment that is required for target protein expression. In addition, it is difficult to simultaneously obtain information on the binding of multiple proteins to the same sequence [236].

4.2.5. CRISPR Affinity Purification In Situ of Regulatory Elements (CAPTURE)

Recently, researchers have developed a new approach, CAPTURE, to isolate chromatin interactions in situ by using the targeting ability of the CRISPR/Cas9 system and high affinity between biotin and streptavidin [237]. CAPTURE includes three key components: an N-terminal FLAG and a biotin-acceptor site (FB)-tagged deactivated Cas9 (dCas9), a biotin ligase (BirA), and a single guide RNA (sgRNA) that serves to direct biotinylated dCas9 to the target genomic sequence. Upon *in vivo* biotinylation of dCas9 by the biotin ligase BirA together with sequence-specific sgRNAs in mammalian cells, the genomic locus-associated macromolecules are isolated by high-affinity streptavidin purification. The purified protein, RNA, and DNA complexes are then identified and analyzed by mass spectrometry (MS)-based proteomics and high-throughput sequencing for the study of native CRE-regulating proteins, RNA, and long-range DNA interactions, respectively. This approach is more specific and sensitive than ChIP, and it does not require protein antibodies and the known TFs. Considering these advantages of CAPTURE, we believe that this method will also be applied for *in vivo* TF–DNA interaction detection in insects in the near future.

4.3. TF Function Verification

4.3.1. CRISPR/Cas9 system

The novel CRISPR/Cas9 technology has been widely used to modify genome sequences in multiple species recently [238]. At present, researchers have begun to apply the CRISPR/Cas9-mediated genome editing system to investigate the regulatory function of TFs in organisms, including insects, plants, and crustaceans [136,239]. The study of TF regulatory function by CRISPR/Cas9 can be divided into two categories. One is to mutagenize the exon of the TF locus through CRISPR/Cas9, and to generate mutants to study TF functions in insects, including *Drosophila*, *B. mori*, and *P. xylostella* [136,239,240], and the other is to knockout the TF-binding site on the promoter of target gene, and then to observe the transcription level of the target gene to study the function of the TF in the crustacean *Daphnia magna* [241].

4.3.2. Yeast Two-Hybrid Assay (Y2H)

TFs often function as homodimers or heterodimers. Hence, understanding the mechanisms of protein–protein interactions is essential for determining the actions of TFs. The Y2H assay has been widely used to identify protein–protein interactions since its appearance in 1989 [242]. As for insects, for example, Bric-a-brac interacting protein 2 (BIP2) has been confirmed to be an ANTP-interacting protein by using this assay in adult *Drosophila* [20]. In the Y2H assay, two proteins are fused into the DBD and AD of Yeast Gal4, respectively. If these two proteins interact with each other, an active Gal4 TF would be generated and induce the transcription of *lacZ* reporter gene in yeast cells. The initial Y2H had some limitations, such as not reflecting complex spatial or temporal interactions *in vivo*. The continuous improvement of Y2H technique has not only overcome the major limitations of the original Y2H system, but also has expanded its application areas. In particular, the development of high-throughput Y2H allows it to be applied for the investigation of complex protein interactions [243–245]. Furthermore, Y2H has also been used to study other types of molecular interactions and to identify domains that stabilize protein–protein interactions [246,247].

4.3.3. Expression Read Depth GWAS (eRD-GWAS)

Many phenotypic changes in organisms are caused by changes in the expression patterns of various regulatory genes, such as genes encoding TFs. eRD-GWAS is a genome-wide association studies based on Bayesian analysis using gene expression level data tested by RNA-Seq as an explanatory variable. This method can identify true relationships between gene expression variation and phenotypic diversity at the genomic level, and it is an effective complement to SNP-based GWAS. Lin et al. applied this method in maize, and revealed that genetic variation in TF expression contributes substantially to phenotypic diversity [248]. Apparently, we can anticipate that this novel and promising method will also be adopted to validate the TF functions in insects.

5. Discussion and Prospects

Evidently, TFs play a central role in the insect genetic regulatory network, as in other organisms. In this review, we integrate vast amounts of TF information in both model and non-model insects, and summarize their vital functions in response to internal signaling and external stimuli.

Probing TFs in the model insect *D. melanogaster* has yielded fruitful results, which provide important insights into the study of TFs in other organisms. In recent years, TF studies in other insects have also achieved great success in the development of genetic tools and next-generation genome sequencing techniques. However, there are still large numbers of unidentified TFs and uncharacterized TF functions in insects. Moreover, understanding the precise regulatory mechanism of TFs still remains a great challenge [249]. Although TF–TF interactions and TF–DNA interactions are prevalent in organisms, both of these interactions are largely undetectable because they depend not only on the opening degree of chromatin, but also on whether the interaction is instantaneous or long-term, and how strong the interactions are. More research is required to understand how TFs interact with specific DNA sites to regulate the spatiotemporal expression of target genes, and how TFs interplay to achieve regulatory functions.

To date, insect resistance to Bt biopesticides and chemical pesticides has seriously threatened pest control in the field. The novel RNAi and CRISPR/Cas9 technologies are promising for insect pest control and resistance management in the near future. However, the RNAi and CRISPR/Cas9-based insect control strategies depend mainly on the selection of safe and efficient target genes, and many insect TFs are suitable candidate targets for lethal genes. For instance, mutations in gap genes such as *kni* can cause serious defects in embryos and impede their normal growth and development during *Drosophila* embryogenesis [17]. In *P. xylostella*, *abd-A* mutagenesis induced by the CRISPR/Cas9 system generated a heritable *abd-A* mutant phenotypes, resulting in severe abdominal morphological defects and significant lethality in the offspring [239]. Moreover, some important TFs, such as CncC and FoxA, have been found to be implicated in insect resistance [200,207], and these TFs can be used as insect lethal targets for pest management. Hence, the identification of these insect TFs will be conducive to developing both new species-specific biopesticides and next-generation transgenic crops combining Bt- and RNAi- or CRISPR/Cas9-based insect control technologies as a pivotal part of integrated pest management (IPM) programs [250].

With the development of high-throughput -omics techniques, the genome-wide identification of insect TFs is becoming easier, and subsequent TF studies will be performed at the genome level. Identifying a more comprehensive TF library in organisms is a major trend in the future. Considering the importance of versatile TFs in the transcriptional regulation of diverse insect physiological processes, undoubtedly, a growing body of entomologists will focus on studying insect TFs in the near future, and the vast range of genome resources and novel genetic methods will greatly propel progress of this area. Collectively, in-depth studies of insect TFs in the future will most likely provide new insights into the intracellular transcriptional regulation network of insects and even humans, which will have important potential for pest control in the field, and protection of human life and health.

Supplementary Materials: Supplementary Materials can be found at http://www.mdpi.com/1422-0067/19/11/3691/s1.

Author Contributions: Conceptualization, Z.G. and Y.Z.; writing—original draft preparation, Z.G., J.Q., and X.Z.; writing—review and editing, Z.G., J.Q., and Y.Z.; supervision, Y.Z.; funding acquisition, Z.G. and Y.Z.

Funding: This research was supported by the National Natural Science Foundation of China (31630059; 31701813), the Beijing Key Laboratory for Pest Control and Sustainable Cultivation of Vegetables, and the Science and Technology Innovation Program of the Chinese Academy of Agricultural Sciences (CAAS-ASTIP-IVFCAAS).

Conflicts of Interest: The authors declare no conflict of interest.

References

1. Harshman, L.G.; James, A.A. Differential gene expression in insects: Transcriptional control. *Annu. Rev. Entomol.* **1998**, *43*, 671–700. [CrossRef] [PubMed]
2. Papantonis, A.; Swevers, L.; Iatrou, K. Chorion genes: A landscape of their evolution, structure, and regulation. *Annu. Rev. Entomol.* **2015**, *60*, 177–194. [CrossRef] [PubMed]
3. Riechmann, J.L.; Ratcliffe, O.J. A genomic perspective on plant transcription factors. *Curr. Opin. Plant Biol.* **2000**, *3*, 423–434. [CrossRef]
4. Riechmann, J.L.; Heard, J.; Martin, G.; Reuber, L.; Jiang, C.; Keddie, J.; Adam, L.; Pineda, O.; Ratcliffe, O.J.; Samaha, R.R.; et al. *Arabidopsis* transcription factors: Genome-wide comparative analysis among eukaryotes. *Science* **2000**, *290*, 2105–2110. [CrossRef] [PubMed]
5. Wingender, E.; Schoeps, T.; Haubrock, M.; Krull, M.; Dönitz, J. TFClass: Expanding the classification of human transcription factors to their mammalian orthologs. *Nucleic Acids Res.* **2018**, *46*, D343–D347. [CrossRef] [PubMed]
6. Spitz, F.; Furlong, E.E. Transcription factors: From enhancer binding to developmental control. *Nat. Rev. Genet.* **2012**, *13*, 613–626. [CrossRef] [PubMed]

7. Reiter, F.; Wienerroither, S.; Stark, A. Combinatorial function of transcription factors and cofactors. *Curr. Opin. Genet. Dev.* **2017**, *43*, 73–81. [CrossRef] [PubMed]

8. Weintraub, A.S.; Li, C.H.; Zamudio, A.V.; Sigova, A.A.; Hannett, N.M.; Day, D.S.; Abraham, B.J.; Cohen, M.A.; Nabet, B.; Buckley, D.L.; et al. YY1 is a structural regulator of enhancer-promoter loops. *Cell* **2017**, *171*, 1573–1588. [CrossRef] [PubMed]

9. King-Jones, K.; Thummel, C.S. Nuclear receptors–a perspective from *Drosophila*. *Nat. Rev. Genet.* **2005**, *6*, 311–323. [CrossRef] [PubMed]

10. Valanne, S.; Wang, J.H.; Rämet, M. The *Drosophila* Toll signaling pathway. *J. Immunol.* **2011**, *186*, 649–656. [CrossRef] [PubMed]

11. Niwa, Y.S.; Niwa, R. Transcriptional regulation of insect steroid hormone biosynthesis and its role in controlling timing of molting and metamorphosis. *Dev. Growth Differ.* **2016**, *58*, 94–105. [CrossRef] [PubMed]

12. Mussabekova, A.; Daeffler, L.; Imler, J.L. Innate and intrinsic antiviral immunity in *Drosophila*. *Cell Mol. Life Sci.* **2017**, *74*, 2039–2054. [CrossRef] [PubMed]

13. Wingender, E.; Schoeps, T.; Dönitz, J. TFClass: An expandable hierarchical classification of human transcription factors. *Nucleic Acids Res.* **2013**, *41*, D165–D170. [CrossRef] [PubMed]

14. Wingender, E.; Schoeps, T.; Haubrock, M.; Dönitz, J. TFClass: A classification of human transcription factors and their rodent orthologs. *Nucleic Acids Res.* **2015**, *43*, D97–D102. [CrossRef] [PubMed]

15. Gehring, W.J.; Qian, Y.Q.; Billeter, M.; Furukubo-Tokunaga, K.; Schier, A.F.; Resendez-Perez, D.; Affolter, M.; Otting, G.; Wuthrich, K. Homeodomain-DNA recognition. *Cell* **1994**, *78*, 211–223. [CrossRef]

16. Gehring, W.J.; Affolter, M.; Bürglin, T. Homeodomain proteins. *Annu. Rev. Biochem.* **1994**, *63*, 487–526. [CrossRef] [PubMed]

17. Latchman, D.S. Families of DNA binding transcription factors. In *Eukaryotic Transcription Factors*; Latchman, D.S., Ed.; Elsevier: Boston, MA, USA, 2008; Volume 4, pp. 96–160.

18. Papadopoulos, D.K.; Skouloudaki, K.; Adachi, Y.; Samakovlis, C.; Gehring, W.J. Dimer formation via the homeodomain is required for function and specificity of Sex combs reduced in *Drosophila*. *Dev. Biol.* **2012**, *367*, 78–89. [CrossRef] [PubMed]

19. Chan, S.K.; Jaffe, L.; Capovilla, M.; Botas, J.; Mann, R.S. The DNA binding specificity of Ultrabithorax is modulated by cooperative interactions with extradenticle, another homeoprotein. *Cell* **1994**, *78*, 603–615. [CrossRef]

20. Prince, F.; Katsuyama, T.; Oshima, Y.; Plaza, S.; Resendez-Perez, D.; Berry, M.; Kurata, S.; Gehring, W.J. The YPWM motif links Antennapedia to the basal transcriptional machinery. *Development* **2008**, *135*, 1669–1679. [CrossRef] [PubMed]

21. Bürglin, T.R.; Affolter, M. Homeodomain proteins: An update. *Chromosoma* **2016**, *125*, 497–521. [CrossRef] [PubMed]

22. Hammonds, A.S.; Bristow, C.A.; Fisher, W.W.; Weiszmann, R.; Wu, S.; Hartenstein, V.; Kellis, M.; Yu, B.; Frise, E.; Celniker, S.E. Spatial expression of transcription factors in *Drosophila* embryonic organ development. *Genome Biol.* **2013**, *14*, R140. [CrossRef] [PubMed]

23. Ma, P.C.; Rould, M.A.; Weintraub, H.; Pabo, C.O. Crystal structure of MyoD bHLH domain-DNA complex: Perspectives on DNA recognition and implications for transcriptional activation. *Cell* **1994**, *77*, 451–459. [CrossRef]

24. Ledent, V.; Paquet, O.; Vervoort, M. Phylogenetic analysis of the human basic helix-loop-helix proteins. *Genome Biol.* **2002**, *3*, research0030.0031. [CrossRef] [PubMed]

25. Katzenberg, D.; Young, T.; Lin, L.; Finn, L.; Mignot, E. A human period gene (*HPER1*) polymorphism is not associated with diurnal preference in normal adults. *Psychiatr. Genet.* **1999**, *9*, 107–109. [CrossRef] [PubMed]

26. Ellis, H.M. Embryonic expression and function of the *Drosophila* helix-loop-helix gene, *extramacrochaetae*. *Mech. Dev.* **1994**, *47*, 65–72. [CrossRef]

27. Van Doren, M.; Ellis, H.M.; Posakony, J.W. The *Drosophila extramacrochaetae* protein antagonizes sequence-specific DNA binding by *daughterless/achaete-scute* protein complexes. *Development* **1991**, *113*, 245–255. [PubMed]

28. Lavista-Llanos, S.; Centanin, L.; Irisarri, M.; Russo, D.M.; Gleadle, J.M.; Bocca, S.N.; Muzzopappa, M.; Ratcliffe, P.J.; Wappner, P. Control of the hypoxic response in *Drosophila melanogaster* by the basic helix-loop-helix PAS protein similar. *Mol. Cell Biol.* **2002**, *22*, 6842–6853. [CrossRef] [PubMed]

29. Dawson, S.R.; Turner, D.L.; Weintraub, H.; Parkhurst, S.M. Specificity for the Hairy/Enhancer of split basic helix-loop-helix (bHLH) proteins maps outside the bHLH domain and suggests two separable modes of transcriptional repression. *Mol. Cell. Biol.* **1995**, *15*, 6923–6931. [CrossRef] [PubMed]

30. Aravind, L.; Koonin, E.V. Gleaning non-trivial structural, functional and evolutionary information about proteins by iterative database searches. *J. Mol. Biol.* **1999**, *287*, 1023–1040. [CrossRef] [PubMed]

31. Crozatier, M.; Valle, D.; Dubois, L.; Ibnsouda, S.; Vincent, A. *Collier*, a novel regulator of *Drosophila* head development, is expressed in a single mitotic domain. *Curr. Biol.* **1996**, *6*, 707–718. [CrossRef]

32. Landschulz, W.H.; Johnson, P.F.; McKnight, S.L. The leucine zipper: A hypothetical structure common to a new class of DNA binding proteins. *Science* **1988**, *240*, 1759–1764. [CrossRef] [PubMed]

33. Miller, J.; McLachlan, A.D.; Klug, A. Repetitive zinc-binding domains in the protein transcription factor IIIA from *Xenopus* oocytes. *EMBO J.* **1985**, *4*, 1609–1614. [CrossRef] [PubMed]

34. Lee, M.S.; Gippert, G.P.; Soman, K.V.; Case, D.A.; Wright, P.E. Three-dimensional solution structure of a single zinc finger DNA-binding domain. *Science* **1989**, *245*, 635–637. [CrossRef] [PubMed]

35. Klug, A. The discovery of zinc fingers and their development for practical applications in gene regulation and genome manipulation. *Proc. Japan Acad. Ser.* **2010**, *B*, 87–102. [CrossRef] [PubMed]

36. Pabo, C.O.; Peisach, E.; Grant, R.A. Design and selection of novel Cys_2His_2 zinc finger proteins. *Annu. Rev. Biochem.* **2001**, *70*, 313–340. [CrossRef] [PubMed]

37. Wolfe, S.A.; Nekludova, L.; Pabo, C.O. DNA recognition by Cys_2His_2 zinc finger proteins. *Annu. Rev. Biophys. Biomol. Struct.* **2000**, *29*, 183–212. [CrossRef] [PubMed]

38. Persikov, A.V.; Osada, R.; Singh, M. Predicting DNA recognition by Cys_2His_2 zinc finger proteins. *Bioinformatics* **2009**, *25*, 22–29. [CrossRef] [PubMed]

39. Nasiadka, A.; Dietrich, B.H.; Krause, H.M. Anterior-posterior patterning in the *Drosophila* embryo. In *Advances in Developmental Biology and Biochemistry*; DePamphilis, M.L., Ed.; Elsevier: Amsterdam, The Netherlands, 2002; Volume 12, pp. 155–204.

40. Gilbert, S.F. The genetics of axis specification in *Drosophila*. In *Developmental Biology (7th ed.)*; Gilbert, S.F., Ed.; Sinauer Associates: Sunderland, MA, USA, 2003; pp. 263–304.

41. Gilbert, S.F. The genetics of axis specification in *Drosophila*. In *Developmental Biology (9th ed.)*; Gilbert, S.F., Ed.; Sinauer Associates: Sunderland, MA, USA, 2010; pp. 203–239.

42. Hang, S.; Gergen, J.P. Different modes of enhancer-specific regulation by Runt and Even-skipped during *Drosophila* segmentation. *Mol. Biol. Cell* **2017**, *28*, 681–691. [CrossRef] [PubMed]

43. Jiang, J.; Levine, M. Binding affinities and cooperative interactions with bHLH activators delimit threshold responses to the dorsal gradient morphogen. *Cell* **1993**, *72*, 741–752. [CrossRef]

44. Shirokawa, J.M.; Courey, A.J. A direct contact between the Dorsal rel homology domain and Twist may mediate transcriptional synergy. *Mol. Cell. Biol.* **1997**, *17*, 3345–3355. [CrossRef] [PubMed]

45. Stathopoulos, A.; Van Drenth, M.; Erives, A.; Markstein, M.; Levine, M. Whole-genome analysis of dorsal-ventral patterning in the *Drosophila* embryo. *Cell* **2002**, *111*, 687–701. [CrossRef]

46. Ratnaparkhi, G.S.; Jia, S.; Courey, A.J. Uncoupling dorsal-mediated activation from dorsal-mediated repression in the *Drosophila* embryo. *Development* **2006**, *133*, 4409–4414. [CrossRef] [PubMed]

47. Zeitlinger, J.; Zinzen, R.P.; Stark, A.; Kellis, M.; Zhang, H.; Young, R.A.; Levine, M. Whole-genome ChIP-chip analysis of Dorsal, Twist, and Snail suggests integration of diverse patterning processes in the *Drosophila* embryo. *Genes Dev.* **2007**, *21*, 385–390. [CrossRef] [PubMed]

48. Dubnicoff, T.; Valentine, S.A.; Chen, G.; Shi, T.; Lengyel, J.A.; Paroush, Z.; Courey, A.J. Conversion of Dorsal from an activator to a repressor by the global corepressor Groucho. *Genes Dev.* **1997**, *11*, 2952–2957. [CrossRef] [PubMed]

49. Flores-Saaib, R.D.; Jia, S.; Courey, A.J. Activation and repression by the C-terminal domain of Dorsal. *Development* **2001**, *128*, 1869–1879. [PubMed]

50. Thor, S. Neuroscience: Stem cells in multiple time zones. *Nature* **2013**, *498*, 441–443. [CrossRef] [PubMed]

51. Ashraf, S.I.; Hu, X.; Roote, J.; Ip, Y.T. The mesoderm determinant Snail collaborates with related zinc-finger proteins to control *Drosophila* neurogenesis. *EMBO J.* **1999**, *18*, 6426–6438. [CrossRef] [PubMed]

52. Ashraf, S.I.; Ip, Y.T. The Snail protein family regulates neuroblast expression of *inscuteable* and *string*, genes involved in asymmetry and cell division in *Drosophila*. *Development* **2001**, *128*, 4757–4767. [PubMed]

53. Lai, S.L.; Miller, M.R.; Robinson, K.J.; Doe, C.Q. The Snail family member Worniu is continuously required in neuroblasts to prevent Elav-induced premature differentiation. *Dev. Cell* **2012**, *23*, 849–857. [CrossRef] [PubMed]

54. Li, X.; Erclik, T.; Bertet, C.; Chen, Z.; Voutev, R.; Venkatesh, S.; Morante, J.; Celik, A.; Desplan, C. Temporal patterning of *Drosophila* medulla neuroblasts controls neural fates. *Nature* **2013**, *498*, 456–462. [CrossRef] [PubMed]

55. Bello, B.C.; Izergina, N.; Caussinus, E.; Reichert, H. Amplification of neural stem cell proliferation by intermediate progenitor cells in *Drosophila* brain development. *Neural Dev.* **2008**, *3*, 5. [CrossRef] [PubMed]

56. Li, X.; Xie, Y.; Zhu, S. Notch maintains *Drosophila* type II neuroblasts by suppressing expression of the Fez transcription factor Earmuff. *Development* **2016**, *143*, 2511–2521. [CrossRef] [PubMed]

57. Zhu, S.; Barshow, S.; Wildonger, J.; Jan, L.Y.; Jan, Y.N. Ets transcription factor Pointed promotes the generation of intermediate neural progenitors in *Drosophila* larval brains. *Proc. Natl. Acad. Sci. USA* **2011**, *108*, 20615–20620. [CrossRef] [PubMed]

58. Xie, Y.; Li, X.; Deng, X.; Hou, Y.; O'Hara, K.; Urso, A.; Peng, Y.; Chen, L.; Zhu, S. The Ets protein Pointed prevents both premature differentiation and dedifferentiation of *Drosophila* intermediate neural progenitors. *Development* **2016**, *143*, 3109–3118. [CrossRef] [PubMed]

59. Janssens, D.H.; Hamm, D.C.; Anhezini, L.; Xiao, Q.; Siller, K.H.; Siegrist, S.E.; Harrison, M.M.; Lee, C.Y. An Hdac1/Rpd3-poised circuit balances continual self-renewal and rapid restriction of developmental potential during asymmetric stem cell division. *Dev. Cell* **2017**, *40*, 367–380. [CrossRef] [PubMed]

60. Koe, C.T.; Li, S.; Rossi, F.; Wong, J.J.; Wang, Y.; Zhang, Z.; Chen, K.; Aw, S.S.; Richardson, H.E.; Robson, P.; et al. The Brm-HDAC3-Erm repressor complex suppresses dedifferentiation in *Drosophila* type II neuroblast lineages. *eLife* **2014**, *3*, e01906. [CrossRef] [PubMed]

61. Pappu, K.S.; Mardon, G. Genetic control of retinal specification and determination in *Drosophila*. *Int. J. Dev. Biol.* **2004**, *48*, 913–924. [CrossRef] [PubMed]

62. Fichelson, P.; Brigui, A.; Pichaud, F. Orthodenticle and Kruppel homolog 1 regulate *Drosophila* photoreceptor maturation. *Proc. Natl. Acad. Sci. USA* **2012**, *109*, 7893–7898. [CrossRef] [PubMed]

63. Mishra, A.K.; Bargmann, B.O.R.; Tsachaki, M.; Fritsch, C.; Sprecher, S.G. Functional genomics identifies regulators of the phototransduction machinery in the *Drosophila* larval eye and adult ocelli. *Dev. Biol.* **2016**, *410*, 164–177. [CrossRef] [PubMed]

64. Sánchez-Higueras, C.; Sotillos, S.; Castelli-Gair Hombria, J. Common origin of insect trachea and endocrine organs from a segmentally repeated precursor. *Curr. Biol.* **2014**, *24*, 76–81. [CrossRef] [PubMed]

65. Fox, R.M.; Vaishnavi, A.; Maruyama, R.; Andrew, D.J. Organ-specific gene expression: The bHLH protein Sage provides tissue specificity to *Drosophila* FoxA. *Development* **2013**, *140*, 2160–2171. [CrossRef] [PubMed]

66. Bachtrog, D.; Mank, J.E.; Peichel, C.L.; Kirkpatrick, M.; Otto, S.P.; Ashman, T.L.; Hahn, M.W.; Kitano, J.; Mayrose, I.; Ming, R. Sex determination: Why so many ways of doing it? *PLoS Biol.* **2014**, *12*, e1001899. [CrossRef] [PubMed]

67. González, A.N.; Lu, H.; Erickson, J.W. A shared enhancer controls a temporal switch between promoters during *Drosophila* primary sex determination. *Proc. Natl. Acad. Sci. USA* **2008**, *105*, 18436–18441. [CrossRef] [PubMed]

68. Verhulst, E.C.; van de Zande, L.; Beukeboom, L.W. Insect sex determination: It all evolves around *transformer*. *Curr. Opin. Genet. Dev.* **2010**, *20*, 376–383. [CrossRef] [PubMed]

69. Lu, H.; Kozhina, E.; Mahadevaraju, S.; Yang, D.; Avila, F.W.; Erickson, J.W. Maternal Groucho and bHLH repressors amplify the dose-sensitive X chromosome signal in *Drosophila* sex determination. *Dev. Biol.* **2008**, *323*, 248–260. [CrossRef] [PubMed]

70. Salz, H.K.; Erickson, J.W. Sex determination in *Drosophila*: The view from the top. *Fly* **2010**, *4*, 60–70. [CrossRef] [PubMed]

71. Verhulst, E.C.; van de Zande, L. Double nexus–*Doublesex* is the connecting element in sex determination. *Brief. Funct. Genomics* **2015**, *14*, 396–406. [CrossRef] [PubMed]

72. Xie, W.; Guo, L.; Jiao, X.; Yang, N.; Yang, X.; Wu, Q.; Wang, S.; Zhou, X.; Zhang, Y. Transcriptomic dissection of sexual differences in *Bemisia tabaci*, an invasive agricultural pest worldwide. *Sci. Rep.* **2014**, *4*, 4088. [CrossRef] [PubMed]

73. Zinna, R.A.; Gotoh, H.; Kojima, T.; Niimi, T. Recent advances in understanding the mechanisms of sexually dimorphic plasticity: Insights from beetle weapons and future directions. *Curr. Opin. Insect Sci.* **2018**, *25*, 35–41. [CrossRef] [PubMed]

74. Goto, S.; Hayashi, S. Specification of the embryonic limb primordium by graded activity of Decapentaplegic. *Development* **1997**, *124*, 125–132. [PubMed]

75. Kubota, K.; Goto, S.; Eto, K.; Hayashi, S. EGF receptor attenuates Dpp signaling and helps to distinguish the wing and leg cell fates in *Drosophila*. *Development* **2000**, *127*, 3769–3776. [PubMed]

76. Dahmann, C.; Basler, K. Opposing transcriptional outputs of Hedgehog signaling and engrailed control compartment cell sorting at the *Drosophila* A/P boundary. *Cell* **2000**, *100*, 411–422. [CrossRef]

77. Milán, M.; Cohen, S.M. A re-evaluation of the contributions of Apterous and Notch to the dorsoventral lineage restriction boundary in the *Drosophila* wing. *Development* **2003**, *130*, 553–562. [CrossRef] [PubMed]

78. Nienhaus, U.; Aegerter-Wilmsen, T.; Aegerter, C.M. In-vivo imaging of the *Drosophila* wing imaginal disc over time: Novel insights on growth and boundary formation. *PLoS ONE* **2012**, *7*, e47594. [CrossRef] [PubMed]

79. Hatini, V.; Kula-Eversole, E.; Nusinow, D.; Del Signore, S.J. Essential roles for *stat92E* in expanding and patterning the proximodistal axis of the *Drosophila* wing imaginal disc. *Dev. Biol.* **2013**, *378*, 38–50. [CrossRef] [PubMed]

80. Recasens-Alvarez, C.; Ferreira, A.; Milán, M. JAK/STAT controls organ size and fate specification by regulating morphogen production and signalling. *Nat. Commun.* **2017**, *8*, 13815. [CrossRef] [PubMed]

81. Terriente, J.; Perea, D.; Suzanne, M.; Díaz-Benjumea, F.J. The *Drosophila* gene *zfh2* is required to establish proximal-distal domains in the wing disc. *Dev. Biol.* **2008**, *320*, 102–112. [CrossRef] [PubMed]

82. Perea, D.; Molohon, K.; Edwards, K.; Díaz-Benjumea, F.J. Multiple roles of the gene *zinc finger homeodomain-2* in the development of the *Drosophila* wing. *Mech. Dev.* **2013**, *130*, 467–481. [CrossRef] [PubMed]

83. Azpiazu, N.; Morata, G. Function and regulation of *homothorax* in the wing imaginal disc of *Drosophila*. *Development* **2000**, *127*, 2685–2693. [PubMed]

84. Casares, F.; Mann, R.S. A dual role for *homothorax* in inhibiting wing blade development and specifying proximal wing identities in *Drosophila*. *Development* **2000**, *127*, 1499–1508. [PubMed]

85. Shukla, V.; Habib, F.; Kulkarni, A.; Ratnaparkhi, G.S. Gene duplication, lineage-specific expansion, and subfunctionalization in the MADF-BESS family patterns the *Drosophila* wing hinge. *Genetics* **2014**, *196*, 481–496. [CrossRef] [PubMed]

86. Vihervaara, T.; Puig, O. dFOXO regulates transcription of a *Drosophila* acid lipase. *J. Mol. Biol.* **2008**, *376*, 1215–1223. [CrossRef] [PubMed]

87. Hossain, M.S.; Liu, Y.; Zhou, S.; Li, K.; Tian, L.; Li, S. 20-Hydroxyecdysone-induced transcriptional activity of FoxO upregulates *brummer* and *acid lipase-1* and promotes lipolysis in *Bombyx* fat body. *Insect Biochem. Mol. Biol.* **2013**, *43*, 829–838. [CrossRef] [PubMed]

88. Baumann, A.A.; Benoit, J.B.; Michalkova, V.; Mireji, P.; Attardo, G.M.; Moulton, J.K.; Wilson, T.G.; Aksoy, S. Juvenile hormone and insulin suppress lipolysis between periods of lactation during tsetse fly pregnancy. *Mol. Cell. Endocrinol.* **2013**, *372*, 30–41. [CrossRef] [PubMed]

89. Kang, P.; Chang, K.; Liu, Y.; Bouska, M.; Birnbaum, A.; Karashchuk, G.; Thakore, R.; Zheng, W.; Post, S.; Brent, C.S.; et al. *Drosophila* Kruppel homolog 1 represses lipolysis through interaction with dFOXO. *Sci. Rep.* **2017**, *7*, 16369. [CrossRef] [PubMed]

90. Zhang, Y.; Lu, Y.X.; Liu, J.; Yang, C.; Feng, Q.L.; Xu, W.H. A regulatory pathway, ecdysone-transcription factor Relish-cathepsin L, is involved in insect fat body dissociation. *PLoS Genet.* **2013**, *9*, e1003273. [CrossRef] [PubMed]

91. Okamura, T.; Shimizu, H.; Nagao, T.; Ueda, R.; Ishii, S. ATF-2 regulates fat metabolism in *Drosophila*. *Mol. Biol. Cell* **2007**, *18*, 1519–1529. [CrossRef] [PubMed]

92. Bond, N.D.; Nelliot, A.; Bernardo, M.K.; Ayerh, M.A.; Gorski, K.A.; Hoshizaki, D.K.; Woodard, C.T. ßFTZ-F1 and Matrix metalloproteinase 2 are required for fat-body remodeling in *Drosophila*. *Dev. Biol.* **2011**, *360*, 286–296. [CrossRef] [PubMed]

93. Vyazunova, I.; Lan, Q. *Yellow fever mosquito sterol carrier protein-2 gene structure and transcriptional regulation*. *Insect Mol. Biol.* **2010**, *19*, 205–215. [CrossRef] [PubMed]

94. Peng, R.; Fu, Q.; Hong, H.; Schwaegler, T.; Lan, Q. THAP and ATF-2 regulated sterol carrier protein-2 promoter activities in the larval midgut of the yellow fever mosquito, *Aedes aegypti. PLoS ONE* **2012**, *7*, e46948. [CrossRef] [PubMed]

95. Liang, L.N.; Zhang, L.L.; Zeng, B.J.; Zheng, S.C.; Feng, Q.L. Transcription factor CAAT/enhancer-binding protein is involved in regulation of expression of sterol carrier protein x in *Spodoptera litura. Insect Mol. Biol.* **2015**, *24*, 551–560. [CrossRef] [PubMed]

96. Glossop, N.R.; Houl, J.H.; Zheng, H.; Ng, F.S.; Dudek, S.M.; Hardin, P.E. VRILLE feeds back to control circadian transcription of *Clock* in the *Drosophila* circadian oscillator. *Neuron* **2003**, *37*, 249–261. [CrossRef]

97. Tataroglu, O.; Emery, P. The molecular ticks of the *Drosophila* circadian clock. *Curr. Opin. Insect Sci.* **2015**, *7*, 51–57. [CrossRef] [PubMed]

98. Gunawardhana, K.L.; Hardin, P.E. VRILLE controls PDF neuropeptide accumulation and arborization rhythms in small ventrolateral neurons to drive rhythmic behavior in *Drosophila. Curr. Biol.* **2017**, *27*, 3442–3453. [CrossRef] [PubMed]

99. Cyran, S.A.; Buchsbaum, A.M.; Reddy, K.L.; Lin, M.C.; Glossop, N.R.; Hardin, P.E.; Young, M.W.; Storti, R.V.; Blau, J. *Vrille*, *Pdp1*, and *dClock* form a second feedback loop in the *Drosophila* circadian clock. *Cell* **2003**, *112*, 329–341. [CrossRef]

100. Matsumoto, A.; Ukai-Tadenuma, M.; Yamada, R.G.; Houl, J.; Uno, K.D.; Kasukawa, T.; Dauwalder, B.; Itoh, T.Q.; Takahashi, K.; Ueda, R.; et al. A functional genomics strategy reveals *clockwork orange* as a transcriptional regulator in the *Drosophila* circadian clock. *Genes Dev.* **2007**, *21*, 1687–1700. [CrossRef] [PubMed]

101. Sivachenko, A.; Li, Y.; Abruzzi, K.C.; Rosbash, M. The transcription factor Mef2 links the *Drosophila* core Clock to *Fas2*, neuronal morphology, and circadian behavior. *Neuron* **2013**, *79*, 281–292. [CrossRef] [PubMed]

102. Jaumouillé, E.; Machado Almeida, P.; Stähli, P.; Koch, R.; Nagoshi, E. Transcriptional regulation via nuclear receptor crosstalk required for the *Drosophila* circadian clock. *Curr. Biol.* **2015**, *25*, 1502–1508. [CrossRef] [PubMed]

103. Kamae, Y.; Uryu, O.; Miki, T.; Tomioka, K. The nuclear receptor genes *HR3* and *E75* are required for the circadian rhythm in a primitive insect. *PLoS ONE* **2014**, *9*, e114899. [CrossRef] [PubMed]

104. Hong, B.; Zhang, Z.F.; Tang, S.M.; Yi, Y.Z.; Zhang, T.Y.; Xu, W.H. Protein-DNA interactions in the promoter region of the gene encoding diapause hormone and pheromone biosynthesis activating neuropeptide of the cotton bollworm, *Helicoverpa armigera. Biochim. Biophys. Acta* **2006**, *1759*, 177–185. [CrossRef] [PubMed]

105. Zhang, T.Y.; Xu, W.H. Identification and characterization of a POU transcription factor in the cotton bollworm, *Helicoverpa armigera. BMC Mol. Biol.* **2009**, *10*, 25. [CrossRef] [PubMed]

106. Hu, C.H.; Hong, B.; Xu, W.H. Identification of an E-box DNA binding protein, activated protein 4, and its function in regulating the expression of the gene encoding diapause hormone and pheromone biosynthesis-activating neuropeptide in *Helicoverpa armigera. Insect Mol. Biol.* **2010**, *19*, 243–252. [CrossRef] [PubMed]

107. Bao, B.; Hong, B.; Feng, Q.L.; Xu, W.H. Transcription factor fork head regulates the promoter of diapause hormone gene in the cotton bollworm, *Helicoverpa armigera*, and the modification of SUMOylation. *Insect Biochem. Mol. Biol.* **2011**, *41*, 670–679. [CrossRef] [PubMed]

108. Zhang, T.Y.; Kang, L.; Zhang, Z.F.; Xu, W.H. Identification of a POU factor involved in regulating the neuron-specific expression of the gene encoding diapause hormone and pheromone biosynthesis-activating neuropeptide in *Bombyx mori. Biochem. J.* **2004**, *380*, 255–263. [CrossRef] [PubMed]

109. Niu, K.; Zhang, X.; Deng, H.; Wu, F.; Ren, Y.; Xiang, H.; Zheng, S.; Liu, L.; Huang, L.; Zeng, B.; et al. BmILF and i-motif structure are involved in transcriptional regulation of *BmPOUM2* in *Bombyx mori. Nucleic Acids Res.* **2018**, *46*, 1710–1723. [CrossRef] [PubMed]

110. Bajgar, A.; Jindra, M.; Dolezel, D. Autonomous regulation of the insect gut by circadian genes acting downstream of juvenile hormone signaling. *Proc. Natl. Acad. Sci. USA* **2013**, *110*, 4416–4421. [CrossRef] [PubMed]

111. Murata, T.; Kageyama, Y.; Hirose, S.; Ueda, H. Regulation of the *EDG84A* gene by FTZ-F1 during metamorphosis in *Drosophila melanogaster. Mol. Cell. Biol.* **1996**, *16*, 6509–6515. [CrossRef] [PubMed]

112. Rebers, J.E.; Niu, J.; Riddiford, L.M. Structure and spatial expression of the *Manduca sexta MSCP14.6* cuticle gene. *Insect Biochem. Mol. Biol.* **1997**, *27*, 229–240. [CrossRef]

113. Deng, H.; Zhang, J.; Li, Y.; Zheng, S.; Liu, L.; Huang, L.; Xu, W.H.; Palli, S.R.; Feng, Q. Homeodomain POU and Abd-A proteins regulate the transcription of pupal genes during metamorphosis of the silkworm, *Bombyx mori*. *Proc. Natl. Acad. Sci. USA* **2012**, *109*, 12598–12603. [CrossRef] [PubMed]

114. He, Y.; Deng, H.; Hu, Q.; Zhu, Z.; Liu, L.; Zheng, S.; Song, Q.; Feng, Q. Identification of the binding domains and key amino acids for the interaction of the transcription factors BmPOUM2 and BmAbd-A in *Bombyx mori*. *Insect Biochem. Mol. Biol.* **2017**, *81*, 41–50. [CrossRef] [PubMed]

115. Wittkopp, P.J.; Beldade, P. Development and evolution of insect pigmentation: Genetic mechanisms and the potential consequences of pleiotropy. *Semin Cell Dev. Biol.* **2009**, *20*, 65–71. [CrossRef] [PubMed]

116. Kronforst, M.R.; Barsh, G.S.; Kopp, A.; Mallet, J.; Monteiro, A.; Mullen, S.P.; Protas, M.; Rosenblum, E.B.; Schneider, C.J.; Hoekstra, H.E. Unraveling the thread of nature's tapestry: The genetics of diversity and convergence in animal pigmentation. *Pigment. Cell Melanoma Res.* **2012**, *25*, 411–433. [CrossRef] [PubMed]

117. Kunte, K.; Zhang, W.; Tenger-Trolander, A.; Palmer, D.H.; Martin, A.; Reed, R.D.; Mullen, S.P.; Kronforst, M.R. *Doublesex* is a mimicry supergene. *Nature* **2014**, *507*, 229–232. [CrossRef] [PubMed]

118. Yoda, S.; Yamaguchi, J.; Mita, K.; Yamamoto, K.; Banno, Y.; Ando, T.; Daimon, T.; Fujiwara, H. The transcription factor Apontic-like controls diverse colouration pattern in caterpillars. *Nat. Commun.* **2014**, *5*, 4936. [CrossRef] [PubMed]

119. Zhao, X.M.; Liu, C.; Jiang, L.J.; Li, Q.Y.; Zhou, M.T.; Cheng, T.C.; Mita, K.; Xia, Q.Y. A juvenile hormone transcription factor Bmdimm-fibroin H chain pathway is involved in the synthesis of silk protein in silkworm, *Bombyx mori*. *J. Biol. Chem.* **2015**, *290*, 972–986. [CrossRef] [PubMed]

120. Zhou, C.; Zha, X.; Shi, P.; Wei, S.; Wang, H.; Zheng, R.; Xia, Q. Multiprotein bridging factor 2 regulates the expression of the *fibroin heavy chain* gene by interacting with Bmdimmed in the silkworm *Bombyx mori*. *Insect Mol. Biol.* **2016**, *25*, 509–518. [CrossRef] [PubMed]

121. Zhao, X.; Liu, C.; Li, Q.; Hu, W.; Zhou, M.; Nie, H.; Zhang, Y.; Peng, Z.; Zhao, P.; Xia, Q. Basic helix-loop-helix transcription factor Bmsage is involved in regulation of *fibroin H-chain* gene via interaction with SGF1 in *Bombyx mori*. *PLoS ONE* **2014**, *9*, e94091. [CrossRef] [PubMed]

122. Hui, C.C.; Matsuno, K.; Suzuki, Y. Fibroin gene promoter contains a cluster of homeodomain binding sites that interact with three silk gland factors. *J. Mol. Biol.* **1990**, *213*, 651–670. [CrossRef]

123. Kimoto, M.; Tsubota, T.; Uchino, K.; Sezutsu, H.; Takiya, S. LIM-homeodomain transcription factor Awh is a key component activating all three fibroin genes, *fibH*, *fibL* and *fhx*, in the silk gland of the silkworm, *Bombyx mori*. *Insect Biochem. Mol. Biol.* **2015**, *56*, 29–35. [CrossRef] [PubMed]

124. Horard, B.; Julien, E.; Nony, P.; Garel, A.; Couble, P. Differential binding of the *Bombyx* silk gland-specific factor SGFB to its target DNA sequence drives posterior-cell-restricted expression. *Mol. Cell. Biol.* **1997**, *17*, 1572–1579. [CrossRef] [PubMed]

125. Grzelak, K. Control of expression of silk protein genes. *Comp. Biochem. Physiol.* **1995**, *110B*, 671–681. [CrossRef]

126. Julien, E.; Bordeaux, M.C.; Garel, A.; Couble, P. Fork head alternative binding drives stage-specific gene expression in the silk gland of *Bombyx mori*. *Insect Biochem. Mol. Biol.* **2002**, *32*, 377–387. [CrossRef]

127. Mach, V.; Takiya, S.; Ohno, K.; Handa, H.; Imai, T.; Suzuki, Y. Silk gland factor-1 involved in the regulation of *Bombyx* sericin-1 gene contains fork head motif. *J. Biol. Chem.* **1995**, *270*, 9340–9346. [CrossRef] [PubMed]

128. Fukuta, M.; Matsuno, K.; Hui, C.C.; Nagata, T.; Takiya, S.; Xu, P.X.; Ueno, K.; Suzuki, Y. Molecular cloning of a POU domain-containing factor involved in the regulation of the *Bombyx sericin-1* gene. *J. Biol. Chem.* **1993**, *268*, 19471–19475. [PubMed]

129. Kimoto, M.; Tsubota, T.; Uchino, K.; Sezutsu, H.; Takiya, S. Hox transcription factor Antp regulates *sericin-1* gene expression in the terminal differentiated silk gland of *Bombyx mori*. *Dev. Biol.* **2014**, *386*, 64–71. [CrossRef] [PubMed]

130. Kimoto, M.; Kitagawa, T.; Kobayashi, I.; Nakata, T.; Kuroiwa, A.; Takiya, S. Inhibition of the binding of MSG-intermolt-specific complex, MIC, to the *sericin-1* gene promoter and *sericin-1* gene expression by POU-M1/SGF-3. *Dev. Genes Evol.* **2012**, *222*, 351–359. [CrossRef] [PubMed]

131. Kayukawa, T.; Jouraku, A.; Ito, Y.; Shinoda, T. Molecular mechanism underlying juvenile hormone-mediated repression of precocious larval-adult metamorphosis. *Proc. Natl. Acad. Sci. USA* **2017**, *114*, 1057–1062. [CrossRef] [PubMed]

132. Yamanaka, N.; Rewitz, K.F.; O'Connor, M.B. Ecdysone control of developmental transitions: Lessons from *Drosophila* research. *Annu. Rev. Entomol.* **2013**, *58*, 497–516. [CrossRef] [PubMed]

133. Uryu, O.; Ou, Q.; Komura-Kawa, T.; Kamiyama, T.; Iga, M.; Syrzycka, M.; Hirota, K.; Kataoka, H.; Honda, B.M.; King-Jones, K.; et al. Cooperative control of ecdysone biosynthesis in *Drosophila* by transcription factors Séance, Ouija board, and Molting defective. *Genetics* **2017**, *208*, 605–622. [CrossRef] [PubMed]

134. Huang, J.; Tian, L.; Peng, C.; Abdou, M.; Wen, D.; Wang, Y.; Li, S.; Wang, J. DPP-mediated TGFβ signaling regulates juvenile hormone biosynthesis by activating the expression of juvenile hormone acid methyltransferase. *Development* **2011**, *138*, 2283–2291. [CrossRef] [PubMed]

135. Cheng, C.; Ko, A.; Chaieb, L.; Koyama, T.; Sarwar, P.; Mirth, C.K.; Smith, W.A.; Suzuki, Y. The POU factor Ventral veins lacking/Drifter directs the timing of metamorphosis through ecdysteroid and juvenile hormone signaling. *PLoS Genet.* **2014**, *10*, e1004425. [CrossRef] [PubMed]

136. Zeng, B.; Huang, Y.; Xu, J.; Shiotsuki, T.; Bai, H.; Palli, S.R.; Huang, Y.; Tan, A. The FOXO transcription factor controls insect growth and development by regulating juvenile hormone degradation in the silkworm, *Bombyx mori*. *J. Biol. Chem.* **2017**, *292*, 11659–11669. [CrossRef] [PubMed]

137. Ashok, M.; Turner, C.; Wilson, T.G. Insect juvenile hormone resistance gene homology with the bHLH-PAS family of transcriptional regulators. *Proc. Natl. Acad. Sci. USA* **1998**, *95*, 2761–2766. [CrossRef] [PubMed]

138. Li, M.; Mead, E.A.; Zhu, J. Heterodimer of two bHLH-PAS proteins mediates juvenile hormone-induced gene expression. *Proc. Natl. Acad. Sci. USA* **2011**, *108*, 638–643. [CrossRef] [PubMed]

139. Zhang, Z.; Xu, J.; Sheng, Z.; Sui, Y.; Palli, S.R. Steroid receptor co-activator is required for juvenile hormone signal transduction through a bHLH-PAS transcription factor, methoprene tolerant. *J. Biol. Chem.* **2011**, *286*, 8437–8447. [CrossRef] [PubMed]

140. Charles, J.P.; Iwema, T.; Epa, V.C.; Takaki, K.; Rynes, J.; Jindra, M. Ligand-binding properties of a juvenile hormone receptor, Methoprene-tolerant. *Proc. Natl. Acad. Sci. USA* **2011**, *108*, 21128–21133. [CrossRef] [PubMed]

141. Shin, S.W.; Zou, Z.; Saha, T.T.; Raikhel, A.S. bHLH-PAS heterodimer of Methoprene-tolerant and Cycle mediates circadian expression of juvenile hormone-induced mosquito genes. *Proc. Natl. Acad. Sci. USA* **2012**, *109*, 16576–16581. [CrossRef] [PubMed]

142. Cui, Y.; Sui, Y.; Xu, J.; Zhu, F.; Palli, S.R. Juvenile hormone regulates *Aedes aegypti* Krüppel homolog 1 through a conserved E box motif. *Insect Biochem. Mol. Biol.* **2014**, *52*, 23–32. [CrossRef] [PubMed]

143. Kayukawa, T.; Shinoda, T. Functional characterization of two paralogous JH receptors, methoprene-tolerant 1 and 2, in the silkworm, *Bombyx mori* (Lepidoptera: Bombycidae). *Appl. Entomol. Zool.* **2015**, *50*, 383–391. [CrossRef]

144. Godlewski, J.; Wang, S.; Wilson, T.G. Interaction of bHLH-PAS proteins involved in juvenile hormone reception in *Drosophila*. *Biochem. Biophys. Res. Commun.* **2006**, *342*, 1305–1311. [CrossRef] [PubMed]

145. Liu, Y.; Sheng, Z.; Liu, H.; Wen, D.; He, Q.; Wang, S.; Shao, W.; Jiang, R.J.; An, S.; Sun, Y.; et al. Juvenile hormone counteracts the bHLH-PAS transcription factors MET and GCE to prevent caspase-dependent programmed cell death in *Drosophila*. *Development* **2009**, *136*, 2015–2025. [CrossRef] [PubMed]

146. Kayukawa, T.; Minakuchi, C.; Namiki, T.; Togawa, T.; Yoshiyama, M.; Kamimura, M.; Mita, K.; Imanishi, S.; Kiuchi, M.; Ishikawa, Y.; et al. Transcriptional regulation of juvenile hormone-mediated induction of Krüppel homolog 1, a repressor of insect metamorphosis. *Proc. Natl. Acad. Sci. USA* **2012**, *109*, 11729–11734. [CrossRef] [PubMed]

147. Kayukawa, T.; Tateishi, K.; Shinoda, T. Establishment of a versatile cell line for juvenile hormone signaling analysis in *Tribolium castaneum*. *Sci. Rep.* **2013**, *3*, 1570. [CrossRef] [PubMed]

148. Smykal, V.; Bajgar, A.; Provaznik, J.; Fexova, S.; Buricova, M.; Takaki, K.; Hodkova, M.; Jindra, M.; Dolezel, D. Juvenile hormone signaling during reproduction and development of the linden bug, *Pyrrhocoris apterus*. *Insect Biochem. Mol. Biol.* **2014**, *45*, 69–76. [CrossRef] [PubMed]

149. Konopova, B.; Jindra, M. Broad-Complex acts downstream of Met in juvenile hormone signaling to coordinate primitive holometabolan metamorphosis. *Development* **2008**, *135*, 559–568. [CrossRef] [PubMed]

150. Belles, X.; Santos, C.G. The MEKRE93 (Methoprene tolerant-Krüppel homolog 1-E93) pathway in the regulation of insect metamorphosis, and the homology of the pupal stage. *Insect Biochem. Mol. Biol.* **2014**, *52*, 60–68. [CrossRef] [PubMed]

151. Urña, E.; Chafino, S.; Manjón, C.; Franch-Marro, X.; Martín, D. The occurrence of the holometabolous pupal stage requires the interaction between E93, Krüppel-Homolog 1 and Broad-Complex. *PLoS Genet.* **2016**, *12*, e1006020.

152. Roy, S.; Saha, T.T.; Zou, Z.; Raikhel, A.S. Regulatory pathways controlling female insect reproduction. *Annu. Rev. Entomol.* **2018**, *63*, 489–511. [CrossRef] [PubMed]

153. Zou, Z.; Saha, T.T.; Roy, S.; Shin, S.W.; Backman, T.W.; Girke, T.; White, K.P.; Raikhel, A.S. Juvenile hormone and its receptor, methoprene-tolerant, control the dynamics of mosquito gene expression. *Proc. Natl. Acad. Sci. USA* **2013**, *110*, E2173–E2181. [CrossRef] [PubMed]

154. Saha, T.T.; Shin, S.W.; Dou, W.; Roy, S.; Zhao, B.; Hou, Y.; Wang, X.L.; Zou, Z.; Girke, T.; Raikhel, A.S. Hairy and Groucho mediate the action of juvenile hormone receptor Methoprene-tolerant in gene repression. *Proc. Natl. Acad. Sci. USA* **2016**, *113*, E735–E743. [CrossRef] [PubMed]

155. Wang, X.; Hou, Y.; Saha, T.T.; Pei, G.; Raikhel, A.S.; Zou, Z. Hormone and receptor interplay in the regulation of mosquito lipid metabolism. *Proc. Natl. Acad. Sci. USA* **2017**, *114*, E2709–E2718. [CrossRef] [PubMed]

156. Guo, W.; Wu, Z.; Song, J.; Jiang, F.; Wang, Z.; Deng, S.; Walker, V.K.; Zhou, S. Juvenile hormone-receptor complex acts on *Mcm4* and *Mcm7* to promote polyploidy and vitellogenesis in the migratory locust. *PLoS Genet.* **2014**, *10*, e1004702. [CrossRef] [PubMed]

157. Wu, Z.; Guo, W.; Xie, Y.; Zhou, S. Juvenile hormone activates the transcription of cell-division-cycle 6 (*Cdc6*) for polyploidy-dependent insect vitellogenesis and oogenesis. *J. Biol. Chem.* **2016**, *291*, 5418–5427. [CrossRef] [PubMed]

158. Luo, M.; Li, D.; Wang, Z.; Guo, W.; Kang, L.; Zhou, S. Juvenile hormone differentially regulates two Grp78 genes encoding protein chaperones required for insect fat body cell homeostasis and vitellogenesis. *J. Biol. Chem.* **2017**, *292*, 8823–8834. [CrossRef] [PubMed]

159. Song, J.; Wu, Z.; Wang, Z.; Deng, S.; Zhou, S. Krüppel-homolog 1 mediates juvenile hormone action to promote vitellogenesis and oocyte maturation in the migratory locust. *Insect Biochem. Mol. Biol.* **2014**, *52*, 94–101. [CrossRef] [PubMed]

160. Gujar, H.; Palli, S.R. Juvenile hormone regulation of female reproduction in the common bed bug, *Cimex lectularius*. *Sci. Rep.* **2016**, *6*, 35546. [CrossRef] [PubMed]

161. Hejnikova, M.; Paroulek, M.; Hodkova, M. Decrease in *Methoprene tolerant* and *Taiman* expression reduces juvenile hormone effects and enhances the levels of juvenile hormone circulating in males of the linden bug *Pyrrhocoris apterus*. *J. Insect Physiol.* **2016**, *93–94*, 72–80. [CrossRef] [PubMed]

162. Uryu, O.; Ameku, T.; Niwa, R. Recent progress in understanding the role of ecdysteroids in adult insects: Germline development and circadian clock in the fruit fly *Drosophila melanogaster*. *Zoological Lett.* **2015**, *1*, 32. [CrossRef] [PubMed]

163. Sourmeli, S.; Papantonis, A.; Lecanidou, R. A novel role for the *Bombyx* Slbo homologue, BmC/EBP, in insect choriogenesis. *Biochem. Biophys. Res. Commun.* **2005**, *337*, 713–719. [CrossRef] [PubMed]

164. Sourmeli, S.; Papantonis, A.; Lecanidou, R. BmCbZ, an insect-specific factor featuring a composite DNA-binding domain, interacts with BmC/EBPγ. *Biochem. Biophys. Res. Commun.* **2005**, *338*, 1957–1965. [CrossRef] [PubMed]

165. Ye, Y.X.; Pan, P.L.; Xu, J.Y.; Shen, Z.F.; Kang, D.; Lu, J.B.; Hu, Q.L.; Huang, H.J.; Lou, Y.H.; Zhou, N.M.; et al. Forkhead box transcription factor L2 activates *Fcp3C* to regulate insect chorion formation. *Open Biol.* **2017**, *7*, 170061. [CrossRef] [PubMed]

166. Hetru, C.; Hoffmann, J.A. NF-κB in the immune response of *Drosophila*. *Cold Spring Harb. Perspect. Biol.* **2009**, *1*, a000232. [CrossRef] [PubMed]

167. Stöven, S.; Ando, I.; Kadalayil, L.; Engström, Y.; Hultmark, D. Activation of the *Drosophila* NF-κB factor Relish by rapid endoproteolytic cleavage. *EMBO Rep.* **2000**, *1*, 347–352. [CrossRef] [PubMed]

168. Lemaitre, B.; Reichhart, J.M.; Hoffmann, J.A. *Drosophila* host defense: Differential induction of antimicrobial peptide genes after infection by various classes of microorganisms. *Proc. Natl. Acad. Sci. USA* **1997**, *94*, 14614–14619. [CrossRef] [PubMed]

169. Michel, T.; Reichhart, J.M.; Hoffmann, J.A.; Royet, J. *Drosophila* Toll is activated by Gram-positive bacteria through a circulating peptidoglycan recognition protein. *Nature* **2001**, *414*, 756–759. [CrossRef] [PubMed]

170. Zambon, R.A.; Nandakumar, M.; Vakharia, V.N.; Wu, L.P. The Toll pathway is important for an antiviral response in *Drosophila*. *Proc. Natl. Acad. Sci. USA* **2005**, *102*, 7257–7262. [CrossRef] [PubMed]

171. Kleino, A.; Silverman, N. The *Drosophila* IMD pathway in the activation of the humoral immune response. *Dev. Comp. Immunol.* **2014**, *42*, 25–35. [CrossRef] [PubMed]

172. Tanji, T.; Yun, E.Y.; Ip, Y.T. Heterodimers of NF-κB transcription factors DIF and Relish regulate antimicrobial peptide genes in *Drosophila*. *Proc. Natl. Acad. Sci. USA* **2010**, *107*, 14715–14720. [CrossRef] [PubMed]

173. Zhong, X.; Rao, X.J.; Yi, H.Y.; Lin, X.Y.; Huang, X.H.; Yu, X.Q. Co-expression of Dorsal and Rel2 negatively regulates antimicrobial peptide expression in the tobacco hornworm *Manduca sexta*. *Sci. Rep.* **2016**, *6*, 20654. [CrossRef] [PubMed]

174. Bhaskar, V.; Courey, A.J. The MADF-BESS domain factor Dip3 potentiates synergistic activation by Dorsal and Twist. *Gene* **2002**, *299*, 173–184. [CrossRef]

175. Ratnaparkhi, G.S.; Duong, H.A.; Courey, A.J. Dorsal interacting protein 3 potentiates activation by *Drosophila* Rel homology domain proteins. *Dev. Comp. Immunol.* **2008**, *32*, 1290–1300. [CrossRef] [PubMed]

176. Junell, A.; Uvell, H.; Pick, L.; Engström, Y. Isolation of regulators of *Drosophila* immune defense genes by a double interaction screen in yeast. *Insect Biochem. Mol. Biol.* **2007**, *37*, 202–212. [CrossRef] [PubMed]

177. Barillas-Mury, C.; Charlesworth, A.; Gross, I.; Richman, A.; Hoffmann, J.A.; Kafatos, F.C. Immune factor Gambif1, a new rel family member from the human malaria vector, *Anopheles gambiae*. *EMBO J.* **1996**, *15*, 4691–4701. [CrossRef] [PubMed]

178. Osta, M.A.; Christophides, G.K.; Vlachou, D.; Kafatos, F.C. Innate immunity in the malaria vector *Anopheles gambiae*: Comparative and functional genomics. *J. Exp. Biol.* **2004**, *207*, 2551–2563. [CrossRef] [PubMed]

179. Meister, S.; Kanzok, S.M.; Zheng, X.L.; Luna, C.; Li, T.R.; Hoa, N.T.; Clayton, J.R.; White, K.P.; Kafatos, F.C.; Christophides, G.K.; et al. Immune signaling pathways regulating bacterial and malaria parasite infection of the mosquito *Anopheles gambiae*. *Proc. Natl. Acad. Sci. USA* **2005**, *102*, 11420–11425. [CrossRef] [PubMed]

180. Shin, S.W.; Kokoza, V.; Bian, G.; Cheon, H.M.; Kim, Y.J.; Raikhel, A.S. REL1, a homologue of *Drosophila* Dorsal, regulates Toll antifungal immune pathway in the female mosquito *Aedes aegypti*. *J. Biol. Chem.* **2005**, *280*, 16499–16507. [CrossRef] [PubMed]

181. Frolet, C.; Thoma, M.; Blandin, S.; Hoffmann, J.A.; Levashina, E.A. Boosting NF-κB-dependent basal immunity of *Anopheles gambiae* aborts development of *Plasmodium berghei*. *Immunity* **2006**, *25*, 677–685. [CrossRef] [PubMed]

182. Xi, Z.; Ramirez, J.L.; Dimopoulos, G. The *Aedes aegypti* toll pathway controls dengue virus infection. *PLoS Pathog.* **2008**, *4*, e1000098. [CrossRef] [PubMed]

183. Shin, S.W.; Kokoza, V.; Ahmed, A.; Raikhel, A.S. Characterization of three alternatively spliced isoforms of the Rel/NF-kappa B transcription factor Relish from the mosquito *Aedes aegypti*. *Proc. Natl. Acad. Sci. USA* **2002**, *99*, 9978–9983. [CrossRef] [PubMed]

184. Shin, S.W.; Kokoza, V.; Lobkov, I.; Raikhel, A.S. Relish-mediated immune deficiency in the transgenic mosquito *Aedes aegypti*. *Proc. Natl. Acad. Sci. USA* **2003**, *100*, 2616–2621. [CrossRef] [PubMed]

185. Antonova, Y.; Alvarez, K.S.; Kim, Y.J.; Kokoza, V.; Raikhel, A.S. The role of NF-κB factor REL2 in the *Aedes aegypti* immune response. *Insect Biochem. Mol. Biol.* **2009**, *39*, 303–314. [CrossRef] [PubMed]

186. Paradkar, P.N.; Duchemin, J.-B.; Voysey, R.; Walker, P.J. Dicer-2-dependent activation of *Culex* Vago occurs via the TRAF-Rel2 signaling pathway. *PLoS Negl. Trop. Dis.* **2014**, *8*, e2823. [CrossRef] [PubMed]

187. Tanaka, H.; Yamamoto, M.; Moriyama, Y.; Yamao, M.; Furukawa, S.; Sagisaka, A.; Nakazawa, H.; Mori, H.; Yamakawa, M. A novel Rel protein and shortened isoform that differentially regulate antibacterial peptide genes in the silkworm *Bombyx mori*. *Biochim. Biophys. Acta* **2005**, *1730*, 10–21. [CrossRef] [PubMed]

188. Tanaka, H.; Matsuki, H.; Furukawa, S.; Sagisaka, A.; Kotani, E.; Mori, H.; Yamakawa, M. Identification and functional analysis of Relish homologs in the silkworm, *Bombyx mori*. *Biochim. Biophys. Acta* **2007**, *1769*, 559–568. [CrossRef] [PubMed]

189. Ryu, J.H.; Nam, K.B.; Oh, C.T.; Nam, H.J.; Kim, S.H.; Yoon, J.H.; Seong, J.K.; Yoo, M.A.; Jang, I.H.; Brey, P.T.; et al. The homeobox gene *Caudal* regulates constitutive local expression of antimicrobial peptide genes in *Drosophila* epithelia. *Mol. Cell. Biol.* **2004**, *24*, 172–185. [CrossRef] [PubMed]

190. Ryu, J.H.; Kim, S.H.; Lee, H.Y.; Bai, J.Y.; Nam, Y.D.; Bae, J.W.; Lee, D.G.; Shin, S.C.; Ha, E.M.; Lee, W.J. Innate immune homeostasis by the homeobox gene caudal and commensal-gut mutualism in *Drosophila*. *Science* **2008**, *319*, 777–782. [CrossRef] [PubMed]

191. Junell, A.; Uvell, H.; Davis, M.M.; Edlundh-Rose, E.; Antonsson, A.; Pick, L.; Engström, Y. The POU transcription factor Drifter/Ventral veinless regulates expression of *Drosophila* immune defense genes. *Mol. Cell Biol.* **2010**, *30*, 3672–3684. [CrossRef] [PubMed]

192. Zhong, X.; Chowdhury, M.; Li, C.F.; Yu, X.Q. Transcription factor forkhead regulates expression of antimicrobial peptides in the tobacco hornworm, *Manduca sexta*. *Sci. Rep.* **2017**, *7*, 2688. [CrossRef] [PubMed]

193. Mai, T.; Chen, S.; Lin, X.; Zhang, X.; Zou, X.; Feng, Q.; Zheng, S. 20-hydroxyecdysone positively regulates the transcription of the antimicrobial peptide, lebocin, via BmEts and BmBR-C Z4 in the midgut of *Bombyx mori* during metamorphosis. *Dev. Comp. Immunol.* **2017**, *74*, 10–18. [CrossRef] [PubMed]

194. Gerardo, N.M.; Altincicek, B.; Anselme, C.; Atamian, H.; Barribeau, S.M.; de Vos, M.; Duncan, E.J.; Evans, J.D.; Gabaldon, T.; Ghanim, M.; et al. Immunity and other defenses in pea aphids, *Acyrthosiphon pisum*. *Genome Biol.* **2010**, *11*, R21. [CrossRef] [PubMed]

195. Peng, T.; Pan, Y.; Gao, X.; Xi, J.; Zhang, L.; Yang, C.; Bi, R.; Yang, S.; Xin, X.; Shang, Q. Cytochrome P450 *CYP6DA2* regulated by *cap 'n'collar isoform C (CncC)* is associated with gossypol tolerance in *Aphis gossypii* Glover. *Insect Mol. Biol.* **2016**, *25*, 450–459. [CrossRef] [PubMed]

196. Kalsi, M.; Palli, S.R. Transcription factor cap n collar C regulates multiple cytochrome P450 genes conferring adaptation to potato plant allelochemicals and resistance to imidacloprid in *Leptinotarsa decemlineata* (Say). *Insect Biochem. Mol. Biol.* **2017**, *83*, 1–12. [CrossRef] [PubMed]

197. Brown, R.P.; McDonnell, C.M.; Berenbaum, M.R.; Schuler, M.A. Regulation of an insect cytochrome P450 monooxygenase gene (*CYP6B1*) by aryl hydrocarbon and xanthotoxin response cascades. *Gene* **2005**, *358*, 39–52. [CrossRef] [PubMed]

198. Peng, T.; Chen, X.; Pan, Y.; Zheng, Z.; Wei, X.; Xi, J.; Zhang, J.; Gao, X.; Shang, Q. Transcription factor *aryl hydrocarbon receptor/aryl hydrocarbon receptor nuclear translocator* is involved in regulation of the xenobiotic tolerance-related cytochrome P450 *CYP6DA2* in *Aphis gossypii* Glover. *Insect Mol. Biol.* **2017**, *26*, 485–495. [CrossRef] [PubMed]

199. Zhao, J.; Liu, X.N.; Li, F.; Zhuang, S.Z.; Huang, L.N.; Ma, J.; Gao, X.W. Yeast one-hybrid screening the potential regulator of CYP6B6 overexpression of *Helicoverpa armigera* under 2-tridecanone stress. *Bull. Entomol. Res.* **2016**, *106*, 182–190. [CrossRef] [PubMed]

200. Wilding, C.S. Regulating resistance: CncC:Maf, antioxidant response elements and the overexpression of detoxification genes in insecticide resistance. *Curr. Opin. Insect Sci.* **2018**, *27*, 89–96. [CrossRef] [PubMed]

201. King-Jones, K.; Horner, M.A.; Lam, G.; Thummel, C.S. The DHR96 nuclear receptor regulates xenobiotic responses in *Drosophila*. *Cell Metab.* **2006**, *4*, 37–48. [CrossRef] [PubMed]

202. Zhao, Q.; Ma, D.; Huang, Y.; He, W.; Li, Y.; Vasseur, L.; You, M. Genome-wide investigation of transcription factors provides insights into transcriptional regulation in *Plutella xylostella*. *Mol. Genet. Genomics* **2018**, *293*, 435–449. [CrossRef] [PubMed]

203. Liu, N.N.; Li, M.; Gong, Y.H.; Liu, F.; Li, T. Cytochrome P450s-their expression, regulation, and role in insecticide resistance. *Pestic. Biochem. Physiol.* **2015**, *120*, 77–81. [CrossRef] [PubMed]

204. Li, T.; Liu, N. Regulation of P450-mediated permethrin resistance in *Culex quinquefasciatus* by the GPCR/Galphas/AC/cAMP/PKA signaling cascade. *Biochem. Biophys. Rep.* **2017**, *12*, 12–19. [PubMed]

205. Adang, M.J.; Crickmore, N.; Jurat-Fuentes, J.L. Diversity of *Bacillus thuringiensis* crystal toxins and mechanism of action. *Adv. Insect Physiol.* **2014**, *47*, 39–87.

206. Guo, Z.J.; Kang, S.; Zhu, X.; Xia, J.X.; Wu, Q.J.; Wang, S.L.; Xie, W.; Zhang, Y.J. Down-regulation of a novel ABC transporter gene (*Pxwhite*) is associated with Cry1Ac resistance in the diamondback moth, *Plutella xylostella* (L.). *Insect Biochem. Mol. Biol.* **2015**, *59*, 30–40. [CrossRef] [PubMed]

207. Li, J.; Ma, Y.; Yuan, W.; Xiao, Y.; Liu, C.; Wang, J.; Peng, J.; Peng, R.; Soberón, M.; Bravo, A.; et al. FOXA transcriptional factor modulates insect susceptibility to *Bacillus thuringiensis* Cry1Ac toxin by regulating the expression of toxin-receptor *ABCC2* and *ABCC3* genes. *Insect Biochem. Mol. Biol.* **2017**, *88*, 1–11. [CrossRef] [PubMed]

208. Cotto, J.J.; Morimoto, R.I. Stress-induced activation of the heat-shock response: Cell and molecular biology of heat-shock factors. *Biochem. Soc. Symp.* **1999**, *64*, 105–118. [PubMed]

209. Guertin, M.J.; Petesch, S.J.; Zobeck, K.L.; Min, I.M.; Lis, J.T. *Drosophila* heat shock system as a general model to investigate transcriptional regulation. *Cold Spring Harb. Symp. Quant. Biol.* **2010**, *75*, 1–9. [CrossRef] [PubMed]

210. Jedlicka, P.; Mortin, M.A.; Wu, C. Multiple functions of *Drosophila* heat shock transcription factor *in vivo*. *EMBO J.* **1997**, *16*, 2452–2462. [CrossRef] [PubMed]

211. Mukherjee, T.; Kim, W.S.; Mandal, L.; Banerjee, U. Interaction between Notch and Hif-α in development and survival of *Drosophila* blood cells. *Science* **2011**, *332*, 1210–1213. [CrossRef] [PubMed]

212. Charoensawan, V.; Wilson, D.; Teichmann, S.A. Genomic repertoires of DNA-binding transcription factors across the tree of life. *Nucleic. Acids. Res.* **2010**, *38*, 7364–7377. [CrossRef] [PubMed]

213. Mathelier, A.; Fornes, O.; Arenillas, D.J.; Chen, C.Y.; Denay, G.; Lee, J.; Shi, W.; Shyr, C.; Tan, G.; Worsley-Hunt, R.; et al. JASPAR 2016: A major expansion and update of the open-access database of transcription factor binding profiles. *Nucleic Acids. Res.* **2016**, *44*, D110–D115. [CrossRef] [PubMed]
214. Hume, M.A.; Barrera, L.A.; Gisselbrecht, S.S.; Bulyk, M.L. UniPROBE, update 2015: New tools and content for the online database of protein-binding microarray data on protein-DNA interactions. *Nucleic Acids. Res.* **2015**, *43*, D117–D122. [CrossRef] [PubMed]
215. Jin, J.; Tian, F.; Yang, D.C.; Meng, Y.Q.; Kong, L.; Luo, J.; Gao, G. PlantTFDB 4.0: Toward a central hub for transcription factors and regulatory interactions in plants. *Nucleic Acids. Res.* **2017**, *45*, D1040–D1045. [CrossRef] [PubMed]
216. Lee, C.; Huang, C.H. LASAGNA-Search: An integrated web tool for transcription factor binding site search and visualization. *BioTechniques* **2013**, *54*, 1417–1453. [CrossRef] [PubMed]
217. Messeguer, X.; Escudero, R.; Farré, D.; Núñez, O.; Martínez, J.; Albà, M.M. PROMO: Detection of known transcription regulatory elements using species-tailored searches. *Bioinformatics* **2002**, *18*, 333–334. [CrossRef] [PubMed]
218. Cartharius, K.; Frech, K.; Grote, K.; Klocke, B.; Haltmeier, M.; Klingenhoff, A.; Frisch, M.; Bayerlein, M.; Werner, T. MatInspector and beyond: Promoter analysis based on transcription factor binding sites. *Bioinformatics* **2005**, *21*, 2933–2942. [CrossRef] [PubMed]
219. Parra, R.G.; Rohr, C.O.; Koile, D.; Perez-Castro, C.; Yankilevich, P. INSECT 2.0: A web-server for genome-wide cis-regulatory modules prediction. *Bioinformatics* **2016**, *32*, 1229–1231. [CrossRef] [PubMed]
220. Wu, C.; Suzuki-Ogoh, C.; Ohmiya, Y. Dual-reporter assay using two secreted luciferase genes. *BioTechniques* **2007**, *42*, 290–292. [CrossRef] [PubMed]
221. Martin, C.S.; Wight, P.A.; Dobretsova, A.; Bronstein, I. Dual luminescence-based reporter gene assay for luciferase and β-galactosidase. *BioTechniques* **1996**, *21*, 520–524. [CrossRef] [PubMed]
222. Paguio, A.; Stecha, P.; Wood, K.V.; Fan, F. Improved dual-luciferase reporter assays for nuclear receptors. *Curr. Chem. Genomics* **2010**, *4*, 43–49. [CrossRef] [PubMed]
223. Wider, D.; Picard, D. Secreted dual reporter assay with *Gaussia* luciferase and the red fluorescent protein mCherry. *PLoS ONE* **2017**, *12*, e0189403. [CrossRef] [PubMed]
224. Garner, M.M.; Revzin, A. A gel electrophoresis method for quantifying the binding of proteins to specific DNA regions: Application to components of the Escherichia coli lactose operon regulatory system. *Nucleic Acids Res.* **1981**, *9*, 3047–3060. [CrossRef] [PubMed]
225. Fried, M.; Crothers, D.M. Equilibria and kinetics of lac repressor-operator interactions by polyacrylamide gel electrophoresis. *Nucleic Acids Res.* **1981**, *9*, 6505–6525. [CrossRef] [PubMed]
226. Hellman, L.M.; Fried, M.G. Electrophoretic mobility shift assay (EMSA) for detecting protein-nucleic acid interactions. *Nat. Protoc.* **2007**, *2*, 1849–1861. [CrossRef] [PubMed]
227. Sun, W.; Wang, C.F.; Zhang, Z. Transcription factor E74A affects the ecdysone titer by regulating the expression of the *EO* gene in the silkworm, *Bomby mori*. *Biochim. Biophys. Acta* **2017**, *1861*, 551–558. [CrossRef] [PubMed]
228. Smith, A.J.; Humphries, S.E. Characterization of DNA-binding proteins using multiplexed competitor EMSA. *J. Mol. Biol.* **2009**, *385*, 714–717. [CrossRef] [PubMed]
229. Pan, Y.; Karns, K.; Herr, A.E. Microfluidic electrophoretic mobility shift assays for quantitative biochemical analysis. *Electrophoresis* **2014**, *35*, 2078–2090. [CrossRef] [PubMed]
230. Reece-Hoyes, J.S.; Diallo, A.; Lajoie, B.; Kent, A.; Shrestha, S.; Kadreppa, S.; Pesyna, C.; Dekker, J.; Myers, C.L.; Walhout, A.J. Enhanced yeast one-hybrid (eY1H) assays for high-throughput gene-centered regulatory network mapping. *Nat. Methods* **2011**, *8*, 1059–1064. [CrossRef] [PubMed]
231. Reece-Hoyes, J.S.; Marian Walhout, A.J. Yeast one-hybrid assays: A historical and technical perspective. *Methods* **2012**, *57*, 441–447. [CrossRef] [PubMed]
232. Mundade, R.; Ozer, H.G.; Wei, H.; Prabhu, L.; Lu, T. Role of ChIP-seq in the discovery of transcription factor binding sites, differential gene regulation mechanism, epigenetic marks and beyond. *Cell Cycle* **2014**, *13*, 2847–2852. [CrossRef] [PubMed]
233. Hanson, B.R.; Tan, M. Intra-ChIP: Studying gene regulation in an intracellular pathogen. *Curr. Genet.* **2016**, *62*, 547–551. [CrossRef] [PubMed]

234. Ren, B.; Robert, F.; Wyrick, J.J.; Aparicio, O.; Jennings, E.G.; Simon, I.; Zeitlinger, J.; Schreiber, J.; Hannett, N.; Kanin, E.; et al. Genome-wide location and function of DNA binding proteins. *Science* **2000**, *290*, 2306–2309. [CrossRef] [PubMed]

235. Nakato, R.; Shirahige, K. Recent advances in ChIP-seq analysis: From quality management to whole-genome annotation. *Brief. Bioinform.* **2017**, *18*, 279–290. [CrossRef] [PubMed]

236. Rusk, N. Reverse ChIP. *Nat. Methods* **2009**, *6*, 187. [CrossRef]

237. Liu, X.; Zhang, Y.; Chen, Y.; Li, M.; Zhou, F.; Li, K.; Cao, H.; Ni, M.; Liu, Y.; Gu, Z.; et al. In situ capture of chromatin interactions by biotinylated dCas9. *Cell* **2017**, *170*, 1028–1043. [CrossRef] [PubMed]

238. Sun, D.; Guo, Z.; Liu, Y.; Zhang, Y. Progress and prospects of CRISPR/Cas systems in insects and other Arthropods. *Front. Physiol.* **2017**, *8*, 608. [CrossRef] [PubMed]

239. Huang, Y.; Chen, Y.; Zeng, B.; Wang, Y.; James, A.A.; Gurr, G.M.; Yang, G.; Lin, X.; Huang, Y.; You, M. CRISPR/Cas9 mediated knockout of the *abdominal-A* homeotic gene in the global pest, diamondback moth (*Plutella xylostella*). *Insect Biochem. Mol. Biol.* **2016**, *75*, 98–106. [CrossRef] [PubMed]

240. Xu, X.S.; Gantz, V.M.; Siomava, N.; Bier, E. CRISPR/Cas9 and active genetics-based trans-species replacement of the endogenous *Drosophila kni*-L2 CRM reveals unexpected complexity. *eLife* **2017**, *6*, e30281. [CrossRef] [PubMed]

241. Mohamad Ishak, N.S.; Nong, Q.D.; Matsuura, T.; Kato, Y.; Watanabe, H. Co-option of the bZIP transcription factor Vrille as the activator of *Doublesex1* in environmental sex determination of the crustacean *Daphnia magna*. *PLoS Genet.* **2017**, *13*, e1006953. [CrossRef] [PubMed]

242. Fields, S.; Song, O. A novel genetic system to detect protein-protein interactions. *Nature* **1989**, *340*, 245–246. [CrossRef] [PubMed]

243. Rain, J.C.; Selig, L.; De Reuse, H.; Battaglia, V.; Reverdy, C.; Simon, S.; Lenzen, G.; Petel, F.; Wojcik, J.; Schächter, V.; et al. The protein-protein interaction map of *Helicobacter pylori*. *Nature* **2001**, *409*, 211–215. [CrossRef] [PubMed]

244. Simonis, N.; Rual, J.F.; Carvunis, A.R.; Tasan, M.; Lemmens, I.; Hirozane-Kishikawa, T.; Hao, T.; Sahalie, J.M.; Venkatesan, K.; Gebreab, F.; et al. Empirically controlled mapping of the *Caenorhabditis elegans* protein-protein interactome network. *Nat. Methods* **2009**, *6*, 47–54. [CrossRef] [PubMed]

245. Rolland, T.; Tasan, M.; Charloteaux, B.; Pevzner, S.J.; Zhong, Q.; Sahni, N.; Yi, S.; Lemmens, I.; Fontanillo, C.; Mosca, R.; et al. A proteome-scale map of the human interactome network. *Cell* **2014**, *159*, 1212–1226. [CrossRef] [PubMed]

246. Silva, J.V.; Freitas, M.J.; Felgueiras, J.; Fardilha, M. The power of the yeast two-hybrid system in the identification of novel drug targets: Building and modulating PPP1 interactomes. *Expert Rev. Proteomics* **2015**, *12*, 147–158. [CrossRef] [PubMed]

247. Wong, J.H.; Alfatah, M.; Sin, M.F.; Sim, H.M.; Verma, C.S.; Lane, D.P.; Arumugam, P. A yeast two-hybrid system for the screening and characterization of small-molecule inhibitors of protein-protein interactions identifies a novel putative Mdm2-binding site in p53. *BMC Biol.* **2017**, *15*, 108. [CrossRef] [PubMed]

248. Lin, H.Y.; Liu, Q.; Li, X.; Yang, J.; Liu, S.; Huang, Y.; Scanlon, M.J.; Nettleton, D.; Schnable, P.S. Substantial contribution of genetic variation in the expression of transcription factors to phenotypic variation revealed by eRD-GWAS. *Genome Biol.* **2017**, *18*, 192. [CrossRef] [PubMed]

249. Lambert, S.A.; Jolma, A.; Campitelli, L.F.; Das, P.K.; Yin, Y.; Albu, M.; Chen, X.; Taipale, J.; Hughes, T.R.; Weirauch, M.T. The human transcription factors. *Cell* **2018**, *172*, 650–665. [CrossRef] [PubMed]

250. Guo, Z.; Kang, S.; Zhu, X.; Xia, J.; Wu, Q.; Wang, S.; Xie, W.; Zhang, Y. The novel ABC transporter ABCH1 is a potential target for RNAi-based insect pest control and resistance management. *Sci. Rep.* **2015**, *5*, 13728. [CrossRef] [PubMed]

© 2018 by the authors. Licensee MDPI, Basel, Switzerland. This article is an open access article distributed under the terms and conditions of the Creative Commons Attribution (CC BY) license (http://creativecommons.org/licenses/by/4.0/).

International Journal of
Molecular Sciences

Article

Pax3 Gene Regulated Melanin Synthesis by Tyrosinase Pathway in *Pteria penguin*

Feifei Yu, Bingliang Qu *, Dandan Lin, Yuewen Deng, Ronglian Huang and Zhiming Zhong

Fishery College, Guangdong Ocean University, 40 East Jiefang Road, Xiashan District, Zhanjiang 524025, China;
yufeifei2000@163.com (F.Y.); molgen512@126.com (D.L.); dengyw@gdou.edu.cn (Y.D.); hrl8849@163.com (R.H.);
qublyuff@163.com (Z.Z.)
* Correspondence: moleculargenetic@163.com; Tel.: +86-759-2383346; Fax: +86-759-2382404

Received: 24 October 2018; Accepted: 17 November 2018; Published: 22 November 2018

Abstract: The paired-box 3 (*Pax3*) is a transcription factor and it plays an important part in melanin synthesis. In this study, a new *Pax3* gene was identified from *Pteria penguin* (*Röding, 1798*) (*P. penguin*) by RACE-PCR (rapid-amplification of cDNA ends-polymerase chain reaction) and its effect on melanin synthesis was deliberated by RNA interference (RNAi). The cDNA of *PpPax3* was 2250 bp long, containing an open reading fragment of 1365 bp encoding 455 amino acids. Amino acid alignment and phylogenetic tree showed *PpPax3* shared the highest (69.2%) identity with *Pax3* of *Mizuhopecten yessoensis*. Tissue expression profile showed that *PpPax3* had the highest expression in mantle, a nacre-formation related tissue. The *PpPax3* silencing significantly inhibited the expression of *PpPax3*, *PpMitf*, *PpTyr* and *PpCdk2*, genes involved in *Tyr*-mediated melanin synthesis, but had no effect on *PpCreb2* and an increase effect on *PpBcl2*. Furthermore, the *PpPax3* knockdown obviously decreased the tyrosinase activity, the total content of eumelanin and the proportion of PDCA (pyrrole-2,3-dicarboxylic acid) in eumelanin, consistent with influence of tyrosinase (*Tyr*) knockdown. These data indicated that *PpPax3* played an important regulating role in melanin synthesis by *Tyr* pathway in *P. penguin*.

Keywords: *Pax3*; *Pteria penguin* (*Röding, 1798*); tyrosinase; melanin; RNA interference; liquid chromatograph-tandem mass spectrometer (LC-MS/MS)

---

1. Introduction

The winged pearl oyster *Pteria penguin* (*Röding, 1798*) (*P. penguin*) is an important marine cultured species that produces high-quality seawater pearl, whose value depends mainly on its color [1,2]. The melanin is the major pigment in *P. penguin* and largely affects the color and value of the pearl [3]. Moreover, the *P. penguin* is considered to be the best research model for melanin synthesis and color reconstruction in bivalve, because of its purely black shell in population. The mantle tissue is the main organ responsible for the formation and secretion of nacre, which is called as "mother of pearl" [4]. Inhibiting the synthesis and secretion of melanin in mantle might change the color of nacre in *P. penguin*.

Melanin plays an important role in a series of physiological processes, including pigmentation, skin photoprotection and aging [5]. In mammals, melanin synthesis is a complex process, and more than 40 genes participate in it [6–8]. Tyrosinase (*Tyr*) is a key rate-limiting enzyme in melanogenesis [9,10], because it catalyzes three different reactions in biosynthetic pathway of melanin [11]. The microphthalmia-associated transcription factor (*Mitf*) is a central regulator of melanogenesis, and it activates the transcription of several important genes, including *Tyr*, *Cdk2* (cyclin-dependent kinase 2) and *Bcl2* (B-cell lymphoma 2), to control melanocyte differentiation, growth and survival [6,8,12].

The paired-box 3 (*Pax3*) is a member of the paired-box family of transcription factors, and it participates in the development of central nervous system, skeletal muscles, and melanocytes [13,14]. Several studies have demonstrated that *Pax3* is frequently expressed in normal melanocytes and aggregated melanomas [15]. PAX3 directly promotes *Mitf* transcription by binding the *Mitf* promoter, functioning with SOX10 (SRY box 10) and synergizing with the CREB (cyclic-AMP responsive element-binding protein) [16]. However, at the same time, PAX3 competes with MITF by occupying the enhancer of dopachrome tautomerase (*Dct*), a downstream enzyme that functions in melanin synthesis [17,18]. Despite extensive investigations about *Pax3* being carried out over recent years, the effect of *Pax3* on melanin synthesis in bivalves is still largely speculative.

The eumelanin, which gave organisms a brown-black color, was the main pigment in *P. penguin* [3]. Natural eumelanin is mainly composed of the monomer units 5,6-dihydroxyindole (DHI) and 5,6-dihydroxyindole-2-carboxylic acid (DHICA), with various ratios of DHI and DHICA [19]. The alkaline hydrogen peroxide oxidation of eumelanin yields pyrrole-2,3-dicarboxylic acid (PDCA) and pyrrole-2,3,5-tricarboxylic acid (PTCA) from DHI- and DHICA-derived units [5]. The PDCA and PTCA can be detected by high-performance liquid chromatography (HPLC), because they are insoluble in both acidic and alkaline solutions. Quantification of PDCA and PTCA has been extensively used to evaluate the amount and composition of eumelanin in a pigment sample using liquid chromatograph-tandem mass spectrometer (LC-MS/MS) [3,20].

In our previous works, we analyzed the crucial function of *Tyr* in menlanin synthesis of *P. penguin* [3]. In this study, a new *Pax3* gene from *P. penguin* was identified, and its exact function in melanin synthesis was deliberated by RNA interference (RNAi) technology. The relative genes involved in the regulation of *PpPax3* to melanin synthesis were enriched. A *Pax3-Tyr*-melanin axis was verified to exist in *P. penguin*.

2. Results

2.1. Cloning and Sequence Analysis of Pax3 cDNA in P. penguin

Based on the cDNA fragment of *Pax3* from transcriptome database of *P. penguin*, the complete cDNA of *PpPax3* gene was obtained by RACE-PCR. The *PpPax3* cDNA consisted of an open reading frame (452–1816) of 1365bp, a 5′-untranslated region (UTR) of 451bp, and a 3′-UTR of 434bp with a typical polyadenylation signal sequence (AATAAA) and a 31bp poly (A) tail (Figure 1). The putative amino acid sequence was 454 amino acids long. No signal peptide and transmembrane domain were found in deduced PAX3 sequence. The predicted molecular mass of *PpPax* protein was 50.7 kDa, and the theoretical isoelectric point (pI) was 8.15. The full-length cDNA sequence of *PpPax3* was submitted to Genebank with the accession no. MH558581.

```
atggggtaacagattagaggcacaggtgagaagacgcgttgtccgatgtgtccgaagtaagctgtgtccaaagca   76
Aatagccacacaattagcatgaaattataattctatatatgacatggagtgttgttgaaaccatttgcatctaag   152
Aggagatttgggacttttacaacgtaatggacataactttggattaatatgaagttttgaaattatgatatactaa   228
Tcaggtatttaaaaatttgggaaaataagtttagctctgatatcattttaaatacatatttgaagacttgcacga   304
Atattgtgaaaacgtagggaaaaatacgtgtacaaatttaggtaagagcaagaggaagccggtaccagcttaaaa   380
ggacaaaataatcgggaaaacaaataaatcactttattatatcattggacattacgacagaaaatatatcgATGTTA   457
                                                                        M  L    2
ACTAGTGGACTTGTAGTAAGCAACATGTTCTCCTATCATCTCGCAGCCTTGGGACTCATGCCGTCATTCCAGATGGAA   535
 T  S  G  L  V  V  S  N  M  F  S  Y  H  L  A  A  L  G  L  M  P  S  F  Q  M  E    28
GGGCGAGGTCGAGTTAATCAGCTAGGAGGTGTTTTCATTAATGGACGGCCACTGCCCAACCATATACGACTGAAGATT   613
 G  R  G  R  V  N  Q  L  G  G  V  F  I  N  G  R  P  L  P  N  H  I  R  L  K  I    54
GTCGAGCTAGCGGCCCAGGGCGTCCGTCCGTGCGTCATCAGTAGACAGCTGCGGGTGTCACATGGCTGTGTCAGTAAA   691
 V  E  L  A  A  Q  G  V  R  P  C  V  I  S  R  Q  L  R  V  S  H  G  C  V  S  K    80
ATACTCCAACGATATCAAGAAACCGGAAGTATCCGACCTGGGGTTATTGGCGGAAGTAAACCAAGGGTCGCAACTCCG   769
 I  L  Q  R  Y  Q  E  T  G  S  I  R  P  G  V  I  G  G  S  K  P  R  V  A  T  P   106
GAAGTTGAGAAAAAGATAGAACAATACAAAAAAGATAATCCGGGAATTTTCAGTTGGGAGATTCGGGATCGGCTGCTG   847
 E  V  E  K  K  I  E  Q  Y  K  K  D  N  P  G  I  F  S  W  E  I  R  D  R  L  L   132
AAGGAGGGGATTTGTGACCGCAGCACCGTGCCAAGTGTGAGCTCCATCAGTCGAGTATTACGGAGCAGGTTCCAGAAA   925
 K  E  G  I  C  D  R  S  T  V  P  S  V  S  S  I  S  R  V  L  R  S  R  F  Q  K   158
TGTGATTCTGATGACAATGACAATGACAATGACAATGAGGACGACGATGGCGATGACGGCAGTAACAGTAGTGTGGCA  1003
 C  D  S  D  D  N  D  N  D  N  D  N  E  D  D  D  G  D  D  G  S  N  S  S  V  A   184
GACAGGTCTGTTAACTTCTCTGTCAGCGGTCTGCTGTCCGACAATAAAAGCGACAAAAGCGACAACGATTCCGATTGT  1081
 D  R  S  V  N  F  S  V  S  G  L  L  S  D  N  K  S  D  K  S  D  N  D  S  D  C   210
GAATCAGAGCCGGGGCTATCTGTAAAACGGAAGCAACGCCGCAGTCGAACTACTTTCACCGCGGAGCAGTTGGAGGAA  1159
 E  S  E  P  G  L  S  V  K  R  K  Q  R  R  S  R  T  T  F  T  A  E  Q  L  E  E   236
CTGGAAAGAGCCTTTGAACGAACTCACTATCCGGATATATATACGCGAGAGGAATTAGCACAAAGAACAAAGCTAACC  1237
 L  E  R  A  F  E  R  T  H  Y  P  D  I  Y  T  R  E  E  L  A  Q  R  T  K  L  T   262
GAGGCAAGAGTCCAAGTATGGTTTAGTAACCGAAGAGCGAGATGGCGGAAACAGATGGGTAGCAATCAGCTGACAGCC  1315
 E  A  R  V  Q  V  W  F  S  N  R  R  A  R  W  R  K  Q  M  G  S  N  Q  L  T  A   288
TTGAACAGTATATTACAAGTGCCACAGGGTATGGGAACGCCCTCTTATATGCTGCACGAGCCTGGGTATCCACTCTCA  1393
 L  N  S  I  L  Q  V  P  Q  G  M  G  T  P  S  Y  M  L  H  E  P  G  Y  P  L  S   314
CATAATGCAGACAATCTTTGGCATAGATCGTCTATGGCCCAGTCATTACAGTCATTTGGTCAGACAATAAAACCAGAG  1471
 H  N  A  D  N  L  W  H  R  S  S  M  A  Q  S  L  Q  S  F  G  Q  T  I  K  P  E   340
AATTCCTACGCCGGTCTTATGGAAAACTATTTATCTCATTCATCACAGCTTCATGGTCTTCCTACACATAGTTCATCC  1549
 N  S  Y  A  G  L  M  E  N  Y  L  S  H  S  S  Q  L  H  G  L  P  T  H  S  S  S   366
GATCCCCTCTCATCCACTTGGTCATCTCCACGTCCATCCTCCGTTCCTGCGGCTAGGATACACCGCCATCTAGTGGCCAT  1627
 D  P  L  S  S  T  W  S  S  P  V  S  T  S  V  P  A  L  G  Y  T  P  S  S  G  H   392
TACCATCATTACTCTGATGTCACCAAAGTACTCTTCATTCATATAACGCTCATATTCCTTCAGTCACAAACATGGAG  1705
 Y  H  H  Y  S  D  V  T  K  S  T  L  H  S  Y  N  A  H  I  P  S  V  T  N  M  E   418
AGATGTTCAGTTGATGACAGTTTGGTTGCTTTACGTATGAAGTCACGTGAGCATTCCGCCGCTCTCAGTTTGATGCAG  1783
 R  C  S  V  D  D  S  L  V  A  L  R  M  K  S  R  E  H  S  A  A  L  S  L  M  Q   444
GTGGCAGACAACAAATGGCTACCTCATTTTGAagaccggagaagtacatatccatgacattgtatgacgtcactaag  1861
 V  A  D  N  K  M  A  T  S  F  *                                                454
Ttccatgtcagtattaatgggttgttgtacgtgtcctgaaccaaataaaatgtatttatcattcggaactgcctcgca  1939
Cttttccttcatttttacggatacacaaatatagggagacgagttccgaatatatatttctgtgttcctgtaaatatgt  2017
Gtataatgtgtcaactttcactgttttggaccaagcgtccacggtgtcgtccatattactaatatttcaatagtatac  2095
Gtgtagttgttatgtatgtcatggtgtttttgacctttgacctatcagtcatagttttacatttcaatgtctattgaca  2173
tcttgtgattttgaagtcgctgaataataaaattgtaatgcattgcaaaaaaaaaaaaaaaaaaaaaaaaaaaaaa  2250
```

Figure 1. Nucleotide and deduced amino acid sequences of *PpPax3*. PD (paried box domain) and HD (homeodomain) were shown in grey boxes. The ORF (open reading frame) and deduced amino acid sequences were shown in uppercase. The 5′-UTR (untranslated region) and 3′-UTR were shown in lowercase. The initiation codon (ATG) and the stop codon (TGA) were boxed. The putative polyadenylation signal (aataaa) was underlined.

2.2. Multiple Sequence Alignment and Phylogenetic Analysis

Amino acid sequence alignments of *Pax3* gene from *P. penguin* and other species were performed. The PpPax3 shared the highest (69.2%) identity with *Pax3* of *Mizuhopecten yessoensis*, 52.5% with *Pax3* of *Aplysia californica*, 45.8% with *Pax3* of *Parasteatoda tepidariorum*, and 45.0% with *Pax3* of *Branchiostoma belcheri*. The deduced amino acid sequence comparison revealed a highly conserved paried box domain (PD) containing 127 amino acids and a homeodomain (HD) containing 59 amino acids (Figure 2) in PpPax3 gene. But the octapeptide motifs located between PD and HD were not obvious in *Pax3* of *P. Pengui*, *A. californica* and *P. tepidariorum*.

The phylogenetic tree analysis was performed to indicate the evolutionary relationships of *Pax3* from different species. As shown in Figure 2B, PpPax3 was close to *Pax3* of *M. yessoensis*, one bivalve, with a support of 89%. Three *Pax3* genes of mollusk referred, including *P. penguin*, *M. yessoensis* and *Aplysia californica*, were grouped into a close cluster. The *Pax3* of *Aplysia californica* showed high homology with that of mollusk, agreed with their taxonomic relationships. All *Pax3* genes of vertebrates referred were classified to a big clade, which exhibited farther distance to *Pax3* of invertebrates.

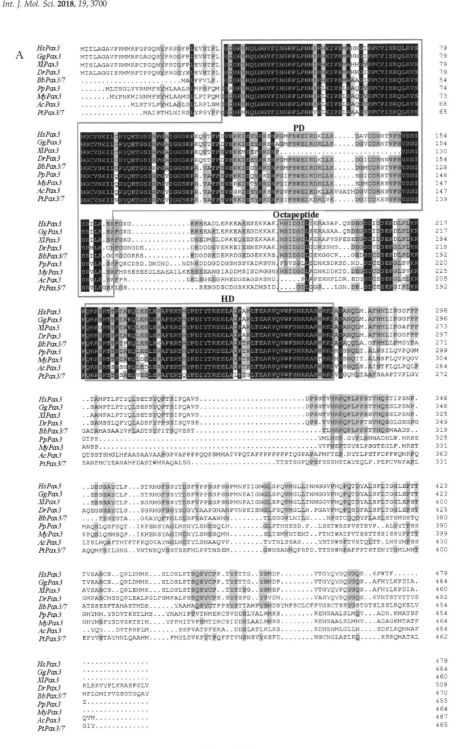

Figure 2. *Cont.*

B

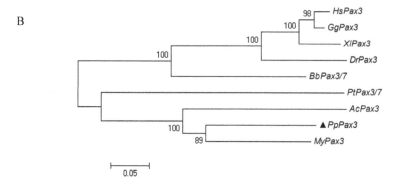

Figure 2. Sequence alignment and phylogenetic analysis. (A) Multiple sequence comparison of *Pax3* among different species including *Pteria penguin(Röding, 1798)* (*PpPax3*, MH558581), *Mizuhopecten yessoensis* (*MyPax3*, XP_021364914.1), *Aplysia californica* (*AcPax3*, XP_012943435.1), *Parasteatoda tepidariorum* (*PtPax3/7*, BBD75270.1), *Branchiostoma belcheri* (*BbPax3/7*, ABK54280.1), *Danio rerio* (*DrPax3*, AAC41253.1), *Homo sapiens* (*HsPax3*, NP_852122.1), *Xenopus laevis* (*XlPax3*, AAI08574.1) and *Gallus gallus* (*GgPax3*, BAB85652.1). The conserved amino acids were written in black background, and similar amino acids were shaded in green and pink. PD, octapeptide motif and HD were indicated in blue boxes. (B) Phylogenetic tree of *Pax3* genes. Numbers in the branches represented the bootstrap values (as a percentage). ▲ meaned the *Pax3* of *P. penguin*.

2.3. PpPax3 mRNA Expression Profile in Different Tissues

The relative expression levels of *PpPax3* gene in various tissues were compared by qRT-PCR (Quantitative real time polymerase chain reaction), with β-actin as an internal control. As shown in Figure 3, the *PpPax3* has constitutive expression in mantle, gill, adductor muscle, digestive diverticulum, foot, testis and ovary. The highest expression level of *PpPax3* was shown in the mantle, followed by the foot, without significant difference between them ($P > 0.05$). The expression of *PpPax3* in the mantle was approximately 2-fold higher than that in adductor muscle, testis and ovary with obvious significant difference ($P < 0.05$). The digestive diverticulum has the lowest expression level. Since *PpPax3* was mainly expressed in the mantle, which was responsible for nacre secretion, the mantle was selected for further studies.

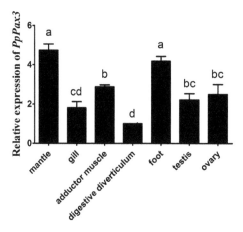

Figure 3. Relative expression of *PpPax3* in various tissues of *P. penguin* estimated by qRT-PCR. Each bar was a mean of 6 pearl oysters. Error bars were the SD. Different letters (a, b, c and d) meaned significant difference ($P < 0.05$).

2.4. PpPax3 Expression Was Inhibited by RNA Interference in P. penguin

To investigate the function of *Pax3* in melanin synthesis of *P. penguin*, RNA interference was performed to inhibit the expression of *PpPax3*, and qRT-PCR was employed to evaluate the silencing effects. As shown in Figure 4A, the *PpPax3* expression was reduced by 65.7% in the *Pax3*-siRNA1 group ($P < 0.05$) and 37.7% in the *Pax3*-siRNA2 group ($P < 0.05$) compared with the negative control (NC) group. This indicated that the expression of *PpPax3* was significantly knocked down by RNA interference.

2.5. PpPax3 Silencing Affected the Expression of PpMitf, PpTyr, PpCreb2, PpBcl2 and PpCdk2 in P. penguin

After *PpPax3* silencing, the transcripts of *PpMitf*, *PpTyr*, *PpCreb2*, *PpBcl2* and *PpCdk2* genes were analyzed. As shown in Figure 4B, after *PpPax3* knockdown, the transcript of *PpMitf*, a central transcription factor of melanogenesis, was significantly decreased by 64.7% in the *Pax3*-siRNA1 (Small interfering RNA) group ($P < 0.01$) and 46.7% in the *Pax3*-siRNA2 group ($P < 0.05$). The expression of *PpTyr*, the key rate-limiting enzyme for melanin synthesis, was obviously reduced by 53.3% ($P < 0.01$) by *Pax3*-siRNA1 and 33.3% ($P < 0.05$) by *Pax3*-siRNA2. The *PpCdk2* mRNA, a melanocyte growth-dependent kinase, was also depressed by 36.6% ($P < 0.05$) and 26.7% ($P < 0.05$). However, no significant difference was observed in *PpCreb2* transcript ($P > 0.05$), a regulatory factor in melanin synthesis pathway. Moreover, the transcript of *PpBcl2*, an apoptosis-related gene, was raised up to 1.9 fold through *Pax3*-siRNA1 interference ($P < 0.05$).

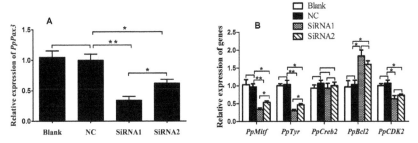

Figure 4. Expression of *PpPax3*, *PpMitf*, *PpTyr*, *PpCreb2*, *PpBcl2* and *PpCdk2* after *PpPax3* RNA interference (RNAi). (**A**) Expression of *PpPax3*. (**B**) Expression of *PpMitf*, *PpTyr*, *PpCreb2*, *PpBcl2* and *PpCdk2*. The qRT-PCR was done with RNA samples from blank group (RNase-free water), NC group (GFP-siRNA), *PpPax3*-siRNA1 group and *PpPax3*-siRNA2 group. The β-actin of *P. penguin* was used as an internal control. Each bar was a mean of 6 individuals. Significant difference was indicated by * ($P < 0.05$) and highly significant difference was indicated by ** ($P < 0.01$).

2.6. PpPax3 Silencing Depressed Tyrosinase Activity in P. penguin

The tyrosinase activity was investigated according to the change in absorbance per minute at 475nm due to dopachrome formation from L-tyrosine of Levodopa (L-DOPA). As respected, after *PpPax3* silencing, the tyrosinase activity was obviously decreased about 48.7% in siRNA1 group ($P < 0.05$) and 31.2% in siRNA2 group ($P < 0.05$) compared to NC group (Figure 5). This indicated that the PAX3 could obviously affect the tyrosinase activity in *P. penguin*.

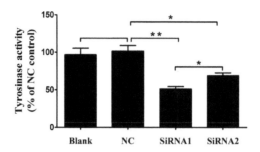

Figure 5. The tyrosinase activity of the control groups and RNA interference groups. The tyrosinase activity was performed with samples from blank group (RNase-free water), NC group (green fluorescent protein (GFP)-siRNA), *PpPax3*-siRNA1 group and *PpPax3*-siRNA2 group. The tyrosinase activity was shown with percentage of every group and NC. Each bar was a mean of 6 pearl oysters. Significant difference was indicated by * ($P < 0.05$) and highly significant difference was indicated by ** ($P < 0.01$).

2.7. PpPax3 Silencing Decreased Melanin Content and Proportion of PDCA

To further investigate the function of *PpPax3* in melanin synthesis in *P. penguin*, we detected the content and composition of melanin in mantle after RNA interference using LC-MS/MS. As expected, by ion spectra examination, the alkaline hydrogen peroxide oxidation products of eumelanin from *P. Penguin* were identified as PDCA and PTCA, because their mass-to-charge ratio values were 156 and 199, respectively, consistent with their molecular weight [3]. The quantity of PDCA and PTCA was calculated according to the special area of peak, which appeared at 2.39 min and 3.58 min, respectively (Figure 6A). As shown in Figure 6B, the total content of PDCA and PTCA was obviously decreased from 674.6 ng/mg to 348.4 ng/mg (by 49.3%) in *PpPax3*-siRNA1 group and to 478.5 ng/mg (by 30.1%) in *PpPax3*-siRNA2 group ($P < 0.05$). The quantity of PDCA was inhibited by 63.5% and 42.6% in *PpPax3*-siRNA1 and *PpPax3*-siRNA2 groups ($P < 0.05$). The quantity of PTCA was inhibited by 45.0% and 26.1% groups ($P < 0.05$). Moreover, after RNA interference, the proportion of PDCA in total oxidation products was obviously decreased from 19.1% to 13.5% (SiRNA1) and 14.6% (SiRNA2) ($P < 0.05$).

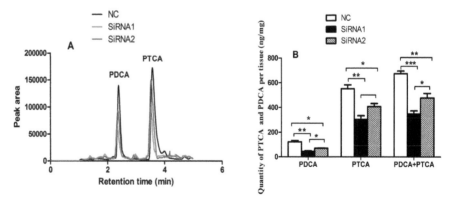

Figure 6. LC-MS/MS analysis of oxidation products of melanin in *P. penguin*. (**A**) HPLC (High Performance Liquid Chromatography) chromatograms of oxidation products of melanin from NC group and RNAi groups. (**B**) The content of PDCA and PTCA from NC group and RNAi groups. Each bar was a mean of 6 pearl oysters. * meaned significant difference ($P < 0.05$); ** meaned very significant difference ($P < 0.01$); *** meaned extremely significant difference ($P < 0.001$).

3. Discussion

Pax gene family encodes transcription factors that are characterized by presence of the paired box domain (PD), an octapeptide motif and homeodomain (HD) [21,22]. As reported [23], after an extensive comparison, both PD and HD of PAX3 proteins among different species were present and highly conserved, but the octapeptide might be absent, such as in Gastropoda, Annelida and in the bivalvia *Pinctada fucata*. Similarly, in this study, the obvious octapeptide only was found in vertebrates, in *B. belcheri* and one bivalve (*M. yessoensis*). No clear homologue of octapeptide could be evidenced in *P. penguin*, *A. californica* and *P. tepidariorum*.

Based on sequence homologies of PD, *Pax* gene family has been classified into four subfamilies, *Pax1/9*, *Pax2/5/8*, *Pax3/7* and *Pax4/6*. The *Pax3/7* subfamily includes *Pax3* gene and *Pax7* genes, which have high similarity in amino acid sequence. The *Pax3* and *Pax7* genes existed in the form of a ancestral gene in protostomes, ascidians and amphioxus, and then were separated into two genes in vertebrates by duplication of the ancestral gene [21]. This might be an explanation for high homology between *Pax3* of *P. penguin* and *Pax7* of other species, which implied the functional diversity of *Pax3* in *P. penguin*.

Our previous studies showed that tyrosinase was a key melanin synthase, and it played a dominant role in melanin synthesis and color formation of *P. penguin* [3]. In this report, the data showed that the knockdown of *PpPax3* caused a significant decrease in *Tyr* expression, tyrosinase activity and melanin content, similar to the influence of *Tyr* silencing [3]. The finding illustrated that PAX3 could affect the melanin synthesis by regulating the expression of *Tyr*. A *Pax3-Tyr*-melanin axis might exist in melanin synthesis pathway of *P. penguin*.

In humans, microphthalmia-associated transcription factor (MITF), known as a master regulator of melanogenesis, binded to the highly conserved binding motif in the regulatory region of the tyrosinase (TYR) promoter, and strongly stimulated the melanocyte-specific transcription of *Tyr* gene [7,24]. *Pax3* directly activated expression of *Mitf* and indirectly affected the expression of *Tyr* in mice [12,20]. However, compared with vertebrates, little is known about whether the MITF also involve in melanin synthesis in bivalve. Our data showed that *PpPax3* silencing significantly cut down the expression of *PpMitf* and obviously depressed the transcription of *PpTyr*. This illustrated that *Mitf* took part in the melanin synthesis of *P. penguin*. The *PpPax3* might affect *Tyr* expression and melanin synthesis through regulation to *Mitf*, similar to that in mammals. Further studies were needed to specify the existence of *Pax3-Mitf-Tyr* pathway in melanin synthesis of *P. penguin*.

As MITF is a multifunctional transcription factor that activates the transcription of various genes involved not only in melanin synthesis, but also in melanocyte proliferation and survival in mammal [12,25], we speculated that *Pax3* might also regulate melanocyte growth and survival by *Mitf* in *P. penguin*. The expressions of *PpCdk2* and *PpBcl2* were analyzed after *PpPax3* interference, because *Cdk2* and *Bcl2* respectively were important genes in melanocyte proliferation and survival in mammals, and both of them were direct MITF target genes [26,27]. In this study, the *PpPax3* silencing obviously decreased the *PpCdk2* expression, which implied *PpPax3* played an important part in control of melanocyte growth in *P. penguin*. Meanwhile, the expression of *PpBcl2* was lightly increased after *PpPax3* silencing, which implied that *PpPax3* played an important part in control of cell survival. It was worth mentioning that the *PpPax3* silencing led to an obvious decrease of *PpMitf*, but a light increase of *PpBcl2*. A possible explanation for this contradiction was that the reduction of tyrosinase protein by *PpPax3* silencing might partly damage of normal cells functions, because tyrosinase involved in several important physiological processes including pigment synthesis [9,28], innate immunity [29] and wound healing [30]. The cell damage led to the up-regulation of cell apoptosis gene *PpBcl2* through another pathway, and antagonized the depression of *PpBcl2* by *PpMitf* decrease.

In human melanocytes, PAX3 partners with SOX10 to induce melanocyte differentiation and melanin synthesis [31]. The CREB (cyclic-AMP responsive element-binding protein), as a cofactor, was inputted to PAX3 and Sox10, so that all three transcription factors induce the expression of *Mitf* [8,11,32,33]. In this report, the silencing of *Pax3* did not affect the expression of *Creb*, which indicated that CREB was not the downstream gene of *Pax3* in *P. penguin*. Further, the alone knockdown of *PpPax3* inhibited the expression of *PpMitf* and *PpTyr*, which implied that the regulation of *PpPax3* to *PpMitf* and *PpTyr* could happen independently of CREB change of in *P. penguin*. The CREB might just work as a cofactor of PAX3 to enhance the regulation effect of *PpPax3* on melanin synthesis in *P. penguin*.

RNA interference is a powerful tool and has been widely used to knock down genes to analyze the genes function, especially in human. For instance, *Pax3* SiRNA was transfected in to human metastatic melanoma to elaborate the function of *Pax3* in melanoma growth and survival [34]. Basing on these new technologies, in human, several widely-believed melanin synthesis pathways were reported, such as the cAMP(cyclic adenosine monophosphate) pathway and the Wnt (wingless-type MMTV integration site family) pathway. The cAMP pathway is a main signal pathway, which goes in the axis: MSH (melanocyte simulating hormone)-MC1R (melanocortin 1 receptor)-cAMP (cycle AMP)–PKA (protein kinase A)–CREB–MITF–tyrosinase. In Wnt pathway, PAX3, partnering with SOX10, induces the expression of *Mitf*, and then affects the expression of *Tyr*. However, in bivalve, the melanin synthesis pathway is still unclear, although some important genes involved melanin synthesis has been cloned and analyzed, such as tyrosinase fom *P. fucata* [35], *Hyriopsis cumingii* [4] and *Crassostrea gigas* [36], *Mitf* from *Meretrix petechialis* [37], *Pax3* from *P. fucata*, *M. yessoensis* and *A. californica* [23]. There has been no clear axis or pathway predicted in bivalve so far. In this research, by functional analysis, we believed that a *Pax3-Mitf-tyr* axis was existent in *P. penguin*, similarly to in humans. The PAX3, by inducing the *Mitf* expression, regulated the melanocyte differentiation, proliferation and survival. However, whether Wnt pathway existed and whether CREB worked as a member of Wnt pathway in *P. penguin* were still problems that deserved further research.

4. Materials and Methods

4.1. Experimental Animals, RNA Isolation and cDNA Synthesis

The *Pteria penguin* (*Röding, 1798*) samples used in this study were obtained from Weizhou Island in Beihai, Guangxi Province, China. All animals were about two years old, with shell length ranging from 12 and 15 cm. They were cultivated with the recirculating seawater at 25–26 °C for one week before the experiment.

Total RNA from mantle (pallial zone and marginal zone), gill, adductor muscle, digestive diverticulum, foot, testis and ovary of *P. penguin* were extracted using RNeasyMini Kit (Qiagen, Gaithersburg, MD, USA), according to the manufacturer's instructions. The integrity and quantity of RNA were detected by electrophoresis on 1% agarose gels and NanoDrop ND1000 Spectrophotometer. The cDNA was synthesized from total RNA using a Superscript II polymerase kit (TransGen, Beijing, China).

4.2. cDNA Cloning and Sequence Analysis

The full-length cDNA sequence of *Pax3* was obtained with SMART RACE cDNA Amplification Kit (Clontech, Mountain View, CA, USA) and Advantage 2 cDNA Polymerase Mix (Clontech, Mountain View, CA, USA) following the manufacturer's protocol. The nested-PCR was employed to enrich the specific DNA band. The test-PCR was used to detect the correctness of linked nucleotide sequence. All used primers were listed in Table 1.

The full-length cDNA of *Pax3* was analyzed by the BLAST program (http://www.ncbi.nlm. nih.gov/). ORF Finder (https://www.ncbi.nlm.nih.gov/orffinder/) was used to characterize the open reading fragment (ORF). Signal 4.1 (http://www.cbs.dtu.dk/services/SignalP/) and TMHMM program (http://www.cbs.dtu.dk/services/TMHMM/) were used to predict signal peptide and transmembrane. Multiple sequence alignments and phylogenetic tree were created using Clustal W and MEGA 6. The protein molecular weight and theoretical pI were analyzed using program tools (http://web.expasy.org/cgibin/protparam/protparam).

4.3. Quantitative Real-Time PCR (qRT-PCR) Analysis

The qRT-PCR assays were performed using Thermo Scientific DyNAmo Flash SYBR Green qPCR Kit (Thermo scientific, Waltham, MA, USA) and the Applied Biosystems 7500/7500 Fast Real-time System (ABI, Carlsbad, CA, USA). Each sample was run in triplicate, along with the internal control gene β-actin. The specific primers were listed in Table 1. The calibration curve was established by several dilutions of standard samples, and used as a linear regression model. The $2^{-\Delta\Delta CT}$ method was applied to calculate the relative expression levels of genes.

Table 1. Primers used in the study.

Primer	Sequence (5′–3′)	Application
PpPax3-outer-F	GGACGGCCACTGCCCAACCATATACG	3′RACE
PpPax3-inner-F	AAGTAAACCAAGGGTCGCAACTCCG	nest-3′RACE
PpPax3-outer-R	GGAACCTGCTCCGTAATACTCGACTGATGG	5′RACE
PpPax3-inner-R	TCAGCAGCCGATCCCGAATCT	nest-5′RACE
UPM (Universal Primer)	TAATACGACTCACTATAGGGCAAGCAGTGGTATC AACGCAGAGT	RACE universal primer
NUP (Nested Universal Primer)	AAGCAGTGGTATCAACGCAGAGT	Nest-RACE universal primer
PpPax3-test-F	GAATGCTCCGTAAACGTTATTG	cDNA test
PpPax3-test-R	GACAACAAAATGGCTACCTCAT	cDNA test
PpPax3-siRNA1-F	GCGTAATACGACTCACTATAGGGGTAAACCAAGG GTCGCAAC	RNAi
PpPax3-siRNA1-R	GCGTAATACGACTCACTATAGGGCGTTGTCGCTTTT GTCGCT	RNAi
PpPax3-siRNA2-F	GCGTAATACGACTCACTATAGGGGATAATCCGGGA ATTTTCAGTTGGG	RNAi
PpPax3-siRNA2-R	GCGTAATACGACTCACTATAGGGGATAGTGAGTTC GTTCAAAGGCTCT	RNAi
GFP-siRNA-F	GATCACTAATACGACTCACTATAGGGATGGTGAGC AAGGGCGAGGA	RNAi
GFP-siRNA-R	GATCACTAATACGACTCACTATAGGGTTACTTGTAC AGCTCGTCCA	RNAi
PpPax3-qPCR-F	TCCGTGCGTCATCAGTAGAC	qRT-PCR
PpPax3-qPCR-R	CCCTTGGTTTACTTCCGCCA	qRT-PCR
PpTyr-qPCR-F	CTCAGGGAAGGGATCAGCTT	qRT-PCR
PpTyr-qPCR-R	AGACCCTCTGCCATTACCAA	qRT-PCR
PpMitf-qPCR-F	TGTTACCTAAATCTGTTGATCCAG	qRT-PCR
PpMitf-qPCR-R	AAATTAGCTGGACAGGAAGAGGAG	qRT-PCR
PpCreb2-qPCR-F	AACTCCCAGTGAAGCAGACA	qRT-PCR
PpCreb2-qPCR-R	GCTCCCCAACAGTAGCCAAT	qRT-PCR
PpBcl2-qPCR-F	TGAGGCACAGTTCCAGGATT	qRT-PCR
PpBcl2-qPCR-R	ACTCTCCACACACCGTACAG	qRT-PCR
PpCdk2-qPCR-F	TGGATTTGCTCGGACACTTG	qRT-PCR
PpCdk2-qPCR-R	TCTACTGCCCTGCCATACTT	qRT-PCR
β-actin-F	CGGTACCACCATGTTCTCAG	qRT-PCR
β-actin-R	GACCGGATTCATCGTATTCC	qRT-PCR

4.4. RNA Interference Experiment

RNA interference was used to analyze the function of *PpPax3* gene. The *PpPax3*-siRNA1 and *PpPax3*-SiRNA2 were synthesized to specially silence the conserved domain of *PpPax3*. The GFP-SiRNA was synthesized from pEGFP-N3 plasmid, a eukaryotic-expression vector encoding green fluorescent protein, as a negative control (NC), and RNase-free water was used as a blank control (primers as Table 1). Six individuals were used in each treatment group. When RNA interference experiment, the experimental individuals were gotten out from seawater and dried in air until the shells were slightly open. Then, 100 μL SiRNAs (small interfering RNA) at a final concentration of 1 μg/μL were gently injected into adductor muscle of experimental individuals, which was then put into seawater and cultivated for 3 days in the lab to have a recovery [38]. At the 4th day, the experimental individuals were injected with same dose of SiRNAs again, and had another recovery for 3 days. At the 7th day, the experimental animals were killed and the tissues were collected for RNA extraction, tyrosinase activity assay and melanin analysis.

4.5. Tyrosinase Activity Assays

Tyrosinase activity assays were performed as described previously with minor modification [19,39]. Briefly, 1g mantle was homogenized in 1 mL of 0.1 mol/L phosphate buffer (pH 6.8) and centrifugated to obtain the tissue supernate. The 0.5 mL of 5 mmol/L L-DOPA (3,4-dihydroxyphenylalanine) was mixed with 2.4 mL of 0.1 mmol/L PBS (phosphate buffer saline), followed by the addition of 0.5 mL of tissue supernate. The mixture was incubated at 37 °C for 30 min, and then the absorbance of the mixture was measured at 475 nm. The total tryosinase activity of every group was represented by the change of absorbance value in 30 min. One unit (U) of the tyrosinase activity was defined as increased or decreased absorbancy per minute at 475 nm. The relative tyrosinase activity was shown using the percentage of every group in NC group.

4.6. Isolation and Oxidation of Total Melanin

The total melanin from mantle of *P. penguin* was isolated and oxidized following our previous report [3]. Briefly, 1 g mantle sample was finely homogenized and incubated in 15 mL phosphate buffer (pH 7.4) with 2% (m/V) papain at 55 °C for 20 h. The mixture was centrifuged at 10,000 rpm for 10min to obtain the precipitate, which was successively washed with 2 mL mineral ether for 3 times, 2 mL ethanol for 3 times and 2 mL water 3 times. Then, the black precipitate was dried and measured as raw melanin production.

The raw melanin production was dissolved in 8.6 mL of 1 mol/L K_2CO_3 and 0.8 mL of 30% H_2O_2. The mixture was heated under reflux at 100 °C for 20 min. After cooling, 0.4 mL of 10% Na_2SO_3 was added to end the reaction. The mixture was acidified to pH 1.0 with 5 mL of 6 mol/L HCl and then was extracted twice with 70 mL of ether. The supernanant was collected and dried to obtain crystalline residue, which was finally redissolved in mobile phase and filtered by 0.45 μm organic membrane for liquid chromatograph-tandem mass spectrometer (LC-MS/MS) analysis.

4.7. LC-MS/MS Assay of Melanin

The content and component of melanin were detected by LC-MS/MS according to previous description [40] with some modification. The chromatographic separation was performed using an Acquity ultraperformance liquid chromatography (UPLC) system (Waters, Milford, MA, USA) consisting of a Waters ACQUITY UPLC HSS T3 (2.1 × 50 mm, 1.7 μm particle size). The mobile phase A and B was 0.1% (*v*/*v*) of formic acid in deionized water and 0.1% (*v*/*v*) of formic acid in methanol, respectively. The ratio of mobile phase A in total mobile phase was gradually decreased from 90% to 0% within 5 min. The cycle time was 5 min per injection. Analyses were performed at 40 °C at a flow rate of 0.3 mL/min. MS/MS detection was performed using a Xevo TQ triple quadrupole mass spectrometer operated in positive electrospray ionization (ESI) mode similar to Yu et al. [3]. The source temperature and desolvation temperature were 150 °C and 550 °C, respectively. The cone gas flow, desolvation gas flow and collision gas flow were 50 L/h, 1100 L/h and 0.14 mL/min (argon), respectively. The analytes were monitored in multireaction monitoring mode (MRM). Specific parameters were given as Table 2.

Table 2. Details of mass spectrometric detection.

Compand	Parent Ion (m/z)	Product Ion (m/z)	Conc Voltage (V)	Collision Energy (eV)	Retention Time (min)
PDCA	155.98	138.01	30	8	2.39
PTCA	199.99	182.09	30	8	3.58

4.8. Statistical Analysis

ANOVA analysis was performed using SPSS 19.0 (IBM, Armonk, NY, USA) to detect the significance of difference among different samples. Significant difference was indicated by * ($P < 0.05$), highly significant difference was indicated by ** ($P < 0.01$) and extremely significant difference was indicated by *** ($P < 0.001$).

5. Conclusions

In this study, we characterized a new *Pax3* gene from *P. penguin*. Tissue expression profile showed that *PpPax3* had the highest expression in mantle, a nacre-formation related tissue. The *PpPax3* silencing significantly inhibited the transcription of *PpPax3*, *PpMitf*, *PpTyr* and *PpCdk2*, genes involved in *Tyr*-mediated melanin synthesis, but had no effect on *PpCreb2* and an increase effect on *PpBcl2*. Furthermore, the *PpPax3* silencing obviously decreased the tyrosinase activity, the total content of eumelanin and the proportion of PDCA in eumelanin, similar to the influence of *Tyr* silencing. Thus, we believed that *PpPax3* played an important role in melanin synthesis by indirectly regulating the expression of *Tyr* in *P. penguin*. The *Pax3-Tyr*-melanin axis was considered a potential strategy in melanin synthesis of *P. penguin*.

Author Contributions: F.Y. and B.Q. conceived and designed the experiments; F.Y., D.L. and Z.Z. performed the experiments; R.H. analyzed the data; Y.D. contributed materials; F.Y. drafted the manuscript; B.Q. made a critical revision of the manuscript.

Funding: This work was supported by the Guangdong Provincial Science and Technology Program (2016A020210115); Guangdong Marine Fishery Development Foundation (B201601-Z08); Doctoral Scientific Research Foundation of Guangdong Ocean University (E15041) and Outstanding Young Teacher Foundation of Guangdong Ocean University (2014004).

Conflicts of Interest: The authors declare no conflict of interest.

Abbreviations

RNAi	RNA interference
UTR	untraslated region
LC-MS/MS	liquid chromatograph-tandem mass spectrometer
PDCA	pyrrole-2,3-dicarboxylic acid
PTCA	pyrrole-2,3,5-tricarboxylic acid
DHI	5,6-dihydroxyindole
DHICA	pyrrole-2,3,5-tricarboxylic acid
CREB	cyclic-AMP responsive element-binding protein
MITF	microphthalmia-associated transcription factor
Tyr	tyrosinase
PAX3	paired-box 3
BCL2	B-cell lymphoma 2
CDK2	cyclin-dependent kinase 2
SOX10	SRY box 10
cAMP	Cyclic Adenosine monophosphate

References

1. Li, H.; Liu, B.; Huang, G.; Fan, S.; Zhang, B.; Su, J.; Yu, D. Characterization of transcriptome and identification of biomineralization genes in winged pearl oyster (Pteria penguin) mantle tissue. *Comp. Biochem. Physiol. D Genom. Proteom.* **2017**, *21*, 67–76. [CrossRef] [PubMed]
2. Naganuma, T.; Hoshino, W.; Shikanai, Y.; Sato, R.; Liu, K.; Sato, S.; Muramoto, K.; Osada, M.; Yoshimi, K.; Ogawa, T. Novel Matrix Proteins of Pteria penguin Pearl Oyster Shell Nacre Homologous to the Jacalin-Related b-Prism Fold Lectins. *PLoS ONE* **2014**, *9*, e112326. [CrossRef]
3. Yu, F.; Pan, Z.; Qu, B.; Yu, X.; Xu, K.; Deng, Y.; Liang, F. Identification of a tyrosinase gene and its functional analysis in melaninsynthesis of *Pteria penguin*. *Gene* **2018**, *656*, 1–8. [CrossRef] [PubMed]
4. Chen, X.; Liu, X.; Bai, Z.; Zhao, L.; Li, J. HcTyr and HcTyp-1 of Hyriopsis cumingii, novel tyrosinase and tyrosinase-related protein genes involved in nacre color formation. *Comp. Biochem. Physiol. B Biochem. Mol. Biol.* **2017**, *204*, 1–8. [CrossRef] [PubMed]

5. Ito, S.; Wakamatsu, K.; Glass, K.; Simon, J.D. High-performance liquid chromatography estimation of cross-linking of dihydroxyindole moiety in eumelanin. *Anal. Biochem.* **2013**, *434*, 221–225. [CrossRef] [PubMed]

6. Cheli, Y.; Ohanna, M.; Ballotti, R.; Bertolotto, C. Fifteen-year quest for microphthalmia-associated transcription factor target genes. *Pigm. Cell Melanoma Res.* **2010**, *23*, 27–40. [CrossRef] [PubMed]

7. Busca, R.; Ballotti, R. Cyclic AMP a key messenger in the regulation of skin pigmentation. *Pigment Cell Res.* **2000**, *13*, 60–69. [CrossRef]

8. Siominski, A.; Tobin, D.J.; Shibahara, S.; Wortsman, J. Melanin pigmentation in mammalian skin and its hormonal regulation. *Physiol. Rev.* **2004**, *84*, 1155–1228. [CrossRef] [PubMed]

9. Hofreiter, M.; Schoneberg, T. The genetic and evolutionary basis of colour variation in vertebrates. *Cell. Mol. Life Sci.* **2010**, *67*, 2591–2603. [CrossRef] [PubMed]

10. Cieslak, M.; Reissmann, M.; Hofreiter, M.; Ludwig, A. Colours of domestication. *Biol. Rev. Camb. Philos. Soc.* **2011**, *86*, 885–899. [CrossRef] [PubMed]

11. Inoue, Y.; Hasegawa, S.; Yamada, T.; Date, Y.; Mizutani, H.; Nakata, S.; Matsunaga, K.; Akamatsu, H. Analysis of the effects of hydroquinone and arbutin on the differentiation of melanocytes. *Biol. Pharm. Bull.* **2013**, *36*, 1722–1730. [CrossRef] [PubMed]

12. Vachtenheim, J.; Borovansky, J. "Transcription physiology" of pigment formation in melanocytes: Central role of MITF. *Exp. Dermatol.* **2010**, *19*, 617–627. [CrossRef] [PubMed]

13. Lee, D.H.; Ahn, S.S.; Kim, J.B.; Lim, Y.; Lee, Y.H.; Shin, S.Y. Downregulation of Melanocyte-Stimulating Hormone-Induced Activation of the Pax3-MITF-Tyrosinase Axis by Sorghum Ethanolic Extract in B16F10 Melanoma Cells. *Int. J. Mol. Sci.* **2018**, *19*, 1640. [CrossRef] [PubMed]

14. Donoghue, P.; Graham, A.; Kelsh, R.N. The origin and evolution of the neural crest. *Bioessay* **2008**, *30*, 530–541. [CrossRef] [PubMed]

15. Medic, S.; Ziman, M. Pax3 expression in normal skin melanocytes and melanocytic lesions (naevi and melanomas). *PLoS ONE* **2010**, *5*, e9977. [CrossRef] [PubMed]

16. Cao, J.; Dai, X.; Wan, L.; Wang, H.; Zhang, J.; Goff, P.S.; Sviderskaya, E.V.; Xuan, Z.; Xu, Z.; Xu, X.; et al. The E ligase APC/CCdh1 promotes ubiquitylation-mediated proteolysis of PAX3 to suppressmelanocyte proliferation and melanoma growth. *Cancer* **2015**, *8*, ra87. [CrossRef]

17. Lang, D.; Lu, M.; Huang, L.; Engleka, K.A.; Zhang, M.; Chu, E.; Lipner, S.; Skoultchi, A.; Millar, S.E.; Epstein, J.A. Epstein1Pax3 functions at a nodal point inmelanocyte stem cell differentiation. *Nature* **2005**, *433*, 884–887. [CrossRef] [PubMed]

18. D'Mello, S.; Finlay, G.J.; Baguley, B.C.; Askarian-Amiri, M.E. Signaling pathways in melanogenesis. *Int. J. Mol. Sci.* **2016**, *17*, 1144. [CrossRef]

19. Szekely-Klepser, G.; Wade, K.; Wooolson, D.; Brown, R.; Fountain, S.; Kindt, E. A validated LC/MS/MS method for the quantification of pyrrole-2,3,5-tricarboxylic acid (PTCA), a eumelanin specific biomarker, in human skin punch biopsies. *J. Chromatogr. B Anal. Technol. Biomed. Life Sci.* **2005**, *826*, 31–40. [CrossRef] [PubMed]

20. Ito, S.; Nakanishi, Y.; Valenzuela, R.K.; Brilliant, M.H.; Kolbe, L.; Wakamatsu, K. Usefulness of alkaline hydrogen peroxide oxidation to analyze eumelanin and pheomelanin in various tissue samples: Application to chemical analysis of human hair melanins. *Pigment Cell Melanoma Res.* **2011**, *24*, 605–613. [CrossRef] [PubMed]

21. Akolkar, D.B.; Asaduzzaman, M.; Kinoshita, S.; Asakawa, S.; Watabe, S. Characterization of Pax3 and Pax7 genes and their expression of patterns during different development and growth stages of Japanese pufferfish Takifugu rubripes. *Gene* **2016**, *575*, 21–28. [CrossRef] [PubMed]

22. Lang, D.; Powell, S.K.; Plummer, R.S.; Young, K.P.; Ruggeri, B.A. PAX genes: Roles in development, pathophysiology, and cancer. *Biochem. Pharmacol.* **2007**, *73*, 1–14. [CrossRef] [PubMed]

23. Navet, S.; Buresi, A.; Baratt, S.; Andouche, A.; Bonnaud-Ponticelli, L.; Bassaglia, Y. The Pax gene family: Highlights from cephalopods. *PLoS ONE* **2017**, *12*, e0172719. [CrossRef]

24. Yasumoto, K.; Yokoyama, K.; Shibata, K.; Tomita, Y.; Shibahara, S. Microphthalmia-associated transcription factor as a regulator for melanocyte-specific transcription of the human tyrosinase gene. *Mol. Cell Biol.* **1994**, *14*, 8058–8070. [CrossRef] [PubMed]

25. Yang, G.; Li, Y.; Nishimura, E.K.; Xin, H.; Zhou, A.; Guo, Y.; Dong, L.; Denning, M.F.; Nickoloff, B.J.; Gui, R. Inhibition of Pax3 by TGF-βmodulates melanocyte viability. *Mol. Cell* **2008**, *32*, 554–563. [CrossRef] [PubMed]

26. Du, J.; Widlund, H.R.; Horstmann, M.A.; Ramaswamy, S.; Ross, K.; Huber, W.E.; Nishimura, E.K.; Golub, T.R.; Fisher, D.E. Critical role of CDK2 for melanoma growth linked to its melanocyte specific transcriptional regulation by MITF. *Cancer Cell* **2004**, *6*, 565–576. [CrossRef] [PubMed]

27. Seiberg, M. Age-induced hair graying-the multiple effects of oxidative stress. *Int. J. Cosmet. Sci.* **2013**, *35*, 532–538. [CrossRef] [PubMed]

28. Kim, T.K.; Lin, Z.; Tidwell, W.J.; Li, W.; Slominski, A.T. Melatonin and its metabolites accumulate in the human epidermis in vivo and inhibit proliferation and tyrosinase activity in epidermal melanocytes in vitro. *Mol. Cell. Endocrinol.* **2015**, *404*, 1–8. [CrossRef] [PubMed]

29. Cerenius, L.; Lee, B.L.; Söderhäll, K. The proPO-system: Pros and cons for its role in invertebrate immunity. *Trends Immunol.* **2008**, *29*, 263–271. [CrossRef] [PubMed]

30. Aguilera, F.; McDougall, C.; Degnan, B.M. Origin, evolution and classification of type-3 copper proteins: Lineage-specific gene expansions and losses across the Metazoa. *BMC Evol. Biol.* **2013**, *13*, 96. [CrossRef]

31. Suzuki, N.; Mutai, H.; Miya, F.; Tsunoda, T.; Terashima, H.; Morimoto, N.; Matsunaga, T. A case report of reversible generalized seizures in a patient with Waardenburg syndrome associated with a novel nonsense mutation in the penultimate exon of SOX10. *BMC Pediatr.* **2018**, *18*, 171. [CrossRef] [PubMed]

32. Moustakas, A. TGF-b Targets PAX3 to Control Melanocyte Differentiation. *Cell* **2008**, *15*, 797–799. [CrossRef] [PubMed]

33. Ortonne, J.P.; Ballotti, R. Melanocyte biology and melanogenesis: What's new? *J. Dermatol. Treat.* **2018**, *9*, 1–28. [CrossRef]

34. He, S.; Li, C.; Slobbe, L.; Glover, A.; Marshall, M.; Baguley, B.C.; Eccles, M.R. Pax3 knockdwon in metastatic melanoma cell lines does not reduce MITF expression. *Melanoma Res.* **2011**, *21*, 24–34. [CrossRef] [PubMed]

35. Takgi, R.; Miyashita, T. A cDNA cloning of a novel alpha-class tyrosinase of Pinctada fucata: Its expression analysis and characterization of the expressed protein. *Enzym. Res.* **2014**, *780549*, 1–9. [CrossRef] [PubMed]

36. Feng, D.; Li, Q.; Yu, H.; Zhao, X.; Kong, L. Comparative transcriptome analysis of the Pacific oyster Crassostrea gigas characterized by shell colors: Identification of genetic bases potentially involved in pigmentation. *PLoS ONE* **2015**, *10*, e0145257. [CrossRef]

37. Zhang, S.; Wang, H.; Yu, J.; Jiang, F.; Yue, X.; Liu, B. Identification of a gene encoding microphthalimia-associated transcription factor and its association with shell color in the clam Memetrix petechialis. *Comp. Biochem. Physiol. B Biochem. Mol. Biol.* **2018**, *225*, 75–83. [CrossRef] [PubMed]

38. Yan, F.; Luo, S.; Jiao, Y.; Deng, Y.; Du, X.; Huang, R.; Wang, Q.; Chen, W. Molecular characterization of the BMP7 gene and its potential role in shell formation in pinctada martensii. *Int. J. Mol. Sci.* **2014**, *15*, 21215–21228. [CrossRef] [PubMed]

39. Choi, T.Y.; Sohn, K.C.; Kim, J.H.; Kim, S.M.; Kim, C.H.; Hwang, J.S.; Lee, J.H.; Kim, C.D.; Yoon, T.J. Impact of NAD(P)H: Quinone oxidoreductase-1 on pigmentation. *J. Investig. Dermatol.* **2010**, *130*, 784–792. [CrossRef] [PubMed]

40. Ito, S.; Wakamatsu, K. Chemical degration of melanins: Application to identification of dopamine-melanin. *Pigment Cell Res.* **1998**, *11*, 120–126. [CrossRef]

 © 2018 by the authors. Licensee MDPI, Basel, Switzerland. This article is an open access article distributed under the terms and conditions of the Creative Commons Attribution (CC BY) license (http://creativecommons.org/licenses/by/4.0/).

Review

Molecular Mechanisms Underlying the Link between Diet and DNA Methylation

Fatma Zehra Kadayifci [1,2], Shasha Zheng [1] and Yuan-Xiang Pan [2,3,4,*]

1   Department of Public Health Sciences, California Baptist University, Riverside, CA 92504, USA;
    fzk@illinois.edu (F.Z.K.); szheng@calbaptist.edu (S.Z.)
2   Department of Food Science and Human Nutrition, University of Illinois at Urbana-Champaign, IL 61801,
    USA
3   Division of Nutritional Sciences, University of Illinois at Urbana-Champaign, IL 61801, USA
4   Illinois Informatics Institute, University of Illinois at Urbana-Champaign, IL 61801, USA
*   Correspondence: yxpan@illinois.edu; Tel.: +1-217-333-3466

Received: 5 November 2018; Accepted: 10 December 2018; Published: 14 December 2018

Abstract: DNA methylation is a vital modification process in the control of genetic information, which contributes to the epigenetics by regulating gene expression without changing the DNA sequence. Abnormal DNA methylation—both hypomethylation and hypermethylation—has been associated with improper gene expression, leading to several disorders. Two types of risk factors can alter the epigenetic regulation of methylation pathways: genetic factors and modifiable factors. Nutrition is one of the strongest modifiable factors, which plays a direct role in DNA methylation pathways. Large numbers of studies have investigated the effects of nutrition on DNA methylation pathways, but relatively few have focused on the biochemical mechanisms. Understanding the biological mechanisms is essential for clarifying how nutrients function in epigenetics. It is believed that nutrition affects the epigenetic regulations of DNA methylation in several possible epigenetic pathways: mainly, by altering the substrates and cofactors that are necessary for proper DNA methylation; additionally, by changing the activity of enzymes regulating the one-carbon cycle; and, lastly, through there being an epigenetic role in several possible mechanisms related to DNA demethylation activity. The aim of this article is to review the potential underlying biochemical mechanisms that are related to diet modifications in DNA methylation and demethylation.

Keywords: epigenetics; gene expression; nutrition; transcription; disorders; mechanisms

1. Introduction

It has been well known that cytosine (C) in the genome, as part of the genetic code, also transfers epigenetic information through the chemical modification of its pyrimidine ring [1,2]. Methylation of the fifth position of cytosine (5mC) is a highly conserved epigenetic modification of DNA that is found in most prokaryotic and eukaryotic models [3], and it has a pivotal impact on genome stability, gene expression, and development [1]. Methylation of the DNA takes place almost completely in the symmetric cytidine–guanine dinucleotide (CpG) context, and is assessed to occur at nearly 70–80% of CpG sites throughout the genome [4]. Additionally, on bacterial and plant DNA, methylation can also occur at an adenine site, which regulates different bacterial and plant DNA functions. Recently, it has been discussed that there is indirect evidence suggesting the presence of adenine site methylation on mammalian DNA. However, the functionality of this base remains unclear on mammals [5].

DNA methylation is a crucial element in the control of the precise expression of genetic information, and both hypermethylation and hypomethylation have been associated with improper gene expression [6]. Irregular changes in genetic methylation patterns or an unusual analysis of DNA methylation signals are associated with many disorders and cancers [7]. Furthermore, the regulation of DNA methylation, crucially, is associated with other metabolic pathways, such as the one-carbon cycle, which have a significant impact on epigenetic regulations [8]. Two types of risk factors can alter the epigenetic regulation of methylation pathways. The first factor is the genetic factors, such as polymorphism and genetic mutations, which can cause aberrant DNA methylation [7,8]. Secondly, there are potentially modifiable factors, such as the modification of essential nutrients that are involved in the metabolism of methyl groups [9].

Nutrition is a strong player not only for its influence on gene expression, but more importantly, because early nutrition alterations could be responsible for the later development of chronic diseases through epigenetic mechanisms [10]. Both animal and human studies have investigated the effects of nutrition on DNA methylation pathways, but to our knowledge, relatively few have focused on the biochemical mechanisms. Understanding the biological mechanisms is important for future studies to clarify how nutrients function in epigenetics. Thus, the aim of this article is to review the underlying biochemical mechanisms of diet-related modifications in DNA methylation and demethylation. We also aim to go through all of the possible nutrient and DNA methylation interactions in more detail, and examine the underlying mechanisms of these relations by including both recent human and animal studies. Additionally, we opt to clarify the effect of diet on the DNA demethylation pathway, which has not been cleared in previous review articles.

2. Mechanisms of DNA Methylation

2.1. What is DNA Methylation?

DNA methylation is a biological process that occurs in the addition of methyl groups to DNA. Methylation marks on DNA occur mainly on the $5'$ position of cytosine residues of a CpG. It contributes to the epigenetics by regulating the gene expression without changing the DNA sequence [1]. In prokaryotes, DNA methylation is essential for transcription, the direction of post-replicative mismatch repair, the regulation of DNA replication, cell-cycle control, bacterial virulence, and differentiating self and non-self DNA [2]. In mammalians, DNA methylation is crucial in many key physiological processes, including the inactivation of the X-chromosome, imprinting, and the silencing of germline-specific genes and repetitive elements [1]. Besides, DNA methylation has been found to be present in actively transcribed gene bodies, and it may play a part in suppressing cryptic transcriptional initiation from the interior of genes [3]. DNA methyltransferase (DNMT) enzymes, which are pivotal for normal development, catalyze the transfer of the methyl group to DNA [4]. Importantly, DNMT's interaction with other components and modifications are required to maintain DNA methylation [6].

The methylation cycle starts with the transportation of a methyl group by tetrahydrofolate, which carries it on its N-5 atom. Since the transfer potential of tetrahydrofolate is not sufficiently high for most biosynthetic methylations, S-adenosyl-L-methionine (SAM) supplies the main activated methyl donors for DNA methylation, which is synthesized by the transfer group from ATP to the sulfur atom of methionine. The positively charged sulfur atoms and the methyl groups become more electrophilic, and thus, the high transfer potential of the S-methyl group enables it to be transferred to a wide variety of acceptors. After SAM transfers the methyl group to an acceptor, S-adenosylhomocysteine (SAH) forms, which then hydrolyzes to homocysteine and adenosine [7].

Methionine can be renewed by the transfer of a methyl group to homocysteine from N^5-methyltetrahydrofolate [7]. Additionally, this reaction is catalyzed by methionine synthase (MS) and requires vitamin B12 as a cofactor in animals. However, the same system in plants is cobalamin-free [8]. In mammalians, not only vitamin B12 has an important cofactor role: vitamin B2, which is a cofactor of methylenetetrahydrofolate reductase (MTHFR), and vitamin B6, which is a cofactor of serine hydroxymethyltransferase (SHMT), also have crucial roles as precursors of SAM [9].

Betaine is also an important methyl donor mediated by betaine homocysteine methyltransferase (BHMT), which is an alternative pathway that supplies the transfer of homocysteine to methionine [10].

Betaine can be produced through the irrecoverable oxidation of choline, and converts into dimethylglycine (DMG) after it provides a methyl group to homocysteine [11]. Therefore, any changes in these cofactors or enzymes may change the activity of folate and the methionine cycle, and thus further DNA methylation (Figure 1).

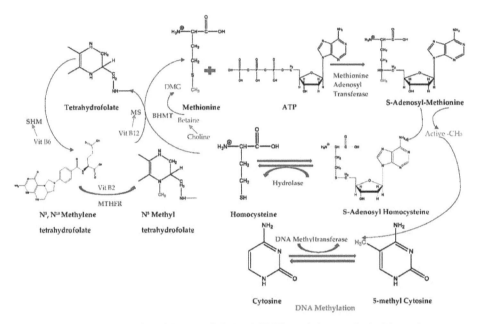

Figure 1. One-Carbon Cycle and DNA Methylation. MTHFR, methylenetetrahydrofolate reductase; SHMT, serine hydroxymethyltransferase; MS, methionine synthase; BHMT, betaine homocysteine methyltransferase; DMG, dimethylglycine.

As it has been revealed, to methylate CpG sites in DNA, methyl groups need to enter the methionine cycle in the conversion of homocysteine into methionine; here, they are made highly reactive by the addition of adenosyl groups, which are specific bases in DNA that are modified by SAM. Furthermore, the CpG base to be methylated is flipped out of the DNA double helix into the active site, where it can accept the methyl group from SAM [7]. Primarily, DNA methyltransferases catalyze the transference of the methyl groups from SAM to methylate cytosine in DNA [2].

2.2. Why Are DNA Methyltransferases Essential for DNA Methylation?

DNMTs are the enzymes that maintain the formation of DNA methylation [12], and have key roles in transcriptional silencing, transcriptional activation, and post-transcriptional gene regulation [13]. Mammalians encode five DNMTs: DNMT1, DNMT2, DNMT3A–DNMT3B (de novo methyltransferases), and DNMTL. DNMT1, DNMT3A, and DNMT3B are the three active enzymes that maintain DNA methylation. DNMT3L has no catalytic activity and functions as a regulator of DNMT3A and DNMT3B, whereas DNMT2 acts as a tRNA transferase rather than a DNA methyltransferase [14]. The coordination of all of the DNMT enzymes is crucial for the regulation of DNA methylation patterns [15]. Although both DNMT1 and DNMT3 enzymes have different and non-redundant functions, they act cooperatively in some respects, such as both enzymes being required for the maintenance of the global hypomethylation patterns in mouse embryonic stem cells [16]. However, the functioning of both enzymes together and the mechanisms that maintain methylation patterns are still debatable [17].

Furthermore, in the past few years, studies have increased their attention on assessing the functional role of DNMTs by combining molecular approaches with a broad analysis of methylation patterns [13] such as looking through the pathways through which DNMTs catalyze the transference of the methyl groups to DNA from SAM, which is required as a cofactor [2]. The research has revealed that DNMTs are mechanistically multi-directional, which supports the notion that these enzymes have a significant role in epigenetic regulations [13].

2.3. How DNMT1 Functions in DNA Methylation

DNMT1 consists of a C-terminal methyltransferase domain and an N-terminal regulatory domain that mediates interactions with proteins, substrates, DNA, etc. [18]. DNMT1 is mainly responsible for replicating pre-existence methylation patterns, from hemimethylated CpG sites to the newly synthesized strands [19]. Additionally, DNMT1 contains functional subdomains that mediate molecular interactions. In order to have a deeper understanding of metabolic pathways, it is crucial to understand the role of these subdomains, which consist of the DNMT1-associated protein 1 (DMAP1) binding domain, replication foci targeting sequence (RFTS) domain, CXXC domain, Bromo-adjacent homology (BAH) domain, and catalytic domain [13]. To give a brief overview of their molecular roles: the DMAP1 binding domain is a protein that links DNMT1 to histone acetylation [20]. The RFTS domain targets the DNMT1 to replication foci, and thus promotes post-replicative maintenance methylation [21]. The CXXC domain is a zinc-finger domain, which mediates binding to unmethylated CpG dinucleotides [22]. Unfortunately, the functions of the BAH domain are still unknown.

Furthermore, the DNMT1 activity can be regulated by other molecular interactions. For instance, the DNMT1-interacting protein E3 ubiquitin-protein ligase (UHRF1), which is essential for methylation, flips the methylated base out of the DNA helix, and thus targets DNMT1 to its physiological substrate. Moreover, both UHRF1 depletion and overexpression indicated a global loss of DNA methylation [23,24], which indicates the importance of the interaction of DNMT1 with proteins in DNA methylation pathways. However, a small number of studies have shown the biochemical mechanisms leading to aberrant DNA methylation when DNMT1's expression is reduced. To illustrate, an epigenetic study has demonstrated that mice with low DNMT1 expression at 10% of the wild-type level established a marked reduction in genome-wide DNA methylation, and revealed a significant increase in genomic instability and the activation of proto-oncogenes [25]. On the other hand, the study did not show whether DNMT reduction causing abnormal DNA methylation was because of the repression of sub-binding domains, the catalytic site, or DNMT-interacting proteins. It is necessary to know the underlying reasons in order to be able to understand the functioning of DNA methylation.

2.4. How DNMT3 Functions in DNA Methylation

DNMT3A and DNMT3B, de novo methyltransferases, are responsible for the methylation of unmodified DNA and the establishment of DNA methylation patterns [19]. These enzymes are mainly essential for de novo methylation, but several studies have shown that DNMT3 enzymes are also crucial for the stable inheritance and active remodeling of DNA methylation patterns in differentiated cells [26,27]. Structurally, both enzymes have a C-terminal catalytic domain that is similar to DNMT1, and a variable region at the N-terminus [28]. Additionally, two subdomains have also been described for DNMT3 enzymes that are important for chromatin interactions, which are the Pro-Trp-Trp-Pro (PWWP) and ATRX-DNMT3-DNMT3L (ADD) domains [29].

The targeted impairment or inactivation of both DNMT3A and DNMT3B in mammal embryonic stem cells blocks de novo methylation [30] and leads to the gradual loss of DNA methylation [14]. It has been indicated that the impaired activity of DNMT3A is a causal factor of tumorigenesis that causes global hypomethylation in specific types of cancer [15,31]. The deletion and overexpression of DNMT3B have been shown, respectively, to suppress and stimulate a specific type of cancer [32,33]. Knockout studies in mice have shown that de novo DNA methylation is pivotal for development, while DNMT3A-deficient mice die several weeks after birth, and DNMT3B-deficient mice die in utero [34].

To sum up, it is assumed that altering the regulation of DNMT3 enzymes may affect DNA methylation activity that results in several diseases, but further studies are needed to clarify the mechanisms between the activity of DNMT3 enzymes and DNA methylation.

2.5. What Is DNA Demethylation?

DNA demethylation is the process of removal of the methyl group. Currently, the DNA demethylation process is not clearly identified because of the multiple different pathways that contribute and act redundantly during this process [35]. DNA methylation has always been an active chemical process, which was originally regarded as an irreversible modification [35], but now it is found that DNA demethylation can occur, which follows either a passive or active process [36].

Mainly, DNA demethylation is passively diluted after DNA replication. However, recently, it has been revealed that DNA demethylation may also occur through the active process [14]. Unfortunately, studies about DNA demethylation mechanisms are conflicting, and its interaction with modifiable factors in mammals is still not well understood [37].

2.6. How Active DNA Demethylation Occurs

It has been proposed that the direct conversion of 5-methylcytosine (5mC) to cytosine does not occur [35]. Instead, active demethylation follows a series of chemical reactions that further transform 5mC to 5-hydroxymethylcytosine (5hmC), 5-hydroxymethyluracil (5hmU), 5-formylcytosine (5fC), 5-carboxylcytosine (5caC), and thymine (Thy), by deamination and/or oxidation reactions. Additionally, ten-eleven translocation (TET) and activation-induced deaminase (AID) enzymes catalyze these reactions. Later, these products are believed to recognized, mainly, by the base excision repair (BER) pathway to replace the modified base with naked cytosine [38]. Similarly, uracil misincorporation is repaired by BER, involving a series of enzymatic steps [39] (Figure 2). However, reserving the reactions in global DNA demethylation has generated conflicting results [40].

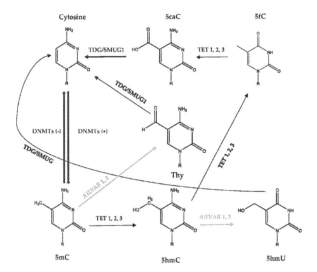

Figure 2. Active DNA demethylation process. 5mC, 5-methylcytosine; 5hmC, 5-hydroxymethylcytosine; 5hmU, 5-hydroxymethyluracil; 5fC, 5-formylcytosine; 5caC, 5-carboxylcytosine; Thy; thymine; DNMT, DNA methyltransferase; TET, ten-eleven translocation; AID, activation-induced deaminase; TDG, thymine DNA glycosylase; SMUG1, single-strand selective monofunctional uracil DNA glycosylase.

Active DNA demethylation is often carried out by members of the ten-eleven translocation (TET) family of enzymes, which functions against the actions of the DNMTs and prevents genome hypermethylation. The three members (TET1, TET2, and TET3) of the TET family oxidize 5mC to promote DNA demethylation [41]. The interaction between TET and demethylation has been shown, as TETs oxidize 5mC and significantly reduce the level of 5mC, which may result in hypomethylation. On the other hand, a loss of TETs may result in hypermethylation [42]. Furthermore, the balance between TET and other demethylation enzymes is also important. For instance, DNMTs and TETs are necessary to define the methylation landscape of gene regulatory regions. The examinations of mice that lack both DNMT3A and TET have suggested that these enzymes act in both a counteractive and synergetic way [43]. Moreover, it has been proposed that the conversion of 5mC to 5hmC by TET1 initiated an oxidative deamination process mediated by the coordinated action of activation-induced deaminase (AID)/APOBEC proteins and the BER pathway, which led to DNA demethylation in the adult brain [44]. AID deaminates cytosine to uracil and, to a smaller extent, 5mC to Thy, by simple hydrolysis [35]. The role of AID in demethylation and expression in embryonic stem cells is controversial. Popp et al. demonstrated that the absence of AID has increased DNA methylation levels, mainly in introns and repetitive elements and also in exons, but not in the promoter regions [45]. However, in vitro findings by Nabel et al., which also apply in vivo, showed that the role of AID in the direct demethylation of 5mC and 5hmC may be limited [46].

Evidence suggests that some of the products of 5mC oxidation could essentially revert back to cytosine [35]. It is known that thiol reagents and DNMT3A/DNMT3B can convert 5hmC (with the loss of formaldehyde) and 5caC (with the loss of CO_2) to cytosine in the absence of SAM [47]. Moreover, uracil generated from cytosine can be excised by thymine DNA glycosylase (TDG) and single-strand selective monofunctional uracil DNA glycosylase (SMUG1). The TDG enzyme is one of the main BER glycosylases involved in the base excision step, which is able to revert 5caC, 5fC, and Thy back to cytosine; it also plays a crucial role in cellular defense against genetic mutation caused by self-induced deamination of 5mC and cytosine [48]. In addition to BER, nucleotide excision repair (NER), Gadd45a, and non-canonical mismatch repair (ncMMR) systems are suggested to have roles in the reverse step of active DNA demethylation [35]. However, both the BER and alternative repair pathways are not well understood, and it is not clear how modified factors may affect the regulation of these pathways.

3. What Are the Underlying Mechanisms of Diet and DNA Methylation?

Numerous studies have focused on the link between diet and DNA methylation in mammalians to elucidate the dietary exposures that may have lifelong consequences on epigenetic marks [12,49,50]. Different types of researchers (in vitro and in vivo) have presented the relationship between nutrition and DNA methylation, including prenatal and postnatal periods, showing that diets deficient in methyl donors and proteins may cause global DNA hypomethylation, or that high-fat diet consumption may result in changes in DNA methylation [1,51–54]. One of the most popular models that has studied the link between diet and DNA methylation is the 'yellow agouti (A^{vy}) mice' model. The agouti gene is responsible for the regulation of brown/black (eumelanin) and yellow (pheomelanin) pigmentation in the mammalian coat. It has been shown that dietary methyl donors' supplementation of dams can change the coat color by correlating with the A^{vy} methylation status [55]. However, the agouti mice model and most of these studies were incapable of showing the underlying epigenetic mechanisms regarding whether the DNA methylation occurred due to the expression or inhibition of special binding sites of methylation enzymes, substrates, cofactors, or something else. Besides, there are other questions that need to be clarified, especially regarding the nutrient doses and the duration of dietary exposure to DNA methylation [56].

Recently, evidence has suggested that nutrition affects the epigenetic regulation of DNA methylation in several possible epigenetic pathways: mainly, by altering the substrates and cofactors that are necessary for proper DNA methylation; additionally, by changing the activity of enzymes regulating the one-carbon cycle; and, lastly, by playing a role in several possible mechanisms related to DNA demethylation activity [1].

3.1. How Diet Influences Methylation Cycle and Methyl Donors

The key methyl donor for DNA and protein methyltransferases, SAM, is synthesized in the methionine cycle while accompanied by various nutrients present in the diet, including methionine, folate, choline, betaine, vitamins B2, B6, and B12 [57]. These nutrients act as precursors and contribute to the production of SAM, although they enter the cycle at different sites [1]. Therefore, any deficiencies in these nutrients may result in changes in the SAM pool, which can influence DNMTS' reaction kinetics and DNA methylation, as well. Taking this opinion into account, many studies have started to investigate the link between SAM availability and its dietary sources, together with endogenous genetic factors [4].

Furthermore, after the removal of the methyl group, SAM is transformed into SAH, which is a strong competitive inhibitor of almost all methylation reactions, and also competes with SAM for the active site on the methyltransferase enzyme [58,59]. Since the supply of SAM and removal of SAH is necessary for DNMT activity, the SAM/SAH has been suggested as a 'methylation index' to show the probability of DNA hypermethylation or hypomethylation [50]. Remarkably, some earlier studies have shown that SAH is an inhibitor of the DNMT-mediated DNA methylation [60,61]. Additionally, moderate elevations in plasma homocysteine concentrations have been shown to be associated with increased levels of SAH, but not SAM, and increased SAH levels have been associated with global DNA hypomethylation [59].

So far, the methyl and folate-deficient diets have been found to be largely associated with reduced levels of SAM, increased levels of SAH, and decreased SAM/SAH ratios in the livers of male rats and mice [62–64]. The changes in SAM and SAH levels also showed irreversible alteration in hepatic DNA methylation [63]. Moreover, a study showed that zinc deficiency has reduced the use of methyl groups from SAM in rat liver and resulted in global DNA hypomethylation [65]. A low-protein diet or undernutrition during gestation in mice and in utero in human studies resulted in both hypomethylation and hypermethylation at specific loci in offspring [66–69]. Although one study hypothesized that the hypomethylation of certain promoters upon protein restriction may be a consequence of decreased methyl group availability [66], most of the studies remained unclear regarding how diet changes the activity of DNA methylation, and they did not observe the upregulation of SAM, SAH, or DNMTs.

On the other side, high dietary methionine intake is believed to increase DNA methylation, and the methyl groups that are transferred in mammalian DNA methylation reactions are believed to eventually derive from methionine [55]. High doses of folate supplementation showed an increase in methylation and normalized gene expression at specific loci, which is believed to induce a substantial increase of the intracellular pool of the SAM and SAM/SAH ratio [70,71]. However, little is known regarding the effect of methionine or methyl donors' supplementation, and the mechanisms of action on DNA methylation are not clear [50]. Very few studies have examined the epigenetic mechanisms of the effects of high methionine intake on DNA methylation. In an epigenetic mouse model, Tremolizzo et al. [72] studied the effect of methionine on SAM, SAH, methylation status, and the expression of the reelin gene in the frontal cortex. The study showed interesting results. For example, after 15 days of methionine treatment, brain SAH was found to be double, whereas SAM was not affected. The reduction in the SAM/SAH ratio would be expected to hypomethylate DNA, but it has been found that specific CpG sites in the reelin promoter were actually hypermethylated in the cortex of methionine-treated mice. Hence, the significant increase in CpG methylation appeared to downregulate reelin expression. A follow-up study from Dong et al. [73] showed that a 15-day methionine (MET) treatment increased the binding of methyl CpG binding protein 2 (MeCP2) to the reelin promoter, which is thought to be the factor behind hypermethylation. However, the same effect was not found in other control genes (*Gad65* and *β-globin*). Another model examining MET-induced alterations in DNA methylation found no significant dietary effects on genome-wide DNA methylation, although methionine supplementation significantly decreased the SAM/SAH ratio in the liver and brain [74]. The problem with determining the SAM, SAH, and their ratio in order to examine nutritional influences on DNA methylation is complicated for several reasons. To start with, each mammalian cell is responsible for synthesizing its own SAM, and SAM cannot cross the plasma membrane. However, SAH does leak from the cell with excessive accumulation. Thus, interpreting the SAM/SAH ratio on a tissue-specific basis and the ratio in plasma may not provide a meaningful indication of systemic methylation [75].

Betaine is an important methyl donor, which can be produced by choline or taken through diet. Betaine converts into dimethylglycine (DMG) after it provides a methyl group to homocysteine [11]. Studies have suggested that plasma DMG is a good indicator of betaine utilization as a methyl donor [76,77]. Moreover, SAM can inhibit BHMT and reduce the usage of betaine as a methyl donor [10], and it is important for SAM to stimulate the BHMT pathway in order to sustain its concentrations [78]. Choline methyl-deficient diets showed reduced hepatic concentrations of SAM and increased levels of SAH in the livers of mice [62]. A rat study evaluating the choline-deficient diet for seven days also showed that the effects of choline deficiency on reducing liver methionine formation by 20–25%, SAM by 60%, and increasing liver SAH by 50% were significant [79]. Plasma SAM levels were found to be significantly correlated with plasma levels of choline and DMG, but not with betaine [80]. To date, evidence has also shown that folate deficiency may lower choline and betaine levels in liver, or that choline deficiency may decrease hepatic folate stores, and thus can affect the methyl transfer of one carbon cycle in the liver [81,82]. On the other hand, a study showed that folic acid-supplemented, BHMT-deleted mice have produced more hepatic SAM compared to BHMT-deleted mice fed a folate-deficient diet or a control diet [83]. It has been a long time since a diet very low in choline and methionine resulted in the decreased methylation of cytosine in the liver [84–87]. However, studies have failed to show the direct interaction between choline, biotin, and DNA methylation through SAM and SAH activities or different mechanisms, if available.

Ultimately, most of the studies did not show the biochemical mechanisms of how methyl donors lead to aberrant DNA methylation. They relied heavily on assumptions. It is not clear how reduced levels of SAM or increased levels of SAH were causing global hypomethylation. Is it because there were not enough methyl donors to bind DNMTs? Alternatively, perhaps SAH was inhibiting the entry of the DNA nucleotide cytosine into enzymes' active sites. It is believed that there is not a simple correlation between methyl donors and DNA methylation. Hence, more studies are warranted to explain the underlying mechanisms in order to contribute to set patterns of DNA methylation in cells.

3.2. What Are the Diet-Related Cofactor and Enzyme Activities in One-Carbon Cycles?

Enzymes taking a role in the folate cycle (MTHFR, MTR, MS, SHMT, etc.) are regulated by micronutrients such as vitamins B2, B6, and B12. It is assumed that supplementing diets with these micronutrients may contribute to the maintenance of DNA methyl marks and therefore regulate DNA methylation [71]. Additionally, it is believed that variations in the bioavailability of these micromolecules may affect DNA methylation by altering the activity of the one-carbon cycle and the production of SAM [1].

MTHFR is an essential enzyme for the maintenance of the folate cycle and methylation of CpG islands [88]. SAM is a strong inhibitor of MTHFR, which also makes it the major regulator of folate-dependent homocysteine remethylation [89]. MTHFR activity may deteriorate due to an excess concentration of methionine and SAM or polymorphisms, or a low concentration of its cofactor vitamin B2, which decreases the synthesis of 5-methyltetrahydrofolate and thus the remethylation of homocysteine [90]. Conversely, when SAM concentrations are low and cofactor levels are high, the remethylation of homocysteine may be favored [89].

Moreover, a reduction of MTHFR activity increases the 5,10-methylenetetrahydrofolate levels while it drops the 5-methyltetrahydrofolate levels, which in return may favor the synthesis of deoxythymidine triphosphate (dTTP) over the methylation of CpG, and therefore alter DNA methylation [88,90]. Additional research has suggested that subjects who are homozygous for the polymorphism (*C677T*) in the *MTHFR* gene exhibited a significantly lower level of methylated DNA, but only under conditions of low folate status [91]. In tissue culture, a study has shown that folic acid, vitamin B2, and *MTHFR*

C677T polymorphism affect genome instability, and that high B2 concentration may increase the activity of MTHFR, which may lead folate to provide methyl groups for the methionine synthesis enzyme instead of for thymidylate synthase [88]. Furthermore, it has been suggested that low vitamin B2 concentration in the presence of low folate may maximize the risk of genome hypomethylation [88]. However, this study did not measure DNA methylation directly. Instead, it measured several markers related to genome stability and linked it with methylation. Unfortunately, most of the evidence from in vivo studies has not clarified the direct link between folate cycle enzymes or cofactors and DNA methylation. More studies are warranted in order to evaluate the interaction between diet-enzyme activities in the one-carbon cycle and DNA methylation.

3.3. How Diet Affects the DNA Methyltransferase Activity

Li et al. were the first scientists showing the *DNMT1* gene leading to the genome-wide loss of DNA methylation and embryonic lethality in mice [92]. Numerous other studies later underlined the link between DNMTs and DNA methylation [93]. Besides, it is believed that those genetic modifications and the DNMT's activity can be modified by nutritional factors. Animal studies reported that feeding methyl-deficient diets for nine weeks or longer caused DNA hypomethylation, which was associated with the suppressed expression of DNMT1 [94,95]. Lillycrop et al. showed a significant decrease of DNA methylation following a protein-restricted diet in pregnant rats, and indicated that altered DNMT1 expression may provide a mechanism for the induction of the hypomethylation of specific genes and individual CpG, although they did not show how such targeting may occur [96]. In this section, potential nutrient-based epigenetic mechanisms mostly involving the inhibition of DNMTs and altered DNA methylation have been evaluated.

The studies outlined in Table 1 suggest that several diet compounds may directly affect the expression of DNMT, or that methyl donors from the diet may indirectly modify DNMT activity by changing the intracellular concentration of SAM [97]. These assumptions have been demonstrated for several bioactive food components such as epigallocatechin-3-gallate (EGCG), genistein, caffeic acid, ascorbate, etc. [1]. A study found that each of the tea polyphenols (catechin, epicatechin, and EGCG) and bioflavonoids (quercetin, fisetin, and myricetin) inhibited SssI DNMT and DNMT1-mediated DNA methylation in a concentration-dependent manner. EGCG was found to be a more potent inhibitor that had direct inhibitory interaction with the DNMTs and the catalytic site of the human DNMT1. Additionally, when epicatechin was used as a model inhibitor, kinetic analyses indicated that this catechol-containing dietary polyphenol inhibited enzymatic DNA methylation (indirect) in vitro, largely by increasing the formation of SAH. [98]. Moreover, the treatment of the human esophageal KYSE 510 cell line with EGCG showed a dose and time-dependent reversal of hypermethylation and the re-expression of mRNA of *p16^INK4a^*, *RARβ*, *MGMT*, and *hMLH1* genes. Reactivation of some methylation-silenced genes by EGCG was also demonstrated in human colon cancer HT-29 cells, prostate cancer PC3 cells, and KYSE cells [99]. Both studies tried to explain the underlying mechanisms between EGCG and DNMT by using the structural model, molecular docking, and binding energy analysis. They revealed that EGCG shows competitive inhibition of DNMT1 by forming hydrogen bonds within the DNMT1 catalytic-binding region, thus blocking the entry of the DNA nucleotide cytosine into its active site, and inhibiting the methylation process [98,99]. Several other studies also revealed that EGCG decreased global DNA methylation levels, and also showed a protective effect by inhibiting the promoter hypermethylation of specific genes. These effects were attributed to the decreased mRNA and protein expression activity of DNMT1 and EGCG inducing the binding domain of DNMT1 to the promoter of the specific genes [100–103].

Genistein also showed a dose-dependent inhibitory effect on recombinant DNMT1 activity, and also decreased DNMT activity in nuclear extracts from KYSE cells, but this activity was found to be weaker than that of EGCG. However, six days of genistein treatment did not affect the mRNA expression levels of DNMTs and the methyl-CpG binding domain 2. Although genistein was found to have a synergistic or additive effect on DNMT inhibitors because it is a weak inhibitor of DNMTs, genomic global hypomethylation was not expected to occur after the dietary intake of soy isoflavones [104]. Another study showed that a genistein diet (300 mg of genistein/kg) was positively correlated with alterations in prostate DNA methylation at CpG islands of specific mouse genes. However, the mechanistic role of genistein was not examined [105].

Lee et al. revealed the effect of several other catechol polyphenols on DNMT activity. It has been shown that quercetin, fisetin, and myricetin may inhibit DNMT activity by transferring SAM to SAH [98]. The same group also showed that two common coffee polyphenols, caffeic acid and chlorogenic acid, have inhibited DNA methylation, which was catalyzed by prokaryotic CpG methylase (M.SssI) DNMT and human DNMT1. The inhibition of DNA methylation by caffeic acid or chlorogenic acid was found to be concentration-dependent, and the inhibition was predominantly through a non-competitive mechanism, which suggested that it was due to the increased formation of SAH [106]. Eventually, caffeic acid/chlorogenic acid treatment in cultured human breast cancer cells showed no significant change in the global methylation status. However, the concentration-dependent inhibition of DNA methylation in the promoter region of the *RARβ* gene was detected, which showed a potential inhibition effect in the promoter region [106].

Curcumin, an antioxidant component of a spice called turmeric, has been investigated by some study groups for its effect on DNA methylation [107]. Liu et al. suggested that curcumin covalently blocks the catalytic thiolate of DNMT1 to exert its inhibitory effect on DNA methylation by using molecular docking [108]. Moreover, a combination of curcumin with the hypomethylating agents increased the response to the drug in breast cancer patients [109]. However, Medina-Franco et al. suggested that curcumin has no significant effect on DNMT inhibition and global hypomethylation after following a multistep docking approach [110]. Thus, more studies are required to detect an interaction between curcumin and DNA methylation.

Parthenolide, a component of a plant called feverfew, has been used for the treatment of several diseases. It has been suggested that parthenolide may have a potential role in inhibiting the activity of DNMT1 by blocking the enzyme's catalytic site, and a study indicated that dose and cell type-dependent parthenolide treatment decreased DNMT1 protein levels and induced a decrease in global DNA methylation. The same study showed that parthenolide inhibited the DNMT1 analog M.SssI by blocking the functional thiolate of the enzyme. Although parthenolide's binding energy is not as strong as EGCG, it has been suggested that it may be an effective DNA methylation inhibitor [111].

Mahanine is found in several Asian herbs and species, and it is an alkaloid from the leaves of the curry leaf tree (*Murraya koenigii*) and lime berry (*Micromelum minutum*). It is mostly studied for its anti-inflammatory and anti-mutagenic activity. [112,113]. Mahanine is thought to have an anti-proliferative activity, which was associated with the inhibition of DNMT activity, and hence, may prevent the hypermethylation of a specific gene in the prostate cancer cell line [114]. However, the mechanisms of action were not clarified.

Eventually, studies evaluating the consumption of polyphenols showed that in general, EGCG and several other polyphenols are promising candidates, especially for future cancer therapies, based on their influence on the epigenetic pathway. Most of these studies showed kinetics and possible mechanisms that alter DNA methylation. These include increasing SAH, inhibiting DNMT's catalytic base, blocking the promoter sites of specific genes, or covalently binding to thiol groups of enzymes/transcriptional factors. However, future studies evaluating the underlying mechanisms are still needed in order to clarify the pathways of epigenetics.

3.4. Is There a Link between Diet and DNA Demethylation?

The reversal of DNA methylation is crucial, and abnormalities are often observed in anomalies and diseases. Genetic and modifiable factors such as diet may affect the regulation of DNA demethylation, and thus genetic regulations. However, DNA demethylation's interaction with modifiable factors in mammals is still not well understood [37]. Recent epigenetic studies have tried to investigate the link between nutrition and active DNA demethylation, which is believed to lead to several modifications in DNA methylation. One study tried to clarify the DNA methylation status of the liver of mice fed the methionine–choline-deficient (MCD) diet (for a week) by measuring the amount of 5mC and investigating the involvement of the active DNA demethylation. The results showed that the expression of DNMT1 and DNMT3a was significantly increased on the MCD diet. In addition, mRNA expression of Tet2 and Tet3 was significantly upregulated on the MCD diet. However, no statistical differences for 5mC content and other demethylation enzymes were found [115]. It is believed that for better epigenetic investigations, long-term studies are necessary. The deletion of Tet2 was found to cause an extensive loss of 5hmC, which was accompanied by enhancer hypermethylation and delayed gene induction in the early steps of differentiation [116]. It is assumed that methyl-deficient diets that alter the expression Tet2 may contribute to hypermethylation in specific areas [115].

Table 1. Studies that have evaluated the interaction between bioactive dietary components [1] and DNMT's activity [2].

Studies	Dietary Components	Enzymes Inhibited or Expressed	Epigenetic Outcomes
Lee, W. J., et al. [98]	EGCG	DNMT1	EGCG inhibited human DNMT1 activity by binding in the catalytic core region
Fang et al. [99]	EGC–EGCG	DNMT	EGC and EGCG showed competitive inhibition of DNMT1 and treatment of the KYSE 510 cell line. EGCG showed a dose and time-dependent reversal of hypermethylation and re-expression of mRNA of $p16^{INK4a}$, *RARβ*, *MGMT*, and *hMLH1* genes
Nandakumar, V., et al. [101]	EGC–EGCG	DNMTs	EGCG reduced the activity of DNMTs by decreasing the mRNA levels and protein expression of DNMTs.
Zhang, B. K., et al. [100]	EGCG	DNMT1	EGCG inhibited the mRNA and protein expression activity of DNMT1 and downregulated binding to the promoter of DDAH2.
Shukla, S., et al. [103]	EGCG	DNMT	EGCG decreased the mRNA and protein expression activity of DNMT1, and increased the expression of unmethylation-specific GSTP1 promoter.
Pandey, M., et al. [102]	Green tea polyphenols, EGCG	DNMT1	A dose and time-dependent inhibition of DNMT activity and protein expression was observed.
Day et al. [105]	Genistein		Genistein diet was positively correlated with alterations in prostate DNA methylation at CpG islands of specific mouse genes.
Fang et al. [104]	Genistein	DNMT1	Genistein showed a dose-dependent inhibitory effect on recombinant DNMT1 activity, and also decreased DNMT activity in nuclear extracts from KYSE cells. However, no effect on the mRNA expression levels of DNMTs and methyl-CpG binding domain 2 was observed.
Lee and Zhu [106]	Caffeic acid, Chlorogenic acid	DNMT1, M.Sssl DNMT	The caffeic acid and chlorogenic acid inhibited the DNA methylation that was catalyzed by prokaryotic M.Sssl DNMT and human DNMT1, and increased levels of SAH.

Table 1. *Cont.*

Studies	Dietary Components	Enzymes Inhibited or Expressed	Epigenetic Outcomes
Liu, Z., et al. [108]	Curcumin	DNMT1,	Curcumin covalently blocks the catalytic thiolate of DNMT1 to exert its inhibitory effect on DNA methylation.
Liu, Z., et al. [111]	Parthenolide	DNMT1, M.Sssl DNMT	Dose-dependent parthenolide treatment decreased DNMT1 protein levels and induced a decrease in global DNA methylation. The same study showed that parthenolide inhibited M.SssI by blocking the functional thiolate of the enzyme.
Minor, E.A., et al. [117]	Ascorbate (Vitamin C)	DNMTs, TET2-TET3	Ascorbate increased the expression of DNMT1, DNMT3a, and mRNA expression of Tet2 and Tet3.
Sheikh, K. D., et al. [114]	Mahanine	DNMT	Mahanine was associated with the inhibition of DNMT activity, and hence, prevented the hypermethylation of a specific gene in the prostate cancer cell line. However, mechanisms are not clarified.

[1] EGCG, epigallocatechin-3-gallate; EGC, epigallocatechin; [2] DNMT, DNA methyltransferase; KYSE 510, oesophageal squamous cell carcinoma; p16^{INK4a}, tumor suppressor protein; RARβ, retinoic acid receptor beta; MGMT, *O-6*-methylguanine-DNA methyltransferase; hMLH1, human mutL homolog 1; DDAH2, dimethylarginine dimethylaminohydrolase; GSTP1, glutathione S-transferase Pi 1; M.Sssl, CpG methylase; SAH, S-adenosylhomocysteine; TET, ten-eleven translocation.

Some studies have shown that the presence of ascorbate (vitamin C) may modify the status of DNA methylation [117,118]. In embryonic stem cells, ascorbate caused the widespread DNA demethylation of nearly 2,000 genes [118]. However, it remains unknown whether the effect of ascorbate on DNA demethylation is due to an enhanced hydroxylation of 5mC. A study showed that ascorbate enhances 5hmC generation, most likely by acting as a cofactor for Tet methylcytosine dioxygenase to hydroxylate 5mC in mouse embryonic fibroblasts [117].

Pogribny et al. evaluated epigenetic changes during hepatocarcinogenesis, which was induced by diets deficient in methyl donors, in his review, and he commented that methyl donors' deficiency sustains the demethylation of genomic DNA that occurs in methyl-deficient animal's cytosine in their liver [87]. Further, the results of past studies have suggested that demethylation may be associated with decreased levels of SAM, increased levels of SAH, a decreased SAM/SAH ratio [86], and the changed activity of DNMTs [119]. However, the latest studies have demonstrated that DNA hypomethylation or demethylation induced by methyl-deficient diets might be attributed to the induction of uracil, 5hmC, and 8-oxodeoxyguanosine [95,120]. The presence of these products may significantly coordinate with DNMT1 and lead to the demethylation of DNA [121].

Less is known about the role of nutrition in the base excision repair system. In one of the few studies that has examined the five genes (*SMUG1, TDG, UNG, MBD4*, and *DUT*) that are involved in the repair system to identify polymorphisms and establish whether one-carbon nutrient status can further alter their effects, single nucleotide polymorphisms in *SMUG1, DUT*, and *UNG* genes showed an association with DNA uracil concentration. However, one-carbon nutrient status was not associated with DNA uracil concentration, and did not modify the effect of the single nucleotide polymorphisms [122]. An older study showed that folate deficiency impairs the DNA excision repair system in rat colonic mucosa [123], and folate status was found to be associated with uracil misincorporation and genomic instability in humans. However, both studies were not linked to DNA demethylation. Together, the evidence suggests that more studies are required in order to understand the demethylation pathways and the part that dietary factors play in demethylation.

4. Conclusions

It is well known that nutrition has an indisputable influence on the epigenome. A great number of studies showed the changes in DNA methylation in specific genes, tissues, hormones, and cell lines after applying different diets [59–61,63,75]. These findings raise important questions about the diet-induced epigenetic pathways, such as: 'What are the underlying regulatory pathways causing hypomethylation or hypermethylation?' Recent evidence makes it clearer that the mechanisms regulating DNA methylation are very complicated, and that there is not one answer to this question. However, several possible assumptions were made for the interaction between diet and DNA methylation (Figure 3). It is suggested that nutrition may affect the epigenetic regulation of DNA methylation by altering the substrates and cofactors that are necessary for proper DNA methylation such as methyl donors, SAM, and SAH. These factors may impair the DNMT's catalytic base, blocking the promoter sites of specific genes or covalently binding to the thiol groups of the enzymes. Likewise, nutrition-based cofactors may change the activity of enzymes regulating the one-carbon cycle and the production of SAM. Lastly, nutrition may have a role in several possible mechanisms related to DNA demethylation activity, which have been suggested to be a new epigenetic approach [11]. For example, changing the expression of Tet family enzymes by methyl–choline-deficient diets is believed to alter DNA methylation [123].

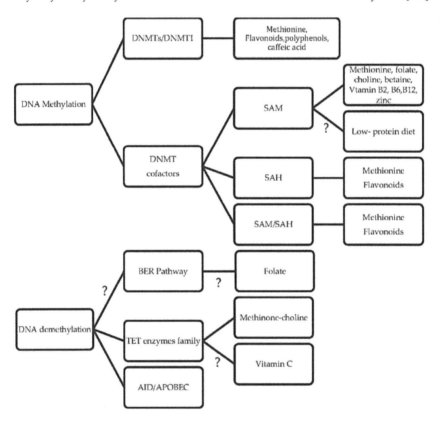

Figure 3. The possible mechanisms that nutrition can stimulate patterns of DNA methylation.

Diet influences organs, body systems, and epigenetics as well. It is extremely important for researchers to study the mechanisms of dietary implications on DNA methylation in order to determine the optimal concentration of macro nutrients and micronutrients for genome stability, which would provide a guide to establishing recommended dietary allowances for the prevention of genome damage and further diseases [124]. However, our knowledge of nutrition and epigenetic mechanisms are still limited. Future studies are required that should focus on the comprehensive understanding of nutrition-epigenetic mechanisms and biochemical pathways, especially those interacting with enzymatic functions rather than just showing general modifications.

Funding: This research received no external funding.

Conflicts of Interest: The authors declare no conflict of interest.

Abbreviations

MTHFR	Methylenetetrahydrofolate reductase
SHMT	Serine hydroxyl methyltransferase
MS	Methionine synthase
BHMT	Betaine homocysteine methyltransferase
DMG	Dimethylglycine
5mC	5-methylcytosine
5hmC	5-hydroxymethylcytosine
5hmU	5-hydroxymethyluracil
5fC	5-formylcytosine
5caC	5-carboxylcytosine
Thy	Thymine
DNMT	DNA methyltransferase
TET	Ten-eleven translocation
AID	Activation-induced deaminase
TDG	Thymine DNA glycosylase
SMUG1	Single-strand selective monofunctional uracil DNA glycosylase
SAM	S-adenosyl-L-methionine
SAH	S-adenosylhomocysteine
BER	Base excision repair
RFTS	Replication foci targeting sequence
BAH	Bromo-adjacent homology
UHRF1	Interacting protein E3 ubiquitin-protein ligase
NER	Nucleotide excision repair
dTTP	Deoxythymidine triphosphate
EGCG	Epigallocatechin-3-gallate

References

1. Zhang, N. Epigenetic modulation of DNA methylation by nutrition and its mechanisms in animals. *Anim. Nutr.* **2015**, *1*, 144–151. [CrossRef] [PubMed]
2. Bheemanaik, S.; Reddy, Y.V.; Rao, D.N. Structure, function and mechanism of exocyclic DNA methyltransferases. *Biochem. J.* **2006**, *399*, 177–190. [CrossRef] [PubMed]
3. Ma, D.K.; Guo, J.U.; Ming, G.L.; Song, H. DNA excision repair proteins and Gadd45 as molecular players for active DNA demethylation. *Cell Cycle* **2009**, *8*, 1526–1531. [CrossRef] [PubMed]
4. Niculescu, M.D. Nutritional epigenetics. *ILAR J.* **2012**, *53*, 270–278. [CrossRef] [PubMed]
5. Ratel, D.; Ravanat, J.-L.; Berger, F.; Wion, D. N6-methyladenine: The other methylated base of DNA. *Bioessays* **2006**, *28*, 309–315. [CrossRef] [PubMed]
6. Jin, B.; Li, Y.; Robertson, K.D. DNA Methylation: Superior or Subordinate in the Epigenetic Hierarchy? *Genes Cancer* **2011**, *2*, 607–617. [CrossRef]
7. Berg, J.M.; Tymoczko, J.L.; Stryer, L. *Biochemistry*, 5th ed.; W.H. Freeman and Co.: New York, NY, USA, 2002.
8. Zydowsky, T.M.; Courtney, L.F.; Frasca, V.; Kobayashi, K.; Shimizu, H.; Yuen, L.D.; Matthews, R.G.; Benkovic, S.J.; Floss, H.G. Stereochemical analysis of the methyl transfer catalyzed by cobalamin-dependent methionine synthase from Escherichia coli B. *J. Am. Chem. Soc.* **1986**, *108*, 3152–3153. [CrossRef]

9. Feil, R.; Fraga, M.F. Epigenetics and the environment: Emerging patterns and implications. *Nat. Rev. Genet.* **2012**, *13*, 97–109. [CrossRef]

10. Obeid, R. The Metabolic Burden of Methyl Donor Deficiency with Focus on the Betaine Homocysteine Methyltransferase Pathway. *Nutrients* **2013**, *5*, 3481–3495. [CrossRef]

11. Finkelstein, J.D.; Martin, J.J.; Harris, B.J.; Kyle, W.E. Regulation of hepatic betaine-homocysteine methyltransferase by dietary betaine. *J. Nutr.* **1983**, *113*, 519–521. [CrossRef]

12. Jaenisch, R.; Bird, A. Epigenetic regulation of gene expression: How the genome integrates intrinsic and environmental signals. *Nat. Genet.* **2003**, *33*, 245–254. [CrossRef] [PubMed]

13. Lyko, F. The DNA methyltransferase family: A versatile toolkit for epigenetic regulation. *Nat. Rev. Genet.* **2018**, *19*, 81–92. [CrossRef]

14. He, X.J.; Chen, T.; Zhu, J.K. Regulation and function of DNA methylation in plants and animals. *Cell Res.* **2011**, *21*, 442–465. [CrossRef] [PubMed]

15. Jia, Y.; Li, P.; Fang, L.; Zhu, H.; Xu, L.; Cheng, H.; Zhang, J.; Li, F.; Feng, Y.; Li, Y.; et al. Negative regulation of DNMT3A de novo DNA methylation by frequently overexpressed UHRF family proteins as a mechanism for widespread DNA hypomethylation in cancer. *Cell Discov.* **2016**, *2*, 16007. [CrossRef] [PubMed]

16. Chen, T.; Li, E. Structure and Function of Eukaryotic DNA Methyltransferases. In *Current Topics in Developmental Biology*; Academic Press: Cambridge, MA, USA, 2004; Volume 60, pp. 55–89.

17. Rhee, I.; Jair, K.-W.; Yen, R.-W.C.; Lengauer, C.; Herman, J.G.; Kinzler, K.W.; Vogelstein, B.; Baylin, S.B.; Schuebel, K.E. CpG methylation is maintained in human cancer cells lacking DNMT1. *Nature* **2000**, *404*, 1003–1007. [CrossRef] [PubMed]

18. Song, J.; Teplova, M.; Ishibe-Murakami, S.; Patel, D.J. Structure-based mechanistic insights into DNMT1-mediated maintenance DNA methylation. *Science* **2012**, *335*, 709–712. [CrossRef] [PubMed]

19. Pradhan, S.; Bacolla, A.; Wells, R.; Roberts, R.J. Recombinant human DNA (cytosine-5) methyltransferase I. Expression, purification, and comparision of de novo and maintenance methylation. *J. Biol. Chem.* **1999**, *274*, 33002–33010. [CrossRef]

20. Rountree, M.R.; Bachman, K.E.; Baylin, S.B. DNMT1 binds HDAC2 and a new co-repressor, DMAP1, to form a complex at replication foci. *Nat. Genet.* **2000**, *25*, 269–277. [CrossRef]

21. Leonhardt, H.; Page, A.W.; Weier, H.U.; Bestor, T.H. A targeting sequence directs DNA methyltransferase to sites of DNA replication in mammalian nuclei. *Cell* **1992**, *71*, 865–873. [CrossRef]

22. Song, J.; Rechkoblit, O.; Bestor, T.H.; Patel, D.J. Structure of DNMT1-DNA complex reveals a role for autoinhibition in maintenance DNA methylation. *Science* **2011**, *331*, 1036–1040. [CrossRef]

23. Bostick, M.; Kim, J.K.; Esteve, P.O.; Clark, A.; Pradhan, S.; Jacobsen, S.E. UHRF1 plays a role in maintaining DNA methylation in mammalian cells. *Science* **2007**, *317*, 1760–1764. [CrossRef] [PubMed]

24. Mudbhary, R.; Hoshida, Y.; Chernyavskaya, Y.; Jacob, V.; Villanueva, A.; Fiel, M.I.; Chen, X.; Kojima, K.; Thung, S.; Bronson, R.T.; et al. UHRF1 overexpression drives DNA hypomethylation and hepatocellular carcinoma. *Cancer Cell* **2014**, *25*, 196–209. [CrossRef] [PubMed]

25. Gaudet, F.; Hodgson, J.G.; Eden, A.; Jackson-Grusby, L.; Dausman, J.; Gray, J.W.; Leonhardt, H.; Jaenisch, R. Induction of tumors in mice by genomic hypomethylation. *Science* **2003**, *300*, 489–492. [CrossRef] [PubMed]

26. Chen, T.; Ueda, Y.; Dodge, J.E.; Wang, Z.; Li, E. Establishment and maintenance of genomic methylation patterns in mouse embryonic stem cells by Dnmt3a and Dnmt3b. *Mol. Cell. Biol.* **2003**, *23*, 5594–5605. [CrossRef] [PubMed]

27. Arand, J.; Spieler, D.; Karius, T.; Branco, M.R.; Meilinger, D.; Meissner, A.; Jenuwein, T.; Xu, G.; Leonhardt, H.; Wolf, V.; et al. In vivo control of CpG and non-CpG DNA methylation by DNA methyltransferases. *PLoS Genet.* **2012**, *8*, e1002750. [CrossRef] [PubMed]

28. Bestor, T. The DNA methyltransferases of mammals. *Hum. Mol. Genet.* **2000**, *9*, 2395–2402. [CrossRef] [PubMed]

29. Du, J.; Johnson, L.M.; Jacobsen, S.E.; Patel, D.J. DNA methylation pathways and their crosstalk with histone methylation. *Nat. Rev. Mol. Cell Biol.* **2015**, *16*, 519–532. [CrossRef]

30. Li, Y.; Tollefsbol, T.O. Impact on DNA methylation in cancer prevention and therapy by bioactive dietary components. *Curr. Med. Chem.* **2010**, *17*, 2141–2151. [CrossRef]

31. Raddatz, G.; Gao, Q.; Bender, S.; Jaenisch, R.; Lyko, F. Dnmt3a protects active chromosome domains against cancer-associated hypomethylation. *PLoS Genet.* **2012**, *8*, e1003146. [CrossRef]

32. Steine, E.J.; Ehrich, M.; Bell, G.W.; Raj, A.; Reddy, S.; van Oudenaarden, A.; Jaenisch, R.; Linhart, H.G. Genes methylated by DNA methyltransferase 3b are similar in mouse intestine and human colon cancer. *J. Clin. Investig.* **2011**, *121*, 1748–1752. [CrossRef]

33. Linhart, H.G.; Lin, H.; Yamada, Y.; Moran, E.; Steine, E.J.; Gokhale, S.; Lo, G.; Cantu, E.; Ehrich, M.; He, T.; et al. Dnmt3b promotes tumorigenesis in vivo by gene-specific de novo methylation and transcriptional silencing. *Genes Dev.* **2007**, *21*, 3110–3122. [CrossRef] [PubMed]

34. Okano, M.; Bell, D.W.; Haber, D.A.; Li, E. DNA methyltransferases Dnmt3a and Dnmt3b are essential for de novo methylation and mammalian development. *Cell* **1999**, *99*, 247–257. [CrossRef]

35. Bochtler, M.; Kolano, A.; Xu, G.L. DNA demethylation pathways: Additional players and regulators. *Bioessays* **2017**, *39*, 1–13. [CrossRef] [PubMed]

36. Chen, Z.X.; Riggs, A.D. DNA methylation and demethylation in mammals. *J. Biol. Chem.* **2011**, *286*, 18347–18353. [CrossRef] [PubMed]

37. Wu, S.C.; Zhang, Y. Active DNA demethylation: Many roads lead to Rome. *Nat. Rev. Mol. Cell Biol.* **2010**, *11*, 607–620. [CrossRef] [PubMed]

38. Bhutani, N.; Burns, D.M.; Blau, H.M. DNA Demethylation Dynamics. *Cell* **2011**, *146*, 866–872. [CrossRef] [PubMed]

39. Blount, B.C.; Mack, M.M.; Wehr, C.M.; MacGregor, J.T.; Hiatt, R.A.; Wang, G.; Wickramasinghe, S.N.; Everson, R.B.; Ames, B.N. Folate deficiency causes uracil misincorporation into human DNA and chromosome breakage: Implications for cancer and neuronal damage. *Proc. Natl. Acad. Sci. USA* **1997**, *94*, 3290–3295. [CrossRef] [PubMed]

40. Jin, C.; Qin, T.; Barton, M.C.; Jelinek, J.; Issa, J.-P.J. Minimal role of base excision repair in TET-induced global DNA demethylation in HEK293T cells. *Epigenetics* **2015**, *10*, 1006–1013. [CrossRef] [PubMed]

41. Rasmussen, K.D.; Helin, K. Role of TET enzymes in DNA methylation, development, and cancer. *Genes Dev.* **2016**, *30*, 733–750. [CrossRef] [PubMed]

42. Dawlaty, M.M.; Breiling, A.; Le, T.; Barrasa, M.I.; Raddatz, G.; Gao, Q.; Powell, B.E.; Cheng, A.W.; Faull, K.F.; Lyko, F.; et al. Loss of Tet enzymes compromises proper differentiation of embryonic stem cells. *Dev. Cell* **2014**, *29*, 102–111. [CrossRef]

43. Zhang, X.; Su, J.; Jeong, M.; Ko, M.; Huang, Y.; Park, H.J.; Guzman, A.; Lei, Y.; Huang, Y.H.; Rao, A.; et al. DNMT3A and TET2 compete and cooperate to repress lineage-specific transcription factors in hematopoietic stem cells. *Nat. Genet.* **2016**, *48*, 1014–1023. [CrossRef] [PubMed]

44. Guo, J.U.; Su, Y.; Zhong, C.; Ming, G.L.; Song, H. Hydroxylation of 5-methylcytosine by TET1 promotes active DNA demethylation in the adult brain. *Cell* **2011**, *145*, 423–434. [CrossRef] [PubMed]

45. Popp, C.; Dean, W.; Feng, S.; Cokus, S.J.; Andrews, S.; Pellegrini, M.; Jacobsen, S.E.; Reik, W. Genome-wide erasure of DNA methylation in mouse primordial germ cells is affected by AID deficiency. *Nature* **2010**, *463*, 1101–1105. [CrossRef]

46. Nabel, C.S.; Jia, H.; Ye, Y.; Shen, L.; Goldschmidt, H.L.; Stivers, J.T.; Zhang, Y.; Kohli, R.M. AID/APOBEC deaminases disfavor modified cytosines implicated in DNA demethylation. *Nat. Chem. Biol.* **2012**, *8*, 751–758. [CrossRef] [PubMed]

47. Chen, C.C.; Wang, K.Y.; Shen, C.K. DNA 5-methylcytosine demethylation activities of the mammalian DNA methyltransferases. *J. Biol. Chem.* **2013**, *288*, 9084–9091. [CrossRef] [PubMed]

48. He, Y.F.; Li, B.Z.; Li, Z.; Liu, P.; Wang, Y.; Tang, Q.; Ding, J.; Jia, Y.; Chen, Z.; Li, L.; et al. Tet-mediated formation of 5-carboxylcytosine and its excision by TDG in mammalian DNA. *Science* **2011**, *333*, 1303–1307. [CrossRef] [PubMed]

49. Sinclair, K.D.; Allegrucci, C.; Singh, R.; Gardner, D.S.; Sebastian, S.; Bispham, J.; Thurston, A.; Huntley, J.F.; Rees, W.D.; Maloney, C.A.; et al. DNA methylation, insulin resistance, and blood pressure in offspring determined by maternal periconceptional B vitamin and methionine status. *Proc. Natl. Acad. Sci. USA* **2007**, *104*, 19351–19356. [CrossRef]

50. Waterland, R.A. Assessing the effects of high methionine intake on DNA methylation. *J. Nutr.* **2006**, *136*, 1706S–1710S. [CrossRef]

51. Amarasekera, M.; Martino, D.; Ashley, S.; Harb, H.; Kesper, D.; Strickland, D.; Saffery, R.; Prescott, S.L. Genome-wide DNA methylation profiling identifies a folate-sensitive region of differential methylation upstream of ZFP57-imprinting regulator in humans. *FASEB J. Off. Publ. Fed. Am. Soc. Exp. Biol.* **2014**, *28*, 4068–4076. [CrossRef]

52. Yu, H.L.; Dong, S.; Gao, L.F.; Li, L.; Xi, Y.D.; Ma, W.W.; Yuan, L.H.; Xiao, R. Global DNA methylation was changed by a maternal high-lipid, high-energy diet during gestation and lactation in male adult mice liver. *Br. J. Nutr.* **2015**, *113*, 1032–1039. [CrossRef]

53. Altmann, S.; Murani, E.; Schwerin, M.; Metges, C.C.; Wimmers, K.; Ponsuksili, S. Dietary protein restriction and excess of pregnant German Landrace sows induce changes in hepatic gene expression and promoter methylation of key metabolic genes in the offspring. *J. Nutr. Biochem.* **2013**, *24*, 484–495. [CrossRef] [PubMed]

54. Zhang, Y.; Wang, H.; Zhou, D.; Moody, L.; Lezmi, S.; Chen, H.; Pan, Y.X. High-fat diet caused widespread epigenomic differences on hepatic methylome in rat. *Physiol. Genom.* **2015**, *47*, 514–523. [CrossRef] [PubMed]

55. Waterland, R.A.; Jirtle, R.L. Transposable elements: targets for early nutritional effects on epigenetic gene regulation. *Mol. Cell. Biol.* **2003**, *23*, 5293–5300. [CrossRef]

56. Cravo, M.L.; Pinto, A.G.; Chaves, P.; Cruz, J.A.; Lage, P.; Nobre Leitao, C.; Costa Mira, F. Effect of folate supplementation on DNA methylation of rectal mucosa in patients with colonic adenomas: Correlation with nutrient intake. *Clin. Nutr.* **1998**, *17*, 45–49. [CrossRef]

57. McKay, J.A.; Mathers, J.C. Diet induced epigenetic changes and their implications for health. *Acta Physiol.* **2011**, *202*, 103–118. [CrossRef]

58. Mato, J.M.; Alvarez, L.; Ortiz, P.; Pajares, M.A. S-adenosylmethionine synthesis: Molecular mechanisms and clinical implications. *Pharmacol. Ther.* **1997**, *73*, 265–280. [CrossRef]

59. Yi, P.; Melnyk, S.; Pogribna, M.; Pogribny, I.P.; Hine, R.J.; James, S.J. Increase in plasma homocysteine associated with parallel increases in plasma S-adenosylhomocysteine and lymphocyte DNA hypomethylation. *J. Biol. Chem.* **2000**, *275*, 29318–29323. [CrossRef] [PubMed]

60. Sibani, S.; Melnyk, S.; Pogribny, I.P.; Wang, W.; Hiou-Tim, F.; Deng, L.; Trasler, J.; James, S.J.; Rozen, R. Studies of methionine cycle intermediates (SAM, SAH), DNA methylation and the impact of folate deficiency on tumor numbers in Min mice. *Carcinogenesis* **2002**, *23*, 61–65. [CrossRef]

61. Bacolla, A.; Pradhan, S.; Roberts, R.J.; Wells, R.D. Recombinant human DNA (cytosine-5) methyltransferase. II. Steady-state kinetics reveal allosteric activation by methylated dna. *J. Biol. Chem.* **1999**, *274*, 33011–33019. [CrossRef]

62. Shivapurkar, N.; Poirier, L.A. Tissue levels of S-adenosylmethionine and S-adenosylhomocysteine in rats fed methyl-deficient, amino acid-defined diets for one to five weeks. *Carcinogenesis* **1983**, *4*, 1051–1057. [CrossRef]

63. Pogribny, I.P.; Ross, S.A.; Wise, C.; Pogribna, M.; Jones, E.A.; Tryndyak, V.P.; James, S.J.; Dragan, Y.P.; Poirier, L.A. Irreversible global DNA hypomethylation as a key step in hepatocarcinogenesis induced by dietary methyl deficiency. *Mutat. Res.* **2006**, *593*, 80–87. [CrossRef]

64. Jhaveri, M.S.; Wagner, C.; Trepel, J.B. Impact of Extracellular Folate Levels on Global Gene Expression. *Mol. Pharmacol.* **2001**, *60*, 1288–1295. [CrossRef] [PubMed]

65. Wallwork, J.C.; Duerre, J.A. Effect of zinc deficiency on methionine metabolism, methylation reactions and protein synthesis in isolated perfused rat liver. *J. Nutr.* **1985**, *115*, 252–262. [CrossRef] [PubMed]

66. Van Straten, E.M.; Bloks, V.W.; Huijkman, N.C.; Baller, J.F.; van Meer, H.; Lutjohann, D.; Kuipers, F.; Plosch, T. The liver X-receptor gene promoter is hypermethylated in a mouse model of prenatal protein restriction. *Am. J. Physiol. Regul. Integr. Comp. Physiol.* **2010**, *298*, R275–282. [CrossRef]

67. Heijmans, B.T.; Tobi, E.W.; Stein, A.D.; Putter, H.; Blauw, G.J.; Susser, E.S.; Slagboom, P.E.; Lumey, L.H. Persistent epigenetic differences associated with prenatal exposure to famine in humans. *Proc. Natl. Acad. Sci. USA* **2008**, *105*, 17046–17049. [CrossRef] [PubMed]

68. Burdge, G.C.; Slater-Jefferies, J.; Torrens, C.; Phillips, E.S.; Hanson, M.A.; Lillycrop, K.A. Dietary protein restriction of pregnant rats in the F_0 generation induces altered methylation of hepatic gene promoters in the adult male offspring in the F_1 and F_2 generations. *Br. J. Nutr.* **2007**, *97*, 435–439. [CrossRef] [PubMed]

69. Tobi, E.W.; Lumey, L.H.; Talens, R.P.; Kremer, D.; Putter, H.; Stein, A.D.; Slagboom, P.E.; Heijmans, B.T. DNA methylation differences after exposure to prenatal famine are common and timing- and sex-specific. *Hum. Mol. Genet.* **2009**, *18*, 4046–4053. [CrossRef] [PubMed]

70. Ingrosso, D.; Cimmino, A.; Perna, A.F.; Masella, L.; De Santo, N.G.; De Bonis, M.L.; Vacca, M.; D'Esposito, M.; D'Urso, M.; Galletti, P.; et al. Folate treatment and unbalanced methylation and changes of allelic expression induced by hyperhomocysteinaemia in patients with uraemia. *Lancet* **2003**, *361*, 1693–1699. [CrossRef]

71. Farias, N.; Ho, N.; Butler, S.; Delaney, L.; Morrison, J.; Shahrzad, S.; Coomber, B.L. The effects of folic acid on global DNA methylation and colonosphere formation in colon cancer cell lines. *J. Nutr. Biochem.* **2015**, *26*, 818–826. [CrossRef]

72. Tremolizzo, L.; Carboni, G.; Ruzicka, W.B.; Mitchell, C.P.; Sugaya, I.; Tueting, P.; Sharma, R.; Grayson, D.R.; Costa, E.; Guidotti, A. An epigenetic mouse model for molecular and behavioral neuropathologies related to schizophrenia vulnerability. *Proc. Natl. Acad. Sci. USA* **2002**, *99*, 17095–17100. [CrossRef]

73. Dong, E.; Agis-Balboa, R.C.; Simonini, M.V.; Grayson, D.R.; Costa, E.; Guidotti, A. Reelin and glutamic acid decarboxylase67 promoter remodeling in an epigenetic methionine-induced mouse model of schizophrenia. *Proc. Natl. Acad. Sci. USA* **2005**, *102*, 12578–12583. [CrossRef]

74. Devlin, A.M.; Arning, E.; Bottiglieri, T.; Faraci, F.M.; Rozen, R.; Lentz, S.R. Effect of Mthfr genotype on diet-induced hyperhomocysteinemia and vascular function in mice. *Blood* **2004**, *103*, 2624–2629. [CrossRef]

75. Finkelstein, J.D. The metabolism of homocysteine: Pathways and regulation. *Eur. J. Pediatr.* **1998**, *157*, S40–S44. [CrossRef]

76. Dominguez-Salas, P.; Moore, S.E.; Cole, D.; da Costa, K.A.; Cox, S.E.; Dyer, R.A.; Fulford, A.J.; Innis, S.M.; Waterland, R.A.; Zeisel, S.H.; et al. DNA methylation potential: Dietary intake and blood concentrations of one-carbon metabolites and cofactors in rural African women. *Am. J. Clin. Nutr.* **2013**, *97*, 1217–1227. [CrossRef] [PubMed]

77. Allen, R.H.; Stabler, S.P.; Lindenbaum, J. Serum betaine, *N,N*-dimethylglycine and *N*-methylglycine levels in patients with cobalamin and folate deficiency and related inborn errors of metabolism. *Metabolism* **1993**, *42*, 1448–1460. [CrossRef]

78. Jacobs, R.L.; Stead, L.M.; Devlin, C.; Tabas, I.; Brosnan, M.E.; Brosnan, J.T.; Vance, D.E. Physiological regulation of phospholipid methylation alters plasma homocysteine in mice. *J. Biol. Chem.* **2005**, *280*, 28299–28305. [CrossRef] [PubMed]

79. Zeisel, S.H.; Zola, T.; daCosta, K.A.; Pomfret, E.A. Effect of choline deficiency on S-adenosylmethionine and methionine concentrations in rat liver. *Biochem. J.* **1989**, *259*, 725–729. [CrossRef] [PubMed]

80. Imbard, A.; Smulders, Y.M.; Barto, R.; Smith, D.E.; Kok, R.M.; Jakobs, C.; Blom, H.J. Plasma choline and betaine correlate with serum folate, plasma S-adenosyl-methionine and S-adenosyl-homocysteine in healthy volunteers. *Clin. Chem. Lab. Med.* **2013**, *51*, 683–692. [CrossRef]

81. Kim, Y.I.; Miller, J.W.; da Costa, K.A.; Nadeau, M.; Smith, D.; Selhub, J.; Zeisel, S.H.; Mason, J.B. Severe folate deficiency causes secondary depletion of choline and phosphocholine in rat liver. *J. Nutr.* **1994**, *124*, 2197–2203. [CrossRef]

82. Horne, D.W.; Cook, R.J.; Wagner, C. Effect of dietary methyl group deficiency on folate metabolism in rats. *J. Nutr.* **1989**, *119*, 618–621. [CrossRef]

83. Teng, Y.W.; Mehedint, M.G.; Garrow, T.A.; Zeisel, S.H. Deletion of betaine-homocysteine S-methyltransferase in mice perturbs choline and 1-carbon metabolism, resulting in fatty liver and hepatocellular carcinomas. *J. Biol. Chem.* **2011**, *286*, 36258–36267. [CrossRef] [PubMed]

84. Wilson, M.J.; Shivapurkar, N.; Poirier, L.A. Hypomethylation of hepatic nuclear DNA in rats fed with a carcinogenic methyl-deficient diet. *Biochem. J.* **1984**, *218*, 987–990. [CrossRef]

85. Zeisel, S.H. Choline, Other Methyl-Donors and Epigenetics. *Nutrients* **2017**, *9*, 445. [CrossRef]

86. Wainfan, E.; Poirier, L.A. Methyl groups in carcinogenesis: Effects on DNA methylation and gene expression. *Cancer Res.* **1992**, *52*, 2071s–2077s. [PubMed]

87. Pogribny, I.P.; James, S.J.; Beland, F.A. Molecular alterations in hepatocarcinogenesis induced by dietary methyl deficiency. *Mol. Nutr. Food Res.* **2012**, *56*, 116–125. [CrossRef] [PubMed]

88. Kimura, M.; Umegaki, K.; Higuchi, M.; Thomas, P.; Fenech, M. Methylenetetrahydrofolate reductase C677T polymorphism, folic acid and riboflavin are important determinants of genome stability in cultured human lymphocytes. *J. Nutr.* **2004**, *134*, 48–56. [CrossRef]

89. Crider, K.S.; Yang, T.P.; Berry, R.J.; Bailey, L.B. Folate and DNA methylation: A review of molecular mechanisms and the evidence for folate's role. *Adv. Nutr.* **2012**, *3*, 21–38. [CrossRef]

90. Hustad, S.; Ueland, P.M.; Vollset, S.E.; Zhang, Y.; Bjørke-Monsen, A.L.; Schneede, J. Riboflavin as a Determinant of Plasma Total Homocysteine: Effect Modification by the Methylenetetrahydrofolate Reductase C677T Polymorphism. *Clin. Chem.* **2000**, *46*, 1065–1071.

91. Friso, S.; Choi, S.-W.; Girelli, D.; Mason, J.B.; Dolnikowski, G.G.; Bagley, P.J.; Olivieri, O.; Jacques, P.F.; Rosenberg, I.H.; Corrocher, R.; et al. A common mutation in the 5,10-methylenetetrahydrofolate reductase gene affects genomic DNA methylation through an interaction with folate status. *Proc. Natl. Acad. Sci. USA* **2002**, *99*, 5606–5611. [CrossRef]

92. Li, E.; Bestor, T.H.; Jaenisch, R. Targeted mutation of the DNA methyltransferase gene results in embryonic lethality. *Cell* **1992**, *69*, 915–926. [CrossRef]

93. Niculescu, M.D.; Haggarty, P. *Nutrition in Epigenetics*; Blackwell: Ames, IA, USA, 2011.

94. James, S.J.; Pogribny, I.P.; Pogribna, M.; Miller, B.J.; Jernigan, S.; Melnyk, S. Mechanisms of DNA damage, DNA hypomethylation, and tumor progression in the folate/methyl-deficient rat model of hepatocarcinogenesis. *J. Nutr.* **2003**, *133*, 3740s–3747s. [CrossRef]

95. Pogribny, I.P.; Shpyleva, S.I.; Muskhelishvili, L.; Bagnyukova, T.V.; James, S.J.; Beland, F.A. Role of DNA damage and alterations in cytosine DNA methylation in rat liver carcinogenesis induced by a methyl-deficient diet. *Mutat. Res.* **2009**, *669*, 56–62. [CrossRef] [PubMed]

96. Lillycrop, K.A.; Phillips, E.S.; Jackson, A.A.; Hanson, M.A.; Burdge, G.C. Dietary protein restriction of pregnant rats induces and folic acid supplementation prevents epigenetic modification of hepatic gene expression in the offspring. *J. Nutr.* **2005**, *135*, 1382–1386. [CrossRef] [PubMed]

97. Mukherjee, N.; Kumar, A.P.; Ghosh, R. DNA Methylation and Flavonoids in Genitourinary Cancers. *Curr. Pharmacol. Rep.* **2015**, *1*, 112–120. [CrossRef] [PubMed]

98. Lee, W.J.; Shim, J.Y.; Zhu, B.T. Mechanisms for the inhibition of DNA methyltransferases by tea catechins and bioflavonoids. *Mol. Pharmacol.* **2005**, *68*, 1018–1030. [CrossRef]

99. Fang, M.Z.; Wang, Y.; Ai, N.; Hou, Z.; Sun, Y.; Lu, H.; Welsh, W.; Yang, C.S. Tea polyphenol (−)-epigallocatechin-3-gallate inhibits DNA methyltransferase and reactivates methylation-silenced genes in cancer cell lines. *Cancer Res.* **2003**, *63*, 7563–7570.

100. Zhang, B.K.; Lai, Y.Q.; Niu, P.P.; Zhao, M.; Jia, S.J. Epigallocatechin-3-gallate inhibits homocysteine-induced apoptosis of endothelial cells by demethylation of the DDAH2 gene. *Planta Med.* **2013**, *79*, 1715–1719. [CrossRef]

101. Nandakumar, V.; Vaid, M.; Katiyar, S.K. (-)-Epigallocatechin-3-gallate reactivates silenced tumor suppressor genes, Cip1/p21 and p16INK4a, by reducing DNA methylation and increasing histones acetylation in human skin cancer cells. *Carcinogenesis* **2011**, *32*, 537–544. [CrossRef]

102. Pandey, M.; Shukla, S.; Gupta, S. Promoter Demethylation and Chromatin Remodeling by Green Tea Polyphenols Leads to Re-expression of GSTP1 in Human Prostate Cancer Cells. *Int. J. Cancer* **2010**, *126*, 2520–2533. [CrossRef]

103. Shukla, S.; Trokhan, S.; Resnick, M.I.; Gupta, S. Epigallocatechin-3-gallate causes demethylation and activation of GSTP1 gene expression in human prostate cancer LNCaP cells. *Cancer Res.* **2005**, *65*, 369.

104. Fang, M.Z.; Chen, D.; Sun, Y.; Jin, Z.; Christman, J.K.; Yang, C.S. Reversal of hypermethylation and reactivation of p16INK4a, RARbeta, and MGMT genes by genistein and other isoflavones from soy. *Clin. Cancer Res.* **2005**, *11*, 7033–7041. [CrossRef] [PubMed]

105. Day, J.K.; Bauer, A.M.; DesBordes, C.; Zhuang, Y.; Kim, B.E.; Newton, L.G.; Nehra, V.; Forsee, K.M.; MacDonald, R.S.; Besch-Williford, C.; et al. Genistein alters methylation patterns in mice. *J. Nutr.* **2002**, *132*, 2419S–2423S. [CrossRef] [PubMed]

106. Lee, W.J.; Zhu, B.T. Inhibition of DNA methylation by caffeic acid and chlorogenic acid, two common catechol-containing coffee polyphenols. *Carcinogenesis* **2006**, *27*, 269–277. [CrossRef] [PubMed]

107. Reuter, S.; Gupta, S.C.; Park, B.; Goel, A.; Aggarwal, B.B. Epigenetic changes induced by curcumin and other natural compounds. *Genes Nutr.* **2011**, *6*, 93–108. [CrossRef]

108. Liu, Z.; Xie, Z.; Jones, W.; Pavlovicz, R.E.; Liu, S.; Yu, J.; Li, P.K.; Lin, J.; Fuchs, J.R.; Marcucci, G.; et al. Curcumin is a potent DNA hypomethylation agent. *Bioorg. Med. Chem. Lett.* **2009**, *19*, 706–709. [CrossRef] [PubMed]

109. Bayet-Robert, M.; Kwiatkowski, F.; Leheurteur, M.; Gachon, F.; Planchat, E.; Abrial, C.; Mouret-Reynier, M.A.; Durando, X.; Barthomeuf, C.; Chollet, P. Phase I dose escalation trial of docetaxel plus curcumin in patients with advanced and metastatic breast cancer. *Cancer Biol. Ther.* **2010**, *9*, 8–14. [CrossRef]

110. Medina-Franco, J.L.; Lopez-Vallejo, F.; Kuck, D.; Lyko, F. Natural products as DNA methyltransferase inhibitors: A computer-aided discovery approach. *Mol. Divers.* **2011**, *15*, 293–304. [CrossRef]

111. Liu, Z.; Liu, S.; Xie, Z.; Pavlovicz, R.E.; Wu, J.; Chen, P.; Aimiuwu, J.; Pang, J.; Bhasin, D.; Neviani, P.; et al. Modulation of DNA Methylation by a Sesquiterpene Lactone Parthenolide. *J. Pharmacol. Exp. Ther.* **2009**, *329*, 505–514. [CrossRef]

112. Nakahara, K.; Trakoontivakorn, G.; Alzoreky, N.S.; Ono, H.; Onishi-Kameyama, M.; Yoshida, M. Antimutagenicity of some edible Thai plants, and a bioactive carbazole alkaloid, mahanine, isolated from Micromelum minutum. *J. Agric. Food Chem.* **2002**, *50*, 4796–4802. [CrossRef]

113. Ramsewak, R.S.; Nair, M.G.; Strasburg, G.M.; DeWitt, D.L.; Nitiss, J.L. Biologically active carbazole alkaloids from Murraya koenigii. *J. Agric. Food Chem.* **1999**, *47*, 444–447. [CrossRef]

114. Sheikh, K.D.; Banerjee, P.P.; Jagadeesh, S.; Grindrod, S.C.; Zhang, L.; Paige, M.; Brown, M.L. Fluorescent epigenetic small molecule induces expression of the tumor suppressor ras-association domain family 1A and inhibits human prostate xenograft. *J. Med. Chem.* **2010**, *53*, 2376–2382. [CrossRef]

115. Takumi, S.; Okamura, K.; Yanagisawa, H.; Sano, T.; Kobayashi, Y.; Nohara, K. The effect of a methyl-deficient diet on the global DNA methylation and the DNA methylation regulatory pathways. *J. Appl. Toxicol.* **2015**, *35*, 1550–1556. [CrossRef]

116. Hon, G.C.; Song, C.X.; Du, T.; Jin, F.; Selvaraj, S.; Lee, A.Y.; Yen, C.A.; Ye, Z.; Mao, S.Q.; Wang, B.A.; et al. 5mC oxidation by Tet2 modulates enhancer activity and timing of transcriptome reprogramming during differentiation. *Mol. Cell* **2014**, *56*, 286–297. [CrossRef] [PubMed]

117. Minor, E.A.; Court, B.L.; Young, J.I.; Wang, G. Ascorbate induces ten-eleven translocation (Tet) methylcytosine dioxygenase-mediated generation of 5-hydroxymethylcytosine. *J. Biol. Chem.* **2013**, *288*, 13669–13674. [CrossRef] [PubMed]

118. Chung, T.L.; Brena, R.M.; Kolle, G.; Grimmond, S.M.; Berman, B.P.; Laird, P.W.; Pera, M.F.; Wolvetang, E.J. Vitamin C promotes widespread yet specific DNA demethylation of the epigenome in human embryonic stem cells. *Stem Cells* **2010**, *28*, 1848–1855. [CrossRef] [PubMed]

119. Ghoshal, K.; Li, X.; Datta, J.; Bai, S.; Pogribny, I.; Pogribny, M.; Huang, Y.; Young, D.; Jacob, S.T. A folate- and methyl-deficient diet alters the expression of DNA methyltransferases and methyl CpG binding proteins involved in epigenetic gene silencing in livers of F344 rats. *J. Nutr.* **2006**, *136*, 1522–1527. [CrossRef]

120. Dahl, C.; Grønbæk, K.; Guldberg, P. Advances in DNA methylation: 5-hydroxymethylcytosine revisited. *Clin. Chim. Acta* **2011**, *412*, 831–836. [CrossRef]

121. Valinluck, V.; Sowers, L.C. Endogenous cytosine damage products alter the site selectivity of human DNA maintenance methyltransferase DNMT1. *Cancer Res.* **2007**, *67*, 946–950. [CrossRef] [PubMed]

122. Chanson, A.; Parnell, L.D.; Ciappio, E.D.; Liu, Z.; Crott, J.W.; Tucker, K.L.; Mason, J.B. Polymorphisms in uracil-processing genes, but not one-carbon nutrients, are associated with altered DNA uracil concentrations in an urban Puerto Rican population. *Am. J. Clin. Nutr.* **2009**, *89*, 1927–1936. [CrossRef]

123. Choi, S.W.; Kim, Y.I.; Weitzel, J.N.; Mason, J.B. Folate depletion impairs DNA excision repair in the colon of the rat. *Gut* **1998**, *43*, 93–99. [CrossRef]

124. Fenech, M. Recommended dietary allowances (RDAs) for genomic stability. *Mutat. Res./Fund. Mol. Mech. Mutagen.* **2001**, *480–481*, 51–54. [CrossRef]

 © 2018 by the authors. Licensee MDPI, Basel, Switzerland. This article is an open access article distributed under the terms and conditions of the Creative Commons Attribution (CC BY) license (http://creativecommons.org/licenses/by/4.0/).

International Journal of
Molecular Sciences

Communication

In Silico Analysis of Pacific Oyster (*Crassostrea gigas*) Transcriptome over Developmental Stages Reveals Candidate Genes for Larval Settlement

Valentin Foulon [1,*], Pierre Boudry [2], Sébastien Artigaud [1], Fabienne Guérard [1] and Claire Hellio [1]

[1] Laboratoire des Sciences de l'Environnement Marin (LEMAR), UMR 6539 CNRS/UBO/IRD/Ifremer, Institut Universitaire Européen de la Mer, Technopole Brest-Iroise, Rue Dumont d'Urville, 29280 Plouzané, France; sebastien.artigaud@univ-brest.fr (S.A.); fabienne.guerard@univ-brest.fr (F.G.); claire.hellio@univ-brest.fr (C.H.)

[2] Ifremer, Laboratoire des Sciences de l'Environnement Marin (LEMAR), UMR 6539 CNRS/UBO/IRD/Ifremer, Centre Bretagne, 29280 Plouzané, France; pierre.boudry@ifremer.fr

* Correspondence: valentin.foulon@univ-brest.fr; Tel.: +33-298-498-662

Received: 26 November 2018; Accepted: 4 January 2019; Published: 8 January 2019

Abstract: Following their planktonic phase, the larvae of benthic marine organisms must locate a suitable habitat to settle and metamorphose. For oysters, larval adhesion occurs at the pediveliger stage with the secretion of a proteinaceous bioadhesive produced by the foot, a specialized and ephemeral organ. Oyster bioadhesive is highly resistant to proteomic extraction and is only produced in very low quantities, which explains why it has been very little examined in larvae to date. In silico analysis of nucleic acid databases could help to identify genes of interest implicated in settlement. In this work, the publicly available transcriptome of Pacific oyster *Crassostrea gigas* over its developmental stages was mined to select genes highly expressed at the pediveliger stage. Our analysis revealed 59 sequences potentially implicated in adhesion of *C. gigas* larvae. Some related proteins contain conserved domains already described in other bioadhesives. We propose a hypothetic composition of *C. gigas* bioadhesive in which the protein constituent is probably composed of collagen and the von Willebrand Factor domain could play a role in adhesive cohesion. Genes coding for enzymes implicated in DOPA chemistry were also detected, indicating that this modification is also potentially present in the adhesive of pediveliger larvae.

Keywords: *Crassostrea gigas*; Pacific oyster; pediveliger larvae; bioadhesive; transcriptome

---

1. Introduction

The majority of bioadhesives secreted by animals are composed of proteins that allow permanent or reversible links to the substrate [1]. In the marine environment, these glues are efficient in wet conditions and could thus potentially represent useful alternatives to synthetic adhesives [2], particularly for biomedical applications [3]. The molecular composition of marine bioadhesives, especially proteins, can sometimes be difficult to characterize, however, due to their high resistance and small quantities, particularly for bioadhesives secreted at the larval stage. This issue has already been reported for bivalve mollusk larvae [4–8].

The Pacific oyster *Crassostrea gigas* (Thunberg 1973) is a benthic mollusc of the bivalve family with a two-phase life cycle. Its pelagic larvae adhere to a surface prior to metamorphosis. Larval settlement occurs at the pediveliger stage by secretion of a bioadhesive [4]. Overall molecular characterization of the adhesive secreted by the pediveliger larvae of *C. gigas* revealed its proteinaceous nature [4] and corroborate previous results published on pediveliger larval adhesive in other species [7–10]. However, the constitutive protein sequences of adhesive from *C. gigas* larvae remain unknown. The identification of genes involved in adhesion could be a useful first step towards protein identification that would enable us to successfully characterize the composition of *C. gigas* larval adhesive.

Numerous transcriptomic studies have recently been carried out on bioadhesive secretory organs. Rodrigues et al. (2016) used transcriptomics and proteomics approaches in cnidarians of the genus *Hydra*, to successfully pinpoint genes, proteins and enzymes potentially involved in adhesive composition and polymerisation [11]. A similar approach was used on the foot and byssus of *Chlamys farreri*, making it possible to understand scallop attachment [12]. Moreover, the transcriptome of the *Mytilus coruscus* foot allowed the identification of sequences with a strong homology to the adhesive sequences of other *Mytilidae* [13]. A transcriptomic study on adhesive glands of polychaetes of the *Sabellariidae* family recently described the phylogenetic evolution of certain adhesion genes and highlighted the importance of post-translational changes in adhesive proteins [14]. Transcriptomic analyses are described as an innovative and effective tool for determining candidate genes in marine organisms, but require validation by other molecular and functional investigations [1,15]. In *C. gigas* pediveliger larvae, the transcriptome of the adhesive gland is difficult to obtain due to the small size of the organism and the complexity of this organ.

However, the development of high-throughput nucleic acid sequencing methods (DNA and RNA) has led to a significant increase in the number of sequences available in generalist or specific databases (for the transcriptome of *C. gigas*: Riviere et al. (2015) [16]). The use of appropriate informatics tools makes it possible to identify sequences of interest in databases, compare them, analyze them and define their potential biological roles [17]. Many studies, known as in silico studies, use the available genomic data to identify genes involved in a defined biological process in order to answer to a working hypothesis. Sequence selection criteria (genomic expression rate, specificity of organs or certain stages of development, functional annotations) are defined according to the biological questions raised. Meta-analyses based on the exploration of published genomes are becoming increasingly common [18–20]. For example, in *C. gigas*, a recent study focusing on adult photosensitivity used this method to identify genes involved in this process [21]. Here, we propose to use a similar method to investigate bioadhesion of pediveliger larvae.

In *C. gigas*, adhesive is synthesized and stored before secretion from glands located in the foot [4]. Morphogenesis of the foot is a rapid process (24 to 48 h), specific to the pediveliger stage. This organ has locomotory, sensory, and secretory roles during the adhesion phase and disappears during metamorphosis (just a few hours after settlement). This indicates that adhesive synthesis is also a rapid process, resulting from significant and episodic cellular activity in the foot.

The presence of an mRNA in an organism at a given time is an indication of protein synthesis. The translation time of any given mRNA into protein is highly variable, however, from a few minutes to a few hours [22,23], and detection of an mRNA is not proof of the presence of the corresponding protein at a given time. However, identifying the genes expressed at a given time can still provide arguments for discussion about its involvement in a biological process. In our study, the first version of the genome assembly of the Pacific oyster *C. gigas*, published in 2012 by Zhang et al., is a particularly interesting resource [24]. The transcriptomic data also published in this article is available in Supplementary Table S14 "Transcriptomic representation of genes (RPKM) at different developmental stages and in different adult organs". This dataset groups the RPKMs (read per kilobase million) of each gene at each stage of development, and in different tissue of adult oysters. These quantitative data make it possible to visualize the expression rate of each genomic sequence. Nine developmental stages defined by 38 sampling time from hours to days after fertilization and 11 different adult organs were analyzed in this article. The pediveliger stages were named P1 and P2 and correspond to larvae of 18 days old larvae (precisely sampled at 18 days and 45 min and at 18 days, 4 h and 35 min after fertilization).

The objective of this study was to identify genes in the transcriptomic data published with the genome of *C. gigas* that could have a potential role in the adhesion of the pediveliger larvae. The identification of these genes could allow us to suggest the probable protein composition of the adhesive and to pinpoint the biosynthesis pathways and molecular cascades involved in their secretion and cross-linking. The sequences specifically expressed at the pediveliger stage and the potential role of the corresponding proteins are presented. After functional annotation of the sequences, those of them with interesting adhesion characteristics can be considered as relevant candidates for future molecular investigations.

2. Results

Fifty-nine sequences were selected as being specifically expressed at the pediveliger stage of *C. gigas* (Table 1) according to the following selection criteria: RPKM [pre-pediveliger stage (LU1 and LU2)]/RPKM [pediveliger stage] > 0.7 * RPKM [pediveliger stage] and RPKM [other stages]/RPKM [pediveliger stage] > 0.2. This selection represents 0.23% of the 27,902 sequences from the Table S14 of Zhang et al. (2012) [24]. sequences had at least one predicted conserved domain and/or one repeat sequence based on analysis with InterPro [25] (Figure 1). Forty-two sequences had extracellular localization according to DeepLoc 1.0 [26]. Twenty-one sequences, or 35.6% of the selected sequences, were annotated as hypothetical proteins, indicating the absence of known functions from the databases. The number of uncharacterized sequences is slightly lower than the 41.8% of sequences annotated as hypothetical proteins in the database used as a whole.

Figure 1. *Cont.*

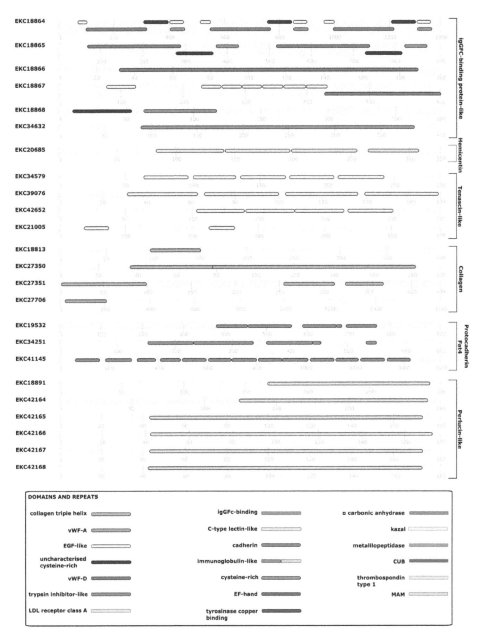

Figure 1. Conserved domains and repeated sequences predicted by the InterPro program (Finn et al., 2016) [25] among 38 sequences specifically expressed at the pediveliger stage in *Crassostrea gigas*, based on transcriptomic data published by Zhang et al. (2012) [24].

Table 1. Genes specifically expressed at the pediveliger stage of *Crassostrea gigas* according to the selection of RPKM from transcriptomic data published by Zhang et al. (2012) [24].

Group	Ensembl Gene ID	Protein ID	Name	Cell. Loc.
Hypothetical protein	CGI_10014580	EKC18206	Hypothetical protein CGI_10014580	Mit 0.58
	CGI_10010208	EKC18972	Hypothetical protein CGI_10010208	Ext 1
	CGI_10004853	EKC21005	Hypothetical protein CGI_10004853	Ext 0.42
	CGI_10002578	EKC22248	Hypothetical protein CGI_10002578	Ext 0.89
	CGI_10001746	EKC22673	Hypothetical protein CGI_10001746	Ext 0.90
	CGI_10013335	EKC23310	Hypothetical protein CGI_10013335	Nuc 0.50
	CGI_10013386	EKC24388	Hypothetical protein CGI_10013386	Ext 0.41
	CGI_10005578	EKC25384	Hypothetical protein CGI_10005578	Ext 0.43
	CGI_10013385	EKC24387	Hypothetical protein CGI_10013385	Ext 0.60
	CGI_10003237	EKC27225	Hypothetical protein CGI_10003237	Ext 0.99
	CGI_10025142	EKC28625	Hypothetical protein CGI_10025142	Ext 0.91
	CGI_10009961	EKC31321	Hypothetical protein CGI_10009961	Ext 0.99
	CGI_10025191	EKC32101	Hypothetical protein CGI_10025191	Cyt 0.38
	CGI_10012470	EKC33059	Hypothetical protein CGI_10012470	Ext 0.99
	CGI_10016093	EKC35263	Hypothetical protein CGI_10016093	Ext 0.52
	CGI_10016094	EKC35264	Hypothetical protein CGI_10016094	Ext 0.52
	CGI_10027526	EKC35968	Hypothetical protein CGI_10027526	ER 0.40
	CGI_10026725	EKC38958	Hypothetical protein CGI_10026725	Ext 0.62
	CGI_10022908	EKC41146	Hypothetical protein CGI_10022908	ER 0.25
	CGI_10008429	EKC41249	Hypothetical protein CGI_10008429	Ext 0.90
	CGI_10013282	EKC42653	Hypothetical protein CGI_10013282	Nuc 0.59
Enzyme	CGI_10009044	EKC19270	Putative tyrosinase-like protein tyr 1	Mem 0.82
	CGI_10014286	EKC25254	Putative tyrosinase-like protein tyr-3	Mem 0.99
	CGI_10006802	EKC29117	Tyrosinase-like protein 1	Ext 0.64
	CGI_10016593	EKC32997	Peroxidase-like protein	Ext 0.49
	CGI_10010889	EKC32754	Carbonic anhydrase 2	Cyt 0.41
	CGI_10011324	EKC18733	Carbonic anhydrase 7	Ext 0.65
	CGI_10003099	EKC28981	Cell surface hyaluronidase-like	Plast 0.39
	CGI_10003100	EKC28982	Cell migration-inducing and hyaluronan-binding protein-like	Cyt 0.29
	CGI_10007190	EKC19955	Metalloendopeptidase	Ext 0.39
	CGI_10007191	EKC19956	Metalloendopeptidase	Ext 0.51
	CGI_10020760	EKC31184	Zinc metalloproteinase nas-15	Ext 0.70
Protease inhibitor	CGI_10010154	EKC18991	Serine protease inhibitor dipetalogastin-like	Ext 0.55
	CGI_10010155	EKC18992	Serine protease inhibitor dipetalogastin-like	Ext 1
Structural protein	CGI_10005627	EKC20685	Hemicentin-1	Ext 0.72
	CGI_10010553	EKC18864	IgGFc-binding protein (zonadhesin-like)	Ext 0.64
	CGI_10010554	EKC18865	IgGFc-binding protein (zonadhesin-like)	Ext 0.72
	CGI_10010555	EKC18866	IgGFc-binding protein (zonadhesin-like)	Ext 0.93
	CGI_10010556	EKC18867	IgGFc-binding protein (zonadhesin-like)	Ext 0.60
	CGI_10010557	EKC18868	IgGFc-binding protein (zonadhesin-like)	Ext 0.91
	CGI_10023170	EKC34632	IgGFc-binding protein	Ext 0.52
	CGI_10010465	EKC34579	Tenascin-X	Ext 0.80
	CGI_10000981	EKC39076	Tenascin-R	Ext 0.84
	CGI_10025295	EKC40994	Multiple EGF-like domains 10	Ext 0.42
	CGI_10013281	EKC42652	Tenascin-R	Ext 0.99
	CGI_10010827	EKC18813	Collagen-like protein 7	Ext 0.89
	CGI_10010374	EKC27350	Collagen-like protein 7	Mem 0.56
	CGI_10010375	EKC27351	Collagen-like protein 7	Mem 0.72
	CGI_10011175	EKC27706	Collagen alpha-5(VI) chain	Ext 0.55
Calcification-related protein and calcium-binding protein	CGI_10010615	EKC18891	Aggrecan core protein	Ext 0.79
	CGI_10006917	EKC42164	Asialoglycoprotein receptor 2	Ext 0.99
	CGI_10006919	EKC42165	Perlucin-like protein	Ext 0.95
	CGI_10006920	EKC42166	C-type mannose receptor 2	Ext 0.90
	CGI_10006921	EKC42167	Perlucin-like protein	Ext 0.99
	CGI_10006922	EKC42168	Perlucin-like protein	Ext 0.99
	CGI_10008331	EKC19532	Protocadherin Fat 4-like	Lys 0.36
	CGI_10018326	EKC34251	Protocadherin Fat 4	Cyt 0.72
	CGI_10022907	EKC41145	Protocadherin Fat 4-like	Cyt 0.44
	CGI_10006247	EKC20329	Putative calmodulin	Cyt 0.48

Cell. Loc. indicates the subcellular localization prediction of the corresponding protein by DeepLoc 1.0. Ext: extracellular, Cyt: cytoplasm, Lys: lysosome, Mit: mitochondria, Pla: plastid, Mem: cell membrane, ER: endoplasmic reticulum, Nuc: nucleus.

3. Discussion

3.1. Sequences Involved in Reduction-Oxidation Reactions (Redox)

Among the fifty-nine selected sequences, three tyrosinase-like proteins (CGI_10009044, CGI_10014286 and CGI_10006802) and one peroxidase-like protein (CGI_10016593) are enzymes involved in reduction-oxidation (Redox) mechanisms. Tyrosinase is an oxidoreductase, also known as phenoloxidase, which allows hydroxylation (addition of an -OH group) of the aromatic part of tyrosine residues. Hydroxylated tyrosine, known as 3,4-dihydroxyphenylalanine (DOPA), can in turn be oxidized by tyrosinase, thus becoming a dopaquinone carrying two =O groups. DOPA-based marine bioadhesives are known to be sensitive to pH variations and to involve coacervation mechanisms [27,28]. Coacervation is a physicochemical mechanism allowing the spontaneous dissociation into two phases of a mixture of molecules due to their opposite charges. The most studied adhesive mechanism is mussel byssus, composed of a filament and a byssal plaque in contact with the substrate. In adult mussels, during the formation of the byssal plaque by the foot, the adhesive is secreted in a coacervated form, and polymerized by redox mechanisms mediated by the pH of the environment [27,29]. This strategy has also been described in the adhesive of the polychaetes *Sabellaria alveolata* and *Phragmatopoma californica* [28,30–32]. After secretion, the described DOPA-based adhesives combined with a coacervation mechanism had a foamy structure. However, the adhesive secreted by *C. gigas* larvae was described as a fibrous structure [4]. Phenoloxidase granules were reported in the main gland of the foot of pediveliger larvae of *O. edulis* by histochemistry [33]. The presence of phenoloxidase granules has not been confirmed in *C. gigas*, although secretion of byssal-like filaments by pediveliger larvae was observed before permanent adhesion at the end of the crawling phase [4]. It is possible that similar byssal secretion strategies could be used by pediveliger oyster larvae and adult mussels. Tyrosinase EKC29117 (CGI_10006802) was predicted for extracellular localization and presented a similarity of 47.95% (*E*-value: 6.5×10^{-75}) to a byssal protein sequence from *Mytilus corsuscus* (ANN45959 | Byssal tyrosinase-like protein 2). The sequence EKC25254 (CGI_10014286) had 46.19% homology (*E*-value: 1.4×10^{-84}) with an analogue protein (AKI87982 | Byssal tyrosinase-like protein-1) [34]. Interestingly, this last sequence had a C-type lectin domain (position 67–187), indicating probable linking to a polysaccharide. The C-type lectin domain could allow the immobilisation of the enzyme in the adhesive after secretion by linkage to a polysaccharide. This binding site could also act as an activation/inactivation site of the enzyme, as observed in tyrosinases implicated in melanin synthesis [35]. The presence of glycosylated active enzymes has also been reported in the adhesive of *Trichopterae* larvae [36].

Another sequence implicated in redox mechanisms, CGI_10016593 | EKC32997 | peroxidase-like protein, was also detected. This sequence had 44.8% similarity (*E*-value: 2.6×10^{-57}) to Byssal peroxidase-like protein 1 of *Mytilus coruscus* [34]. Byssal peroxidases have also been detected in the foot of *Limnoperna fortunei* [37] and byssus of *Pinctada fucata* [38]. The role of these enzymes remains uncertain, but byssal peroxidases could be involved in the protection of byssus from oxidizing environments and degradation brought about by microorganisms. The sequence EKC32997 (CGI_10016593) had a signal peptide indicating the extracellular secretion of this protein. These enzymes could also be directly involved in cross-linking of the adhesive, as described in the larvae of *Hysperophylax occidentalis*, allowing the establishment of di-tyrosine bonds that stabilize adhesive fibers [36]. The presence of peroxidases was also reported in the basal disc of the polyp *Hydra magnipapillata* [11], at the interface of the adhesive plaque in the barnacle *Amphibalanus amphitrite* [39,40], and in the parathorax of the polychaetes *Sabellaria alveolata* and *Phragmatopoma caudata* [14]. In algae, haloperoxidases are also involved during adhesion, catalyzing the redox reactions of phenolic compounds in adhesive mucilages [41,42]. All these observations indicate an important role of this enzyme, particularly in the biosynthesis of DOPA and its derivatives.

The presence of tyrosinase and peroxidase coding sequences in our *C. gigas* pediveliger larva sequence selection could indicate a use of redox mechanisms in its adhesive with the presence of DOPA or phenol groups.

Two other sequences coding for enzymes involved in redox processes were selected: CGI_10010889 and CGI_10011324, annotated as carbonic anhydrase. These enzymes are closely related to calcification, are important regulators of the acid–base balance, and could be involved in the cellular regulation of CO_2 at different cellular levels [43,44]. In molluscks, carbonic anhydrase activity is associated with shell calcification and active domains have been reported in nacrein protein [45]. This enzyme is strongly expressed in the mollusck mantle [46]. To date, carbonic anhydrase has never been reported in bioadhesive studies. It is therefore very likely that these two mRNA are expressed at the pediveliger stage in order to prepare the rapid calcification observed after metamorphosis. Also, it should be noted that the adhesion of pediveliger larvae in bivalve molluscks involves the shell as the upper interface. This is unique in mollusck adhesives since, for adult bivalves secreting byssal secretions, the binding between the adhesive and the body is provided by the tissues of the foot [27,47]. The binding between the shell and the adhesive in the pediveliger larvae of *C. gigas* could be strengthened by the action of carbonic anhydrase mobilizing carbonate from the shell. This selected set of genes specifically expressed at the pediveliger stage, encoding proteins involved in redox mechanisms, could therefore play a major role in the adhesion of *C. gigas*.

3.2. Proteases and Enzyme Inhibitors

Proteases were also detected in our selection. The sequence CGI_10007190 (EKC19955 | metalloproteinase) contained two Epidermal Growth-Factor (EGF) domains (positions 277–312 and 438–476) and a complement domain C1r/C1s, Uegf, Bmp1 (CUB) (position 320–438). The sequence EKC19956 has the same architecture as two EGF domains (positions 447–485 and 654–684), one CUB domain (position 329–447) and a meprin, an A-5 protein, and a receptor protein-tyrosine phosphatase mu domain (MAM) at the end of the sequence (position 920–1074). The sequence CGI_10020760 was annotated as Zinc metalloproteinase nas-15 (EKC31184), with a ZnMc domain (position 110–239). All these proteases were extracellular metalloproteinases which could play a role in the remodeling of the extracellular matrix during metamorphosis. Metalloproteinases have also been reported to play a role in synaptic systems and neural development. The larval transition to the pediveliger stage is accompanied by the development of the nervous system of the foot, which is largely innervated. It is probable that the presence of these sequences is related to this phenomenon. In addition, immediately after adhesion, the foot disappears during metamorphosis [48]; this tissue remodeling probably involves protease action.

Protease inhibitor sequences were also selected. The sequence CGI_10005578 (EKC25384 | hypothetical protein) was annotated with Gene Ontology indicating a metalloendoproteinase inhibitor molecular function. Recently, a protein with a similar function was identified in the foot and byssus of *Chlamys farreri* [12,49]. In this species, the protein Sbp8-1, which was described as an atypical metalloproteinase inhibitor, is a component of the byssus that is probably involved in the binding between the different byssal proteins. Two sequences annotated as serine protease inhibitors (CGI_10010154 and CGI_10010155) were specific to the pediveliger stage. The three sequences CGI_10010553, CGI_10010554 and CGI_10010557 had a trypsin inhibitor domain, and the sequences CGI_10010556 and CGI_10010557 had a serine protease inhibitor domain. Gene expression of serine protease inhibitor has been detected in the foot of *Mytilisepta virgata* [50] and *Chlamys farreri* [12].

In *C. gigas*, enzyme inhibitors could be involved in protecting the adhesive from degradation after secretion. Indeed, the presence of the adhesive secreted at the pediveliger stage (14 days post-fertilization) was still observable 72 days post-fertilization, indicating the robustness of this biomaterial.

3.3. Sequences Related to the Extracellular Matrix

Two hyaluronidase-related sequences were selected as specific to the pediveliger stage: CGI_10003099 and CGI_10003100. The role of these sequences was difficult to determine. Hyaluronidases had hydrolytic activity on hyaluronic acid and some forms of chondroitin sulfate, and could allow the remodeling of the extracellular matrix.

Six of the selected sequences were annotated as Fc receptors (IgGFc-binding protein): CGI_10010553, CGI_10010554, CGI_10010555, CGI_10010556, CGI_10010557 and CGI_10023170. These proteins are generally associated with immunity as they bind antigens from pathogens. In humans, some IgGFc-binding proteins are associated with mucus composition [51]. This type of protein has been identified in the mucus of *Crassostrea virginica* [52], where it was involved in the structure of the mucus via interactions with mucins. Mucus could be secreted as reversible adhesive by pediveliger larvae of *C. gigas* during the crawling phase. Indeed, contents of foot glands A and B, implicated in crawling, could potentially be related to mucus [4].

It is probable that IgGFc-binding protein sequences expressed at the pediveliger stage would play a role during the crawling phase of *C. gigas*.

These proteins have adhesion properties and could also be implicated in the structure of the final adhesive. Indeed, some structures present in these sequences are common to adhesive proteins. IgGFc-binding protein sequences have been identified in adhesive footprints of starfish *Asterias rubens* [53]. The sequence EKC18867 (CGI_10010556) has seven EGF domain sites. This type of repetition has been observed in byssal plaque protein (mfp2) in mussels, which has a repetition of 11 EGF domains [54]. Sequences EKC18864, EKC18865 and EKC18868 (CGI_10010553, CGI_10010554 and CGI_10010557) have cysteine-rich regions, indicating the potential ability to establish disulfide bonds. These domains may also indicate a folding conformation in these proteins. These sequences also contain von Willebrand Factor type D (vWF-D) and EGF domains. This type of domain have been described in byssal proteins of mussels and in adhesive proteins of starfish [54,55]. The sequence EKC20685 (CGI_10005627 | Hemicentin 1) has one LDL receptor domain (low density lipoproteins) and three TSP-1 domains (type 1 thrombospondin), indicating a probable extracellular localization. The central role of a protein containing three TSP-1 domains has been reported in the byssus of *P. fucata* [56], and is probably related to the elastic properties of the distal part of the filament.

Four sequences containing an EGF-like and tenascin-related domain were selected (CGI_10010465, CGI_10000981, CGI_10025295 and CGI_10013281). Tenascins are generally extracellular, glycosylated, and have elastic properties [57]. Tenascin R may play a role in the development of the nervous system [58]. Tenascin X is a protein with essential architectural functions, implicated in the structural properties of many tissues by binding to other constitutive proteins [59]. This extracellular protein is particularly involved in cell-matrix adhesion.

Among the specific sequences of the pediveliger stage of *C. gigas*, four have a collagen triple helix domain: CGI_10010827, CGI_10010374, CGI_10010375 and CGI_10011175. Collagen is a structural protein forming fibrous structures. In marine bioadhesives, a component of the byssal filament secreted by mussels [60,61]. Sequences EKC18813 (CGI_10010827), EKC27350 (CGI_10010374) and EKC27351 (CGI_10010375) have high glycine contents (25.7%, 27.2%, and 23.3% respectively), close to the levels observed in the byssus of *P. fucata*, *M. californianus* [60] and *M. edulis* [62]. These sequences also have high proline contents, close to 11%. Glycine and proline are essential amino acids for the establishment of the triple collagen helix for fiber formation [63]. The sequence EKC27706 (CGI_10011175) has a von Willebrand Factor type A domain (vWF-A), which is a glycoprotein-binding site. The vWF-A domain is also involved in collagen binding, as previously described in mussel byssus [56,64]. According to the Phyre2 program [65], this sequence has 25% identity (99.4% confidence) with the proximal thread matrix protein (ptmp-1 | AAL17974.1) from *Mytilus galloprovincialis* [64].

The filamentous structure of the adhesive of *C. gigas* pediveliger larvae fits with a collagen-rich composition, and the cohesion of the adhesive could result from the presence of von Willebrand Factor type domains.

3.4. Calcifying Sequences

Redox enzymes such as tyrosinases (CGI_10009044 | EKC19270, CGI_10014286 | EKC25254 and CGI_10006802 | EKC29117), peroxidase (CGI_10016593 | EKC32997), and the two carbonic anhydrases (CGI_10010889 | EKC18733 and CGI_10011324 | EKC32754) could be related to bio-calcification. Indeed, these enzymes have been classically reported in shell synthesis processes in molluscs and could coincide with the beginning of calcite layer production, which occurs after metamorphosis, at the spat stage. These sequences are poorly expressed at the spat stage, however, compared with the pediveliger stage, although no quantitative relationship has yet been established between the RPKM value observed for any mRNA of sequences in the dataset and the abundance of the corresponding protein. Nevertheless, these sequences are also poorly expressed at the adult stage, particularly in the mantle, the organ responsible for shell synthesis. In contrast, a recent study on *C. virginica* showed the succession of two adhesion strategies during the transition of pediveliger larvae to spat [9]: the secretion of the organic adhesive by the pediveliger larva is followed by the secretion of an adhesive containing a larger inorganic fraction, allowing adherence to the substrate of the growing shell. It also appears that adhesive proteins in molluscs have similar molecular domains and functions to so-called calcifying proteins. Thus, the detection of tyrosinase, DOPA chemistry-related proteins, polysaccharide-binding domains, vWF domains and EGF domains are often common to adhesion proteins [66], including those involved in bivalve mollusck byssus [27,54] and calcification [46].

Calcium-binding sequences were selected: three protocadherin Fat 4-like proteins (CGI_10008331, CGI_10018326 and CGI_10022907) and a putative calmodulin (CGI_10006247) containing multiple EF-hand domains. An EF-hand domain consists of two alpha helix forming a loop by interaction with a Ca^{2+} ion. A sequence annotated as hypothetical protein (CGI_10025191) also presents two EF-hand domains. Extracellular proteins containing an EF-hand domain were reported in the calcification process in pearl oysters [67]. Calmodulin is a ubiquitous protein, involved in calcium metabolism. In oysters, this protein plays an important role in calcification [68]. Protocadherin Fat 4 proteins contain multiple cadherin domains, which could be involved in cell binding [69]. Many protocadherin Fat 4 proteins have been described as involved in the development of the nervous system in cephalopods [70]. In contrast, proteins containing a Ca^{2+}-binding site play a role in many cellular processes (homeostasis maintenance, muscle contraction, cell differentiation, cell adhesion, immunity, signal transmission) [71].

To date, no proteins containing EF-hand domains have been reported in bioadhesives composition. However, calmodulins were detected in the adhesive organs of urchins and could play a role in exocytosis of adhesive [72]. The role of these sequences, specifically expressed at the pediveliger stage in *C. gigas*, remains unknown. It could coincide with the morphogenesis of the foot and the development of the nervous system, but also with metamorphosis or with the secretion of the shell and its adhesive matrix after metamorphosis [9].

Six sequences (CGI_10010615, CGI_10006917, CGI_10006919, CGI_10006920, CGI_10006921, and CGI_10006922) coding for proteins with a C-type lectin domain (EKC18891, EKC42164, EKC42165, EKC42166, EKC42167 and EKC42168) were selected. A C-type lectin domain is a calcium-dependent polysaccharide-binding domain [73]. These six sequences are also related to perlucin. Perlucins are proteins involved in calcification, implicated in the nucleation of calcium carbonate crystals [74]. However, the expression profile of these sequences raises doubts about their true function in *C. gigas* larvae. The sequences CGI_10010615 (EKC18891), CGI_10006917 (EKC42164) and CGI_10006920 (EKC42166) were annotated as aggrecan core protein, asialoglycoprotein receptor 2 and C-type mannose receptor 2, respectively. In vertebrates, aggrecan core protein is a constitutive protein of cartilage [75]. Perlucin-like sequences were identified in the foot of *Chlamys farreri* [12]. It is possible that these sequences are directly involved in the composition of the adhesive. Indeed, C-type lectin domains are present in many bioadhesives [47,55,76–78]. In addition, the sequences EKC18891, EKC42164, EKC42165, EKC42165, EKC42166, EKC42167, and EKC42168 have 46%, 48%, 44%, 47%, 44%, and 48% homology (*E*-values: 1×10^{-47}, 2×10^{-47}, 7×10^{-41}, 7×10^{-41}, 4×10^{-45}, 2×10^{-39} and

3×10^{-48}), respectively, with the foot protein 1 (AIWO4139) from *Atrina pectinata* [47]. The multiple alignment of the sequences indicates that the homology comes from the C-type lectin domain (Figure 1).

The shell of mollusck larvae is composed of a polymorphic inorganic matrix of $CaCO_3$ and an organic matrix, the periostracum. The link between the adhesive and the shell could involve bonds with the periostracum. The periostracum is composed of glycoproteins and polysaccharides such as chitin [79]. It is highly likely that the C-type lectin domain present in perlucin-like sequences allows a link between the periostracum and the adhesive.

3.5. Hypothetical Sequences

Twenty-one sequences were annotated as hypothetical proteins, fifteen of which had extracellular localization according to DeepLoc 1.0 (CGI_10010208, CGI_10004853, CGI_10002578, CGI_10001746, CGI_10013386, CGI_10005578, CGI_10013385, CGI_10003237, CGI_10025142, CGI_10009961, CGI_10012470, CGI_10016093, CGI_10016094, CGI_10026725, and CGI_10008429). The sequence CGI_10022908 had a transmembrane domain and a predicted localization in the endoplasmic reticulum. When further information is available about these genes, it will be possible to suggest the roles they could play in adhesion.

4. Hypothetical Model of Molecular Interactions within *C. gigas* Adhesive

In silico analysis of transcriptomic data on the *C. gigas* oyster [24] made it possible to select transcripts that were over-expressed at the pediveliger stage. After analysis of the conserved domains and repeat sequences of the 59 selected transcripts, it appeared that the majority probably had an extracellular localization and potential roles in adhesion. These results must be treated with caution, however, based on the analysis of the transcriptome. They do not validate the presence of the proteins encoded by the identified genes. Hypothetical involvement in adhesion of some protein domains identified in genes specifically expressed in the pediveliger stage of *C. gigas* is shown in Figure 2.

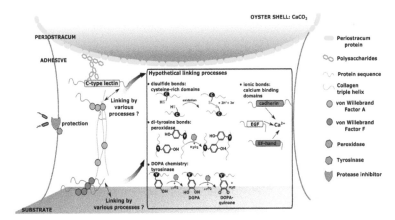

Figure 2. Schematic representation of the hypothetical molecular interactions involved in the adhesion of *C. gigas* pediveliger larvae, based on the selection of genes specifically expressed at the pediveliger stage.

Thus, the presence of structural proteins such as collagen and proteins rich in vWF domains is in accordance with the fibrous structure of the adhesive secreted by *C. gigas* pediveliger larvae. These proteins can be proposed as the structural components of the fibers and central matrix of the adhesive. In addition, the presence of vWF-A domains in an extracellular collagen sequence and vWF-D domains within an IgGFc-binding protein may indicate that these domains could be involved in the adhesive structure by mediation of protein aggregation.

The adhesive of *C. gigas* has the particularity of having its upper part in contact with the shell of the left larval valve. This shell is composed of calcium carbonate and an organic matrix, the periostracum. Since the periostracum is composed of glycosylated proteins and polysaccharides, it is likely that perlucin-like proteins containing a C-type lectin domain could mediate linkage between the adhesive and the periostracum.

The secreted adhesive needs to be resistant to the environment and not degraded until the shell of the spat has reached a sufficient size for the oyster to remain attached to the substrate. Sequences selected in this study and annotated with protease and peroxidase inhibitory domains could have a role in protecting the adhesive against bacterial degradation. In addition, the distribution of *C. gigas* on the foreshore means that the adhesive will potentially be exposed to UV radiation, desiccation, fresh water and high temperatures, which could be stress factors for this material.

The chemical bonds involved in the structure of the adhesive, and in linkage to the substrate, could be of a diversity of types. The cysteine-rich domains observed in some sequences (EKC18864 | CGI_10010553, EKC18865 | CGI_10010554, and EKC18868 | CGI_10010557) may be involved in the structural conformation of these proteins, but also in the establishment of disulfide bonds. Disufide bonds can also be established between other proteins containing cysteines in their sequences. The selection of sequences coding for enzymes involved in redox mechanisms (tyrosinases, peroxidase) could potentially indicate the biosynthesis of DOPA and these derivatives. This makes it possible to consider adhesion mechanisms in pediveliger larvae similar to those described in the byssus of adult bivalve molluscks. Peroxidases and tyrosinases can thus allow the formation of dityrosine bonds, and DOPA or DOPA-quinone groups, respectively. DOPA can mediate linking between adhesive and substrate, but also between the proteins in the adhesive, by covalent bonds or ionic interactions. Ionic interactions could also occur within the adhesive, particularly due to the presence of sequences with domains having a calcium affinity (cadherin, EGF, EF-hand). It is possible that the difference in structure observed between the inner and outer zone of the adhesive in *C. gigas* [4] may result from the presence in the outer zone of proteins or chemical groups allowing the establishment of a greater number of covalent bonds or ionic interactions, resulting in a tightening of the adhesive mesh.

5. Materials and Methods

Table S14 "Transcriptomic representation of genes (RPKM) at different developmental stages and in different adult organs" was downloaded from the *C. gigas* genome publication on the Nature website [24]. Stages were defined by the number of days after fertilization of the larval cohort. Thus, P1 and P2 were defined as the pediveliger stage at 18 days post-fertilization (precisely sampled at 18 days and 45 min and at 18 days, 4h and 35 min after fertilization). In order to identify genes potentially involved in adhesion, sequences with "strong" RPKM at stages P1 and P2 relative to the other stages were selected according to the following thresholds:

- RPKM of stages E to U6, stages S and J, and adult organs less than 20% that of P1 or P2.
- RPKM of stages LU1, LU2 below 70% that of P1 or P2.

LU1 and LU2 correspond to "later umbo larvae 1" and "later umbo larva 2" at 14 and 15 days post-fertilization, respectively. The RPKM selection threshold of these two stages was higher than for other larval stages according to the potential heterogeneity of the larval cohort and potential RNA synthesis before settlement at the pediveliger stage.

The protein sequences corresponding to the selected genomic sequences were then searched for on the NCBI database. Functional annotation of protein sequences was performed with BLAST2GO 4.0.7 software. A BLASTP search was performed with BLASTP2.7.1 + (NCBI) from BLAST2GO in the non-redundant protein database (nr) on 23 August 2018. Conserved domains and repetitive sequences were predicted with InterPro [25], and subcellular localization was predicted with DeepLoc 1.0 program [26]. Sequences of interest were then also submitted to Phyre2 to search for protein similarities based on protein structure prediction [65].

Int. J. Mol. Sci. **2019**, *20*, 197

6. Conclusions

Our in silico analysis successfully identified genes specifically expressed at the pediveliger stage in *C. gigas*. The majority of these genes were annotated for proteins containing conserved domains potentially involved in adhesion of *C. gigas*. Hypotheses formulated during this analysis advanced our understanding of larval adhesion in *C. gigas*. The set of 59 selected transcripts found by this method are interesting candidates that could be functionally explored by localization approaches (RNA in situ hybridization, antibody hybridization) and phenotyping approaches (interfering RNA, CRISPR). Thus, the localization of these transcripts and/or of the corresponding proteins within the pediveliger larvae and/or adhesive imprints is needed in order to validate their respective involvement in adhesion. The localization of these proteins would also make it possible to define molecular interactions and bonds involved in the structure of this adhesive. However, our results are based on gene expression and are thus not a perfect reflection of the real protein composition of the adhesive. Indeed, despite the precautions taken in the sequence selection protocol (time window tightened around the pediveliger stage), regulatory mechanisms between the mRNA and its translation into protein are multiple. Larval adhesives are difficult to describe and characterize because of the small amount of material and the resistance of these matrices. Our results represent significant progress about oyster larval adhesive, and open new possibilities for further analysis on this bioadhesive.

Author Contributions: V.F., S.A., P.B., F.G. and C.H. wrote the main manuscript text. V.F. and S.A. performed the transcriptome analysis. V.F. made figures. All authors reviewed the manuscript.

Funding: This research received no external funding.

Acknowledgments: This work was supported by the European Institute for Marine Studies (IUEM, France), University of Western Brittany, the Brittany local government (CR Bretagne) and by the "Laboratoire d'Excellence" LabexMER (ANR-10-LABX-19). It was co-funded by a grant from the French government under the program "Investissements d'Avenir".

Conflicts of Interest: The authors declare no conflicts of interest.

References

1. Hennebert, E.; Maldonado, B.; Ladurner, P.; Flammang, P.; Santos, R. Experimental strategies for the identification and characterization of adhesive proteins in animals: A review. *Interface Focus* **2014**, *5*, 20140064. [CrossRef] [PubMed]
2. North, M.A.; Del Grosso, C.A.; Wilker, J.J. High strength underwater bonding with polymer mimics of mussel adhesive proteins. *ACS Appl. Mater. Interfaces* **2017**, *9*, 7866–7872. [CrossRef] [PubMed]
3. Foster, L.J.R. Bioadhesives as surgical sealants: A Review. *Bioadhesive. Biomim. Nat. Appl.* **2015**, 203–234.
4. Foulon, V.; Artigaud, S.; Buscaglia, M.; Bernay, B.; Fabioux, C.; Petton, B.; Elies, P.; Boukerma, K.; Hellio, C.; Guérard, F.; et al. Proteinaceous secretion of bioadhesive produced during crawling and settlement of *Crassostrea gigas* larvae. *Sci. Rep.* **2018**, *8*, 15298. [CrossRef] [PubMed]
5. Petrone, L.; Ragg, N.L.C.; McQuillan, A.J. In situ infrared spectroscopic investigation of *Perna canaliculus* mussel larvae primary settlement. *Biofouling* **2008**, *24*, 405–413. [CrossRef] [PubMed]
6. Petrone, L.; Ragg, N.L.C.; Girvan, L.; McQuillan, J.A. Scanning electron microscopy and energy dispersive X-Ray microanalysis of *Perna canaliculus* mussel larvae adhesive secretion. *J. Adhes.* **2009**, *85*, 78–96. [CrossRef]
7. Cranfield, H.J. The ultrastructure and histochemistry of the larval cement of *Ostrea edulis* L. *J. Mar. Biol. Assoc. UK* **1975**, *55*, 497–503. [CrossRef]
8. Gruffydd, L.D.; Lane, D.J.W.; Beaumont, A.R. The glands of the larval foot in *Pecten maximus* L. and possible homologues in other bivalves. *J. Mar. Biol. Assoc. UK* **1975**, *55*, 463–476. [CrossRef]
9. Tibabuzo Perdomo, A.M.; Alberts, E.M.; Taylor, S.D.; Sherman, D.M.; Huang, C.-P.; Wilker, J.J. Changes in cementation of reef building oysters transitioning from larvae to adults. *ACS Appl. Mater. Interfaces* **2018**, *10*, 14248–14253. [CrossRef] [PubMed]
10. Lane, D.; Nott, J. A study of the morphology, fine structure and histochemistry of the foot of the pediveliger of *Mytilus edulis* L. *J. Mar. Biol. Assoc. UK* **1975**, *55*, 477–495. [CrossRef]

11. Rodrigues, M.; Ostermann, T.; Kremeser, L.; Lindner, H.; Beisel, C.; Berezikov, E.; Hobmayer, B.; Ladurner, P. Profiling of adhesive-related genes in the freshwater cnidarian *Hydra magnipapillata* by transcriptomics and proteomics. *Biofouling* **2016**, *32*, 1115–1129. [CrossRef]

12. Miao, Y.; Zhang, L.; Sun, Y.; Jiao, W.; Li, Y.; Sun, J.; Wang, Y.; Wang, S.; Bao, Z.; Liu, W. Integration of transcriptomic and proteomic approaches provides a core set of genes for understanding of scallop attachment. *Mar. Biotechnol.* **2015**, *17*, 523–532. [CrossRef]

13. Pan, Q.; Qi, Q.; Bao, L.-F.; Qin, C.-L.; He, J.-Y.; Fan, M.-H.; Liao, Z. Illumina-based De novo sequencing and characterization of *Mytilus coruscus* foot transcriptome. *Chin. J. Biochem. Mol. Biol.* **2015**, *8*, 014.

14. Buffet, J.-P.; Corre, E.; Duvernois-Berthet, E.; Fournier, J.; Lopez, P.J. Adhesive gland transcriptomics uncovers a diversity of genes involved in glue formation in marine tube-building polychaetes. *Acta Biomater.* **2018**, *72*, 316–328. [CrossRef]

15. Pennati, R.; Rothbächer, U. Bioadhesion in ascidians: A developmental and functional genomics perspective. *Interface Focus* **2015**, *5*, 20140061. [CrossRef] [PubMed]

16. Riviere, G.; Klopp, C.; Ibouniyamine, N.; Huvet, A.; Boudry, P.; Favrel, P. GigaTON: An extensive publicly searchable database providing a new reference transcriptome in the pacific oyster Crassostrea gigas. *BMC Bioinform.* **2015**, *16*, 401. [CrossRef] [PubMed]

17. Rodrigues, M.; Lengerer, B.; Ostermann, T.; Ladurner, P. Molecular biology approaches in bioadhesion research. *Beilstein J. Nanotechnol.* **2014**, *5*, 983. [CrossRef]

18. Makarev, E.; Schubert, A.D.; Kanherkar, R.R.; London, N.; Teka, M.; Ozerov, I.; Lezhnina, K.; Bedi, A.; Ravi, R.; Mehra, R. In silico analysis of pathways activation landscape in oral squamous cell carcinoma and oral leukoplakia. *Cell Death Discov.* **2017**, *3*, 17022. [CrossRef] [PubMed]

19. Mitchell, A.; Guerra, D.; Stewart, D.; Breton, S. In silico analyses of mitochondrial ORFans in freshwater mussels (Bivalvia: Unionoida) provide a framework for future studies of their origin and function. *BMC Genom.* **2016**, *17*, 597. [CrossRef]

20. Alkhalili, R.; Canbäck, B. Identification of Putative Novel Class-I Lanthipeptides in Firmicutes: A Combinatorial In Silico Analysis Approach Performed on Genome Sequenced Bacteria and a Close Inspection of Z-Geobacillin Lanthipeptide Biosynthesis Gene Cluster of the Thermophilic *Geobacillus* sp. Strain ZGt-1. *Int. J. Mol. Sci.* **2018**, *19*, 2650. [CrossRef]

21. Wu, C.; Jiang, Q.; Wei, L.; Cai, Z.; Chen, J.; Yu, W.; He, C.; Wang, J.; Guo, W.; Wang, X. A Rhodopsin-Like gene may be associated with the light-sensitivity of adult Pacific oyster *Crassostrea gigas*. *Front. Physiol.* **2018**, *9*, 221. [CrossRef] [PubMed]

22. Dermit, M.; Dodel, M.; Mardakheh, F.K. Methods for monitoring and measurement of protein translation in time and space. *Mol. Biosyst.* **2017**, *13*, 2477–2488. [CrossRef] [PubMed]

23. Morisaki, T.; Lyon, K.; DeLuca, K.F.; DeLuca, J.G.; English, B.P.; Zhang, Z.; Lavis, L.D.; Grimm, J.B.; Viswanathan, S.; Looger, L.L. Real-time quantification of single RNA translation dynamics in living cells. *Science* **2016**, *352*, 1425–1429. [CrossRef] [PubMed]

24. Zhang, G.; Fang, X.; Guo, X.; Li, L.; Luo, R.; Xu, F.; Yang, P.; Zhang, L.; Wang, X.; Qi, H.; et al. The oyster genome reveals stress adaptation and complexity of shell formation. *Nature* **2012**, *490*, 49–54. [CrossRef] [PubMed]

25. Finn, R.D.; Attwood, T.K.; Babbitt, P.C.; Bateman, A.; Bork, P.; Bridge, A.J.; Chang, H.-Y.; Dosztányi, Z.; El-Gebali, S.; Fraser, M. InterPro in 2017—Beyond protein family and domain annotations. *Nucleic Acids Res.* **2016**, *45*, D190–D199. [CrossRef] [PubMed]

26. Almagro Armenteros, J.J.; Sønderby, C.K.; Sønderby, S.K.; Nielsen, H.; Winther, O. DeepLoc: Prediction of protein subcellular localization using deep learning. *Bioinformatics* **2017**, *33*, 3387–3395. [CrossRef] [PubMed]

27. Waite, J.H. Mussel adhesion–essential footwork. *J. Exp. Biol.* **2017**, *220*, 517–530. [CrossRef] [PubMed]

28. Stewart, R.J.; Wang, C.S.; Song, I.T.; Jones, J.P. The role of coacervation and phase transitions in the sandcastle worm adhesive system. *Adv. Colloid Interface Sci.* **2017**, *239*, 88–96. [CrossRef] [PubMed]

29. Martinez Rodriguez, N.R.; Das, S.; Kaufman, Y.; Israelachvili, J.N.; Waite, J.H. Interfacial pH during mussel adhesive plaque formation. *Biofouling* **2015**, *31*, 221–227. [CrossRef] [PubMed]

30. Becker, P.T.; Lambert, A.; Lejeune, A.; Lanterbecq, D.; Flammang, P. Identification, characterization, and expression levels of putative adhesive proteins from the tube-dwelling polychaete *Sabellaria alveolata*. *Biol. Bull.* **2012**, *223*, 217–225. [CrossRef] [PubMed]

31. Stewart, R.J.; Weaver, J.C.; Morse, D.E.; Waite, J.H. The tube cement of *Phragmatopoma californica*: A solid foam. *J. Exp. Biol.* **2004**, *207*, 4727–4734. [CrossRef] [PubMed]

32. Wang, C.S.; Stewart, R.J. Multipart copolyelectrolyte adhesive of the sandcastle worm, *Phragmatopoma californica* (Fewkes): Catechol oxidase catalyzed curing through peptidyl-DOPA. *Biomacromolecules* **2013**, *14*, 1607–1617. [CrossRef] [PubMed]

33. Cranfield, H.J. A study of the morphology, ultrastructure, and histochemistry of the foot of the pediveliger of *Ostrea edulis*. *Mar. Biol.* **1973**, *22*, 187–202. [CrossRef]

34. Qin, C.; Pan, Q.; Qi, Q.; Fan, M.; Sun, J.; Li, N.; Liao, Z. In-depth proteomic analysis of the byssus from marine mussel *Mytilus coruscus*. *J. Proteom.* **2016**, *144*, 87–98. [CrossRef]

35. Mikami, M.; Sonoki, T.; Ito, M.; Funasaka, Y.; Suzuki, T.; Katagata, Y. Glycosylation of tyrosinase is a determinant of melanin production in cultured melanoma cells. *Mol. Med. Rep.* **2013**, *8*, 818–822. [CrossRef]

36. Wang, C.-S.; Pan, H.; Weerasekare, G.M.; Stewart, R.J. Peroxidase-catalysed interfacial adhesion of aquatic caddisworm silk. *J. R. Soc. Interface* **2015**, *12*, 20150710. [CrossRef]

37. Li, S.; Xia, Z.; Chen, Y.; Gao, Y.; Zhan, A. Byssus structure and protein composition in the highly invasive fouling mussel *Limnoperna fortunei*. *Front. Physiol.* **2018**, *9*, 418. [CrossRef]

38. Liu, C.; Xie, L.; Zhang, R. Ca^{2+} mediates the self-assembly of the foot proteins of *Pinctada fucata* from the nanoscale to the microscale. *Biomacromolecules* **2016**, *17*, 3347–3355. [CrossRef]

39. So, C.R.; Scancella, J.M.; Fears, K.P.; Essock-Burns, T.; Haynes, S.E.; Leary, D.H.; Diana, Z.; Wang, C.; North, S.; Oh, C.S. Oxidase activity of the barnacle adhesive interface involves peroxide-dependent catechol oxidase and lysyl oxidase enzymes. *ACS Appl. Mater. Interfaces* **2017**, *9*, 11493–11505. [CrossRef]

40. Zhang, G.; He, L.; Wong, Y.-H.; Xu, Y.; Zhang, Y.; Qian, P. Chemical component and proteomic study of the *Amphibalanus* (=Balanus) amphitrite shell. *PLoS ONE* **2015**, *10*, e0133866. [CrossRef]

41. Berglin, M.; Delage, L.; Potin, P.; Vilter, H.; Elwing, H. Enzymatic cross-linking of a phenolic polymer extracted from the marine alga *Fucus serratus*. *Biomacromolecules* **2004**, *5*, 2376–2383. [CrossRef] [PubMed]

42. Vreeland, V.; Waite, J.H.; Epstein, L. Minireview—Polyphenols and oxidases in substratum adhesion by marine algae and mussels. *J. Phycol.* **1998**, *34*, 1–8. [CrossRef]

43. Hopkinson, B.M.; Tansik, A.L.; Fitt, W.K. Internal carbonic anhydrase activity in the tissue of scleractinian corals is sufficient to support proposed roles in photosynthesis and calcification. *J. Exp. Biol.* **2015**. [CrossRef] [PubMed]

44. Wang, X.; Wang, M.; Jia, Z.; Qiu, L.; Wang, L.; Zhang, A.; Song, L. A carbonic anhydrase serves as an important acid–base regulator in pacific oyster *Crassostrea gigas* exposed to elevated CO_2: Implication for physiological responses of mollusk to ocean acidification. *Mar. Biotechnol.* **2017**, *19*, 22–35. [CrossRef] [PubMed]

45. Zhang, C.; Zhang, R. Matrix proteins in the outer shells of molluscks. *Mar. Biotechnol.* **2006**, *8*, 572–586. [CrossRef] [PubMed]

46. McDougall, C.; Degnan, B.M. The evolution of mollusck shells. *Wiley Interdiscip. Rev. Dev. Biol.* **2018**, *7*, e313. [CrossRef] [PubMed]

47. Yoo, H.Y.; Iordachescu, M.; Huang, J.; Hennebert, E.; Kim, S.; Rho, S.; Foo, M.; Flammang, P.; Zeng, H.; Hwang, D. Sugary interfaces mitigate contact damage where stiff meets soft. *Nat. Commun.* **2016**, *7*, 11923. [CrossRef]

48. Bayne, B.L. *Biology of Oysters*; Developments in Aquaculture and Fisheries Science; Elsevier Science: Amsterdam, The Netherlands, 2017; ISBN 978-0-12-803500-9.

49. Zhang, X.; Dai, X.; Wang, L.; Miao, Y.; Xu, P.; Liang, P.; Dong, B.; Bao, Z.; Wang, S.; Lyu, Q. Characterization of an atypical metalloproteinase inhibitors like protein (Sbp8-1) from scallop byssus. *Front. Physiol.* **2018**, *9*, 597. [CrossRef]

50. Gerdol, M.; Fujii, Y.; Hasan, I.; Koike, T.; Shimojo, S.; Spazzali, F.; Yamamoto, K.; Ozeki, Y.; Pallavicini, A.; Fujita, H. The purplish bifurcate mussel *Mytilisepta virgata* gene expression atlas reveals a remarkable tissue functional specialization. *BMC Genom.* **2017**, *18*, 590. [CrossRef]

51. Harada, N.; Iijima, S.; Kobayashi, K.; Yoshida, T.; Brown, W.R.; Hibi, T.; Oshima, A.; Morikawa, M. Human IgGFc binding protein (FcγBP) in colonic epithelial cells exhibits mucin-like structure. *J. Biol. Chem.* **1997**, *272*, 15232–15241. [CrossRef]

52. Espinosa, E.P.; Koller, A.; Allam, B. Proteomic characterization of mucosal secretions in the eastern oyster, *Crassostrea virginica*. *J. Proteom.* **2016**, *132*, 63–76. [CrossRef] [PubMed]

53. Hennebert, E.; Leroy, B.; Wattiez, R.; Ladurner, P. An integrated transcriptomic and proteomic analysis of sea star epidermal secretions identifies proteins involved in defense and adhesion. *J. Proteom.* **2015**, *128*, 83–91. [CrossRef] [PubMed]

54. Inoue, K.; Takeuchi, Y.; Miki, D.; Odo, S. Mussel adhesive plaque protein gene is a novel member of epidermal growth factor-like gene family. *J. Biol. Chem.* **1995**, *270*, 6698–6701. [CrossRef] [PubMed]

55. Hennebert, E.; Wattiez, R.; Demeuldre, M.; Ladurner, P.; Hwang, D.S.; Waite, J.H.; Flammang, P. Sea star tenacity mediated by a protein that fragments, then aggregates. *Proc. Natl. Acad. Sci. USA* **2014**, *111*, 6317–6322. [CrossRef] [PubMed]

56. Liu, C.; Li, S.; Huang, J.; Liu, Y.; Jia, G.; Xie, L.; Zhang, R. Extensible byssus of *Pinctada fucata*: Ca²⁺-stabilized nanocavities and a thrombospondin-1 protein. *Sci. Rep.* **2015**, *5*, 15018. [CrossRef] [PubMed]

57. Oberhauser, A.F.; Marszalek, P.E.; Erickson, H.P.; Fernandez, J.M. The molecular elasticity of the extracellular matrix protein tenascin. *Nature* **1998**, *393*, 181.

58. Anlar, B.; Gunel-Ozcan, A. Tenascin-R: Role in the central nervous system. *Int. J. Biochem. Cell Biol.* **2012**, *44*, 1385–1389. [CrossRef]

59. Valcourt, U.; Alcaraz, L.B.; Exposito, J.-Y.; Lethias, C.; Bartholin, L. Tenascin-X: Beyond the architectural function. *Cell Adhes. Migr.* **2015**, *9*, 154–165. [CrossRef]

60. Pasche, D.; Horbelt, N.; Marin, F.; Motreuil, S.; Macías-Sánchez, E.; Falini, G.; Hwang, D.S.; Fratzl, P.; Harrington, M.J. A new twist on sea silk: The peculiar protein ultrastructure of fan shell and pearl oyster byssus. *Soft Matter* **2018**, *14*, 5654–5664. [CrossRef]

61. Suhre, M.H.; Scheibel, T. Structural diversity of a collagen-binding matrix protein from the byssus of blue mussels upon refolding. *J. Struct. Biol.* **2014**, *186*, 75–85. [CrossRef]

62. Waite, J.H. The formation of mussel byssus: Anatomy of a natural manufacturing process. In *Structure, Cellular Synthesis and Assembly of Biopolymers*; Springer: Berlin, Germany, 1992; pp. 27–54.

63. Rich, A.; Crick, F. The Structure of Collagen. *Nature* **1955**, *176*, 915–916. [CrossRef] [PubMed]

64. Suhre, M.H.; Gertz, M.; Steegborn, C.; Scheibel, T. Structural and functional features of a collagen-binding matrix protein from the mussel byssus. *Nat. Commun.* **2014**, *5*, 3392. [CrossRef] [PubMed]

65. Kelley, L.A.; Mezulis, S.; Yates, C.M.; Wass, M.N.; Sternberg, M.J. The Phyre2 web portal for protein modeling, prediction and analysis. *Nat. Protoc.* **2015**, *10*, 845. [CrossRef] [PubMed]

66. Smith, A.M. *Biological Adhesives*, 2nd ed.; Springer International Publishing: Berlin, Germany, 2016; Volume VIII, p. 378, ISBN 978-3-319-46082-6.

67. Huang, J.; Zhang, C.; Ma, Z.; Xie, L.; Zhang, R. A novel extracellular EF-hand protein involved in the shell formation of pearl oyster. *Biochim. Biophys. Acta BBA Gen. Subj.* **2007**, *1770*, 1037–1044. [CrossRef]

68. Li, X.-X.; Yu, W.-C.; Cai, Z.-Q.; He, C.; Wei, N.; Wang, X.-T.; Yue, X.-Q. Molecular cloning and characterization of full-length cDNA of calmodulin gene from Pacific oyster *Crassostrea gigas*. *BioMed Res. Int.* **2016**, *2016*, 5986519. [PubMed]

69. Takeichi, M. Cadherin cell adhesion receptors as a morphogenetic regulator. *Science* **1991**, *251*, 1451–1455. [CrossRef] [PubMed]

70. Albertin, C.B.; Simakov, O.; Mitros, T.; Wang, Z.Y.; Pungor, J.R.; Edsinger-Gonzales, E.; Brenner, S.; Ragsdale, C.W.; Rokhsar, D.S. The octopus genome and the evolution of cephalopod neural and morphological novelties. *Nature* **2015**, *524*, 220. [CrossRef] [PubMed]

71. Zhao, X.; Yu, H.; Kong, L.; Li, Q. Transcriptomic responses to salinity stress in the Pacific oyster *Crassostrea gigas*. *PLoS ONE* **2012**, *7*, e46244. [CrossRef] [PubMed]

72. Lebesgue, N.; Da Costa, G.; Ribeiro, R.M.; Ribeiro-Silva, C.; Martins, G.G.; Matranga, V.; Scholten, A.; Cordeiro, C.; Heck, A.J.R.; Santos, R. Deciphering the molecular mechanisms underlying sea urchin reversible adhesion: A quantitative proteomics approach. *J. Proteom.* **2016**, *138*, 61–71. [CrossRef]

73. Drickamer, K. C-type lectin-like domains. *Curr. Opin. Struct. Biol.* **1999**, *9*, 585–590. [CrossRef]

74. Blank, S.; Arnoldi, M.; Khoshnavaz, S.; Treccani, L.; Kuntz, M.; Mann, K.; Grathwohl, G.; Fritz, M. The nacre protein perlucin nucleates growth of calcium carbonate crystals. *J. Microsc.* **2003**, *212*, 280–291. [CrossRef] [PubMed]

75. Kiani, C.; Liwen, C.; Wu, Y.J.; Albert, J.Y.; Burton, B.Y. Structure and function of aggrecan. *Cell Res.* **2002**, *12*, 19. [CrossRef] [PubMed]

76. De Gregorio, B.T.; Stroud, R.M.; Burden, D.K.; Fears, K.P.; Everett, R.K.; Wahl, K.J. Shell structure and growth in the base plate of the Barnacle *Amphibalanus amphitrite*. *ACS Biomater. Sci. Eng.* **2015**, *1*, 1085–1095. [CrossRef]

77. Flammang, P.; Demeuldre, M.; Hennebert, E.; Santos, R. Adhesive secretions in echinoderms: A review. In *Biological Adhesives*; Springer: Berlin, Germany, 2016; pp. 193–222.

78. Peng, Y.Y.; Glattauer, V.; Skewes, T.D.; McDevitt, A.; Elvin, C.M.; Werkmeister, J.A.; Graham, L.D.; Ramshaw, J.A. Identification of proteins associated with adhesive prints from *Holothuria dofleinii* Cuvierian tubules. *Mar. Biotechnol.* **2014**, *16*, 695–706. [CrossRef] [PubMed]

79. Peters, W. Occurrence of chitin in Molluscka. *Comp. Biochem. Physiol. Part B Comp. Biochem.* **1972**, *41*, 541–550. [CrossRef]

© 2019 by the authors. Licensee MDPI, Basel, Switzerland. This article is an open access article distributed under the terms and conditions of the Creative Commons Attribution (CC BY) license (http://creativecommons.org/licenses/by/4.0/).

Article

Global Transcriptional Insights of Pollen-Pistil Interactions Commencing Self-Incompatibility and Fertilization in Tea [*Camellia sinensis* (L.) O. Kuntze]

Romit Seth [1,2], Abhishek Bhandawat [1], Rajni Parmar [1,3], Pradeep Singh [1,2], Sanjay Kumar [1,3] and Ram Kumar Sharma [1,3,*]

[1] Biotechnology Department, CSIR-Institute of Himalayan Bioresource Technology, Palampur, Himachal Pradesh 176061, India; romit_seth18@yahoo.com (R.S.); abhishek.bhandawat@gmail.com (A.B.); rajni.parmar03@gmail.com (R.P.); ps111186@gmail.com (P.S.); sanjayplp1@gmail.com (S.K.)

[2] Department of Biotechnology, Guru Nanak Dev University, Amritsar 143005, India

[3] Academy of Scientific and Innovative Research, CSIR-Institute of Himalayan Bioresource Technology, Palampur, Himachal Pradesh 176061, India

* Correspondence: rksharma.ihbt@gmail.com or ramsharma@ihbt.res.in

Received: 26 November 2018; Accepted: 9 January 2019; Published: 28 January 2019

Abstract: This study explicates molecular insights commencing Self-Incompatibility (SI) and CC (cross-compatibility/fertilization) in self (SP) and cross (CP) pollinated pistils of tea. The fluorescence microscopy analysis revealed ceased/deviated pollen tubes in SP, while successful fertilization occurred in CP at 48 HAP. Global transcriptome sequencing of SP and CP pistils generated 109.7 million reads with overall 77.9% mapping rate to draft tea genome. Furthermore, concatenated de novo assembly resulted into 48,163 transcripts. Functional annotations and enrichment analysis (KEGG & GO) resulted into 3793 differentially expressed genes (DEGs). Among these, de novo and reference-based expression analysis identified 195 DEGs involved in pollen-pistil interaction. Interestingly, the presence of 182 genes [PT germination & elongation (67), S-locus (11), fertilization (43), disease resistance protein (30) and abscission (31)] in a major hub of the protein-protein interactome network suggests a complex signaling cascade commencing SI/CC. Furthermore, tissue-specific qRT-PCR analysis affirmed the localized expression of 42 DE putative key candidates in stigma-style and ovary, and suggested that LSI initiated in style and was sustained up to ovary with the active involvement of *cs*RNS, SRKs & SKIPs during SP. Nonetheless, COBL10, RALF, FERONIA-rlk, LLG and MAPKs were possibly facilitating fertilization. The current study comprehensively unravels molecular insights of phase-specific pollen-pistil interaction during SI and fertilization, which can be utilized to enhance breeding efficiency and genetic improvement in tea.

Keywords: gene expression; interactome; microscopy; fertilization; self-incompatibility; transcriptome; tea

1. Introduction

The purpose of pollination is fertilization and seed production to secure future survivability. Charles Darwin pioneered studies on the phenomenon of self-incompatibility in flowering plants "which are completely sterile with their own pollen, but fertile with that of any other individual of same species" [1]. This incapacity for self-pollination impeding self-fertilization is defined as self-incompatibility (SI). It is a genetically controlled mechanism that predominantly exists in flowering plants to overcome inbreeding depression and provides a high level of heterozygosity [2]. Self-incompatible plants have evolved genetic systems to prevent self-fertilization by recognition and rejection of pollen/pollen tube (PT) expressing the same allelic specificity either with pistils (pollen-pistil incompatibility) or ovular vicinity (ovular incompatibility/late-acting incompatibility), and post-fertilization mortality (post-zygotic incompatibility), inhibiting seed set [3]. Depending on the genetic control system, SI may be homomorphic or heteromorphic under the control of sporophytic or gametophytic conditions, and is categorized into three mechanisms, namely homomorphic sporophytic, homomorphic gametophytic and heteromorphic self-incompatibility [4,5].

Although SI has been widely studied in various angiosperms, nevertheless, molecular insights remained limited to Brassicaceae, Plantaginaceae, Rosaceae, Solanaceae and Papaveraceae. Among these, Brassicaceae possesses Sporophytic Self-Incompatibility (SSI), wherein, S-alleles of both the parents determine pollen's compatibility [6]. The mechanism is controlled by a tightly linked allele of stigma-specific S-receptor kinase (SRK) and pollen-specific S-locus cysteine-rich protein (SCR)/s-locus protein 11 (SP11), often referred as S haplotype [7]. The pollen germination in plants with similar S-haplotype is obstructed by inhibition of a stigmatic compatibility factor, Exo70A1 by regulating the pollen hydration via water transport from papilla cells in stigma to facilitate the pollen germination [8].

The members of Plantaginaceae, Rosaceae & Solanaceae exhibit Gametophytic Self-Incompatibility (GSI), wherein the female determinant S-RNase acts as a cytotoxin inhibiting pollen with similar S-allele. A group of pollen determinant S-locus F-box (SLF/SCF complex) found in the vicinity of S-RNase gene in *Petunia* was controlling the pollen specificity commencing for either GSI or fertilization/cross-compatibility (CC) [9]. Furthermore, non-self S-RNase were targeted by pollen specific SCF complex and undergoes ubiquitin-mediated degradation inside the cross PT, while self S-RNase were not blocked by SCFs, subsequently degrading the pollen's RNA and arresting PT growth [10]. Additionally, the roles of Pectin methyl esterase (PME) and pectin methyl esterase inhibitors (PMEI) were also reported in GSI in *Solanum species* [11]. Another type of GSI is reported in Papaveraceae, wherein Ca^{2+} mediated programmed cell death (PCD) occurs in self PT, preventing fertilization [12]. A recent transcriptome study in *Pyrus* species indicated a role of ATPase in SI through the calcium signaling pathway during the onset of pollination [13]. Moreover, late acting pre-zygotic SI or ovarian SI has been predominantly reported in Winteraceae, Theaceae, Malvaceae, Apocynaceae and Bignoniaceae families (eudicots); and Velloziaceae, Iridaceae, Amaryllidaceae and Xanthorrhoeaceae in monocots [4,14]. In some plant species like *Melaleuca alternifolia*, *Acacia retinodes* and *Theobroma cacao*, the PT normally grows up to ovary but failed to penetrate the ovule; while *Asclepiassyriaca* and *Spathodea campanulate* have been reported with post-zygotic LSI having abnormal/no seed set [15,16].

Tea (*Camellia sinensis* (L) Kuntze), indigenous to India and China, has been among the most profitable cash-crop across the globe. It is chiefly used as a 'health/energy drink' due to its ability to accumulate beneficial ingredients (mainly polyphenols) [17,18]. Belonging to family Theaceae, commercially important tea species have been classified into Chinese (*Camellia sinensis* var. *sinensis*), Assam (*Camellia sinensis* var. *assamica*) and Cambod (*Camellia sinensis* var. *assamica* subssp. *lasiocalyx*) forms of tea [19]. Due to tea's high economic value, breeding efforts have been made for its genetic improvement, though these efforts are still incomplete due to certain bottlenecks such as a high outcrossing nature (allogamy), profuse phenotypic variation, perennial, long gestation periods, high inbreeding depression and self-incompatibility contributing to tremendous heterozygosity in tea [20,21]. Hence, conventional clonal propagation is preferred over natural propagation to maintain the quality lines. Considering the multiple advantages of cost-effective next-generation sequencing (NGS) technologies for molecular dissection of complex traits [22,23], an earlier study suggested involvement of the SCF complex and S-RNase during SI in the style [24]. Furthermore, investigations of ion components in self and cross pollinated pistils indicated the role of Ca^{+2} and K^+ signal during SI [25]. Additionally, microscopy studies revealed LSI or ovarian sterility with pollen tube growth arrest in the SP ovary [26]. However, being a novel SI system, limited information is available regarding molecular insights regulating LSI response due to unidentified pollen/pistil factor having an important role in SI/CC reactions in tea [25,26]. In the current study, novel candidates involved in pollen-pistil interaction (LSI & fertilization) were identified by comparing the transcriptome of self-(SP) and cross-pollinated (CP) pistils in tea using high-throughput NGS technology. Furthermore, tissue-specific relative expression (style vs. ovary) of key genes provides a better understanding of the spatial transcriptional changes throughout the pistil during LSI. The results generated in this study elucidates important insights to understand the molecular mechanisms of LSI in light of fertilization in tea.

2. Results

2.1. Field Study and Microscopy Analysis

Pistil of both accessions (SA-6 and T78) possess wet type stigma with an ascending type style and syncarpous superior ovary [27]. The 24 h after Pollination (HAP) pistils were observed with PTs elongation up to the terminal region of style towards ovary in each case (Figure 1A). At 48 HAP Cross Pollinated pistils (CP), higher abundance of PT density and embryo sac with infiltrating PTs was observed in style and ovary, respectively (Figure 1B,C). In contrast, 48 HAP "Self-pollinated SA-6" (SP_S) and "Self-pollinated T78" (SP_T) exhibited less PT density in style with ceased/deviated PT towards integuments or other connective tissues in ovary (Figure 1B,C). A significant number of fertilized ovules (~97%) were recorded in reciprocal crosses of CP ovaries (SxT & TxS) at 48 HAP, while being insignificant in SP_S (1.1%) and SP_T (1.6%). However, a significant number of ovaries with abnormal PT behavior (ceased/deviated) near the micropyle in SP_S (98.8%) and SP_T (98.4%) was observed (Figure 1E and Table S1). Furthermore, a field study revealed ~60% fruit set at 180 Days after Pollination (DAP), and a seed set was observed at 360 DAP in both CP pistils (Figure 1D). In contrast, abortive ovules were also observed at 144 HAP in SP pistils [Figure 1C(c,f)]. Considering the microscopy inferences, 48 HAP was found to be an appropriate time to capture both fertilization and self-incompatible interactions for molecular analysis in our study. Additionally, a significant number of fertilization events with a strong positive correlation was recorded in both the reciprocal crosses (SxT and TxS) at 48 HAP, therefore, a single cross SxT of CP was utilized for transcriptome analysis.

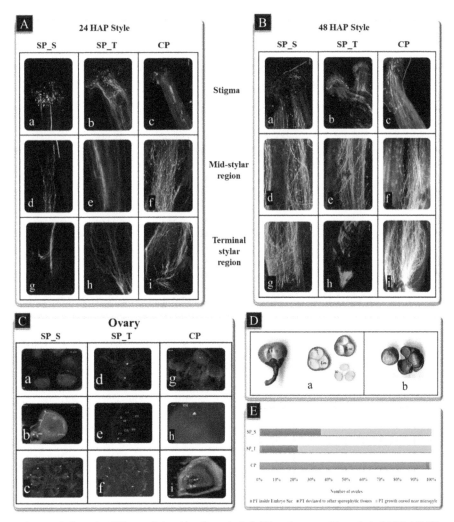

Figure 1. Pollen tube (PT) growth in self-pollinated pistil (SP) and cross-pollinated pistil (CP). (**A**) PT elongation in stigma (**a–c**), mid stylar region (**d–f**) and terminal stylar region (**g–i**) at 24 HAP SP and CP style. (**B**) PT growth in stigma (**a–c**), mid stylar region (**d–f**) and terminal stylar region (**g–i**) at 48 HAP SP and CP style. (**C**) PTs cessation (**a,d,e**) and deviation (**b**) at 48 HAP with abortive ovules (**c,f**) at 144 HAP in SP ovaries; PT (callose fluorescence) inside ovules (**g**), PTs infiltrating embryo sac (**h**), fertilized ovule with degenerated synergid (**i**) at 48 HAP CP ovaries. (**D**) 180 DAP fruit morphology and anatomy in CP pistil (**a**), 360 DAP seed morphology in CP pistil (**b**). nu represents nucellus, EA: Egg apparatus, in: integuments, ii: inner integument, oi: outer integument, mi: micropyle, sy: synergid, dsy: degenerated synergid, EC: Egg Cell, LEN: liquid endosperm, loc: locule, SC: seed coat (**E**) Graphical representation of microscopy inferences showing number fertilized ovules, number of ovules with PT deviation and number of ovules with PT cessation near micropyle at 48 HAP SP and CP pistils.

2.2. Illumina Sequencing, Sequence Assembly and Functional Annotation

Based on microscopy inferences, cDNA libraries of self (SP) and cross-pollinated CP pistils were sequenced to surmise the global molecular insights of pollen tube-pistil interaction. Overall,

91.2 million filtered reads were obtained after quality filtering of 109.7 million raw reads (Figure S1). The de novo assembly of high-quality reads yielded 51,489 (average length: 543 bp; N_{50}:719 bp) and 68,176 (average length: 776 bp; N_{50}:960 bp) transcripts using CLC genomic workbench and TRINITY, respectively (Table S2). Furthermore, the assembled transcripts obtained from both assemblers were concatenated and clustered into 48,163 high-quality non-redundant (NR) transcripts. Additionally, reference-based assembly resulted in a 77.9% overall mapping rate of filtered reads (SP_S, 81.1%; CP, 77.9% and SP_T, 74.7%) with the tea draft genome [28].

To obtain the global functional insights of assembled transcripts, sequence homology search (BLASTx) was performed with various publicly available protein databases annotating 35,136 (73%), 33,017 (68.56%), 26,945 (55.9%) and 31,798 (66.02%) transcripts with NCBI's nr, EggNOG, Swiss-Prot and TAIR10, respectively. The gene ontology (GO) annotation identified 23,996 transcripts assigned with 82,326 GO terms and classified them into the biological process (52%; 17 sub-categories), molecular function (22%; 7 sub-categories) & cellular component (26%; 8 sub-categories) (Table S3 and Figure S2a). Furthermore, a sequence search with Plant-TFDB resulted into 17,760 (36.56%) transcripts representing 58 transcription factors families. Among these, transcripts encoding basic helix-loop-helix transcription factor (bHLH) were the most abundant (2429 transcripts) followed by NAC (1663), MYB-related (1584), ERF (1278) and C2H2 (1038), (Figure S2b). Moreover, 378 pathways representing "metabolism" (44.5%), "genetic information & processing" (46.7%) and "signaling & cellular processes" (8.8%) exhibited significant enrichment in the KEGG pathway (Figure S2c).

2.3. Global Transcripts Expression Dynamics and Gene Ontology Enrichment Analysis

To elucidate molecular insights and key regulators involved in SI and fertilization, differential gene expression (DGE) of self (SP_S and SP_T) and cross-pollinated (CP: SxT) pistils resulted into 3793 (SP_S vs. CP), 3530 (SP_T vs. CP) and 3423 (SP_S vs. SP_T) differentially expressed (DE) genes in de novo DGE analysis (p-value & FDR $\leq$ 0.05) (Table S4). While the reference genome based DGE yielded 1847 (SP_S vs. CP), 1919 (SP_T vs. CP) and 1298 (SP_S vs. SP_T) DE genes with p-value & FDR $\leq$ 0.05 (Table S5). Moreover, the gene ontology (GO) enrichment analysis revealed a maximum enrichment of GO categories in CP followed by SP_S and SP_T, respectively (Figure S3). The categories: "signal transduction", "pollen-pistil interaction", "embryonic and post-embryonic development" of biological process and "hydrolase", "transferase", "kinase"; "signal transducer & receptor activity"; and "proteasome & its regulatory complexes" of molecular function exhibited significantly higher enrichment in CP (Figure S4). However, "cell death" and "response to stress" showed significantly higher enrichment in SP pistils (SP_S and SP_T) (Figures S5c and S6c).

2.4. Phase Specific Differentially Expressed Transcripts Involved in Pollen-Pistil Interaction

Based on the global expression and GO enrichment analysis, 195 significantly DE transcripts (considering both de novo and reference-based DGE along with their functional relevance in SI & fertilization) were extracted and categorized into five phases during pollen-pistil interactions [29]. These phases include pollen germination in stigma region (Phase I), PT elongation in the upper stylar region (Phase II), PT elongation and incompatible interactions in the style transmitting tract (Phase III), PT ovular guidance and LSI interactions (Phase IV) and ovarian region encompassing genes involved in fertilization (Phase V) (Table S6). The transcripts corresponding to genes involved in the pollen germination of phase I (Exo70A1, SRK, CER4) along with gametophytic self-incompatibility of phase II-III [S-RNase (*c*sRNS), SKIP (ABI1 and EBF1), F-box like (FBL), Pectin lyase (polygalacturonase, PGLR; Exo-polygalacturonase, ExoPG)] and some disease resistance proteins (DRPs) were significantly upregulated in SP. Meanwhile, transcripts involved in normal PT elongation in style of Phase III (ANXUR-rlk, 26s proteasome, LAT52, Root hair defective (RHD), Lipid transfer proteins (LTP), Arabinogalactan protein (AGP)); PT-ovular guidance of phase IV [Rapid alkalization factor (RALF), COBL10, SETH, K^+ transporters] and fertilization of phase V [FERONIA-rlk, LORELEI like glycoprotein (LLG), PMEI, GEX and ECP] along with auxin biosynthesis and auxin response factors (ARF) exhibited higher expression in CP (Figure 2).

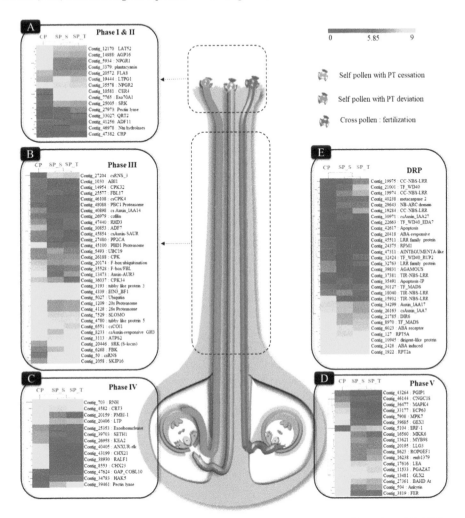

Figure 2. Schematic representation of PT elongation inside self and cross-pollinated pistil [Self PT: deviated (red) and ceased (brown), cross PT fertilization (purple)] as observed in microscopy, along with expression pattern of transcripts involved in different phases of pollen-pistil interaction. The heatmap represents expression pattern (log₂ transformed FPKM) in yellow-blue scale. (**A**) Transcripts expression of genes involved in pollen germination and PT elongation (Phase I & II); (**B**) PT elongation in mid-stylar region and incompatible interactions (Phase III); (**C**) PT ovular guidance/rejection (Phase IV); (**D**) fertilization in cross-pollinated (Phase V) and; (**E**) disease resistance proteins.

2.5. Protein-Protein Interactome Network Analysis

To identify the key regulatory genes and their involvement in complex signaling pathways during pollen-pistil interactions, a predetermined *At*PIN (*Arabidopsis thaliana* protein interaction network) was used [30]. The 195 DE transcripts showed direct interactions with 330 first neighbors (average number of neighbors: 27.170; network heterogeneity: 0.941 and clustering coefficient: 0.452).

Interestingly, 182 nodes (1953 edges) were present in the major hub representing PT germination & elongation (67), S-locus related (13), Fertilization (43), disease resistance protein (DRPs, 30) and abscission (31) (Figure 3A and Table S7).

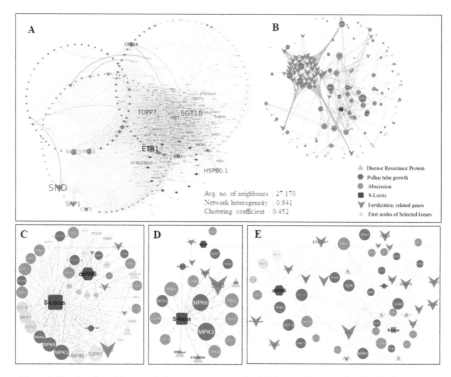

Figure 3. Predicted protein-protein interactome network of DE transcripts involved in fertilization or Self-Incompatibility in tea. (**A**) A major hub of 182 genes interacting with 343 first neighbors (6598 edges), (**B**) Co-expression network of 169 genes (1417 edges) extracted from 182 genes. (**C**) Gene specific predicted PPI-interactome network of S-Locus group (SRKs) and *cs*RNS (*C. sinensis*). (**D**) Direct interactions of S-locus related group and S-RNase. (**E**) Direct and indirect interactions of transcripts encoding genes involved in SI and Fertilization.

Furthermore, co-expression analysis revealed 148 genes (105 nodes in major hub) interacting with 211 first neighbors (2943 edges), displaying 129 incoming and 161 outgoing interactions (Figure 3B and Table S8). The degree of outgoing edges from node/gene (outgoing interactions) represents its regulatory function towards the node/gene receiving edges (incoming interaction) [31]. The intra-interactome network among five categories revealed that transcripts belonging to PT germination & elongation showed maximum outgoing interactions to the disease resistance proteins (DRP, 29) and abscission (16). Thus, transcripts involved in PT germination & elongation may have a role in pollen-pistil interaction by regulating DRPs and abscission-related genes. Furthermore, higher outgoing interactions of fertilization related genes with S-locus related (11), PT germination & elongation (57) and abscission (26) putatively suggested their major role in regulating PT growth to undergo fertilization or LSI. Higher outgoing interactions of S-locus related transcripts with the abscission-related genes, put forward their putative involvement in regulating PT abscission during LSI (Table 1).

Table 1. Intra-interactome network analysis among five categories showing a number of outgoing and incoming interactions.

Outgoing Interactions	Incoming Interactions				
	PT Germination & Elongation	S-Locus Related	Fertilization	DRP	Abscission
PT germination & elongation	67	3	26	29	16
S-locus related	2	11	5	3	8
Fertilization	57	11	43	5	26
DRP	7	2	6	30	4
abscission	27	10	20	2	31

The direct interactions of S-locus related transcripts with the ovular guidance & fertilization, abscission, DRP, PT elongation; and indirect interactions with SI related transcripts (csRNS & Exo70A1) and ovular guidance cysteine rich proteins (RALF) also suggest their regulatory function during SI and CC. Furthermore, direct interaction of csRNS with AGP8A (autophagy 8A), peroxidase (PAP17), pectin lyase; and indirect interactions with actin depolymerization factor (ADF) & PMEI indicates its key role during incompatible interactions. Moreover, the ExoPG recorded direct interactions with the genes involved in PT growth arrest (PMEI & CPK24) may also have a role in self-incompatibility. A gene belonging to family receptor-like kinase (ANXUR-rlk) exhibited direct interactions with the genes involved in normal PT elongation and abscission, which probably suggests its role in normal PT elongation, and was also recorded with higher expression in CP. Moreover, the genes involved in ovular guidance GPI-Anchored proteins (COBL10) were found to be directly interacting with Rapid alkalization factor (RALF), arabinogalactan protein (AGP), Ca^{++} mediated signal transduction (csCPK), SETH and ROPGEF. This indicates their role in regulating PT ovular guidance for successful fertilization. Additionally, another receptor-like kinases (FERONIA-rlk) with significantly upregulated expression in CP, recorded direct/indirect interactions with fertilization related genes (ROPGEF, LLG, SETH, MPKs), thus it probably has a role in regulating fertilization (CC) (Figure 3C–E).

2.6. RNA-Seq Data Validation by qRT-PCR

To confirm DGE inferences, qRT-expression validation of 12 key genes involved in pollen-pistil interaction during SP and CP showed a strong positive correlation with RNA-Seq expression data using GAPDH as an internal control (Figure 4A,B; Tables S9 and S10).

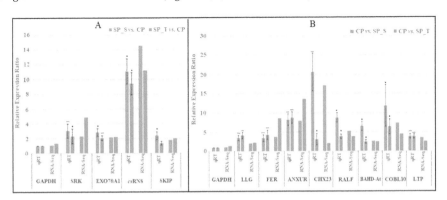

Figure 4. qRT-PCR validation of RNA-seq data using GAPDH as internal control. (**A**) Significantly upregulated SI related transcripts in SP pistils. (**B**) Significantly upregulated fertilization related transcripts in CP pistils. The bar represents standard deviation (SD) of relative expression for three replicated, and significance level is represented as stars: *p*-values (0.001, 0.01, 0.05) <=> symbols ("***", "**", "*").

Interestingly, 9 of 12 fertilization related genes were significantly up-regulated in both CP with respect to their SP pistils (SxT vs. SP_S and TxS vs. SP_T), and recorded a strong positive correlation [R squared correlation coefficient (R^2) = 0.8292] between the two reciprocal crosses of CP pistils (SxT & TxS) (Figure 5 and Table S10).

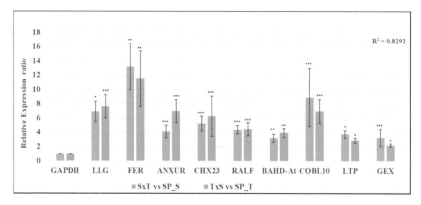

Figure 5. Relative expression analysis depicting strong positive correlation of reciprocal crosses (SxT & TxS) in CP pistils using GAPDH as internal control having strong positive correlation (R^2 = 0.8292) between them. The bar represents SD of relative expression for three replicated experiments, significance level is represented as symbols ("***", "**", "*") <=> *p*-values (0.001, 0.01, 0.05).

2.7. Tissue-Specific qRT-PCR Expression Dynamics during Pollen-Pistil Interaction

To study tissue and event specific expression, 42 key regulatory transcripts [pollen germination & elongation (9); ubiquitin-mediated protein degradation (6), ovular guidance (8), fertilization (12) and disease resistance (7)] were utilized for qRT-PCR relative expression analysis in style and ovary during SP and CP condition using GAPDH as an internal control (Table S9). A strong positive correlation in the expression pattern between SP genotypes in stylar (SP_S_style &. SP_T_style; R^2 = 0.83) and ovary (SP_S_ovary & SP_T_ovary; R^2 = 0.75) tissues possibly suggests a similar molecular behavior of incompatibility in both the SP pistils (Figure 6A and Figure S7A). However, an insignificant correlation in expression pattern between SP and CP possibly suggests a contrasting molecular mechanism commencing with SI and CC (Figure 6B,C; Figure S7B,C and Table S10).

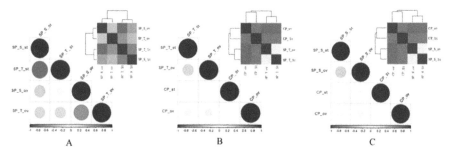

Figure 6. Tissue specific (style vs. ovary) and event specific (SP vs. CP) qRT-PCR expression correlation plot (correlation matrix and correlogram) of 42 key genes involved self-incompatibility and fertilization. (**A**) Tissue-specific correlation between self-pollinated tissues. (**B**) Event specific correlation between SP_S style and ovaries and (**C**) SP_T style and ovaries with respect to CP style and ovaries. The relative expression pattern is depicted in red-blue scale. Color intensity and size of the circle are proportional to the correlation coefficients. The legend color in the bottom represents the scale of correlation coefficients.

The transcripts involved in SI (*cs*RNS & SRK) and pollen tube growth regulator (PMEI, PGLR, ExoPG) were upregulated in SP_style, while the transcripts participating in PT elongation (LAT52, cofilin, RHD, FBL) along with Ubiquitin mediated protein degradation (20s, 26s) [29] were highly expressed in CP_style. However, genes involved in PT-ovular guidance from stylar transmitting tract to ovule (RALF, LLG & COBL10) and fertilization (FER, GEX, hapless2, MAPKs and ECP) [29,32] exhibited upregulated expression in CP ovaries, and suggested a higher probability of PT ovular guidance commencing fertilization (Figure 7 and Figure S8).

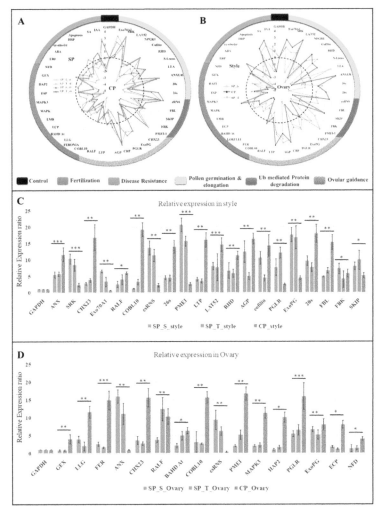

Figure 7. qRT-PCR expression analysis (log$_2$ fold change) of 42 key genes using GAPDH as internal control in event specific (**A**) and tissue specific (**B**) manner. The positive values (periphery) represent genes upregulated in SP and style, while negative (center) represents upregulation in CP and ovaries. (**C**) Significant relative expression of genes in self and cross-pollinated style and; (**D**) ovaries with respect to unpollinated style and ovaries respectively. The error bar in the graph represents SD of relative expression for three replicated experiments and significance level is represented as symbols ("***", "**", "*") <=> *p*-values (0.001, 0.01, 0.05).

Interestingly, a significantly higher expression of *cs*RNS, PMEI, polygalactuonase (PGLR) and Exo-polygalaturonase (ExoPG) in SP style [33,34] suggests its role in lower PT growth rate in SP style. Furthermore, higher expression of ANXUR-rlk in CP_style and SP_ovary, possibly associated with normal PT elongation in CP_style, while deviation in SP_ovary. Additionally, up-regulated expression of PMEI in CP_ovary and SP_style might involve in slackening PT elongation to assist PT burst [35].

3. Discussion

The journey to fertilization is arbitrated by a series of complex signaling mechanisms from stigma to an ovary, wherein PT growth in style can stimulate changes within the ovary [36]. In the present study, the phenological, microscopic and genome-wide expressions forms of analysis have been comprehensively explored to unravel the complexity of SI/CC in tea. 48 HAP as implicated in this study was also appropriately reported with the pollen tubes elongation up to ovary commencing successful fertilization in earlier studies in tea [26]. The concatenated *de novo* assembly using two assemblers (CLC and TRINITY) resulted in high-quality non-redundant transcripts in this study [37]. Furthermore, ≥ 77 % mapping of reads with the reference genome of tea suggested quality transcriptome data in this study [28,38]. The higher enrichment of 'signal transduction', 'post-embryonic development' and 'pollen-pistil interactions' putatively suggests successful commencement of fertilization in CP (Figure S4), while, 'cell death' and 'response to stress (endogenous and biotic)' enrichment indicates the occurrence of SI in SP (Figures S5 and S6) [24]. Additionally, qRT validation of key genes of pollen-pistil interaction suggests the reliability of the RNA-Seq expression data. Significant abundance of fertilized ovules with a strong positive correlation in the expression pattern of fertilization related genes in reciprocal crosses (SxT & TxS) suggests the rare probability of unilateral incompatibility (UI) in tea [39].

Most of the SI related earlier studies have been focused on molecular dynamics between pollen and style, with limited attention given to the ovary specific events. Hence, the tissue-specific relative expression of 42 key candidates obtained in the current study were further investigated in a phase-specific manner (Phase I to V) [40] using stigma-style and ovary to gain a better understanding of the LSI response in the light of fertilization in tea. Considering an evolution of SI from pathogen defense mechanisms, the higher expression pattern of defense-related genes (CC-NBS-LRR; NB-ARC domain) and transcription factors (WD40) in SP suggests their possible involvement in incompatible PT arrest in tea [16,41].

3.1. Pollen Germination & PT Elongation (Phase I-III)

As reported in Brassicaceae, the pollens were physically adhered to stigmatic papilla cells by pollen coat proteins and hydrated via Exo70A1 in stigma, wherein pollen coat lipids assist in pollen hydration to undergo germination [42]. The higher expression of Exo70A1 in SP_style is possibly responsible for the wet type of stigma with higher stigma receptivity in SP than CP at 48 HAP [27]. Furthermore, lower PT density in SP_style can be attributed by an upregulated expression of SI related transcripts (*cs*RNS, SRK, SKIP, ADF, pectin lyase, PGLR and Exo-PG) [43]. Moreover, *cs*RNS and S-locus related transcripts can be considered as key regulators due to their interactions with many compatibility and incompatibility factors in PPI network analysis. Additionally, indirect interaction of *cs*RNS with ADF suggests its possible role in programmed cell death (PCD) by depolymerization of actin cytoskeletons, hence arresting the self PT growth during GSI [10,44]. Considering an indicator of self-incompatibility, a significantly higher expression of Ca^{+2} transporters recorded in SP pistils may be responsible for higher concentration of Ca^{+2} ions in SP [25]. Nonetheless, the upregulated expression of transcripts involved in normal PT elongation (ANXUR-rlk, LAT52, cysteine rich proteins and RHD) and Ubiquitin-mediated S-RNase degradation (20s, 26s proteasome and SCF complex) may be attributed to higher PT density in the CP style (Figure 8A,B) [29,43].

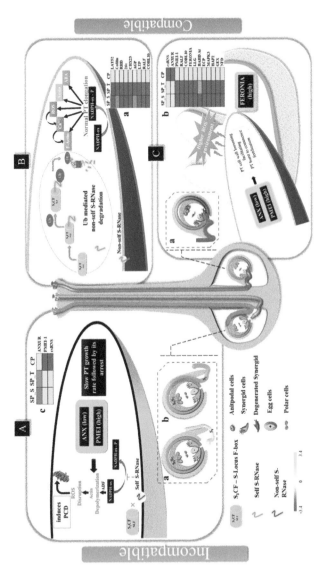

Figure 8. Summarized illustration representing self-incompatibility and cross-compatibility with tissue specific expression. The PT elongation within self and cross-pollinated pistil [Self PT: deviated (red) and ceased (brown), cross PT fertilization (purple)]. (**A**) Ceased (**a**) and deviated pollen tubes (**b**) representing incompatible interactions in style; (**c**) Heat map showing tissue specific qPCR expression of ANXUR-rlk, PMEI and csRNS in style revealing upregulated expression (yellow) of NADPH-ox, PMEI and csRNS in both SP coupled with downregulated expression (blue) of ANXUR-rlk; The self S-RNase (csRNS) in SP style inhibits phosphorylation of NADPH-ox, resultantly inducing programmed cell death (PCD) via depolymerization of actin cytoskeleton [44] (**B**) Normal PT elongation in style in CP as non-self S-RNase undergoes ubiquitin mediated protein degradation [10], (**a**) qPCR expression pattern showing up-regulated expression of genes involved in normal PT elongation in CP style. (**C**) Cross PT growth arrest followed by its burst within synergids commencing fertilization (**a**); qPCR expression pattern (**b**) of PT-ovular guidance and fertilization related genes exhibiting significantly up-regulated expression in CP ovaries. The lower expression of ANX coupled with higher expression of PMEI as observed in SP_style and CP_ovary suggests their putative role in inhibiting self-PT growth in SP style leading to SI and cross-PT inhibition in CP ovary to facilitate PT burst during fertilization. The yellow-blue scale represents fold change obtained in tissue specific relative expression analysis.

3.2. PT-Ovular Guidance (PHASE IV & V)

The higher tissue-specific expression of ANXUR-rlk, cofilin and RHD involved in PT elongation may be correlated with PT deviation in SP ovaries [45,46]. Additionally, higher expression of *cs*RNS, PGLR, ExoPG and PMEI in SP (style & ovary) possibly associated with anomalous PT behavior, suggesting the initiation of LSI in style and its sustenance up to ovary [24]. Nevertheless, higher expression RALF, GPI-APs (COBL10 and LLG) and SETH in CP ovaries suggests their involvement during normal PT-ovular guidance (Phase IV) [42]. Furthermore, indirect interaction of S-locus (SRK) with COBL10 via SETH in PPI network probably suggests its regulation by SRK during compatible and incompatible interactions. Also, the interaction of SETH with GPI-APs (COBL10 & LLG), calcium channels (*cs*CPKs) and ROPGEF involved in downstream activation of NADPH-oxidase (increase ROS level) leads to PT-synergid cell burst, thereby commencing fertilization [25,32]. Additionally, COBL10 is reportedly involved in regulating PTs cell wall organization *via* pectin modifications by activating PMEI causing PT burst during fertilization and is governed by the ovular guiding signals [42]. The higher expression of PMEIs coupled with lower expression of ANXUR-rlk in SP_style and CP_ov suggests their putative role in inhibiting self-PT elongation in SP style leading to LSI, and cross-PT inhibition in CP ovary commencing successful fertilization [35,47] (Figure 8).

During fertilization (Phase V), the female "FERONIA dependent signaling pathway" is activated within synergid, while the male "ANXUR dependent signaling pathway" is deactivated in compatible PT [48]. In the current study, ANXUR-rlk and PME were found to be co-expressed in network analysis with significantly higher expression in CP_style and SP_ovary, which can be correlated with normal PT elongation by regulating PME. Meanwhile higher PMEI expression coupled with low ANXUR-rlk in CP ovary were possibly involved in the commencement of fertilization (Figure 8C) [35,47]. Additionally, the presence of FERONIA-rlk in the major hub having direct interactions with transcripts involved in the fertilization suggests its key regulatory role in commencing fertilization. Moreover, upregulated expression of FERONIA-rlk along with genes involved in double fertilization (GEX, HAP2 and BAHD acyltransferase) and transcription factor MAPK3 (Mitogen-activated Protein Kinase 3) can be correlated with higher frequency of fertilized ovules in CP ovaries as observed in microscopy [32].

4. Materials and Methods

4.1. Plant Material

Two self-incompatible tea accessions, SA-6 and Tukdah (T)-78 with high level of cross-compatibility [19,49] were selected in this study. These accessions were maintained at the CSIR-Institute of Himalayan Bioresource Technology, Palampur, India (1300 m altitude; 32°06' N, 76°33' E). Controlled pollination was carried out at the balloon stage (flowering) during October to December in three subsequent years (2013-2015). Enlarged and about to open floral buds with maximal stigmatic receptivity were emasculated, bagged and pollinated next day between 8:30 to 10:00 AM, followed by immediate re-bagging after pollination. The experimental analysis was performed in three combinations as "Self-pollinated SA-6 (SP_S)"; "Self-pollinated T78 (SP_T)" and "Cross-pollinated SA-6 x T78 and T78 x SA6 (CP)". Pistils at 24 and 48 HAP were fixed for microscopy. A total 320 pollinated pistils (40 each for SP_S, SP_T and CP at 24 and 48 HAP) were collected for the microscopy, while some were leftover in the field to monitor the subsistent fruit and seeds set.

4.2. Microscopic Analysis

Twenty-four HAP and 48 HAP pistils were harvested and fixed in FAA fixative solution (Formaldehyde: Acetic acid:Alcohol::1:1:18) to target the PTs localization inside female gametophyte (pistil). Of the forty pistils, twenty each were used to trace the PTs inside stigma-stylar region using squash method and for targeting the PTs inside the ovary using microtome. For squash method, the pistils were fixed in F.A.A. for 24 h and stained using the aniline blue staining protocol [50]. Furthermore, 10 μm thin transverse sections of paraffin wax embedded ovaries were performed using microtome (Thermo Shandon Finesse microtome, Thermo Fisher Scientific, Waltham, MA, USA). Sections were mounted and stained using 0.1% aniline blue staining solution. The mounted stained samples and squashed samples were scanned and captured using Fluorescence microscope with AxioCam Zeiss MR Lenses (Oberkochen, Germany). Chi-square test was used to assess significance level of microscopy data to affirm the distinctness ($p < 0.05$) among collected samples.

4.3. RNA Extraction, cDNA Library Preparation And Illumina Sequencing

Based on microscopy inferences, 48 HAP pistils of SP_S, SP_T and CP (SxT) in ten biological replicates were collected and snap-frozen to liquid nitrogen for total RNA extraction using IRIS method [51]. The RNA was quantified on NanoDrop 2000 (Thermo Scientific, Waltham, MA, USA), and quality was assessed on 1% formaldehyde agarose gel (MOPS) and Agilent Bioanalyzer with RNA 7500 series II Chip (Agilent Technologies, CA, USA). The RNA samples with RIN (RNA Integrity Number) value greater than 8 and the final concentration of 4.0 µg were used for cDNA library preparation.

Eight cDNA libraries in biological replicates SP_T (3), CP (3) and SP_S (2) were constructed using the illumina Truseq RNA Sample prep v2 LS Protocol (Illumina Inc., CA, USA). The libraries were quantified on Qubit 2.0 fluorometer (Invitrogen, USA), while quality was assessed using an Agilent 2100 Bioanalyzer (Agilent Technologies, CA, USA). The paired-end (PE) (2×72 bp) sequencing was performed using Illumina GAIIx.

4.4. Quality Filtering, Sequence Assembly and Differential Expression Dynamics

The base calling and demultiplexing of raw data obtained from GAIIx run was performed using Illumina Casava 1.8.2 pipeline (http://support.illumina.com/). The demultiplexed raw reads were filtered using NGS QC Toolkit [52]. Filtered fastq reads were de novo assembled using both CLC Genomics Workbench 6.5 (CLC Bio-Qiagen, Aarhus, Denmark) and TRINITY RNA-Seq ver. 2.3.0 [53] with default parameters. Both of the assemblies were combined independently to optimize the coding region of transcriptome as discussed by Cerveau and Jackson (2016) [37]. The intra-assembly clustering of both the de novo assembled transcripts was performed using CD-HIT-ESTver4.6 clustering tool [54]. The unique transcripts derived from both the assemblies were concatenated and ORFs were detected using TransDecoder ver.3.0.1. These ORFs were further re-clustered based on their sequence similarity, yielding non-redundant high-quality transcripts. Individual sample reads were then separately mapped to the concatenated transcripts using Bowtie 2 and normalized to estimate transcript abundance and DE. The Transcript abundance was estimated using RPKM (Reads Per Kilobase of transcript per Million mapped reads) [55,56]. The differential gene expression between self-pollination (SP_S and SP_T) and cross-pollination (CP) events were estimated using the edgeR tool [57,58]. The p-values of DE transcripts were adjusted for multiple testing by the Benjamini-Hochberg false discovery rate (FDR) method [59]. The transcripts with FDR ≤ 0.05 and $\log_2$ FC ≥ 1 & ≤ -1 were extracted for downstream analysis. Transcripts abundance (RPKM) was illustrated as a heatmap using MeV package v.4.9.0. Furthermore, with the advent of draft tea genome [28], reference-based DGE was also performed using Tuxedo reference genome based assembly pipeline with default parameters [60]. The sample-specific filtered reads were mapped to reference genome using TOPHAT ver2.1.0. Cufflink was used to assemble the transcriptome and estimate transcript abundance followed by Cuffmerge, to merge all the assemblies and estimate expression level. The DE transcripts between CP and SP conditions were compared using Cuffdiff. The TransDecoder ver.3.0.1 was used to extract the longest coding sequence using the merged GTF file obtained as an output from cuffmerge.

4.5. Transcripts Homology, Functional Classification and GO Enrichment Analysis

The de novo assembled transcripts were subjected to blastx analysis against the protein sequences in NCBI's nr, Swiss-Prot, TAIR10, EggNOG v4.5 (http://eggnogdb.embl.de/), KEGG (http:

//www.kegg.jp/kegg/tool/annotate_sequence.html) and Plant Transcription Factor Database (http://planttfdb.cbi.pku.edu.cn/) considering *e*-value $\leq 1 \times 10^{-5}$ to retrieve the top hits with functional attributes showing highest sequence similarity with the assembled transcripts. Gene enrichment was estimated using AgriGO toolkit. TAIR orthologous ID of DE transcripts was retrieved for GO enrichment using singular enrichment analysis (SEA) in AgriGO toolkit [61]. Plant GO slim was performed using Fischer statistical analysis (Hochberg-FDR adjustment cut-off <0.01) for optimal gene enrichment and represented in a hierarchical semantic similarity based scattered model and treemap (Figure S3). The in-silico enrichment analyses were computed using Bioconductor R package version 3.2.3. The GO terms were grouped into three categories: molecular function, biological processes, and cellular component. The over and under-represented GO terms were reduced and visualized on the Revigo tool using the Fisher-exact test.

4.6. Protein-Protein Interactome Network Analysis

A protein-protein interactome network was built to identify key regulatory genes involved in incompatible and compatible interactions. The sequences homologues of DE transcripts were extracted from nr, TAIR and Swiss-Prot protein database and subjected to the STRING interactome public database for network analysis [62]. A correlation edge was considered as conserved when the selected tea genes had a significant correlation edge with its respective orthologs in the *Arabidopsis thaliana* PPI network. First neighbors of the mapped IDs were selected for predicting their interaction. Subsequently, a regulatory network was built based on phylogenetic co-occurrence, the number of directed edges, homology and co-expression of values. This network was visualized on Cytoscape ver. 3.4.0 [63]. Genes of selected categories were represented in circular layouts using a number of directed edges as an attribute.

4.7. RNA-Seq Expression Pattern Validation Using Real-Time PCR

Differential Gene expression of 12 DE transcripts from RNA-Seq data were validated utilizing Real time PCR (RT-PCR). The RNA of whole pistil from each SP_S, SP_T and CP was considered in RNA-Seq validation as utilized in RNA-Seq analysis. Additionally, RNA from SxT & TxS pistils was also extracted to scrutinize the expression pattern of 9 fertilization related genes between two reciprocal crosses (Table S9). The first strand cDNA was synthesized using 2 μg of total RNA by Revert Aid First strand cDNA synthesis kit (Thermo Scientific, USA). Gene-specific primers from selected transcripts were designed with BatchPrimer3 (http://probes.pw.usda.gov/batchprimer3/). Reactions were performed in 20 μL reaction volume containing 200 ng template cDNA with FG-POWER SYBR® Green PCR Master Mix Applied Biosystem (Foster City, CA, USA) and gene-specific primers (Table S9) in StepOne™ Real-Time PCR System (Applied Biosystem). Specific GAPDH primers were used as an internal control. The expression analysis of all the genes were performed in three replicates and relative expression was calculated using comparative Ct values [59,64].

4.8. Tissue-Specific Transcript Expression Analysis Using qRT-PCR

42 putative key candidate genes involved in compatible/incompatible interactions were selected based on their functional annotation, enrichment and PPI network analysis to assess tissue specific (style vs. ovary) and event specific (SP vs. CP) relative expression analysis using qRT-PCR. Total RNA was extracted from both 48 HAP style and ovaries, separately from each SP_S, SP_T and CP along with their respective controls (un-pollinated style and ovary) using IRIS method [51]. The cDNA preparation and qRT-analysis were performed as mentioned in the previous section (Section 4.7) using GAPDH as a reference gene (Table S9). The expression analysis of all the genes were performed in three replicates and relative expression was calculated using comparative Ct values [59,64]. The relative expression ratio of SP and CP, style and ovaries were obtained with respect to unpollinated style and ovaries. Furthermore, ovaries and CP were considered as control in tissue specific and event specific fold change analysis respectively. Pearson's correlation coefficient along with their significance were

computed based on candidate genes specific relative expression ratio to find tissue specific and event specific correlation and were plotted using the R package.

5. Conclusions

The current study provides a comprehensive atlas of genes and pathways involved in pollen pistil interaction leading to LSI in light of fertilization in tea. Combined inferences drawn based on microscopy, genome-wide transcriptome, interactome network and tissue specific qRT-expression analysis suggests a pre-zygotic type of LSI, which probably initiates in style and sustains up to ovary with the active involvement of potential candidates belonging to categories cysteine-rich proteins (RALF), receptor-like kinases (FER-rlk, ANXUR-rlk), GPI-Aps (COBL10, LLG), enzyme (csRNS, PME & PMEI) and transcription factors (MAPK). The valuable genomic resources and putative master regulators obtained in this study will promote a better understanding of the molecular mechanism of pollen-pistil interaction that commences LSI and fertilization in tea. These resources can be employed to enhance breeding efficiency and genetic improvement in tea and other perennial plant species.

Supplementary Materials: The following are available online at http://www.mdpi.com/1422-0067/20/3/539/s1, Figures S1–S8; Tables S1–S10. Figure S1: Quality check and filtering of RNA-seq Data. [a] Overall Filtering of Data. [b] Sample wise Filtering of Data. Figure S2: (a) Overall GO annotation of transcripts categorized into cellular component, biological process and molecular function, (b) Transcripts annotation with 58 Plant transcription factors represented in the form of scattered plot, top 20 TF represented in the form of pie chart and (c) KEGG pathway classification of overall transcripts. Figure S3: Scattered plot of GO terms associated with upregulated transcripts in (a) SP_S, (b) SP_T and (c) CP condition exhibiting clusters representatives. The matrix was created using pairwise semantic similarities of GO terms. The semantically similar GO terms were clustered together. Revigo tree map representation of Upregulated transcripts in (d) SP_S, (e) SP_T and (f) CP condition showing GO enrichment. The treemap was built using absolute log10 P-value, the blocks with larger area showing categories with higher GO enrichment and the blocks with similar semantic values are represented with similar colours. Figure S4: GO enrichment analysis of upregulated transcripts in CP condition [a] Biological Processes; [b] Molecular Function; [c] Cellular Components. Figure S5: GO enrichment analysis of upregulated transcripts in SP_S condition [a] Biological Processes; [b] Molecular Function; [c] Cellular Components. Figure S6: GO enrichment analysis of upregulated transcripts in SP_T condition [a] Biological Processes; [b] Molecular Function; [c] Cellular Components. Figure S7: Correlation plot representing correlation between event specific and condition specific RT expression analysis (a) Correlation among SP_S_ss, SP_T_ss, SP_S_ov and SP_T_ov; (b) Correlation among SP_S_ss, CP_ss, SP_S_ov and CP_ov; (c) Correlation among SP_T_ss, CP_ss, SP_T_ov and CP_ov. The distribution of each variable is shown on the diagonal with bottom showing bivariate scatter plots with a fitted line top, showing the value of the correlation. The significance level is represented as stars, each significance level is associated to a symbol: p-values(0, 0.001, 0.01, 0.05, 0.1, 1) <=> symbols("***", "**", "*", ".", " "). Figure S8: Graph representing log2 fold change expression of 42 key genes using GAPDH as internal control in event specific (A) and tissue specific (B) manner. The error bar represents SD between three replicated samples. Table S1: Chi-square estimation of microscopy results. Table S2: Summary details of assembled sequences produced after filtered reads de novo assembly using both CLC work bench and TRINITY. Table S3: Annotation of assembled sequences (Transcripts) based on five public databases TAIR, swissprot, nr, KOG (EGGNOG) and Transcription Factors. Table S4: Differential Gene Expression extracted from de novo assembly using EdgeR tools. Tab 1 showing DGE between SP_S and CP, Tab 2 DGE between SP_T and CP. Table S5: Differential Gene Expression extracted from tea reference genome using tuxedo pipeline. The fold change was calculated between SP_S vs CP and SP_T vs CP using cuffdiff. Table S6: 408 Differentially expressed transcripts encapsulating 195 genes related to fertilization and incompatibility categorized into five different phases. Table S7: Gene specific interactome network analysis of differentially expressed genes. Table S8: Gene specific coexpression analysis of differentially expressed genes. Table S9: Primer details of selected transcripts sequences to assess relative expression analysis using quantitative Real Time-polymerase chain reaction (qRT-PCR) analysis. Table S10: RNA-Seq validation of DGE with raw Ct values of reciprocal crosses and tissue specific relative expression analysis with significance test.

Data availability and material: The raw reads were deposited to NCBI SRA database with the following accession numbers: SRR7037029, SRR7037030, SRR7037031, SRR7037032, SRR7037033, SRR7037034, SRR7037035, SRR7037036.

Author Contributions: R.K.S., R.S. conceived and designed the study. R.S., A.B., R.P., P.S. conducted field work. R.S., P.S. performed microscopy. R.S., A.B., R.K.S. analysed data. R.S., R.P. performed qRT expression analysis. R.S., R.K.S. wrote and edited the manuscript. S.K. helped in editing the manuscript. R.K.S. approved the final version of the manuscript. All authors read and approved the manuscript.

Funding: This research was funded by Council of Scientific & Industrial Research, New Delhi, grant numbers BSC-0301 and MLP-0146, and DST/INT/SL/P-16/2016.

Acknowledgments: RS and PS acknowledge GNDU, Amritsar for Ph. D registration. Mohit K. Swarnkar is acknowledged for assisting in Illumina GAIIx sequencing. This is IHBT communication No. 3992.

Conflicts of Interest: The authors declare no conflict of interest.

Abbreviations

SI	Self-incompatibility
LSI	Late-acting gametophytic self-incompatibility
CC	Cross-compatibility (Fertilization)
SP	Self-Pollinated
CP	Cross-pollinated
PT	Pollen tube
HAP	Hours after pollination
DAP	Days after pollination
KEGG	Kyoto encyclopedia of genes and genomes
GO	Gene ontology
DGE	Differential gene expression
NGS	next generation sequencing
SLF/SCF	S-locus F-box protein
SRK	S-receptor kinase
CPK	Calcium-dependent protein kinases
TLP	Tubby like proteins
RHD	Root hair defective
PMEI	Pectin methylesterase inhibitor
GEX	Gamete expressed
ARF	auxin response factors
DRP	Disease resistance proteins
RALF	Rapid alkalization factors
LLG	LORELLEI like glycoprotein
GPI-Ap	Glycosylphosphatidylinositol anchored protein
CRP	Cystein rich protein
qRT-PCR	Quantitative Real-Time PCR

References

1. Darwin, C. *The Effects of Cross and Self Fertilisation in the Vegetable Kingdom*; Cambridge University Press: Cambridge, UK, 1876.
2. De Nettancourt, D. Incompatibility in angiosperms. *Sex. Plant Reprod.* **1997**, *10*, 185–199. [CrossRef]
3. Chapman, L.A.; Goring, D.R. Pollen-pistil interactions regulating successful fertilization in the Brassicaceae. *J. Exp. Bot.* **2010**, *61*, 1987–1999. [CrossRef] [PubMed]
4. Gibbs, P.E. Late-acting self-incompatibility-the pariah breeding system in flowering plants. *New Phytol.* **2014**, *203*, 717–734. [CrossRef] [PubMed]
5. Lewis, D. Comparative Incompatibility in Angiosperms and Fungi. *Adv. Genet.* **1954**, *6*, 235–285. [CrossRef]
6. Bateman, A.J. Self-incompatibility systems in angiosperms II. Iberis amara. *Heredity* **1954**, *8*, 305–332. [CrossRef]
7. Haasen, K.E.; Goring, D.R. The recognition and rejection of self-incompatible pollen in the Brassicaceae. *Bot. Stud.* **2010**, *51*, 1–6.
8. Samuel, M.A.; Chong, Y.T.; Haasen, K.E.; Aldea-Brydges, M.G.; Stone, S.L.; Goring, D.R. Cellular pathways regulating responses to compatible and self-incompatible pollen in Brassica and Arabidopsis stigmas intersect at Exo70A1, a putative component of the exocyst complex. *Plant Cell* **2009**, *21*, 2655–2671. [CrossRef] [PubMed]
9. Sun, L.; Williams, J.S.; Li, S.; Wu, L.; Khatri, W.A.; Stone, P.G.; Keebaugh, M.D.; Kao, T. S-Locus F-Box Proteins Are Solely Responsible for Pollen Function in S-RNase-Based Self-Incompatibility of Petunia. *Plant Cell* **2018**, tpc.00615. [CrossRef]
10. McClure, B.A.; Franklin-Tong, V. Gametophytic self-incompatibility: Understanding the cellular mechanisms involved in "self" pollen tube inhibition. *Planta* **2006**, *224*, 233–245. [CrossRef] [PubMed]

11. Pease, J.B.; Guerrero, R.F.; Sherman, N.A.; Hahn, M.W. Molecular mechanisms of postmating prezygotic reproductive isolation uncovered by transcriptome analysis. *Mol. Ecol.* **2016**, *25*, 2592–2608. [CrossRef]

12. Wilkins, K.A.; Bosch, M.; Haque, T.; Teng, N.; Poulter, N.S.; Franklin-Tong, V.E. Self-incompatibility-induced programmed cell death in field poppy pollen involves dramatic acidification of the incompatible pollen tube cytosol. *Plant Physiol.* **2015**, *167*, 766–779. [CrossRef] [PubMed]

13. Chen, Q.; Meng, D.; Gu, Z.; Li, W.; Yuan, H.; Duan, X.; Yang, Q.; Li, Y.; Li, T. SLFL Genes Participate in the Ubiquitination and Degradation Reaction of S-RNase in Self-compatible Peach. *Front. Plant Sci.* **2018**, *9*, 227. [CrossRef] [PubMed]

14. Seavey, S.R.; Bawa, K.S. Late-acting self-incompatibility in angiosperms. *Bot. Rev.* **1986**, *52*, 195–219. [CrossRef]

15. Liao, T.; Yuan, D.-Y.; Zou, F.; Gao, C.; Yang, Y.; Zhang, L.; Tan, X.-F. Self-sterility in Camellia oleifera may be due to the prezygotic late-acting self-incompatibility. *PLoS ONE* **2014**, *9*, e99639. [CrossRef]

16. Allen, A.M.; Hiscock, S.J. Evolution and phylogeny of self-incompatibility systems in angiosperms. In *Self-Incompatibility in Flowering Plants*; Springer Verlag: Berlin, Germany, 2008; pp. 73–101.

17. Preedy, V.R. *Tea in Health and Disease Prevention*; Academic Press: Cambrage, MA, USA, 2012; ISBN 0123849381.

18. Singh, S.; Sud, R.K.; Gulati, A.; Joshi, R.; Yadav, A.K.; Sharma, R.K. Germplasm appraisal of western Himalayan tea: A breeding strategy for yield and quality improvement. *Genet. Resour. Crop Evol.* **2013**, *60*, 1501–1513. [CrossRef]

19. Sharma, R.K.; Negi, M.S.; Sharma, S.; Bhardwaj, P.; Kumar, R.; Bhattachrya, E.; Tripathi, S.B.; Vijayan, D.; Baruah, A.R.; Das, S.C. AFLP-based genetic diversity assessment of commercially important tea germplasm in India. *Biochem. Genet.* **2010**, *48*, 549–564. [CrossRef] [PubMed]

20. Raina, S.N.; Ahuja, P.S.; Sharma, R.K.; Das, S.C.; Bhardwaj, P.; Negi, R.; Sharma, V.; Singh, S.S.; Sud, R.K.; Kalia, R.K. Genetic structure and diversity of India hybrid tea. *Genet. Resour. Crop Evol.* **2012**, *59*, 1527–1541. [CrossRef]

21. Sharma, H.; Kumar, R.; Sharma, V.; Kumar, V.; Bhardwaj, P.; Ahuja, P.S.; Kumar, R. Identification and cross-species transferability of 112 novel unigene-derived microsatellite markers in tea (Camellia sinensis). *Am. J. Bot.* **2011**, *98*, 133–138. [CrossRef]

22. Unamba, C.I.N.; Nag, A.; Sharma, R.K. Next Generation Sequencing technologies: The doorway to the unexplored genomics of non-model plants. *Front. Plant Sci.* **2015**, *6*. [CrossRef]

23. Jayaswall, K.; Mahajan, P.; Singh, G.; Parmar, R.; Seth, R.; Raina, A.; Swarnkar, M.K.; Singh, A.K.; Shankar, R.; Sharma, R.K. Transcriptome Analysis Reveals Candidate Genes involved in Blister Blight defense in Tea (*Camellia sinensis* (L.) Kuntze). *Sci. Rep.* **2016**, *6*. [CrossRef]

24. Zhang, C.; Wang, L.; Wei, K.; Wu, L.; Li, H.; Zhang, F.; Cheng, H. Transcriptome analysis reveals self-incompatibility in the tea plant (Camellia sinensis) might be under gametophytic control. *BMC Genom.* **2016**, *17*, 1–15. [CrossRef] [PubMed]

25. Ma, Q.; Chen, C.; Zeng, Z.; Zou, Z.; Li, H.; Zhou, Q.; Chen, X.; Sun, K.; Li, X. Transcriptomic analysis between self-and cross-pollinated pistils of tea plants (Camellia sinensis). *BMC Genom.* **2018**, *19*, 289. [CrossRef] [PubMed]

26. Chen, X.; Hao, S.; Wang, L.; Fang, W.; Wang, Y.; Li, X. Late-acting self-incompatibility in tea plant (*Camellia sinensis*). *Biol. Sect. Bot.* **2012**, *672*, 347–351. [CrossRef]

27. Heslop-Harrison, Y.; Shivanna, K.R. The receptive surface of the angiosperm stigma. *Ann. Bot.* **1977**, *41*, 1233–1258. [CrossRef]

28. Xia, E.-H.; Zhang, H.-B.; Sheng, J.; Li, K.; Zhang, Q.-J.; Kim, C.; Zhang, Y.; Liu, Y.; Zhu, T.; Li, W. The Tea Tree Genome Provides Insights into Tea Flavor and Independent Evolution of Caffeine Biosynthesis. *Mol. Plant* **2017**, *10*, 866–877. [CrossRef] [PubMed]

29. Dresselhaus, T.; Franklin-Tong, N. Male-female crosstalk during pollen germination, tube growth and guidance, and double fertilization. *Mol. Plant* **2013**, *6*, 1018–1036. [CrossRef] [PubMed]

30. Bhandawat, A.; Singh, G.; Seth, R.; Singh, P.; Sharma, R.K.R.K. Genome-wide transcriptional profiling to elucidate key candidates involved in bud burst and rattling growth in a subtropical bamboo (Dendrocalamus hamiltonii). *Front. Plant Sci.* **2016**, *7*. [CrossRef] [PubMed]

31. Kusonmano, K. Gene Expression Analysis Through Network Biology: Bioinformatics Approaches. In *Network Biology*; Springer Nature: Basingstoke, UK, 2016; pp. 15–32.

32. Dresselhaus, T.; Sprunck, S.; Wessel, G.M. Fertilization mechanisms in flowering plants. *Curr. Biol.* **2016**, *26*, R125–R139. [CrossRef] [PubMed]

33. Muschietti, J.; Dircks, L.; Vancanneyt, G.; McCormick, S. LAT52 protein is essential for tomato pollen development: Pollen expressing antisense LAT52 RNA hydrates and germinates abnormally and cannot achieve fertilization. *Plant J.* **1994**, *6*, 321–338. [CrossRef] [PubMed]

34. Boisson-Dernier, A.; Roy, S.; Kritsas, K.; Grobei, M.A.; Jaciubek, M.; Schroeder, J.I.; Grossniklaus, U. Disruption of the pollen-expressed FERONIA homologs ANXUR1 and ANXUR2 triggers pollen tube discharge. *Development* **2009**, *136*, 3279–3288. [CrossRef]

35. Li, S.; Ge, F.R.; Xu, M.; Zhao, X.Y.; Huang, G.Q.; Zhou, L.Z.; Wang, J.G.; Kombrink, A.; McCormick, S.; Zhang, X.S.; Zhang, Y. Arabidopsis COBRA-LIKE 10, a GPI-anchored protein, mediates directional growth of pollen tubes. *Plant J.* **2013**, *74*, 486–497. [CrossRef] [PubMed]

36. Zheng, Y.-Y.; Lin, X.-J.; Liang, H.-M.; Wang, F.-F.; Chen, L.-Y. The Long Journey of Pollen Tube in the Pistil. *Int. J. Mol. Sci.* **2018**, *19*, 3529. [CrossRef] [PubMed]

37. Cerveau, N.; Jackson, D.J. Combining independent de novo assemblies optimizes the coding transcriptome for nonconventional model eukaryotic organisms. *BMC Bioinform.* **2016**, *17*, 525. [CrossRef] [PubMed]

38. Kovi, M.R.; Amdahl, H.; Alsheikh, M.; Rognli, O.A. De novo and reference transcriptome assembly of transcripts expressed during flowering provide insight into seed setting in tetraploid red clover. *Sci. Rep.* **2017**, *7*, 44383. [CrossRef] [PubMed]

39. Qin, X.; Li, W.; Liu, Y.; Tan, M.; Ganal, M.; Chetelat, R.T. A farnesyl pyrophosphate synthase gene expressed in pollen functions in S-RNase-independent unilateral incompatibility. *Plant J.* **2018**, *93*, 417–430. [CrossRef] [PubMed]

40. Park, S.-W.; Kang, S.-W.; Goo, T.-W.; Kim, S.-R.; Lee, G.; Paik, S.-Y. Tissue-specific gene expression analysis of silkworm (Bombyx mori) by quantitative real-time RT-PCR. *BMB Rep.* **2010**, *43*, 480–484. [CrossRef] [PubMed]

41. Shi, D.; Tang, C.; Wang, R.; Gu, C.; Wu, X.; Hu, S.; Jiao, J.; Zhang, S. Transcriptome and phytohormone analysis reveals a comprehensive phytohormone and pathogen defence response in pear self-/cross-pollination. *Plant Cell Rep.* **2017**, *36*, 1785–1799. [CrossRef] [PubMed]

42. Higashiyama, T.; Takeuchi, H. The Mechanism and Key Molecules Involved in Pollen Tube Guidance. *Annu. Rev. Plant Biol.* **2015**, *66*, 393–413. [CrossRef] [PubMed]

43. McClure, B. Darwin's foundation for investigating self-incompatibility and the progress toward a physiological model for S.-RNase-based SI. *J. Exp. Bot.* **2009**, *60*, 1069–1081. [CrossRef] [PubMed]

44. Onelli, E.; Idilli, A.I.; Moscatelli, A. Emerging roles for microtubules in angiosperm pollen tube growth highlight new research cues. *Front. Plant Sci.* **2015**, *6*, 51. [CrossRef]

45. Boisson-Dernier, A.; Lituiev, D.S.; Nestorova, A.; Franck, C.M.; Thirugnanarajah, S.; Grossniklaus, U. ANXUR Receptor-Like Kinases Coordinate Cell Wall Integrity with Growth at the Pollen Tube Tip Via NADPH Oxidases. *PLoS Biol.* **2013**, *11*. [CrossRef]

46. Geitmann, A.; Palanivelu, R. Fertilization Requires Communication: Signal Generation and Perception During Pollen Tube Guidance. *Floric. Ornam. Biotechnol.* **2007**, *1*, 77–89.

47. Woriedh, M.; Wolf, S.; Márton, M.L.; Hinze, A.; Gahrtz, M.; Becker, D.; Dresselhaus, T. External application of gametophyte-specific ZmPMEI1 induces pollen tube burst in maize. *Plant Reprod.* **2013**, *26*, 255–266. [CrossRef] [PubMed]

48. Qu, L.J.; Li, L.; Lan, Z.; Dresselhaus, T. Peptide signalling during the pollen tube journey and double fertilization. *J. Exp. Bot.* **2015**, *66*, 5139–5150. [CrossRef] [PubMed]

49. Bhardwaj, P.; Kumar, R.; Sharma, H.; Tewari, R.; Ahuja, P.S.; Sharma, R.K. Development and utilization of genomic and genic microsatellite markers in Assam tea (Camellia assamica ssp. assamica) and related Camellia species. *Plant Breed.* **2013**, *132*, 748–763. [CrossRef]

50. Martin, F.W. Staining and observing pollen tubes in the style by means of fluorescence. *Stain Technol.* **1959**, *34*, 125–128. [CrossRef]

51. Ghawana, S.; Paul, A.; Kumar, H.; Kumar, A.; Singh, H.; Bhardwaj, P.K.; Rani, A.; Singh, R.S.; Raizada, J.; Singh, K.; Kumar, S. An RNA isolation system for plant tissues rich in secondary metabolites. *BMC Res. Notes* **2011**, *4*, 85. [CrossRef]

52. Patel, R.K.; Jain, M. NGS QC Toolkit: A toolkit for quality control of next generation sequencing data. *PLoS ONE* **2012**, *7*, e30619. [CrossRef]

53. Haas, B.J.; Papanicolaou, A.; Yassour, M.; Grabherr, M.; Blood, P.D.; Bowden, J.; Couger, M.B.; Eccles, D.; Li, B.; Lieber, M. De novo transcript sequence reconstruction from RNA-seq using the Trinity platform for reference generation and analysis. *Nat. Protoc.* **2013**, *8*, 1494–1512. [CrossRef]
54. Fu, L.; Niu, B.; Zhu, Z.; Wu, S.; Li, W. CD-HIT: Accelerated for clustering the next-generation sequencing data. *Bioinformatics* **2012**, *28*, 3150–3152. [CrossRef]
55. Mortazavi, A.; Williams, B.A.; McCue, K.; Schaeffer, L.; Wold, B. Mapping and quantifying mammalian transcriptomes by RNA-Seq. *Nat. Methods* **2008**, *5*, 621–628. [CrossRef] [PubMed]
56. Langmead, B.; Salzberg, S.L. Fast gapped-read alignment with Bowtie 2. *Nat. Methods* **2012**, *9*, 357–359. [CrossRef] [PubMed]
57. Chen, Y.; Lun, A.T.L.; Smyth, G.K. Differential expression analysis of complex RNA-seq experiments using edgeR. In *Statistical Analysis of Next Generation Sequencing Data*; Datta, S., Nettleton, D., Eds.; Springer International Publishing: New York, NY, USA, 2014; pp. 51–74.
58. Robinson, M.D.; McCarthy, D.J.; Smyth, G.K. edgeR: A Bioconductor package for differential expression analysis of digital gene expression data. *Bioinformatics* **2010**, *26*, 139–140. [CrossRef] [PubMed]
59. Benjamini, Y.; Hochberg, Y. Controlling the false discovery rate: A practical and powerful approach to multiple testing. *J. R. Stat. Soc. Ser. B* **1995**, *57*, 289–300. [CrossRef]
60. Trapnell, C.; Roberts, A.; Goff, L.; Pertea, G.; Kim, D.; Kelley, D.R.; Pimentel, H.; Salzberg, S.L.; Rinn, J.L.; Pachter, L. Differential gene and transcript expression analysis of RNA-seq experiments with TopHat and Cufflinks. *Nat. Protoc.* **2012**, *7*, 562–578. [CrossRef] [PubMed]
61. Du, Z.; Zhou, X.; Ling, Y.; Zhang, Z.; Su, Z. agriGO: A GO analysis toolkit for the agricultural community. *Nucleic Acids Res.* **2010**, *45*, gkq310. [CrossRef] [PubMed]
62. Szklarczyk, D.; Franceschini, A.; Wyder, S.; Forslund, K.; Heller, D.; Huerta-Cepas, J.; Simonovic, M.; Roth, A.; Santos, A.; Tsafou, K.P. STRING v10: Protein–protein interaction networks, integrated over the tree of life. *Nucleic Acids Res.* **2014**, *43*, gku1003. [CrossRef] [PubMed]
63. Shannon, P.; Markiel, A.; Ozier, O.; Baliga, N.S.; Wang, J.T.; Ramage, D.; Amin, N.; Schwikowski, B.; Ideker, T. Cytoscape: A software environment for integrated models of biomolecular interaction networks. *Genome Res.* **2003**, *13*, 2498–2504. [CrossRef] [PubMed]
64. Pfaffl, M.W.; Horgan, G.W.; Dempfle, L. Relative expression software tool (REST©) for group-wise comparison and statistical analysis of relative expression results in real-time PCR. *Nucleic Acids Res.* **2002**, *30*, e36. [CrossRef] [PubMed]

© 2019 by the authors. Licensee MDPI, Basel, Switzerland. This article is an open access article distributed under the terms and conditions of the Creative Commons Attribution (CC BY) license (http://creativecommons.org/licenses/by/4.0/).

International Journal of
Molecular Sciences

Review

Long Non-Coding RNA and Acute Leukemia

Gabriela Marisol Cruz-Miranda [1], Alfredo Hidalgo-Miranda [2], Diego Alberto Bárcenas-López [1],
Juan Carlos Núñez-Enríquez [3], Julian Ramírez-Bello [4], Juan Manuel Mejía-Aranguré [5,*]
and Silvia Jiménez-Morales [2,*]

[1] Programa de Doctorado, Posgrado en Ciencias Biológicas, Universidad Nacional Autónoma de México,
     Mexico City 04510, Mexico; gmcm611@hotmail.com (G.M.C.-M.); d.a.barcenas@outlook.com (D.A.B.-L.)
[2] Laboratorio de Genómica del Cáncer, Instituto Nacional de Medicina Genómica, Mexico City 14610, Mexico;
     ahidalgo@inmegen.gob.mx
[3] Unidad de Investigación Médica en Epidemiología Clínica, UMAE Hospital de Pediatría
     "Dr. Silvestre Frenk Freund", Centro Médico Nacional Siglo XXI, Instituto Mexicano del Seguro Social,
     Mexico City 06720, Mexico; jcarlos_nu@hotmail.com
[4] Unidad de Investigación en Enfermedades Metabólicas y Endócrinas, Hospital Juárez de México,
     Mexico City 07760, Mexico; dr.julian.ramirez.hjm@gmail.com
[5] Coordinación de Investigación en Salud, Instituto Mexicano del Seguro Social, Mexico City 06720, Mexico
*   Correspondence: juan.mejiaa@imss.gob.mx (J.M.M.-A.); sjimenez@inmegen.gob.mx (S.J.-M.);
     Tel.: +52-55-5350-1900 (ext. 1155) (S.J.-M.)

Received: 6 October 2018; Accepted: 22 October 2018; Published: 9 February 2019

Abstract: Acute leukemia (AL) is the main type of cancer in children worldwide. Mortality by this
disease is high in developing countries and its etiology remains unanswered. Evidences showing the
role of the long non-coding RNAs (lncRNAs) in the pathophysiology of hematological malignancies
have increased drastically in the last decade. In addition to the contribution of these lncRNAs in
leukemogenesis, recent studies have suggested that lncRNAs could be used as biomarkers in the
diagnosis, prognosis, and therapeutic response in leukemia patients. The focus of this review is
to describe the functional classification, biogenesis, and the role of lncRNAs in leukemogenesis,
to summarize the evidence about the lncRNAs which are playing a role in AL, and how these genes
could be useful as potential therapeutic targets.

Keywords: long non-coding RNAs; cancer; acute leukemia; therapeutic targets

---

1. Introduction

Leukemia is a group of hematological malignancies characterized by an oligoclonal expansion of
abnormally differentiated, and sometimes poorly differentiated hematopoietic cells which infiltrate
the bone marrow, and could also invade the blood and other extramedullary tissues. In general, AL
can be divided into acute or chronic, and lymphoid or myeloid, according to their progression and
affected lineage, respectively. Thus, we can identify the following subtypes: acute lymphoblastic
leukemia (ALL), chronic lymphoblastic leukemia (CLL), acute myeloid leukemia (AML), and chronic
myeloid leukemia (CML). AL is the main type of cancer in children worldwide [1,2]. In recent
years, it has reported a trend of increase in the incidence AL; notwithstanding, the causes are still
unclear. Studies conducted to identify the etiology of this disease have reported that a genetic
background interacting with environmental factors (i.e., high doses of ionizing radiation, infections,
parental occupational exposures, etc.) could explain this phenomenon [3]; however, the molecular
mechanisms involved are not fully understood. To date, growing data have shown that different
non-coding RNAs (ncRNAs) might be the link between the genome and the environment because
they are closely related to normal physiological and pathological processes [4,5]. ncRNAs, also known
as non-protein-coding RNAs (npcRNAs), non-messenger RNAs (nmRNAs) or functional RNAs

(fRNAs), are functional RNA molecules which are not translated into proteins [6]. These RNAs consist of several distinct families which include microRNAs (miRNAs), small nuclear RNAs (snRNAs), PIWI-interacting RNAs (piRNAs), and long non-coding RNAs (lncRNAs), among others. LncRNAs are one of the most studied ncRNA types, and play an important role as gene expression modulators at the epigenetic, transcriptional, and post-transcriptional level. In fact, it has been suggested that various miRNAs and lncRNAs could act as tumor suppressors genes or oncogenes, because they regulate directly or indirectly the expression of genes involved in molecular mechanisms as cell proliferation/differentiation, apoptosis, and metastasis [4,5]. In comparison with miRNAs, the lncRNAs are more numerous and represents the 41% of the overall ncRNAs. Over the last years, massive technological tools have been useful to increase the knowledge about lncRNAs that are abnormally expressed or mutated in AL and the list of relevant lncRNAs in leukemogenesis is growing rapidly. Moreover, it has reported a distinctive lncRNAs expression signature associated with AL prognosis, suggesting the potential application of these genes to make treatment decisions. Here, we review the most recent findings about lncRNAs in AL pathogenesis and their role as potential biomarkers. We also are pointing out the lncRNAs as promising druggable molecules in the development of new treatments for leukemia [7]. An electronic search strategy using the biomedical database of the National Center for Biotechnology Information (NCBI) was conducted. Studies that combined the keywords lncRNAs with acute leukemia, or acute lymphoblastic leukemia, or acute myeloid leukemia or hematopoiesis were enclosed.

2. Genetic Features of Acute Leukemia

AL has been recognized as a highly genetically heterogeneous disease, where chromosomal abnormalities, either numerical (hyperdiploidy and hypodiploidy) or structural alterations (translocations, amplifications, DNA copy number alterations, insertions/deletions, and punctual mutations) are usually observed; thus, these alterations are the hallmarks of the leukemic cells and represent the major class of oncogenic drivers to the disease. Indeed, due to the fact many childhood ALL cases carry specific fusion genes (*MLL* gene fusions, *ETV6/RUNX1*, *E2A/PBX1*, etc.) and *AML* (*AML1/ETO*, *PML/RARα*, *CBFβ/MYH11*, etc.), this gives more evidence that childhood AL is initiated in utero during fetal hematopoiesis [8]. In addition to the numerical alterations and common targets of translocations in ALL, this disease is characterized by mutations in transcriptional factors (*AML1, ETS, PAX5, IKZF1, EBF1, ETV6*, and *STAT*), suppressor genes (*TP53, RB1, CDKN2A/CDKN2B*, etc.), oncogenes (*ABL1, ABL2, CSF1R, JAK2, PDGFRB*, and *CRLF2*), B lymphoid cell differentiators (*IKZF1, TCF3, EBF1, PAX5*, and *VPREB1*), chromatin remodelers, or epigenetic modifiers (*DNMT3A, CREBBP, MLL2, NSD2, EP300, ARID1A, TET2*, and *CHD6*) [9–12]. Data from the St. Jude/Washington Pediatric Cancer Genome Project (PCGP), that has characterized pediatric cancer genomes by whole-genome or whole-exome sequencing, revealed that the somatic mutation rate in childhood ALL ranges from 7.30×10^{-8} per base [13]. In spite of the fact that chromosomal changes detectable by cytogenetic techniques are present in nearly 75% of the precursor B (pre-B) cell ALL cases, the gene expression profiling and genome-wide sequencing analyses have showed that B cell leukemogenesis is more complex [14]. Meanwhile, mutations in *nRAS, RUNX1, FLT3, KIT*, etc., abnormalities of DNA methylation, biogenesis of ribosomes, activated signaling pathways, myeloid transcription factors, chromatin remodeling, and cohesion complex processes are very common in AML [15].

The discovery of frequent mutations in epigenetic modifiers genes in AL show that epigenetic alterations also play a critical role in leukemogenesis. In this regard, it is known that most of the genes involved in epigenetic process do not code for proteins, and many of them are classified as lncRNAs, which regulate gene expression through different mechanisms.

3. LncRNAs Characteristics

lncRNAs comprise a highly functionally heterogeneous group of RNA molecules with sizes are greater than 200 nucleotides, and, as all the mRNAs usually have more than one exon, most of them are transcribed by RNA polymerase II (RNA pol II), are capped, may be polyadenylate, and can be located within the nucleus or cytoplasm. LncRNAs genes differ from mRNAs because lncRNAs lack protein-coding potential, are mostly expressed in low levels, and show poor species conservation compared to protein-coding genes (mRNAs). Additionally, lncRNAs display tissue-specific and development stage-specific expression showing their important role in cell differentiation mechanisms [16].

The number of lncRNAs is larger than the number of protein-coding RNAs. To date, the GENCODE project lncRNAs catalog consists of 15,779 transcripts (there are potentially more than 28,000 distinct transcripts) in the human genome (https://www.gencodegenes.org); nevertheless, this number could increase, since many primary long non-coding transcripts are often processed into smaller ncRNAs [17]. ncRNA detection led to a solution for the G-value paradox that states that there is no correlation between the amount of coding genes and the complexity of the organism, while we observe a correlation between the complexity of the organism and the ratio of the number of non-coding genes to total genomic DNA. Nowadays, cumulative evidence exhibits that lncRNAs are relevant players in many cellular processes either in physiological as well as pathological conditions. In cancer, the lncRNAs could have oncogenic function and tumor suppressive function since they have been found as upregulated or downregulated in several types of tumors in comparison to healthy tissues [18].

4. Biogenesis and Classification

It has hypothesized that most of lncRNAs are originated from (1) the incorporation of the fragments of original protein-coding genes; (2) juxtaposition of two transcribed and previously well-separated sequence regions of chromosomes giving rise a multi-exon ncRNA; (3) duplication of non-coding genes through retrotransposition; (4) tandem duplication events of neighboring repeats within a ncRNA; and (5) insertion of transcription factor, which is inserted into a sequence.

LncRNAs are transcribed and processed by the RNA pol II transcriptional machinery, thus many of them undergo post-transcriptional modifications such as 5′ capping, splicing, and polyadenylation. Nevertheless, there are also nonpolyadenylated lncRNAs that derive from RNA pol III promoters and snoRNA-related lncRNAs (sno-lncRNAs) expressed from introns via the snoRNP machinery (with the supplementary production of two snoRNAs). LncRNAs have been mapped into a wide range of regions, including coding and non-coding regions (intergenic regions, promoters, enhancers, and introns) [19–27].

To date, there is not a unique system to classify lncRNAs; however, different classifications have been proposed based on their size, genome localization, RNA mechanism of action, and function [28]. According to their location (Figure 1a), orientation (Figure 1b), and transcription direction (Figure 1c) relative to protein-coding genes, an lncRNA can be placed into one or more broad categories. Thus, lncRNAs can be intronic, when they lie into a intron of a second transcript (*COLDAIR*, located in the first intron of the flowering repressor locus C or *FLC*), intergenic (lincRNA) if it is located between two genes without any overlap at least 5 kb from both sides (exemplified by *H19*, *XIST*, and *lincRNA-p21*), exonic if lncRNA is encoded within a exon, or overlapping, which includes those lncRNA located within one or two genes [4,13,29,30]. Based on the orientation, lncRNAs can be transcribed from either the same strand or antisense in a divergent or convergent manner. LncRNAs can be also classified as enhancer-associated RNAs (eRNAs) and promoter-associated long RNAs (or PROMPTs) if they are produced from enhancer or promoter regions, respectively [31].

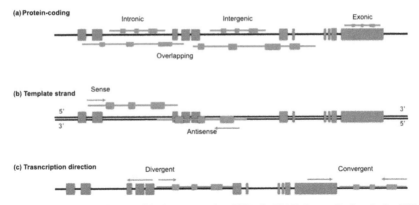

Figure 1. Positional classification of the long non-coding RNAs (lncRNA). Carton displays the LncRNA (red) classification base on (**a**) the location between two coding genes (intronic, exonic, intergenic, or overlapping), (**b**) the template strand (sense, antisense), and (**c**) transcription direction when coding genes and lncRNA are transcribed in the same strand (divergent, convergent). Gray arrow indicates in which direction transcription is proceed. Green and blue boxes represent exons of two different genes.

Although lncRNAs show a spatiotemporal expression pattern during proliferation, differentiation, and cell death; these genes are classified based on their function as guide, decoy, signaling, scaffold, or enhancer lncRNAs [32]. Guide lncRNAs interact with transcription factors or proteins and recruit them to their gene target or their genomic loci regulating downstream signaling events and gene expression. Decoy lncRNAs mimic and compete with their consensus DNA-binding motifs for binding nuclear receptors or transcriptional factors in the nucleus, facilitating gene activation or silencing. These genes can also "sponge" proteins such as chromatin modifiers, adding an extra level transcriptome regulation. Signaling lncRNAs are associated with signaling pathways to regulate transcription in response to various stimuli. Scaffold lncRNAs act as a central platform where many protein complexes tying and get directed to specific genomic loci or target gene promoter [17]. Enhancer lnRNAs are *cis*-encoded DNA elements that bind with mediator complex to regulate transcription genes located within their own chromosome (Table 1) [33]. However, this classification is too simple to cover the whole lncRNAome, cases such as pseudogenes and telomerase RNA (*TERC*) still lie outside the list [20,32].

Table 1. Classification of lncRNAs according to their function.

Functional Type	Cellular Location	Mechanism of Action	Examples	Reference
Guide	Nucleus	Essential for the proper localization of proteins to their site-specific reaction.	XIST, ANRIL	[34]
Decoys	Plasma membrane, nucleus and cytosol	Sequestering regulatory factors (transcription factors, catalytic proteins subunits, chromatin modifiers, etc.) to modulate transcription	GAS5, MALAT1	[35–37]
Scaffold	Nucleus	Providing platforms for assembly of multiple-component complexes such as the polycomb repressive complexes and ribonucleoprotein complex.	CDKN2B-AS1, HOTAIR	[35,36]
Signaling	Nucleus	Serving as a molecular signal to regulate transcription in response to various stimuli	TP53COR1, PANDAR	[35,36]
Enhancer	Nucleus	Binding with mediator complex to enhance transcription	HOTTIP, CCAT1-L, LUNAR1	[25,33]

In terms of size, lincRNAs often range from hundreds of nucleotides to several kilobases [20]. Nevertheless, there are exceptionally long lncRNAs (macroRNAs) and very long intergenic non-coding RNAs (vlincRNAs), stretching 10 kb and 1 Mb, respectively [30].

In addition, lncRNAs have regulatory roles in gene expression at both, the transcriptional, and post-transcriptional levels in mostly biological mechanisms and pathophysiological processes. These molecules can regulate the expression of neighboring genes (*cis*) or affect genes located at different chromosomes (*trans*) [38]. In this way, lncRNAs can regulate gene expression via transcription factor and chromatin-modifiers complex recruitment to their DNA targets, acting as enhancers to activate genes, as part of the heterogeneous nuclear ribonucleoprotein (hnRNP) complex, interacting with RNA and DNA by base paring, etc. [38].

5. LncRNAs in Normal Hematopoiesis

Hematopoietic cell lineage differentiation involves the regulation of gene expression at different levels that can occur to activate lineage specific genes and repress those genes that are not specific to that lineage. This activation/suppression is mediated by transcription factors and chromatin remodeling that act as determinants of the intrinsic cell lineage. However, these factors are reactivated in different lines and stages of differentiation, so that the choice of the final lineage reflects the particular combination of elements interacting in a certain stage of cell differentiation [39]. LncRNAs are involved in regulating different steps in hematopoiesis, immune system development, and activation. In fact, several lncRNAs have been identified in the blood cells either in animal models or human samples. For example, over 1109 poliA+ lncRNAs were detected in murine megakaryocytes, erythroblast, and megakaryocyte-erythroid precursors, of which 15% are expressed in humans [40]. The Eosinophil Granule Ontogeny (*EGO*) was one of the first lncRNAs related with the human normal hematopoiesis process. *EGO* is nested within an intron of inositol triphosphate receptor type 1 (*ITPR1*) and was found to be highly expressed in human bone marrow and in mature eosinophils. Despite that the molecular mechanism of their actions is not well known, experimental evidences show that *EGO* is involved in the eosinophil differentiation of CD34+ hematopoietic progenitor cells by regulating eosinophil granule protein expression at the transcription level [41]. *PU.1-As*, which is antisense to the master hematopoietic transcriptional factor *PU.1*, negatively regulates the expression of *PU.1*, repressing myeloid cells and B cells differentiation [42]. Other examples include dendritic cell-specific lncRNA (*lnc-DC*), non-coding RNA repressor of *NFAT* (*NRON*), and *lincRNA-Cox2*. *lnc-DC* was identified from extensive profiling of lncRNAs expression during differentiation of monocytes into dendritic cells (DCs). Mechanistic studies suggest that *lnc-DC* contributes to prevent *STAT3* (signal transducer and activator of transcription 3) dephosphorylation by Src homology region 2 domain-containing phosphatase-1 (*SHP1*) by directly binding to *STAT* in the cytoplasm [43]. *NRON* plays a relevant role in the adaptive immune response through sequestering transcription factors in the cytoplasm, such as the nuclear factor of activated T cells (NFAT). *LincRNA-Cox2* contributes with the regulation of the innate immune response by repressing the expression of critical immune-response regulators and by the coordinating the assembly, location and orientation of the complexes that specify the cellular fate [39].

Studying twelve distinct blood cell population purified by multicolor flow cytometry, Schwarzer et al. [44] established a human ncRNA hematopoietic expression atlas per blood cell population, finding *LINC00173*, *LINC000524*, *RP11-1029J19*, and *HOTAIRM1* among the lncRNAs that characterize cells of the different human blood lineages. *LINC00173* exhibited the most specific expression, with critical regulatory circuits involved in blood homeostasis and myeloid differentiation. In vitro models showed that suppression of *LINC00173* in human CD34+ hematopoietic stem and progenitor cells (HSPCs) specifically affects granulocyte differentiation and decreases its phagocytic capacity (which is associated with perturbed maturation). Additional studies reported that *LINC00173* is highly expressed in granulocytes [45]. *H19*, *XIST*, *lncHSC-1*, and *lncHSC-2*, which maintain long-term hematopoietic stem cell (HSC) quiescence and self-renewal, have also been involved in normal hematopoiesis [46].

6. LncRNAs in Acute Leukemia

Although many studies have implicated lncRNAs in many cancer types, little is known about the functional impact of lncRNAs in AL etiology, progression, and treatment response [44]. Several lncRNAs have been reported to be exclusively involved in specific ALL lineages but few of these are abnormally expressed in ALL and AML [47,48]. For instance, *CASC15*, involved in cellular survival proliferation and the expression of *SOX4* (*cis* regulation), was detected to be upregulated in t(12;21) (p13;q22) (*ETV6/RUNX1*) B cell ALL and in AML patients with the (8;21) translocation. In both cases, upregulation of *CASC15* was associated with a good prognosis [48]. To date, a large number of lncRNAs have been identified in AL; however, their molecular mechanisms remains elusive. Table 2 includes some examples of lncRNAs which have been reported as implicated in acute leukemia in children [49–77].

Table 2. Examples of lncRNAs described in acute leukemia.

LncRNAs	Classification	Function	Target Genes	Expression Level in Leukemia	Reference
		Myeloblastic Leukemia			
IRAIN	Intronic	Intrachromosomal interactions	IGF1R	Downregulated in leukemia cell lines and in patients with high risk AML	[49]
UCA1	Intergenic	Proliferation of AML cells. Oncofetal gene	CDKN1B	Upregulated	[50–52]
MEG3	Intergenic	Tumor suppressor gene	P53	Downregulated	[52,53]
RUNXOR	Sense	Chromosomal translocations	RUNX1	Upregulated	[54]
NEAT1	Intergenic	Myeloid differentiation cells	Unknown in AML	Downregulated	[50,52,55]
LLEST		Tumor suppressor	BCL-2	Downregulated or even not expressed.	
HOTAIRM1	Antisense	Myeloid differentiation cells, autophagy mechanisms, chromatin remodeling and architecture	HOXA1, HOXA4, CD11b and CD18	Upregulated	[52,56–60]
HOXA-AS2	Antisense	Apoptotic repressor in NB4 promyelocytic leukemia cells	Unknown	Upregulated	[61]
PU.1-AS	Antisense	Involved in the translation of PU.1	PU.1	Downregulated	[62]
WT1-AS	Antisense	WT1 expression	WT1		[63]
EGO	Intronic	MBP and EDN expression			[41]
BGL3	Intergenic	Apoptosis and DNA methylation	miR-17, miR-93, miR-20a, miR-20b, miR-106a and miR-106b	Upregulated	[50,52,64]
CCAT1	Intergenic	Monocytic cell differentiation	miR-155		[9,52,65]
CCDC26	Intergenic	AML cell proliferation	c-Kit		[66]
HOTAIR	Intergenic	Apoptosis inhibitor	miR-193a and c-Kit	Upregulated	[67]

Table 2. *Cont.*

LncRNAs	Classification	Function	Target Genes	Expression Level in Leukemia	Reference
PVT1	Intergenic	Proliferation of promyelocytes	*MYC*	Upregulated	[52,68]
ZNF571-AS1	Antisense	Regulator of JAK/STAT signaling pathway	*KIT* and *STAT5*		[69]
		Lymphoblastic Leukemia			
BALR-2	Uncharacterized	Unknown	Unknown	Overexpressed in prednisone-resistant B-ALL patients	[70]
BALR-1	Unknown	Unknown	Unknown	Upregulated	[70]
BARL-6	Unknown	Promotes cell survival and inhibits apoptosis	Unknown	Upregulated	[70]
LINC00958	Intergenic	Unknown	Unknown	Upregulated in t(12;21) preB cALL	[70,71]
DBH-AS1	Antisense	Cell proliferation and cell survival	Unknown	Upregulated	
RP11-137H2.4	Uncharacterized	Apoptosis, proliferation, cell migration	Unknown	Upregulated. Glucocorticoids resistance	[72]
ANRIL	Antisense	Cellular proliferation and apoptosis	*CDKN2A/B. CBX7, SUZ12*	Upregulated	[52]
T-ALL-R-LncR1	Unknown	Promotor of the formation of Par-4/THAP1 protein complex, and the activity of caspase-3	Unknown	Upregulated in children with T-ALL	[73]
LUNAR1	Enhancer-like	Promotor of T-ALL proliferation by inducing IGF1R expression.	*IGF1R*	Downregulated	[50,52,74]
MALAT1	Intergenic	Alternative splicing and epigenetic modification	Unknown	Upregulated Downregulated in vincristine-resistant ALL	[50,52,75–77]
CASC15	Intergenic	Cellular survival and proliferation	*SOX4*	Upregulated	[48]

7. LncRNAs in Acute Myeloid Leukemia

Regarding the association between lncRNA and hematopoietic cancer, AML has been the most investigated, and has been reported to be an important lncRNA in the biological and pathological processes of the disease. For example, insulin-like growth factor type I receptor antisense imprinted non-protein RNA (*IRAIN*), which is transcribed antisense to insulin-like growth factor type I receptor (*IGF1R*) gene, is downregulated in leukemia cell lines and in patients with high-risk AML. *IRAIN* is involved in the formation of a long-range intrachromosomal interaction between the *IGF1R* promoter and a distant intragenic enhancer [49]. *ZNF571-AS1* is another lncRNA that has been suggested as a relevant player in AML. Based on co-expression correlation analysis across all AML samples with lncRNA–lncRNA pairs, this lncRNA was identified as potential regulator of the Janus Kinase (JAK)/signal transducer and activator of transcription (STAT) 5A and tyrosine-protein kinase Kit (KIT) expression. Thus their participation in AML was suggested via the JAK/STAT signaling pathway [69]. As well, Urothelial carcinoma-associated 1 (*UCA1*), an oncofetal gene that has been involved in embryonic development and carcinogenesis, was found to be upregulated in myeloid cell lines promoting cell viability, migration, invasion, and apoptosis processes [78–80]. A significant upregulation of *UCA1* expression in AML with *CEBPA* (a crucial component during myeloid differentiation) mutations and its relation with chemoresistance in pediatric AML was documented [51,81]. The maternally expressed 3 non-protein-coding gene (*MEG3*), a tumor suppressor, has also been associated with significantly reduced overall survival rate in AML patients. This gene is related to a variety of human tumors and data point out that directly enhance the anticancer effect through p53 [82,83]. Benetatos et al. [53] evaluated the aberrant promoter methylation of *MEG3* in 42 AML patients, and found that *MEG3* hypermethylation was present in 47.6% AML cases and might be associated with significantly reduced overall survival rate in these patients [53]. LncRNAs have also been profiled from AML patients cytogenetically normal (CN) and with specific translocation. For example, AML patients carrying *NPM1*, *CEBPA*, *IDH2*, *ASXL1*, and *RUNX1* mutations and internal tandem duplication mutations in *FLT3* (FLT3/ITD) gene exhibited specific lncRNA expression signature. As well, Diaz-Beya et al. [84], studying AML cases with t(15;17), t(8;21), inv(16), t(6;9), t(3;3), t(9;11), t(8;16), FLT3/ITD, and monosomal karyotype, found a specific lncRNA profile in t(15;17), t(6;9), and t(8;16) positive cases. That study also revealed a correlation between t(8;16) and *linc-HOXA11*, *HOXA11-AS*, *HOTTIP*, and NR_038120 expression, and suggested that GAT2 is an important transcription factor to these lncRNAs. Otherwise, lncRNAs expression correlated with treatment response and survival. One of the lncRNAs that is specifically upregulated in CN-AML cases with *CEBPA* mutation is the lncRNA *UCA1* [85]. Taurine-upregulated gene 1 (*TUG1*) expression was reported to be associated with higher white blood cell counts, monosomal karyotype, FLT3/ITD mutation, and worse prognosis in AML adults. In vitro studies in AML cells indicates that *TUG1* induces cell proliferation but suppressing cell apoptosis via targeting *AURKA* [86].

Schwarzer et al. [44] made a high-density reconstruction of the human coding and non-coding hematopoietic landscape to identify an ncRNA fingerprint associated with lineage specification, HSPC maintenance, and cellular differentiation. They define a core ncRNA stem cell signature in normal HSCs and AML blast, which can serve as a prognostic marker in a different cohort of AML patients and may pave the way for novel therapeutic interventions targeting the non-coding transcriptome [44].

8. LncRNAs in Acute Lymphoblastic Leukemia

Data regarding lncRNA playing a role in ALL are still scarce. One of the first clinicopathological correlations with lncRNA expression data in ALL was performed by Fernando et al. [70] who studied 160 children with B-ALL observing that *BALR-2* correlates with overall survival and with response to prednisone. These authors also demonstrated a putative mechanism in regulating cell survival in B-ALL that it is downregulated by glucocorticoid receptor engagement, and that its downregulation results in the activation of the glucocorticoid receptor signaling pathway [70]. Loie et al. [71] also reports that lncRNA expression patterns can classify ALL disease by subtypes as well as protein-coding genes. In addition to lncRNA, *BARL-2*, which is also correlated with resistance to prednisone treatment, these authors found that lncRNAs *BALR-1*, *BRL-6*, and *LINC0098* were overexpressed in pre-B ALL cases and that all of these genes correlated with cytogenetic abnormalities, disease subtypes, and survivals of B-ALL patients [71]. In that study, they also observed that diverse coding genes adjacent to several of those lncRNAs showed unique overexpression profile in *ETV6/RUNX1* positive BCP-ALLS suggesting a possible *cis* regulatory relationship. Furthermore, Ghazavi et al. [47] identified an *ETV6/RUNX1*-specific lncRNA signature in a 64 children cohort and in 13 BCP-ALL cell lines. Five-hundred-and-ninty-six lncRNA transcripts (434 up- and 162 downregulated) showed significant differential expression between *ETV6/RUNX1*-positive BCP-ALL and other genetic BCP-ALL subclasses. However, 16 lncRNAs, of which 14 were upregulated and two were found downregulated, overlapped with the *ETV6/RUNX1*-specific lncRNA signature, including *NKX2-3-1*, *lncRTN4R-1*, *lncGIP-1*, *lnc-LRP8-3*, *lnc-TCF12-2*, *lncC8ort4-1*, *lnc-C8orf4-2*, *lnc-TINAGL1-1*, *lnc-LSM11-4*, and *lnc-SARDH-1* (also known as *DBH-AS1*). *Lnc-SARDH-1* is known to possess an oncogenic role promoting cell proliferation and cell survival through activation of MAPK signaling in the context of hepatocellular carcinoma [87]. Furthermore, the H3K27ac epigenetic mark (associated to enhancers) was found in nine loci of the rest of the lncRNAs and their adjacent coding genes, which, in addition to the finding of a unique expression signature of these coding genes in *ETV6/RUNX1* pre-B ALL, suggests a *cis* interaction between the lncRNAs and their neighboring coding genes [47]. In another study, Ouimet et al. performed a whole transcriptome analysis in a 56 pre-B ALL children cohort finding five lncRNAs specifically overexpressed in pre-B ALL. These genes may have impact in cancer traits such a cell proliferation, migration, apoptosis and treatment response. Specifically, lncRNA *RP11-137H2.4* had a considerable impact on apoptosis, proliferation, and cell migration and its silencing is sufficient to restore a NR3C1-independent cellular response to glucocorticoid (GC) in GC-resistant pre-B ALL cells, leading to GC-induced apoptosis [72]. Further to this study, Gioia et al. functionally characterized three lncRNAs—*RP-11-624C23.1*, *RP11-203E8*, and *RP11-446E9*—specifically repressed in pre-B ALL, restoring their expression in a pre-B ALL cell line. All the lncRNAs promoted tumor suppressor-like phenotypes: apoptosis induction in response to DNA damaging agents and a reduction in cell proliferation and migration [88]. Additionally, Garitano-Trojaola et al., while analyzing ALL samples and peripheral blood samples obtained from healthy donors, found 43 lncRNAs abnormally expressed in ALL. *Linc-PINT* was downregulated both in T- and B-ALL cases [89]. Studies in T-ALL cells found a significant difference in expression of *LUNAR1* and *lnc-FAM120AOS-1* between *NOTCH1* wild type and mutant cases [68]. The use of bioinformatics tools identified that *lnc-OAZ3-2:7*—located near the RORC gene—was repressed in this leukemia subtype [90]. These studies suggest that lncRNAs might be utilized as diagnostic and prognostic markers in leukemia, but additional analyses are needed.

9. Future Outlooks: Potential Clinical Implications on LncRNAs in Acute Leukemia

It is suggested that more than 97% of the transcribed genome does not encode for proteins. The discovery of the biological role of these non-coding genes took place in 1990, when *XIST* was reported to be involved in X chromosome inactivation (XCI) and gene dosage compensation. Subsequently, *HOTAIR* was identified as a repressor of *HOX* family gene transcription [91]. Most recently, high-throughput expression analyses have been conducted to identify thousands of expressed lncRNA genes either in normal or tumor tissues, showing the potential of lncRNAs as biomarkers for different types of cancer [37,44,52].

Deciphering the molecular mechanisms involved in hematological malignancies addresses new routes to improve diagnosis, prognosis, and treatment of patients with leukemia. In fact, abnormal expression of specific lncRNAs have been reported to be associated with some clinicopathological parameters and molecular subtypes in AL. As example, *BALR-1* and *LINC0098* have been identified as correlating with poor overall survival and diminished response to prednisone treatment in B cell ALL cases [70,71]. Regarding AML, *HOTAIR*, *IRAIN*, and *SNHG5* have been suggested as biomarkers for diagnosis [92]; meanwhile, *UCA1* overexpression was associated with chemoresistance of pediatric cases [81]. *SNHG5* upregulation, which was detected in bone marrow and plasma, was correlated with unfavorable cytogenetics and shorter overall patient survival and was suggested as an independent factor to predict prognosis in AML [93].

Notwithstanding, few of these genes have been replicated across cohorts, probably evidencing biases due to different sample collection and processing techniques, but also as a consequence of AL biological complexity, which is characterized by a wide range of interactions among coding and non-coding genome and spatiotemporal relationships. *HOTAIR*, a proliferation promotor of leukemic blast and leukemia stem cells [94], is one of the most consistently found in AL. A high-expression level defines a subgroup of AL patients with high white blood cell counts at the time of diagnosis and low survival rates [95,96]. Recently, *HOTAIR* high-expression was associated with acquired resistance to antileukemic drugs such as doxorubicin and immatinib [97,98], making this gene as a potential therapeutic target molecule that could contribute to solve a tremendous problem in leukemia chemotherapy, the drug-resistance. On the other hand, experimental data suggest that *HOTAIR* low-expression could be mediated by small interference RNA (siRNA), but still no evidences exist regarding its potential benefit in humans [98]. The development of new molecular strategies as CRISPR/Cas9 to edit the mutated genome or nanotechnology approaches to deliver drugs specifically to leukemia cells prognosticate high applicability of lncRNA as a target to develop new treatments to leukemia [99,100]. Additionally, the high specificity and feasible detection in tissues, serum, plasma, urine, and saliva of the lncRNAs led us to think that lncRNAs could be useful as signals of specific cellular states or read-outs of active cellular pathologies such as leukemia, being promising as predictive biomarkers and potential therapeutic targets in cancer [19].

There is no doubt of the role of lncRNAs in hematopoietic cell transformation, disease evolution, or drug resistance; nevertheless, due to the limited number of studies in hematological entities, these applications are still inconclusive. In fact, before their use as biomarkers in childhood AL, prospective and well-designed cohort studies with adequate sample sizes and further validation of the results in independent cohorts are needed to confirm their clinical usefulness. Therefore, translating this knowledge into the clinical practice still represents a big challenge.

10. Conclusions

At this time, we know that lncRNAs are playing a relevant role in cancer development, including leukemia. However, the knowledge regarding molecular mechanisms underlying the pathogenesis of these diseases remains limited. Massive parallel analysis techniques and, likewise, transcriptome expression analysis and RNA sequencing technologies are increasing the possibility to identify those lncRNAs potentially involved in the pathogenesis of AL and other hematopoietic malignancies. To date, large improvements of the surveillance of AL cases have been achieved; nevertheless, cases still die during the AL treatment. Thus, it is necessary to find suitable biomarkers for early diagnosis and accurate risk stratification in AL patients. The association of lncRNAs with several subtypes of leukemia, such as *MEG3*, *IRAIN*, and *UCA1* related to AML and *ANRIL*, *LUNAR1*, in ALL, increase the possibility to use them as biomarkers for the diagnosis, prognosis, and treatment (to provide a target) for the different subtypes of this disease. In addition, further investigation of the function of aberrant expressed lncRNAs may help to understand the pathogenesis of hematological malignancies and provide an important insight in childhood leukemia therapy.

Author Contributions: G.M.C.-M. and S.J.-M. drafted the work. A.H.-M., D.A.B.-L., J.C.N.-E., J.R.-B., and J.M.M.-A. substantively revised the manuscript and contributed intellectually. S.J.-M. conceived the review. All authors read and approved the submitted version.

Funding: This work was supported by the Consejo Nacional de Ciencia y Tecnología (CONACyT), grant numbers: Investigación en Fronteras de la Ciencia (IFC)-2016-01-2119, PDCPN2013-01-215726, SALUD-2010-1-141026, SALUD-2015-1-262190, FONCICYT/37/2018, and CB-2015-1-258042; and by the Instituto Mexicano del Seguro Social, grant numbers: FIS/IMSS/PROT/PRIO/11/017, FIS/IMSS/PROT/G12/1134, FIS/IMSS/PROT/PRIO/14/031, FIS/IMSS/PROT/PRIO/15/048, FIS/IMSS/PROT/MD15/1504, FIS/IMSS/PROT/G15/1477, FIS/IMSS/PROT/895, FIS/IMSS/PROT/1364, FIS/IMSS/PROT/1533, FIS/IMSS/PROT/1782 and FIS/IMSS/PROT/1548. Gabriela Marisol Cruz-Miranda was supported by CONACyT (Scholarship) 2018-000012-01NACF.

Conflicts of Interest: The authors declare no conflicts of interest.

Abbreviations

1.	ABL1	2.	ABL protoconcogene 1
3.	ABL2	4.	ABL protooncogene 2
5.	AL	6.	Acute leukemia
7.	ALL	8.	Acute lymphoblastic leukemia
9.	AML	10.	Acute myeloblastic leukemia
11.	ANRIL	12.	Antisense non-coding RNA in the INK4-ARF locus B-ALL B cell Acute lymphoblastic leukemia
13.	ARID1A	14.	AT-rich interaction domain 1A
15.	AURKA	16.	Aurora kinase A gene
17.	BALR	18.	B-ALL-associated long non-coding RNAs BL Burkitt Lymphoma
19.	CAS9	20.	CRISPR associated protein 9
21.	CBF	22.	Core-binding factor subunit beta
23.	CCAT1	24.	Colon cancer associated transcript 1 ceRNA Competing endogenous RNA
25.	CDKN2A	26.	Cyclin dependent kinase inhibitor 2A

27. CDKN2B   28. Cyclin dependent kinase inhibitor 2B
29. CDKN2B-AS1   30. CDKN2B antisense RNA 1
31. CEBPA   32. CCAAT enhancer binding protein alpha
33. CHD6   34. Chromodomain helicase DNA binding protein 6
35. circRNA   36. Circular RNA
37. CLL   38. Chronic lymphocytic leukemia
39. CML   40. Chronic myeloblastic leukemia
41. CN   42. Cytogenetically normal
43. COLDAIR   44. COLD assisted intronic non-coding RNA
45. CREBBP   46. CREB binding protein
47. CRISPR   48. Clustered regularly interspaced short palindromic repeats
49. CRLF2   50. Cytokine receptor like factor 2
51. CSF1R   52. Colony stimulating factor 1 receptor
53. DCs   54. Dendritic Cells
55. DNMT3A   56. DNA methyltransferase 3α
57. EBF1   58. Early B cell factor 1
59. EGO   60. Eosinophil granule ontogeny
61. EP300   62. E1A binding protein P300
63. eRNAs   64. Enhancer RNAs
65. ETS1   66. ETS proto-oncogene 1 transcription factor
67. ETV6   68. ETS Variant6
69. FLC   70. Flowering repressor locus
71. FLT3   72. Fms related tyrosine kinase 3
73. fRNAs   74. Functional RNAs
75. GAS5   76. Growth specific 5
77. GEO   78. Gene expression omnibus
79. H19   80. Imprinted maternally expressed transcript
81. hnRNP   82. Heterogenous nuclear ribonucleoprotein
83. HOTAIR   84. The HOX transcript antisense intergenic RNA
85. HOTTIP   86. HOXA distal transcript antisense RNA
87. IGFR1   88. Insuline-like growth factor type 1
89. IKZF1   90. IKAROS family zinc finger 1
91. IRAIN   92. IGFR1 antisense imprinted non protein RNA
93. ITPR1   94. Inositol1,4,5-triophosphate receptor type 1
95. JAK2   96. Janus kinase 2
97. KIT   98. Tyrosine protein kinase
99. LincRNA   100. Long intergenic non-coding RNA
101. LncRNA   102. Long non-coding RNA
103. lnc-DC   104. Dendritic cell-specificit lncRNA
105. lincRNA-p21   106. Large intergenic non-coding RNA p21
107. lncRNA   108. Long non-coding RNA
109. LUNAR1   110. Leukemia-associated non-coding IGF1R
111. MALAT1   112. Metastasis associated lung adenocarcinoma transcript 1 MCL Mantle cell lymphoma
113. MEG3   114. Maternally expressed 3
115. miRNA   116. MicroRNA
117. mRNA   118. Messenger RNA
119. NCBI   120. National center of biotechnology information
121. ncRNA   122. Non-coding RNA
123. NFAT   124. Nuclear factor activated T cells
125. nmRNA   126. Non messengers RNA
127. npcRNA   128. Non protein-coding RNA
129. NRAS   130. NRAS proto-oncogene
131. NRON   132. Non-protein-coding RNA Repressor of NFAT

133.	NSD2	134.	Nuclear receptor binding SET domain protein 2
135.	PANDAR	136.	Promoter of CDKN1A antisense DNA damage activated RNA
137.	PAX5	138.	Paired box 5
139.	PBX1	140.	PBX Homeobox 1
141.	PCGP	142.	Pediatric cancer genome project
143.	PDGFRB	144.	Platelet derived growth factor receptor beta
145.	piRNAs	146.	PIWI-interacting RNAs
147.	PML	148.	Promyelocytic Leukemia gene
149.	PROMPTs	150.	Promoter-associated long RNAs
151.	RB1	152.	RB transcriptional corepressor 1
153.	RBPs	154.	RNA-binding proteins
155.	RUNX1	156.	Runt related transcription factor 1
157.	SHP1	158.	Scr homology region 2 domain containing phosphatase-1
159.	siRNA	160.	Small interference RNA
161.	snRNAs	162.	Small nuclear RNA
163.	snoRNAs	164.	Small nucleolar RNA
165.	STAT3	166.	Signal transducer and activator of transcription 3
167.	TCF3	168.	Transcription Factor 3Ç
169.	TERC	170.	Telomerase RNA component
171.	TET2	172.	Tet methylcytosine dioxygenase 2
173.	TLR	174.	Tool-like receptor
175.	TP53	176.	Tumor protein P53
177.	TP53COR1	178.	Tumor protein P53 pathway corepressor 1
179.	TUG1	180.	Taurine-up regulated gene 1
181.	UCA1	182.	Urothelial carcinoma associated 1
183.	vlincRNA	184.	Very long intergenic RNA
185.	XIST	186.	X inactive specific transcript

References

1. Mejia-Arangure, J.M.; McNally, R.J.Q. Acute Leukemia in Children. *Biomed. Res. Int.* **2015**. [CrossRef] [PubMed]

2. Linet, M.S.; Brown, L.M.; Mbulaiteye, S.M.; Check, D.; Ostroumova, E.; Landgren, A.; Devesa, S.S. International long-term trends and recent patterns in the incidence of leukemias and lymphomas among children and adolescents ages 0–19 years. *Int. J. Cancer* **2016**, *138*, 1862–1874. [CrossRef] [PubMed]

3. Schuz, J.; Erdmann, F. Environmental Exposure and Risk of Childhood Leukemia: An Overview. *Arch. Med. Res.* **2016**, *47*, 607–614. [CrossRef]

4. Beltran-Anaya, F.O.; Cedro-Tanda, A.; Hidalgo-Miranda, A.; Romero-Cordoba, S.L. Insights into the Regulatory Role of Non-coding RNAs in Cancer Metabolism. *Front. Physiol.* **2016**, *7*, 342. [CrossRef] [PubMed]

5. Perez-Saldivar, M.L.; Fajardo-Gutierrez, A.; Bernaldez-Rios, R.; Martinez-Avalos, A.; Medina-Sanson, A.; Espinosa-Hernandez, L.; Flores-Chapa, J.D.; Amador-Sanchez, R.; Penaloza-Gonzalez, J.G.; Alvarez-Rodriguez, F.J.; et al. Childhood acute leukemias are frequent in Mexico City: Descriptive epidemiology. *BMC Cancer* **2011**, *11*, 355. [CrossRef] [PubMed]

6. Wright, M.; Bruford, E.A. Naming 'junk': Human non-protein coding RNA (ncRNA) genome nomenclature. *Hum. Genom.* **2011**, *5*, 90–98. [CrossRef]

7. Connelly, C.M.; Moon, M.H.; Schneekloth, J.S. The Emerging Role of RNA as a Therapeutic Target for Small Molecules. *Cell Chem. Biol.* **2016**, *23*, 1077–1090. [CrossRef]

8. Greaves, M. In utero origins of childhood leukaemia. *Early Hum. Dev.* **2005**, *81*, 123–129. [CrossRef]

9. Mullighan, C.G. Genomic profiling of B-progenitor acute lymphoblastic leukemia. *Best Pract. Res. Clin. Haematol.* **2011**, *24*, 489–503. [CrossRef]

10. Janczar, S.; Janczar, K.; Pastorczak, A.; Harb, H.; Paige, A.J.W.; Zalewska-Szewczyk, B.; Danilewicz, M.; Mlynarski, W. The Role of Histone Protein Modifications and Mutations in Histone Modifiers in Pediatric B-Cell Progenitor Acute Lymphoblastic Leukemia. *Cancers* **2017**, *9*, 2. [CrossRef]

11. Roberts, K.G.; Gu, Z.H.; Payne-Turner, D.; McCastlain, K.; Harvey, R.C.; Chen, I.M.; Pei, D.Q.; Iacobucci, I.; Valentine, M.; Pounds, S.B.; et al. High Frequency and Poor Outcome of Philadelphia Chromosome-Like Acute Lymphoblastic Leukemia in Adults. *J. Clin. Oncol.* **2017**, *35*, 394. [CrossRef] [PubMed]

12. Nordlund, J.; Kiialainen, A.; Karlberg, O.; Berglund, E.C.; Goransson-Kultima, H.; Sonderkaer, M.; Nielsen, K.L.; Gustafsson, M.G.; Behrendtz, M.; Forestier, E.; et al. Digital gene expression profiling of primary acute lymphoblastic leukemia cells. *Leukemia* **2012**, *26*, 1218–1227. [CrossRef] [PubMed]

13. Andersson, A.K.; Ma, J.; Wang, J.M.; Chen, X.; Gedman, A.L.; Dang, J.J.; Nakitandwe, J.; Holmfeldt, L.; Parker, M.; Easton, J.; et al. The landscape of somatic mutations in infant MLL-rearranged acute lymphoblastic leukemias. *Nat. Genet.* **2015**, *47*, 330–337. [CrossRef] [PubMed]

14. Zhang, X.H.; Rastogi, P.; Shah, B.; Zhang, L. B lymphoblastic leukemia/lymphoma: New insights into genetics, molecular aberrations, subclassification and targeted therapy. *Oncotarget* **2017**, *8*, 66728–66741. [CrossRef] [PubMed]

15. Aziz, H.; Ping, C.Y.; Alias, H.; Ab Mutalib, N.S.; Jamal, R. Gene Mutations as Emerging Biomarkers and Therapeutic Targets for Relapsed Acute Myeloid Leukemia. *Front. Pharmacol.* **2017**, *8*, 897. [CrossRef] [PubMed]

16. Ponting, C.P.; Oliver, P.L.; Reik, W. Evolution and Functions of Long Noncoding RNAs. *Cell* **2009**, *136*, 629–641. [CrossRef] [PubMed]

17. Wang, X.T.; Song, X.Y.; Glass, C.K.; Rosenfeld, M.G. The Long Arm of Long Noncoding RNAs: Roles as Sensors Regulating Gene Transcriptional Programs. *Cold Spring Harb. Perspect. Biol.* **2011**, *3*, a003756. [CrossRef] [PubMed]

18. Qi, P.; Du, X. The long non-coding RNAs, a new cancer diagnostic and therapeutic gold mine. *Mod. Pathol.* **2013**, *26*, 155–165. [CrossRef] [PubMed]

19. Kapranov, P.; St Laurent, G.; Raz, T.; Ozsolak, F.; Reynolds, C.P.; Sorensen, P.H.B.; Reaman, G.; Milos, P.; Arceci, R.J.; Thompson, J.F.; et al. The majority of total nuclear-encoded non-ribosomal RNA in a human cell is 'dark matter' un-annotated RNA. *BMC Biol.* **2010**, *8*, 149. [CrossRef] [PubMed]

20. Guttman, M.; Amit, I.; Garber, M.; French, C.; Lin, M.F.; Feldser, D.; Huarte, M.; Zuk, O.; Carey, B.W.; Cassady, J.P.; et al. Chromatin signature reveals over a thousand highly conserved large non-coding RNAs in mammals. *Nature* **2009**, *458*, 223–227. [CrossRef]

21. Kapranov, P.; Cheng, J.; Dike, S.; Nix, D.A.; Duttagupta, R.; Willingham, A.T.; Stadler, P.F.; Hertel, J.; Hackermuller, J.; Hofacker, I.L.; et al. RNA maps reveal new RNA classes and a possible function for pervasive transcription. *Science* **2007**, *316*, 1484–1488. [CrossRef] [PubMed]

22. Dieci, G.; Fiorino, G.; Castelnuovo, M.; Teichmann, M.; Pagano, A. The expanding RNA polymerase III transcriptome. *Trends Genet.* **2007**, *23*, 614–622. [CrossRef] [PubMed]

23. Yin, Q.F.; Yang, L.; Zhang, Y.; Xiang, J.F.; Wu, Y.W.; Carmichael, G.G.; Chen, L.L. Long Noncoding RNAs with snoRNA Ends. *Mol. Cell* **2012**, *48*, 219–230. [CrossRef]

24. Hung, T.; Wang, Y.L.; Lin, M.F.; Koegel, A.K.; Kotake, Y.; Grant, G.D.; Horlings, H.M.; Shah, N.; Umbricht, C.; Wang, P.; et al. Extensive and coordinated transcription of noncoding RNAs within cell-cycle promoters. *Nat. Genet.* **2011**, *43*, 621–629. [CrossRef] [PubMed]

25. Orom, U.A.; Derrien, T.; Beringer, M.; Gumireddy, K.; Gardini, A.; Bussotti, G.; Lai, F.; Zytnicki, M.; Notredame, C.; Huang, Q.H.; et al. Long Noncoding RNAs with Enhancer-like Function in Human Cells. *Cell* **2010**, *143*, 46–58. [CrossRef] [PubMed]

26. Salzman, J.; Gawad, C.; Wang, P.L.; Lacayo, N.; Brown, P.O. Circular RNAs Are the Predominant Transcript Isoform from Hundreds of Human Genes in Diverse Cell Types. *PLoS ONE* **2012**, *7*, e30733. [CrossRef]

27. Yang, L.; Duff, M.O.; Graveley, B.R.; Carmichael, G.G.; Chen, L.L. Genomewide characterization of non-polyadenylated RNAs. *Genome Biol.* **2011**, *12*, R16. [CrossRef]

28. St Laurent, G.; Wahlestedt, C.; Kapranov, P. The Landscape of long noncoding RNA classification. *Trends Genet.* **2015**, *31*, 239–251. [CrossRef]

29. Han, D.; Wang, M.; Ma, N.; Xu, Y.; Jiang, Y.T.; Gao, X. Long noncoding RNAs: Novel players in colorectal cancer. *Cancer Lett.* **2015**, *361*, 13–21. [CrossRef]

30. Di Gesualdo, F.; Capaccioli, S.; Lulli, M. A pathophysiological view of the long non-coding RNA world. *Oncotarget* **2014**, *5*, 10976–10996. [CrossRef]

31. Morlando, M.; Ballarino, M.; Fatica, A. Long Non-Coding RNAs: New Players in Hematopoiesis and Leukemia. *Front. Med.* **2015**, *2*, 23. [CrossRef] [PubMed]

32. Ulitsky, I.; Bartel, D.P. lincRNAs: Genomics, Evolution, and Mechanisms. *Cell* **2013**, *154*, 26–46. [CrossRef] [PubMed]

33. Quinn, J.J.; Chang, H.Y. Unique features of long non-coding RNA biogenesis and function. *Nat. Rev. Genet.* **2016**, *17*, 47–62. [CrossRef] [PubMed]

34. Jeon, Y.; Lee, J.T. YY1 Tethers Xist RNA to the Inactive X Nucleation Center. *Cell* **2011**, *146*, 119–133. [CrossRef] [PubMed]

35. Rinn, J.L.; Chang, H.Y. Genome regulation by long noncoding RNAs. *Annu. Rev. Biochem.* **2012**, *81*, 145–166. [CrossRef] [PubMed]

36. Binder, J.; Frankild, S.; Tsafou, K.; Stolte, C.; O'Donoghue, S.; Schneider, R.; Jensen, L.J. COMPARTMENTS. Available online: https://compartments.jensenlab.org/Search (accessed on 11 July 2018).

37. Fang, K.; Han, B.W.; Chen, Z.H.; Lin, K.Y.; Zeng, C.W.; Li, X.J.; Li, J.H.; Luo, X.Q.; Chen, Y.Q. A distinct set of long non-coding RNAs in childhood MLL-rearranged acute lymphoblastic leukemia: Biology and epigenetic target. *Hum. Mol. Genet.* **2014**, *23*, 3278–3288. [CrossRef]

38. Chen, L.L. Linking Long Noncoding RNA Localization and Function. *Trends Biochem. Sci.* **2016**, *41*, 761–772. [CrossRef]

39. Xia, F.; Dong, F.L.; Yang, Y.; Huang, A.F.; Chen, S.; Sun, D.; Xiong, S.D.; Zhang, J.P. Dynamic Transcription of Long Non-Coding RNA Genes during CD4+ T Cell Development and Activation. *PLoS ONE* **2014**, *9*, e101588. [CrossRef]

40. Paralkar, V.R.; Mishra, T.; Luan, J.; Yao, Y.; Kossenkov, A.V.; Anderson, S.M.; Dunagin, M.; Pimkin, M.; Gore, M.; Sun, D.; et al. Lineage and species-specific long noncoding RNAs during erythro-megakaryocytic development. *Blood* **2014**, *123*, 1927–1937. [CrossRef]

41. Wagner, L.A.; Christensen, C.J.; Dunn, D.M.; Spangrude, G.J.; Georgelas, A.; Kelley, L.; Esplin, M.S.; Weiss, R.B.; Gleich, G.J. EGO, a novel, noncoding RNA gene, regulates eosinophil granule protein transcript expression. *Blood* **2007**, *109*, 5191–5198. [CrossRef]

42. Imperato, M.R.; Cauchy, P.; Obier, N.; Bonifer, C. The RUNX1-PU.1 axis in the control of hematopoiesis. *Int. J. Hematol.* **2015**, *101*, 319–329. [CrossRef]

43. Wang, P.; Xue, Y.Q.; Han, Y.M.; Lin, L.; Wu, C.; Xu, S.; Jiang, Z.P.; Xu, J.F.; Liu, Q.Y.; Cao, X.T. The STAT3-Binding Long Noncoding RNA lnc-DC Controls Human Dendritic Cell Differentiation. *Science* **2014**, *344*, 310–313. [CrossRef] [PubMed]

44. Schwarzer, A.; Emmrich, S.; Schmidt, F.; Beck, D.; Ng, M.; Reimer, C.; Adams, F.F.; Grasedieck, S.; Witte, D.; Kabler, S.; et al. The non-coding RNA landscape of human hematopoiesis and leukemia. *Nat. Commun.* **2017**, *8*. [CrossRef] [PubMed]

45. Hon, C.C.; Ramilowski, J.A.; Harshbarger, J.; Bertin, N.; Rackham, O.J.; Gough, J.; Denisenko, E.; Schmeier, S.; Poulsen, T.M.; Severin, J.; et al. An atlas of human long non-coding RNAs with accurate 5' ends. *Nature* **2017**, *543*, 199–204. [CrossRef]

46. Berg, J.S.; Lin, K.K.; Sonnet, C.; Boles, N.C.; Weksberg, D.C.; Nguyen, H.; Holt, L.J.; Rickwood, D.; Daly, R.J.; Goodell, M.A. Imprinted Genes That Regulate Early Mammalian Growth Are Coexpressed in Somatic Stem Cells. *PLoS ONE* **2011**, *6*, e26410. [CrossRef] [PubMed]

47. Ghazavi, F.; De Moerloose, B.; Van Loocke, W.; Wallaert, A.; Helsmoortel, H.H.; Ferster, A.; Bakkus, M.; Plat, G.; Delabesse, E.; Uyttebroeck, A.; et al. Unique long non-coding RNA expression signature in ETV6/RUNX1-driven B-cell precursor acute lymphoblastic leukemia. *Oncotarget* **2016**, *7*, 73769–73780. [CrossRef]

48. Fernando, T.R.; Contreras, J.R.; Zampini, M.; Rodriguez-Malave, N.I.; Alberti, M.O.; Anguiano, J.; Tran, T.M.; Palanichamy, J.K.; Gajeton, J.; Ung, N.M.; et al. The lncRNA CASC15 regulates SOX4 expression in RUNX1-rearranged acute leukemia. *Mol. Cancer* **2017**, *16*, 126. [CrossRef]

49. Sun, J.N.; Li, W.; Sun, Y.P.; Yu, D.H.; Wen, X.; Wang, H.; Cui, J.W.; Wang, G.J.; Hoffman, A.R.; Hu, J.F. A novel antisense long noncoding RNA within the IGF1R gene locus is imprinted in hematopoietic malignancies. *Nucleic Acids Res.* **2014**, *42*, 9588–9601. [CrossRef]

50. Chen, S.Y.; Liang, H.R.; Yang, H.; Zhou, K.R.; Xu, L.M.; Liu, J.X.; Lai, B.; Song, L.; Luo, H.; Peng, J.M.; et al. Long non-coding RNAs: The novel diagnostic biomarkers for leukemia. *Environ. Toxicol. Pharmacol.* **2017**, *55*, 81–86. [CrossRef]

51. Hughes, J.M.; Legnini, I.; Salvatori, B.; Masciarelli, S.; Marchioni, M.; Fazi, F.; Morlando, M.; Bozzoni, I.; Fatica, A. C/EBPα-p30 protein induces expression of the oncogenic long non-coding RNA UCA1 in acute myeloid leukemia. *Oncotarget* **2015**, *6*, 18534–18544. [CrossRef]

52. Bhan, A.; Soleimani, M.; Mandal, S.S. Long Noncoding RNA and Cancer: A New Paradigm. *Cancer Res.* **2017**, *77*, 3965–3981. [CrossRef] [PubMed]

53. Benetatos, L.; Hatzimichael, E.; Dasoula, A.; Dranitsaris, G.; Tsiara, S.; Syrrou, M.; Georgiou, I.; Bourantas, K.L. CpG methylation analysis of the MEG3 and SNRPN imprinted genes in acute myeloid leukemia and myelodysplastic syndromes. *Leuk. Res.* **2010**, *34*, 148–153. [CrossRef] [PubMed]

54. Wang, H.; Li, W.; Guo, R.; Sun, J.; Cui, J.; Wang, G.; Hoffman, A.R.; Hu, J.F. An intragenic long noncoding RNA interacts epigenetically with the RUNX1 promoter and enhancer chromatin DNA in hematopoietic malignancies. *Int. J. Cancer* **2014**, *135*, 2783–2794. [CrossRef] [PubMed]

55. Zeng, C.; Xu, Y.; Xu, L.; Yu, X.; Cheng, J.; Yang, L.; Chen, S.; Li, Y. Inhibition of long non-coding RNA NEAT1 impairs myeloid differentiation in acute promyelocytic leukemia cells. *BMC Cancer* **2014**, *14*, 693. [CrossRef] [PubMed]

56. Zhang, X.; Lian, Z.; Padden, C.; Gerstein, M.B.; Rozowsky, J.; Snyder, M.; Gingeras, T.R.; Kapranov, P.; Weissman, S.M.; Newburger, P.E. A myelopoiesis-associated regulatory intergenic noncoding RNA transcript within the human HOXA cluster. *Blood* **2009**, *113*, 2526–2534. [CrossRef] [PubMed]

57. Zhang, X.; Weissman, S.M.; Newburger, P.E. Long intergenic non-coding RNA HOTAIRM1 regulates cell cycle progression during myeloid maturation in NB4 human promyelocytic leukemia cells. *RNA Biol.* **2014**, *11*, 777–787. [CrossRef] [PubMed]

58. Chen, Z.H.; Wang, W.T.; Huang, W.; Fang, K.; Sun, Y.M.; Liu, S.R.; Luo, X.Q.; Chen, Y.Q. The lncRNA HOTAIRM1 regulates the degradation of PML-RARA oncoprotein and myeloid cell differentiation by enhancing the autophagy pathway. *Cell Death Differ.* **2017**, *24*, 212–224. [CrossRef]

59. Wang, X.Q.; Dostie, J. Reciprocal regulation of chromatin state and architecture by HOTAIRM1 contributes to temporal collinear HOXA gene activation. *Nucleic Acids Res.* **2017**, *45*, 1091–1104. [CrossRef]

60. Díaz-Beyá, M.; Brunet, S.; Nomdedéu, J.; Pratcorona, M.; Cordeiro, A.; Gallardo, D.; Escoda, L.; Tormo, M.; Heras, I.; Ribera, J.M.; et al. The lincRNA HOTAIRM1, located in the HOXA genomic region, is expressed in acute myeloid leukemia, impacts prognosis in patients in the intermediate-risk cytogenetic category, and is associated with a distinctive microRNA signature. *Oncotarget* **2015**, *6*, 31613–31627. [CrossRef]

61. Zhao, H.; Zhang, X.; Frazão, J.B.; Condino-Neto, A.; Newburger, P.E. HOX antisense lincRNA HOXA-AS2 is an apoptosis repressor in all trans retinoic acid treated NB4 promyelocytic leukemia cells. *J. Cell. Biochem.* **2013**, *114*, 2375–2383. [CrossRef]

62. Ebralidze, A.K.; Guibal, F.C.; Steidl, U.; Zhang, P.; Lee, S.; Bartholdy, B.; Jorda, M.A.; Petkova, V.; Rosenbauer, F.; Huang, G.; et al. PU.1 expression is modulated by the balance of functional sense and antisense RNAs regulated by a shared *cis*-regulatory element. *Genes Dev.* **2008**, *22*, 2085–2092. [CrossRef]

63. McCarty, G.; Loeb, D.M. Hypoxia-sensitive epigenetic regulation of an antisense-oriented lncRNA controls WT1 expression in myeloid leukemia cells. *PLoS ONE* **2015**, *10*, e0119837. [CrossRef] [PubMed]

64. Guo, G.; Kang, Q.; Zhu, X.; Chen, Q.; Wang, X.; Chen, Y.; Ouyang, J.; Zhang, L.; Tan, H.; Chen, R.; et al. A long noncoding RNA critically regulates Bcr-Abl-mediated cellular transformation by acting as a competitive endogenous RNA. *Oncogene* **2015**, *34*, 1768–1779. [CrossRef] [PubMed]

65. Chen, L.; Wang, W.; Cao, L.; Li, Z.; Wang, X. Long Non-Coding RNA CCAT1 Acts as a Competing Endogenous RNA to Regulate Cell Growth and Differentiation in Acute Myeloid Leukemia. *Mol. Cells* **2016**, *39*, 330–336. [CrossRef] [PubMed]

66. Hirano, T.; Yoshikawa, R.; Harada, H.; Harada, Y.; Ishida, A.; Yamazaki, T. Long noncoding RNA, CCDC26, controls myeloid leukemia cell growth through regulation of KIT expression. *Mol. Cancer* **2015**, *14*, 90. [CrossRef] [PubMed]

67. Xing, C.Y.; Hu, X.Q.; Xie, F.Y.; Yu, Z.J.; Li, H.Y.; Bin-Zhou; Wu, J.B.; Tang, L.Y.; Gao, S.M. Long non-coding RNA HOTAIR modulates c-KIT expression through sponging miR-193a in acute myeloid leukemia. *FEBS Lett.* **2015**, *589*, 1981–1987. [CrossRef]

68. Zeng, C.; Yu, X.; Lai, J.; Yang, L.; Chen, S.; Li, Y. Overexpression of the long non-coding RNA PVT1 is correlated with leukemic cell proliferation in acute promyelocytic leukemia. *J. Hematol. Oncol.* **2015**, *8*, 126. [CrossRef]

69. Pan, J.Q.; Zhang, Y.Q.; Wang, J.H.; Xu, P.; Wang, W. lncRNA co-expression network model for the prognostic analysis of acute myeloid leukemia. *Int. J. Mol. Med.* **2017**, *39*, 663–671. [CrossRef]

70. Fernando, T.R.; Rodriguez-Malave, N.I.; Waters, E.V.; Yan, W.H.; Casero, D.; Basso, G.; Pigazzi, M.; Rao, D.S. LncRNA Expression Discriminates Karyotype and Predicts Survival in B-Lymphoblastic Leukemia. *Mol. Cancer Res.* **2015**, *13*, 839–851. [CrossRef]

71. Lajoie, M.; Drouin, S.; Caron, M.; St-Onge, P.; Ouimet, M.; Gioia, R.; Lafond, M.H.; Vidal, R.; Richer, C.; Oualkacha, K.; et al. Specific expression of novel long non-coding RNAs in high-hyperdiploid childhood acute lymphoblastic leukemia. *PLoS ONE* **2017**, *12*, e174124. [CrossRef]

72. Ouimet, M.; Drouin, S.; Lajoie, M.; Caron, M.; St-Onge, P.; Gioia, R.; Richer, C.; Sinnett, D. A childhood acute lymphoblastic leukemia-specific lncRNA implicated in prednisolone resistance, cell proliferation, and migration. *Oncotarget* **2017**, *8*, 7477–7488. [CrossRef] [PubMed]

73. Zhang, L.; Xu, H.G.; Lu, C. A novel long non-coding RNA T-ALL-R-LncR1 knockdown and Par-4 cooperate to induce cellular apoptosis in T-cell acute lymphoblastic leukemia cells. *Leuk. Lymphoma* **2014**, *55*, 1373–1382. [CrossRef] [PubMed]

74. Melo, C.P.D.; Campos, C.B.; Rodrigues, J.D.; Aguirre-Neto, J.C.; Atalla, A.; Pianovski, M.A.D.; Carbone, E.K.; Lares, L.B.Q.; Moraes-Souza, H.; Octacilio-Silva, S.; et al. Long non-coding RNAs: Biomarkers for acute leukaemia subtypes. *Br. J. Haematol.* **2016**, *173*, 318–320. [CrossRef] [PubMed]

75. Romero-Barrios, N.; Legascue, M.F.; Benhamed, M.; Ariel, F.; Crespi, M. Splicing regulation by long noncoding RNAs. *Nucleic Acids Res.* **2018**, *46*, 2169–2184. [CrossRef] [PubMed]

76. Akbari Moqadam, F.; Lange-Turenhout, E.A.; Ariës, I.M.; Pieters, R.; Den Boer, M.L. MiR-125b, miR-100 and miR-99a co-regulate vincristine resistance in childhood acute lymphoblastic leukemia. *Leuk. Res.* **2013**, *37*, 1315–1321. [CrossRef]

77. Zhang, X.; Hamblin, M.H.; Yin, K.J. The long noncoding RNA Malat1: Its physiological and pathophysiological functions. *RNA Biol.* **2017**, *14*, 1705–1714. [CrossRef]

78. Fan, Y.; Shen, B.; Tan, M.; Mu, X.; Qin, Y.; Zhang, F.; Liu, Y. Long non-coding RNA UCA1 increases chemoresistance of bladder cancer cells by regulating Wnt signaling. *FEBS J.* **2014**, *281*, 1750–1758. [CrossRef]

79. Han, Y.; Yang, Y.N.; Yuan, H.H.; Zhang, T.T.; Sui, H.; Wei, X.L.; Liu, L.; Huang, P.; Zhang, W.J.; Bai, Y.X. UCA1, a long non-coding RNA up-regulated in colorectal cancer influences cell proliferation, apoptosis and cell cycle distribution. *Pathology* **2014**, *46*, 396–401. [CrossRef]

80. Sun, M.D.; Zheng, Y.Q.; Wang, L.P.; Zhao, H.T.; Yang, S. Long noncoding RNA UCA1 promotes cell proliferation, migration and invasion of human leukemia cells via sponging miR-126. *Eur. Rev. Med. Pharmacol. Sci.* **2018**, *22*, 2233–2245.

81. Zhang, Y.; Liu, Y.; Xu, X. Knockdown of LncRNA-UCA1 suppresses chemoresistance of pediatric AML by inhibiting glycolysis through the microRNA-125a/hexokinase 2 pathway. *J. Cell. Biochem.* **2018**, *119*, 6296–6308. [CrossRef]

82. Miyoshi, N.; Wagatsuma, H.; Wakana, S.; Shiroishi, T.; Nomura, M.; Aisaka, K.; Kohda, T.; Surani, M.A.; Kaneko-Ishino, T.; Ishino, F. Identification of an imprinted gene, Meg3/Gtl2 and its human homologue MEG3, first mapped on mouse distal chromosome 12 and human chromosome 14q. *Genes Cells* **2000**, *5*, 211–220. [CrossRef] [PubMed]

83. Wang, P.; Ren, Z.; Sun, P. Overexpression of the long non-coding RNA MEG3 impairs in vitro glioma cell proliferation. *J. Cell. Biochem.* **2012**, *113*, 1868–1874. [CrossRef] [PubMed]

84. Diaz-Beya, M.; Navarro, A.; Cordeiro, A.; Pratcorona, M.; Castellano, J.; Torrente, M.A.; Nomdedeu, M.; Risueño, R.; Rozman, M.; Monzo, M.; et al. Exploring the Expression Profile of Long Non-Coding RNA (lncRNA) in Different Acute Myeloid Leukemia (AML) Subtypes: t(8;16)(p11;p13)/MYST3-Crebbp AML Harbors a Distinctive LncRNA Signature. *Blood* **2015**, *126*, 1397.

85. Garzon, R.; Volinia, S.; Papaioannou, D.; Nicolet, D.; Kohlschmidt, J.; Yan, P.S.; Mrozek, K.; Bucci, D.; Carroll, A.J.; Baer, M.R.; et al. Expression and prognostic impact of lncRNAs in acute myeloid leukemia. *Proc. Natl. Acad. Sci. USA* **2014**, *111*, 18679–18684. [CrossRef] [PubMed]

86. Wang, X.; Zhang, L.; Zhao, F.; Xu, R.; Jiang, J.; Zhang, C.; Liu, H.; Huang, H. Long non-coding RNA taurine-upregulated gene 1 correlates with poor prognosis, induces cell proliferation, and represses cell apoptosis via targeting aurora kinase A in adult acute myeloid leukemia. *Ann. Hematol.* **2018**, *97*, 1375–1389. [CrossRef] [PubMed]

87. Huang, J.L.; Ren, T.Y.; Cao, S.W.; Zheng, S.H.; Hu, X.M.; Hu, Y.W.; Lin, L.; Chen, J.; Zheng, L.; Wang, Q. HBx-related long non-coding RNA DBH-AS1 promotes cell proliferation and survival by activating MAPK signaling in hepatocellular carcinoma. *Oncotarget* **2015**, *6*, 33791–33804. [CrossRef] [PubMed]

88. Gioia, R.; Drouin, S.; Ouimet, M.; Caron, M.; St-Onge, P.; Richer, C.; Sinnett, D. LncRNAs downregulated in childhood acute lymphoblastic leukemia modulate apoptosis, cell migration, and DNA damage response. *Oncotarget* **2017**, *8*, 80645–80650. [CrossRef] [PubMed]

89. Garitano-Trojaola, A.; José-Enériz, E.S.; Ezponda, T.; Unfried, J.P.; Carrasco-León, A.; Razquin, N.; Barriocanal, M.; Vilas-Zornoza, A.; Sangro, B.; Segura, V.; et al. Deregulation of linc-PINT in acute lymphoblastic leukemia is implicated in abnormal proliferation of leukemic cells. *Oncotarget* **2018**, *9*, 12842–12852. [CrossRef]

90. Ngoc, P.C.T.; Tan, S.H.; Tan, T.K.; Chan, M.M.; Li, Z.; Yeoh, A.E.J.; Tenen, D.G.; Sanda, T. Identification of novel lncRNAs regulated by the TAL1 complex in T-cell acute lymphoblastic leukemia. *Leukemia* **2018**, *32*, 2138–2151. [CrossRef]

91. Ransohoff, J.D.; Wei, Y.N.; Khavari, P.A. The functions and unique features of long intergenic non-coding RNA. *Nat. Rev. Mol. Cell Biol.* **2018**, *19*, 143–157. [CrossRef]

92. Sayad, A.; Hajifathali, A.; Hamidieh, A.A.; Roshandel, E.; Taheri, M. HOTAIR Long Noncoding RNA is not a Biomarker for Acute Myeloid Leukemia (AML) in Iranian Patients. *Asian Pac. J. Cancer Prev.* **2017**, *18*, 1581–1584. [PubMed]

93. Li, J.; Sun, C.K. Long noncoding RNA SNHG5 is up-regulated and serves as a potential prognostic biomarker in acute myeloid leukemia. *Eur. Rev. Med. Pharmacol. Sci.* **2018**, *22*, 3342–3347.

94. Gao, S.; Zhou, B.; Li, H.; Huang, X.; Wu, Y.; Xing, C.; Yu, X.; Ji, Y. Long noncoding RNA HOTAIR promotes the self-renewal of leukemia stem cells through epigenetic silencing of p15. *Exp. Hematol.* **2018**. [CrossRef] [PubMed]

95. Wu, S.H.; Zheng, C.P.; Chen, S.Y.; Cai, X.P.; Shi, Y.J.; Lin, B.J.; Chen, Y.M. Overexpression of long non-coding RNA HOTAIR predicts a poor prognosis in patients with acute myeloid leukemia. *Oncol. Lett.* **2015**, *10*, 2410–2414. [CrossRef] [PubMed]

96. Zhang, Y.Y.; Huang, S.H.; Zhou, H.R.; Chen, C.J.; Tian, L.H.; Shen, J.Z. Role of HOTAIR in the diagnosis and prognosis of acute leukemia. *Oncol. Rep.* **2016**, *36*, 3113–3122. [CrossRef] [PubMed]

97. Wang, H.; Li, Q.; Tang, S.; Li, M.; Feng, A.; Qin, L.; Liu, Z.; Wang, X. The role of long noncoding RNA HOTAIR in the acquired multidrug resistance to imatinib in chronic myeloid leukemia cells. *Hematology* **2017**, *22*, 208–216. [CrossRef] [PubMed]

98. Shang, C.; Guo, Y.; Zhang, H.; Xue, Y.X. Long noncoding RNA HOTAIR is a prognostic biomarker and inhibits chemosensitivity to doxorubicin in bladder transitional cell carcinoma. *Cancer Chemother. Pharmacol.* **2016**, *77*, 507–513. [CrossRef] [PubMed]

99. Tabassum, N.; Verma, V.; Kumar, M.; Kumar, A.; Singh, B. Nanomedicine in cancer stem cell therapy: From fringe to forefront. *Cell Tissue Res.* **2018**. [CrossRef] [PubMed]

100. Sakuma, T.; Yamamoto, T. Acceleration of cancer science with genome editing and related technologies. *Cancer Sci.* **2018**. [CrossRef] [PubMed]

© 2018 by the authors. Licensee MDPI, Basel, Switzerland. This article is an open access article distributed under the terms and conditions of the Creative Commons Attribution (CC BY) license (http://creativecommons.org/licenses/by/4.0/).

International Journal of
Molecular Sciences

Review

Adiponectin as Link Factor between Adipose Tissue and Cancer

Erika Di Zazzo [1,2]**, Rita Polito** [3,4]**, Silvia Bartollino** [1]**, Ersilia Nigro** [3,5]**, Carola Porcile** [1]**,
Andrea Bianco** [5]**, Aurora Daniele** [3,4,*] **and Bruno Moncharmont** [1,*]

[1]   Department of Medicine and Health Sciences "V. Tiberio", University of Molise, Campobasso 86100, Italy;
       erika.dizazzo@unimol.it (E.D.Z.); silvia.bartollino@unimol.it (S.B.); carola.porcile@unimol.it (C.P.)
[2]   Department of Precision Medicine, University of Campania "Luigi Vanvitelli", Naples 80131, Italy
[3]   Dipartimento di Scienze e Tecnologie Ambientali Biologiche Farmaceutiche, Università degli Studi della
       Campania "Luigi Vanvitelli", Caserta 81100, Italy; rita.polito@unicampania.it (R.P.);
       nigro@ceinge.unina.it (E.N.)
[4]   CEINGE-Biotecnologie Avanzate Scarl, Napoli 80145, Italy
[5]   Dipartimento di Scienze Cardio-Toraciche e Respiratorie, Università degli Studi della Campania "Luigi
       Vanvitelli", Napoli 80131, Italy; andrea.bianco@unimol.it
*   Correspondence: aurora.daniele@unicampania.it (A.D.); moncharmont@unimol.it (B.M.)

Received: 30 November 2018; Accepted: 11 February 2019; Published: 15 February 2019

Abstract: Adipose tissue is a key regulator of energy balance playing an active role in lipid storage as well as in synthesizing several hormones directly involved in the pathogenesis of obesity. Obesity represents a peculiar risk factor for a growing list of cancers and is frequently associated to poor clinical outcome. The mechanism linking obesity and cancer is not completely understood, but, amongst the major players, there are both chronic low-grade inflammation and deregulation of adipokines secretion. In obesity, the adipose tissue is pervaded by an abnormal number of immune cells that create an inflammatory environment supporting tumor cell proliferation and invasion. Adiponectin (APN), the most abundant adipokine, shows anti-inflammatory, anti-proliferative and pro-apoptotic properties. Circulating levels of APN are drastically decreased in obesity, suggesting that APN may represent the link factor between obesity and cancer risk. The present review describes the recent advances on the involvement of APN and its receptors in the etiology of different types of cancer.

Keywords: Adiponectin; cancer; Adiponectin receptors; obesity; inflammatory response; inflammation; nutritional status

---

1. Introduction

Obesity, characterized by an excessive and chronic fat accumulation harmful to health, is defined by the World Health Organization (WHO) as a body mass index (BMI) of 30 or more and represents a worldwide emergent public health problem [1]. Several epidemiological studies revealed an alarming increase in the number of obese individuals in recent decades, reaching epidemic distribution in many areas of the World [2]. Currently, more than 1.5 billion adults are overweight and about 600 million people are classified as obese worldwide and these rates are estimated to increase in the future [3]. Obesity is a major risk factor for the development of metabolic and cardiovascular diseases as well as for several malignancies, frequently associated to a poor clinical outcome [4–6]. Cancer is the second leading cause of death worldwide; consequently, new efforts are needed to understand how obesity induce the cancer onset and affects its outcomes. The hypothesis that adipose tissue is involved in tumorigenesis is now called "adiponcosis" [7]. Molecular mechanisms linking obesity and cancer are complex and still not fully clarified. A low-grade chronic inflammation, deregulation of growth signaling pathways, chronic hyperinsulinemia and obesity-associated hypoxia are widely accepted to be pivotal factors in cancer pathogenesis [8]. Adipose tissue, originally defined as a passive fat depot, is a heterogeneous tissue, strongly committed to energy substrate homeostasis but endowed with complex secretory functions related to the nutritional status, thereby recognized as an active endocrine organ [9]. It produces and secretes different bioactive molecules, called adipokines. In obesity, the alteration of endocrine functions of adipose tissue negatively affects the secretion of different adipocytokines. Among them, Adiponectin (APN), the most adipocyte secretory protein, shows a reduced expression levels in obesity [10]. The reduction of APN expression levels observed in obese patients, has been related to an increase of tumor onset risk. Several studies demonstrated that APN, beyond its actions in metabolic responses such as energy metabolism regulation and insulin-sensitivity, has pleiotropic effects in cancer. Although literature data on the role of APN in carcinogenesis is conflicting, the most accredited hypothesis is that APN has a protective role, such as anti-inflammatory, anti-proliferative and pro-apoptotic effects, avoiding the development and progression of several malignancies, such as breast, colon, prostate, liver and endometrial cancers [11–14].

The aim of this review is to appraise the role of APN and its molecular pathways in "big killer" cancers, such as breast, colon, thyroid and lung cancer.

2. Adiponectin Structure and Function

First described in 1995, APN, also named adipocyte complement-related protein (Acrp30, *ADIPOQ*, apM1, GBP28), represents the most relevant insulin-sensitizing adipokine, primarily controlling glucose uptake as well as stimulating fatty acids oxidation [15]. APN is encoded by the *ADIPOQ* gene, which spans approximately 15.8 kb and is structured in three exons on chromosome 3q27; this region has been linked to a susceptibility locus for metabolic syndrome, type 2 diabetes and cardiovascular disease [16]. APN is mainly produced from adipose tissue but it is released at much lower concentration from other tissues [14]. Full-length APN is a 30-kDa protein with a primary sequence of 244 amino acids, composed of four domains: a signal sequence (aa 1–18), a non-conserved N-terminal domain (aa 19–41), followed by a 22 collagen-like repeat domain (aa 42–107) and a C-terminal globular domain (aa 108–244). By the cleavage of full-length APN, the globular APN (gAPN), containing only the C-terminal domain is obtained. APN can exist as different oligomers: trimers (approx. 90 kDa basic unit; Low Molecular Weight, LMW), hexamers (approx. 180 kDa, Medium Molecular Weight, MMW) and multimers (approx. 360–400 kDa, High Molecular Weight, HMW) [14].

The APN correct folding starts with trimers formation that, through the collagenous domains, assemble into hexamers (MMW); subsequently, these primordial complexes associate into multimers, (HMW), the most biologically active form [14]. APN biological activity depends strictly on its structure assembly, determined by post-translational modifications [17]. In particular, post-translational modifications of the oligomeric forms, involving hydroxylation and glycosylation of four conserved lysine residues at positions 68, 71, 80 and 104 on the collagenous domain, determine the formation of the high-molecular weight (HMW) complex APN [18]. Impairment of APN oligomers formation has an impact on insulin concentration, liver gluconeogenesis and can induce severe cardio-metabolic dysfunctions [17,18]. Furthermore, Arg112Cys and Ile164Thr mutations in the APN protein, preventing the trimer assembly, cause an impaired cellular secretion and are clinically associated with hypoadiponectinemia [19]. In physiological conditions, APN is an abundant protein in systemic circulation, representing about 0.01% of the total serum protein, with a concentration range of 5–50 µg/mL [14,20]. The APN serum concentration is inversely related to BMI and to insulin resistance [10,21]. However, in pathological conditions characterized by a chronic inflammation, such as type 2 diabetes, obesity and atherosclerosis, a lowering in APN serum concentrations is observed [10,20,22]. APN mediates most of its biological effects by binding to its classical receptors, AdipoR1 and AdipoR2, belonging to seven-transmembrane domains receptor family. Both receptors have been detected in almost normal and cancer tissues. AdipoR1 shows higher affinity for the globular protein than the full-length APN molecule, while AdipoR2 has a similar affinity for both forms. In obese individuals, a reduction in AdipoR1 and AdipoR2 expression levels seems to lead to a decreased sensitivity to APN [23]. Additionally, hexameric and multimeric APN bind the third non- classical receptor recognized, the glycosylphosphatidyl inositol (GPI)-anchored T-cadherin receptor [24].

3. Adiponectin Signaling Pathways

Several lines of evidence suggest that APN upon binding to its receptors, induces the recruitment of the adaptor protein APPL1, thereby activating a plethora of downstream signaling pathways controlling cell survival, cell growth and apoptosis. APN effects are mostly mediated via AMPK, mTOR, PI3K/AKT, MAPK, STAT3 and NF-kB [12]. APN induces the activation of AMPK, a central sensor and regulator of cellular energy, that in turn stimulates the expression of p21 and p53 and phosphorylates p53 to initiate cell cycle arrest, senescence and apoptosis. Additionally, studies point toward the inhibitory effects of APN on the PI3K/AKT/mTOR pathways, which leads to a cascade of events resulting in a blockade of cell survival, growth and proliferation. APN signaling also activates the MAPK cascade, which involves cJNK, p38 and ERK1/2. The cJNK and p38 action on proliferation and apoptosis depend on the cell type, whereas ERK1/2 have frequently a mitogenic effect. APN inhibits STAT3 activation that increases tumor cell proliferation, survival, angiogenesis and invasion, as well as inhibiting anti-tumor immunity. APN, through the suppression of inhibitor of NF-kB phosphorylation, suppresses the pro-inflammatory and anti-apoptotic NF-kB pathway [14].

4. Adipose Tissue, Adiponectin and Low Chronic Inflammation

In adipose tissue there is a perfect balance between adipocytes and immune cells that is lost in obesity, leading to a local chronic low inflammation associated with increased cancer risk. Immune cells infiltrating the adipose tissue of obese patients regulate the local immune responses, by increasing the levels of pro-inflammatory cytokines and adipokines thus supporting tumor development. Clusters of enlarged adipocytes become distant from the blood vessels, leading to a local area of hypoxia that underlies the inflammatory response [25,26]. Several immune cell types are involved in the development of adipose tissue inflammation: neutrophils and mast cells have been implicated in promoting inflammation and insulin resistance in obesity, whereas eosinophils and myeloid-derived suppressor cells have been suggested to play a protective role. In addition, a prominent role of B- and T-lymphocytes and natural-killer cells in adipose tissue inflammation recently emerged [27]. The cross-talk between adipocytes and cancer cells is mediated by cytokines (specifically IL-1, IL-6 and TNF-α), adipokines, including APN and other molecules, released by adipose tissue, able to control proliferation and invasion of different cancer cell types [28]. APN suppresses immune cell proliferation and, in particular, proliferation and polarization of type 1 macrophages (pro-inflammatory phenotype) while inducing proliferation and polarization of type 2 macrophages (anti-inflammatory phenotype). APN additionally reduces B-cells lymphopoiesis and T-cells responsiveness. Finally, APN acting on inflammatory response suppresses the expression of several pro-inflammatory mediators, such as TNF-α [29].

5. Adiponectin in Cancer

Although literature data on the role of APN in carcinogenesis are conflicting, it is recognized that APN is able to reduce development and progression of several malignancies, such as breast, colon, lung, thyroid and other cancers, through different molecular mechanisms, which are described below (see Figure 1).

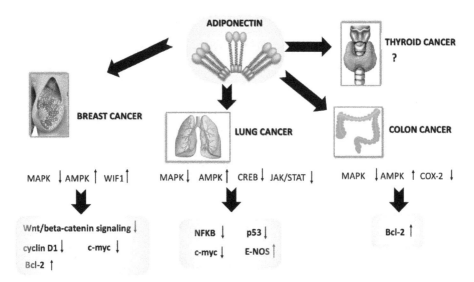

Figure 1. Summary of the molecular mechanisms affected by APN in cancers.

5.1. Adiponectin in Breast Cancer

Breast cancer (BC) is a well-known obesity-related cancer [30]. Adipose tissue may increase BC risk via a dual mechanism: (i) aromatization of adrenal androgens to estrogens in the adipocyte increases estrogen circulating levels and consequently promoting proliferation of mammary epithelial cells; (ii) deregulation of adipokine's expression and secretion, thus reducing the APN anti-proliferative effect on breast cells [31]. Recent evidence demonstrates that low APN serum levels are associated with increased BC risk but the precise APN mechanism of action is not completely understood [32]. Depending on tumor phenotype (ERα-positive or negative) and through the cooperation with circulating or locally-produced growth factors, APN affects BC cell growth, aggressiveness and behavior [33]. In the ERα-negative human BC cell line MDA-MB-231, APN elicits an anti-proliferative effect by modulating the expression of genes controlling cell cycle progression, such as p53, Bax, Bcl-2, c-myc and cyclin D1. In these cells APN inhibits PI3K/AKT pathway and activates AMPK, that in turn phosphorylates Sp1 protein. Phosphorylated Sp1 binds cyclin D1 promoter, causing the displacement of RNA Polymerase II and the recruitment of a co-repressor complex containing SMRT, NCoR and HDAC1, with a consequent repression of cyclin D1 expression and BC growth blockage [33]. In MDA-MB-231 cells and in nude mice APN also negatively controls Wnt/beta-catenin signaling, by positively regulating the expression of the Wnt inhibitory factor-1 (WIF1), a Wnt antagonist, at gene and protein levels [34].

On the other hand, conflicting results have been reported on the effects induced by APN in ERα-positive BC cells. In MCF-7 BC cell line (ERα-positive), APN is able to activate ERα that in turn stimulates cell growth [33]. Moreover, ADP400, a synthetic peptide modulating cellular APN receptor responses, induces mitogenic effects in MCF-7 cells, probably antagonizing endogenous APN actions or acting as an inverse agonist [35]. Emerging evidence shows the existence of a cross-talk between APN/AdipoR1, IGF-IR and ERα in BC [36]. Notoriously, insulin stimulates proliferation of BC cells through the IGF-1 receptors by activating PI3K [37–39]. In BC an increase in circulating insulin and estrogen concentration are observed together with a reduction in APN expression level [33]. Furthermore, low APN levels increase the risk of postmenopausal BC and of ER-positive breast tumors through a combined mitogenic effect of hyperinsulinemia and increased IGFs and estrogen levels. It has been demonstrated that incubation of MCF-7 cells with low APN concentrations enhances the association of IGF-1R with AdipoR1, APPL1, ERα, IGF-IR and c-Src; leading, via c-Src, to MAPK activation. The activated MAPK phosphorylates both Sp1 and ERα, allowing their recruitment on cyclin D1 promoter together with an enhanced association of RNA Polymerase II and pCAF. This leads to an increased cyclin D1 expression, inducing BC growth. Such effect is abrogated in the presence of specific RNA silencers targeting ERα or IGF-1R [36,40]. On the other hand, in ERα-negative MDA-MB-231 cells, APN is unable to induce MAPK phosphorylation. Additional data demonstrates that APN is able to regulate BC cell migration and invasion [41,42]. Indeed, a positive correlation between lympho-vascular and vascular invasion and AdipoR2 but not AdipoR1, expression has been reported [43]. Additionally, the expression of both APN and AdipoRs was significantly higher in invasive BC than noninvasive cases [42].

5.2. Adiponectin in Lung Cancer

Lung cancer is a highly-prevalent malignant carcinoma, representing one of the principal causes of morbidity and death worldwide [44].

To date, the molecular association between obesity and lung cancer remains still unclear and somewhat contradictory [45,46]. In fact, a meta-analysis suggested that overweight and obesity are protective factors against lung cancer, whereas other evidence indicates that obesity, in particular at visceral level, represents an important risk factor for lung cancer [45]. A variation of APN expression level was measured in patients affected by lung cancer, even though with contrasting results [47]. Some studies reported that there is no significant association between APN levels and lung cancer [48,49]. Other studies described a lowering in APN concentrations during lung cancer progression [50]. A further study revealed that APN deficiency significantly inhibited tumor vascularization and increased apoptosis and hypoxia, while APN-null mice showed a higher number of pulmonary metastases [51]. Additionally, two APN-gene promoter polymorphisms, Rs266730 and Rs2241766, have been associated with lung cancer risk and poor prognosis after surgery [52].

The APN inhibitory effects on lung cancer cell proliferation and invasion accompanied by an apoptosis rate increase has been mainly linked to the activation of pAMPK/mTOR pathways [53]. Moreover, APN, may exert an anti-proliferative effects through CREB down-regulation [54]. Indeed, in this recent paper, the authors reported that physiological concentrations of APN significantly reduced cell proliferation of human lung adenocarcinoma cell line A549, mainly by altering cell cycle kinetics and through the inhibition of CREB [54].

Beyond the inhibition of cell proliferation, a role of APN in regulating inflammation, cell growth and oxidative stress could also be observed in lung cancer cell lines. For instance, APN was effective in reducing the activation of inflammatory pathways, especially through the NF-κB-AdipoR1 pathway. Additionally, APN increased the levels of the anti-inflammatory IL-10 without influencing the expression of pro-inflammatory IL-6, IL-8 and MCP-1 in both IL-1β and TNF-α-treated A549 cells [53]. However, a pro-inflammatory role for APN has also been proposed; in fact, a report showed that APN promoted lung inflammation, via up-regulating cPLA2 and COX-2 expression together with intracellular ROS production [55].

Taken together, the current evidence indicates that APN and its receptors may act as molecular mediators in lung cancer at multiple levels although their role is controversial and far from being fully defined.

5.3. Adiponectin in Colon Cancer

Colorectal cancer (CRC) is one of the most common obesity-related cancer [56]. To date, several epidemiological and in vivo studies have investigated the role of APN in CRC [57]. Low APN serum concentration has been strongly associated to an increased risk of colorectal adenoma or early CRC [58–61]. In addition, patients affected by colon carcinoma or advanced adenoma showed markedly low APN serum level but no difference in serum APN concentration was observed between patients with advanced adenoma and patients with CRC [62].

AdipoR1 and AdipoR2 are expressed in CRC and are associated with lymph node involvement. Furthermore, AdipoR1 expression is correlated to tumor size during the early stages of CRC [61,63].

On the contrary, T-cadherin gene is not expressed in this cancer type since its promoter is frequently methylated [64,65].

Choe et al. using The Cancer Genome Atlas for CRC, investigated the association between the adipokine gene family (*ADIPOQ, ADIPOR1, ADIPOR2, LEP, LEPR, RETN, RETNLB, RBP4, SFRP5, NAMPT, SPP1*) mRNA expression levels and the survival rate of CRC patients, observing that a high expression level of the *ADIPOR1* and *SPP1* genes had unfavorable outcomes on CRC patients. Also, the *SPP1* mRNA expression level was significantly associated to the T- and N-stage, overall stage and mortality [66].

Evidence demonstrated that APN may affect CRC cell proliferation, adhesion, invasion as well as inflammation. Most of the studies demonstrated that APN could reduce cell proliferation rate [67,68]. Specifically, APN directly inhibited CRC cell proliferation via AdipoR1- and AdipoR2-mediated AMPK activation [30,67]. Activated AMPK, in turn, regulated many molecular mechanisms responsible for the APN inhibition of cell proliferation. Indeed, AMPK upregulated T-cadherin mRNA expression in a dose-dependent manner in HCT116 cells [69–71].

In vitro assay and immunohistochemical staining suggested that APN could prevent CRC carcinogenesis and proliferation by downregulating COX-2 expression [61,72]. APN treatment reduced the survival rate of both CaCo-2 and HCT116 cell lines in a time- and dose-dependent manner by inducing the phosphorylation of ERK1/2 and the cleavage of Caspase-3 thus activating programmed cell death [73].

In a mouse model, Saxena et al. proposed the use of APN as a therapeutic compound to decrease the severity of the symptoms caused by chronic inflammation-induced by CRC [74]. Additionally, a very recent study in a CRC patient cohort strongly suggested that the mRNA expression levels of APN and its receptors could be used as biomarkers for the prediction of CRC survival prediction [66]. On the other hand, Ogunwobi et al. reported a pro-proliferative and pro-inflammatory APN action on CRC cells [75]. This discrepancy might be explained looking at recent-published data showing that the effect of APN on cancer cell proliferation is glucose-dependent, whereby APN supports CRC survival in a low glucose medium but inhibits proliferation under a high glucose conditions [76].

Taken together these results suggest that, although APN could be an attractive target for obesity-associated colon cancer, further investigations are needed to completely elucidate the potential actions of APN in these cancers.

5.4. Adiponectin in Thyroid Cancer

The incidence of thyroid cancer (TC) underwent such a remarkable worldwide increase that it becomes the second most commonly diagnosed cancer in young women; nevertheless, the mechanisms underlying the development and progression of TC are poorly understood. Epidemiologic studies reported that an increase in BMI and obesity with low levels of circulating APN, are positively associated to TC risk [77]. On the contrary, Abooshahab et al. found no differences in APN levels between TC patients and healthy controls [78]. TC tissues and cells lines express both AdipoR1 and AdipoR2 [79]. A weak expression of AdipoR1 and a moderate expression of AdipoR2 were observed in both K1 and B-CPAP TC cell lines where APN stimulates AMPK phosphorylation. BHP7 and SW579 cell lines express both AdipoR1 and AdipoR2 but are not responsive to APN [79]. APN increases the synthesis of thyroid hormones, especially free thyroxine (fT4), through the interaction between the C-terminal globular domain of APN and the gC1q receptor [80].

Several molecular pathways link obesity to TC. An in vivo study performed in a TC mouse model reported that a high-fat diet (HFD) could increase cell proliferation via two main molecular pathways: i. increasing the protein levels of cyclin D1 and retinoblastoma protein (pRb) phosphorylation; ii. through chronic activation of the JAK2/STAT3 signaling pathway and induction of STAT3 gene expression [81]. In HFD mice, the JAK2-STAT3 signaling pathway was also associated with a higher occurrence of anaplastic *foci*. Interestingly, an activation of the STAT3 pathway by APN has been discovered in fibroblasts, hepatocytes and adipocytes but, to our knowledge, there is no proof of STAT3 involvement by APN in thyroid cells [82,83].

Altogether this evidence suggests that APN might have an important role in the development and progression of TC, even if additional studies are necessary to fully clarify the usefulness of APN as a plausible marker or therapeutic target for TC.

5.5. Adiponectin in Other Cancers

Recently, in a systematic review and meta-analysis, weight loss and a physically active lifestyle were associated to a lower risk of meningiomas; in the same study, the authors also observed a correlation between obesity and glioma risk [84]. Interestingly, our research group showed that APN negatively modulated cell proliferation of human glioblastomas U87-MG and U251 cell lines, by inducing growth arrest with a G1-phase delay and a slow but persistent activation of a specific subset of ERK1/2 proteins. Moreover, we observed that APN negatively regulated Insulin-like Growth Factor 1 IGF-1 action abolishing the IGF-1-induced proliferation of U251 cells [85]. Prostate cancer is one of the most commonly diagnosed cancers in men [86]. The role of APN in this type of cancer is contradictory. In a large retrospective study, low APN expression levels were related to the onset of prostate cancer. Furthermore, the APN expression level was associated to the stage and the grade of the disease. However, other studies have found no association between APN expression level and prostate cancer [12,14].

6. Conclusions

Obesity is a highly prevalent public health problem that has been associated with increased cancers risk in multiple organs. Several mechanisms have been proposed to explain the link between obesity and cancer and, among them, deregulation of adipocyte-secreted factors is critical. An involvement of APN, the most represented circulating adipokine, in the etiology of different cancer types has been proposed (see Figure 2). APN shows multifaceted functions in tumorigenesis. Nevertheless, the anti-proliferative and tumor-suppressor role of APN remains elusive and data collected so far are controversial. In vitro studies suggested that in a number of cancers APN may promote neoplastic growth, while in others it may suppress it. Consequently, APN likely could act both as a tumor-suppressor or as a tumor-promoting factor. The discrepancy is coherent with the evidence that APN exerts different functions depending on environmental factors, such as tissue/organ type and inflammatory state. Moreover, the available cultured-cell models probably are not suitable to reproduce the complex tumor microenvironment existing in vivo. Another level of complexity that could influence experimental results derives from the existence of different APN oligomers. Indeed, the major number of studies do not indicate which APN isoform is involved; probably this is due to the difficulty of discerning them. Moreover, alterations not only in circulating levels of total and isoform-specific APN but also in APN tumor microenvironment concentrations, remain to be explored in human specimens. Additional clues may facilitate the development of new strategies for the successful treatment of many obesity- related malignancies, such as by increasing APN serum concentration or antagonizing its receptors, and/or by targeting APN signaling pathways.

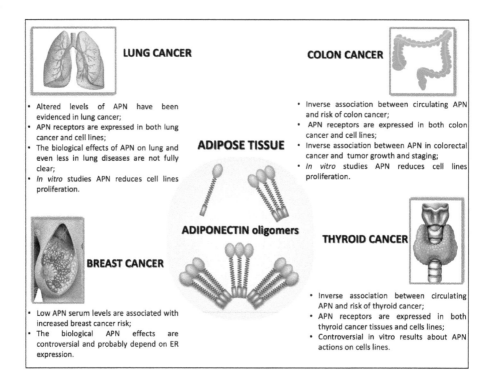

Figure 2. Summary of the biological functions exerted by APN in lung, colon, breast and thyroid cancers.

Author Contributions: Supervision, funding acquisition, review and editing: A.D., A.B., B.M. Conceptualization, Writing—Original Draft: E.D.Z. Writing—Review and Editing: E.D.Z., R.P., E.N., C.P., S.B.

Funding: This work was funded by the National Ministry of University and Research, PRIN.

Acknowledgments: This work was supported by grants from the Italian Ministry of University and Scientific Research, PRIN.

Conflicts of Interest: The authors declare that they have no conflict of interest.

Abbreviations

AdipoR	Adiponectin receptor
AMPK	5′-adenosine monophosphate-activated protein kinase
APN	Adiponectin
BC	Breast cancer
BMI	Body Mass Index
COX-2	Cyclooxygenase
CRC	Colorectal cancer
cPLA2	Cytosolic phospholipase A2
CREB	cAMP response element-binding protein
ER	Estrogen receptor
ERK1/2	Extracellular signal-regulated protein kinases 1 and 2
fT4	Free thyroxine
HFD	High-fat diet
IGF-1	Insulin-like growth factor 1

IL-10	Interleukin-10
IL-6	Interleukin-6
IL-8	Interleukin-8
JAK2	Janus kinase 2
MAPK	Mitogen-activated protein kinase
MCP-1	Monocyte chemoattractant protein 1
mTOR	Mammalian target of rapamycin
NF-κB	Nuclear factor-κB
PI3K	Phosphatidylinositol-4,5-bisphosphate 3-kinase
PPAR	Peroxisome proliferator-activated receptors
PP2A	Protein phosphatase 2 A
pRb	Phosphorylated retinoblastoma protein
ROS	Reactive oxygen species
STAT3	Signal transducer and activator of transcription 3
TC	Thyroid cancer
TNF-α	Tumor necrosis factor-α

References

1. Clinical guidelines on the identification, evaluation and treatment of overweight and obesity in adults: Executive summary. Expert Panel on the Identification, Evaluation and Treatment of Overweight in Adults. *Am. J. Clin. Nutr.* **1998**, *68*, 899–917. [CrossRef] [PubMed]
2. Ng, M.; Fleming, T.; Robinson, M.; Thomson, B.; Graetz, N.; Margono, C.; Mullany, E.C.; Biryukov, S.; Abbafati, C.; Abera, S.F.; et al. Global, regional and national prevalence of overweight and obesity in children and adults during 1980–2013: A systematic analysis for the Global Burden of Disease Study 2013. *Lancet* **2014**, *384*, 766–781. [CrossRef]
3. NCD Risk Factor Collaboration (NCD-RisC). Trends in adult body-mass index in 200 countries from 1975 to 2014: A pooled analysis of 1698 population-based measurement studies with 19·2 million participants. *Lancet* **2016**, *387*, 1377–1396. [CrossRef]
4. Global Burden of Metabolic Risk Factors for Chronic Diseases Collaboration (BMI Mediated Effects); Lu, Y.; Hajifathalian, K.; Ezzati, M.; Woodward, M.; Rimm, E.B.; Danaei, G. Metabolic mediators of the effects of body-mass index, overweight and obesity on coronary heart disease and stroke: A pooled analysis of 97 prospective cohorts with 1·8 million participants. *Lancet* **2014**, *383*, 970–983. [CrossRef]
5. Global BMI Mortality Collaboration; Di Angelantonio, E.; Bhupathiraju, S.; Wormser, D.; Gao, P.; Kaptoge, S.; Berrington de Gonzalez, A.; Cairns, B.; Huxley, R.; Jackson, C.; et al. Body-mass index and all-cause mortality: Individual-participant-data meta-analysis of 239 prospective studies in four continents. *Lancet* **2016**, *388*, 776–786. [PubMed]
6. Renehan, A.G.; Tyson, M.; Egger, M.; Heller, R.F.; Zwahlen, M. Body-mass index and incidence of cancer: A systematic review and meta-analysis of prospective observational studies. *Lancet* **2008**, *371*, 569–578. [CrossRef]
7. Bifulco, M.; Ciaglia, E. Updates on "adiponcosis": More new incoming evidence strengthening the obesity-cancer link. *Eur. J. Intern. Med.* **2017**, *41*, e19–e20. [CrossRef]
8. Stone, T.W.; McPherson, M.; Gail Darlington, L. Obesity and Cancer: Existing and New Hypotheses for a Causal Connection. *EBioMedicine* **2018**, *30*, 14–28. [CrossRef]
9. Galic, S.; Oakhill, J.S.; Steinberg, G.R. Adipose tissue as an endocrine organ. *Mol. Cell. Endocrinol.* **2010**, *316*, 129–139. [CrossRef]
10. Arita, Y.; Kihara, S.; Ouchi, N.; Takahashi, M.; Maeda, K.; Miyagawa, J.; Hotta, K.; Shimomura, I.; Nakamura, T.; Miyaoka, K.; et al. Paradoxical decrease of an adipose-specific protein, adiponectin, in obesity. *Biochem. Biophys. Res. Commun.* **1999**, *257*, 79–83. [CrossRef]
11. Nishida, M.; Funahashi, T.; Shimomura, I. Pathophysiological significance of adiponectin. *Med. Mol. Morphol.* **2007**, *40*, 55–67. [CrossRef] [PubMed]
12. Dalamaga, M.; Diakopoulos, K.N.; Mantzoros, C.S. The role of adiponectin in cancer: A review of current evidence. *Endocr. Rev.* **2012**, *33*, 547–594. [CrossRef] [PubMed]

13. Hebbard, L.; Ranscht, B. Multifaceted roles of adiponectin in cancer. *Best Pract. Res. Clin. Endocrinol. Metab.* **2014**, *28*, 59–69. [CrossRef] [PubMed]

14. Katira, A.; Tan, P.H. Evolving role of adiponectin in cancer-controversies and update. *Cancer Biol. Med.* **2016**, *13*, 101–119. [CrossRef] [PubMed]

15. Berg, A.H.; Combs, T.P.; Du, X.; Brownlee, M.; Scherer, P.E. The adipocyte-secreted protein Acrp30 enhances hepatic insulin action. *Nat. Med.* **2001**, *7*, 947–953. [CrossRef] [PubMed]

16. Bermúdez, V.J.; Rojas, E.; Toledo, A.; Rodríguez-Molina, D.; Vega, K.; Suárez, L.; Pacheco, M.; Canelón, R.; Arráiz, N.; Rojas, J.V.M. Single-nucleotide polymorphisms in adiponectin, AdipoR1 and AdipoR2 genes: Insulin resistance and type 2 diabetes mellitus candidate genes. *Am. J. Ther.* **2013**, *20*, 414–421. [CrossRef] [PubMed]

17. Simpson, F.; Whitehead, J.P. Adiponectin-It's all about the modifications. *Int. J. Biochem. Cell Biol.* **2010**, *42*, 785–788. [CrossRef]

18. Wang, Y.; Lam, K.S.L.; Chan, L.; Kok, W.C.; Lam, J.B.B.; Lam, M.C.; Hoo, R.C.L.; Mak, W.W.N.; Cooper, G.J.S.; Xu, A. Post-translational modifications of the four conserved lysine residues within the collagenous domain of adiponectin are required for the formation of its high molecular weight oligomeric complex. *J. Biol. Chem.* **2006**, *281*, 16391–16400. [CrossRef]

19. Waki, H.; Yamauchi, T.; Kamon, J.; Ito, Y.; Uchida, S.; Kita, S.; Hara, K.; Hada, Y.; Vasseur, F.; Froguel, P.; et al. Impaired multimerization of human adiponectin mutants associated with diabetes. Molecular structure and multimer formation of adiponectin. *J. Biol. Chem.* **2003**, *278*, 40352–40363. [CrossRef]

20. Chiarugi, P.; Fiaschi, T. Adiponectin in health and diseases: From metabolic syndrome to tissue regeneration. *Expert Opin. Ther. Targets* **2010**, *14*, 193–206. [CrossRef]

21. Yamauchi, T.; Kamon, J.; Waki, H.; Terauchi, Y.; Kubota, N.; Hara, K.; Mori, Y.; Ide, T.; Murakami, K.; Tsuboyama-Kasaoka, N.; et al. The fat-derived hormone adiponectin reverses insulin resistance associated with both lipoatrophy and obesity. *Nat. Med.* **2001**, *7*, 941–946. [CrossRef] [PubMed]

22. Nigro, E.; Scudiero, O.; Monaco, M.L.; Palmieri, A.; Mazzarella, G.; Costagliola, C.; Bianco, A.; Daniele, A. New insight into adiponectin role in obesity and obesity-related diseases. *Biomed. Res. Int.* **2014**, *2014*, 658913. [CrossRef] [PubMed]

23. Yamauchi, T.; Iwabu, M.; Okada-Iwabu, M.; Kadowaki, T. Adiponectin receptors: A review of their structure, function and how they work. *Best Pract. Res. Clin. Endocrinol. Metab.* **2014**, *28*, 15–23. [CrossRef]

24. Hug, C.; Wang, J.; Ahmad, N.S.; Bogan, J.S.; Tsao, T.-S.; Lodish, H.F. T-cadherin is a receptor for hexameric and high-molecular-weight forms of Acrp30/adiponectin. *Proc. Natl. Acad. Sci. USA* **2004**, *101*, 10308–10313. [CrossRef] [PubMed]

25. Wu, L.; Van Kaer, L. Contribution of lipid-reactive natural killer T cells to obesity-associated inflammation and insulin resistance. *Adipocyte* **2013**, *2*, 12–16. [CrossRef] [PubMed]

26. Catalán, V.; Gómez-Ambrosi, J.; Rodríguez, A.; Frühbeck, G. Adipose tissue immunity and cancer. *Front. Physiol.* **2013**, *4*, 275. [CrossRef] [PubMed]

27. Sell, H.; Eckel, J. Adipose tissue inflammation: Novel insight into the role of macrophages and lymphocytes. *Curr. Opin. Clin. Nutr. Metab. Care* **2010**, *13*, 366–370. [CrossRef]

28. Lengyel, E.; Makowski, L.; DiGiovanni, J.; Kolonin, M.G. Cancer as a Matter of Fat: The Crosstalk between Adipose Tissue and Tumors. *Trends Cancer* **2018**, *4*, 374–384. [CrossRef]

29. Ouchi, N.; Walsh, K. Adiponectin as an anti-inflammatory factor. *Clin. Chim. Acta* **2007**, *380*, 24–30. [CrossRef]

30. James, F.R.; Wootton, S.; Jackson, A.; Wiseman, M.; Copson, E.R.; Cutress, R.I. Obesity in breast cancer–what is the risk factor? *Eur. J. Cancer* **2015**, *51*, 705–720. [CrossRef]

31. Avgerinos, K.I.; Spyrou, N.; Mantzoros, C.S.; Dalamaga, M. Obesity and cancer risk: Emerging biological mechanisms and perspectives. *Metabolism* **2018**, *92*, 121–135. [CrossRef] [PubMed]

32. Ye, J.; Jia, J.; Dong, S.; Zhang, C.; Yu, S.; Li, L.; Mao, C.; Wang, D.; Chen, J.; Yuan, G. Circulating adiponectin levels and the risk of breast cancer: A meta-analysis. *Eur. J. Cancer Prev.* **2014**, *23*, 158–165. [CrossRef] [PubMed]

33. Mauro, L.; Pellegrino, M.; De Amicis, F.; Ricchio, E.; Giordano, F.; Rizza, P.; Catalano, S.; Bonofiglio, D.; Sisci, D.; Panno, M.L.; et al. Evidences that estrogen receptor α interferes with adiponectin effects on breast cancer cell growth. *Cell Cycle* **2014**, *13*, 553–564. [CrossRef] [PubMed]

34. Liu, J.; Lam, J.B.B.; Chow, K.H.M.; Xu, A.; Lam, K.S.L.; Moon, R.T.; Wang, Y. Adiponectin stimulates Wnt inhibitory factor-1 expression through epigenetic regulations involving the transcription factor specificity protein 1. *Carcinogenesis* **2008**, *29*, 2195–2202. [CrossRef] [PubMed]

35. Otvos, L.; Knappe, D.; Hoffmann, R.; Kovalszky, I.; Olah, J.; Hewitson, T.D.; Stawikowska, R.; Stawikowski, M.; Cudic, P.; Lin, F.; et al. Development of second generation peptides modulating cellular adiponectin receptor responses. *Front. Chem.* **2014**, *2*, 93. [CrossRef] [PubMed]

36. Mauro, L.; Naimo, G.D.; Ricchio, E.; Panno, M.L.; Andò, S. Cross-Talk between Adiponectin and IGF-IR in Breast Cancer. *Front. Oncol.* **2015**, *5*, 157. [CrossRef] [PubMed]

37. Di Zazzo, E.; Feola, A.; Zuchegna, C.; Romano, A.; Donini, C.F.; Bartollino, S.; Costagliola, C.; Frunzio, R.; Laccetti, P.; Di Domenico, M.; et al. The p85 regulatory subunit of PI3K mediates cAMP-PKA and insulin biological effects on MCF-7 cell growth and motility. *Sci. World J.* **2014**, *2014*, 565839. [CrossRef]

38. Donini, C.F.; Coppa, A.; Di Zazzo, E.; Zuchegna, C.; Di Domenico, M.; D'Inzeo, S.; Nicolussi, A.; Avvedimento, E.V.; Coppa, A.; Porcellini, A. The p85α regulatory subunit of PI3K mediates cAMP-PKA and retinoic acid biological effects on MCF7 cell growth and migration. *Int. J. Oncol.* **2012**, *40*, 1627–1635.

39. Di Zazzo, E.; Bartollino, S.; Moncharmont, B. The master regulator gene PRDM2 controls C2C12 myoblasts proliferation and Differentiation switch and PRDM4 and PRDM10 expression. *Insights Biol. Med.* **2017**, *1*, 75–91.

40. Mauro, L.; Pellegrino, M.; Giordano, F.; Ricchio, E.; Rizza, P.; De Amicis, F.; Catalano, S.; Bonofiglio, D.; Panno, M.L.; Andò, S. Estrogen receptor- α drives adiponectin effects on cyclin D1 expression in breast cancer cells. *FASEB J.* **2015**, *29*, 2150–2160. [CrossRef]

41. Jia, Z.; Liu, Y.; Cui, S. Adiponectin Induces Breast Cancer Cell Migration and Growth Factor Expression. *Cell Biochem. Biophys.* **2014**, *70*, 1239–1245. [CrossRef] [PubMed]

42. Jeong, Y.J.; Bong, J.G.; Park, S.H.; Choi, J.H.; Oh, H.K. Expression of leptin, leptin receptor, adiponectin and adiponectin receptor in ductal carcinoma in situ and invasive breast cancer. *J. Breast Cancer* **2011**, *14*, 96–103. [CrossRef] [PubMed]

43. Pfeiler, G.; Treeck, O.; Wenzel, G.; Goerse, R.; Hartmann, A.; Schmitz, G.; Ortmann, O. Influence of insulin resistance on adiponectin receptor expression in breast cancer. *Maturitas* **2009**, *63*, 253–256. [CrossRef] [PubMed]

44. De Groot, P.; Munden, R.F. Lung cancer epidemiology, risk factors and prevention. *Radiol. Clin. N. Am.* **2012**, *50*, 863–876. [CrossRef]

45. Hidayat, K.; Du, X.; Chen, G.; Shi, M.; Shi, B. Abdominal Obesity and Lung Cancer Risk: Systematic Review and Meta-Analysis of Prospective Studies. *Nutrients* **2016**, *8*, 810. [CrossRef] [PubMed]

46. Yang, Y.; Dong, J.; Sun, K.; Zhao, L.; Zhao, F.; Wang, L.; Jiao, Y. Obesity and incidence of lung cancer: A meta-analysis. *Int. J. Cancer* **2013**, *132*, 1162–1169. [CrossRef] [PubMed]

47. Daniele, A.; De Rosa, A.; Nigro, E.; Scudiero, O.; Capasso, M.; Masullo, M.; de Laurentiis, G.; Oriani, G.; Sofia, M.; Bianco, A. Adiponectin oligomerization state and adiponectin receptors airway expression in chronic obstructive pulmonary disease. *Int. J. Biochem. Cell Biol.* **2012**, *44*, 563–569. [CrossRef]

48. Petridou, E.T.; Mitsiades, N.; Gialamas, S.; Angelopoulos, M.; Skalkidou, A.; Dessypris, N.; Hsi, A.; Lazaris, N.; Polyzos, A.; Syrigos, C.; et al. Circulating adiponectin levels and expression of adiponectin receptors in relation to lung cancer: Two case-control studies. *Oncology* **2007**, *73*, 261–269. [CrossRef]

49. Gulen, S.T.; Karadag, F.; Karul, A.B.; Kilicarslan, N.; Ceylan, E.; Kuman, N.K.; Cildag, O. Adipokines and Systemic Inflammation in Weight-Losing Lung Cancer Patients. *Lung* **2012**, *190*, 327–332. [CrossRef]

50. Karapanagiotou, E.M.; Tsochatzis, E.A.; Dilana, K.D.; Tourkantonis, I.; Gratsias, I.; Syrigos, K.N. The significance of leptin, adiponectin and resistin serum levels in non-small cell lung cancer (NSCLC). *Lung Cancer* **2008**, *61*, 391–397. [CrossRef]

51. Denzel, M.S.; Scimia, M.-C.; Zumstein, P.M.; Walsh, K.; Ruiz-Lozano, P.; Ranscht, B. T-cadherin is critical for adiponectin-mediated cardioprotection in mice. *J. Clin. Investig.* **2010**, *120*, 4342–4352. [CrossRef] [PubMed]

52. Cui, E.; Deng, A.; Wang, X.; Wang, B.; Mao, W.; Feng, X.; Hua, F. The role of adiponectin (ADIPOQ) gene polymorphisms in the susceptibility and prognosis of non-small cell lung cancer. *Biochem. Cell Biol.* **2011**, *89*, 308–313. [CrossRef] [PubMed]

53. Nigro, E.; Scudiero, O.; Sarnataro, D.; Mazzarella, G.; Sofia, M.; Bianco, A.; Daniele, A. Adiponectin affects lung epithelial A549 cell viability counteracting TNFα and IL-1ß toxicity through AdipoR1. *Int. J. Biochem. Cell Biol.* **2013**, *45*, 1145–1153. [CrossRef] [PubMed]

54. Illiano, M.; Nigro, E.; Sapio, L.; Caiafa, I.; Spina, A.; Scudiero, O.; Bianco, A.; Esposito, S.; Mazzeo, F.; Pedone, P.V.; et al. Adiponectin down-regulates CREB and inhibits proliferation of A549 lung cancer cells. *Pulm. Pharmacol. Ther.* **2017**, *45*, 114–120. [CrossRef] [PubMed]

55. Chen, H.-M.; Yang, C.-M.; Chang, J.-F.; Wu, C.-S.; Sia, K.-C.; Lin, W.-N. AdipoR-increased intracellular ROS promotes cPLA2 and COX-2 expressions via activation of PKC and p300 in adiponectin-stimulated human alveolar type II cells. *Am. J. Physiol. Lung Cell. Mol. Physiol.* **2016**, *311*, L255–L269. [PubMed]

56. Tarasiuk, A.; Mosińska, P.; Fichna, J. The mechanisms linking obesity to colon cancer: An overview. *Obes. Res. Clin. Pract.* **2018**, *12*, 251–259. [CrossRef] [PubMed]

57. Riondino, S.; Roselli, M.; Palmirotta, R.; Della-Morte, D.; Ferroni, P.; Guadagni, F. Obesity and colorectal cancer: Role of adipokines in tumor initiation and progression. *World J. Gastroenterol.* **2014**, *20*, 5177–5190. [CrossRef]

58. Moon, H.-S.; Liu, X.; Nagel, J.M.; Chamberland, J.P.; Diakopoulos, K.N.; Brinkoetter, M.T.; Hatziapostolou, M.; Wu, Y.; Robson, S.C.; Iliopoulos, D.; et al. Salutary effects of adiponectin on colon cancer: In vivo and in vitro studies in mice. *Gut* **2013**, *62*, 561–570. [CrossRef]

59. Saetang, J.; Boonpipattanapong, T.; Palanusont, A.; Maneechay, W.; Sangkhathat, S. Alteration of Leptin and Adiponectin in Multistep Colorectal Tumorigenesis. *Asian Pac. J. Cancer Prev.* **2016**, *17*, 2119–2123. [CrossRef]

60. Xu, X.T.; Xu, Q.; Tong, J.L.; Zhu, M.M.; Huang, M.L.; Ran, Z.H.; Xiao, S.D. Meta-analysis: Circulating adiponectin levels and risk of colorectal cancer and adenoma. *J. Dig. Dis.* **2011**, *12*, 234–244. [CrossRef]

61. Tae, C.H.; Kim, S.-E.; Jung, S.-A.; Joo, Y.-H.; Shim, K.-N.; Jung, H.-K.; Kim, T.H.; Cho, M.-S.; Kim, K.H.; Kim, J.S. Involvement of adiponectin in early stage of colorectal carcinogenesis. *BMC Cancer* **2014**, *14*, 811. [CrossRef] [PubMed]

62. Kumor, A.; Daniel, P.; Pietruczuk, M.; Małecka-Panas, E. Serum leptin, adiponectin and resistin concentration in colorectal adenoma and carcinoma (CC) patients. *Int. J. Colorectal Dis.* **2009**, *24*, 275–281. [CrossRef] [PubMed]

63. Ayyildiz, T.; Dolar, E.; Ugras, N.; Adim, S.B.; Yerci, O. Association of adiponectin receptor (Adipo-R1/-R2) expression and colorectal cancer. *Asian Pac. J. Cancer Prev.* **2014**, *15*, 9385–9390. [CrossRef] [PubMed]

64. Ren, J.-Z.; Huo, J.-R. Correlation between T-cadherin gene expression and aberrant methylation of T-cadherin promoter in human colon carcinoma cells. *Med. Oncol.* **2012**, *29*, 915–918. [CrossRef] [PubMed]

65. Ye, M.; Huang, T.; Li, J.; Zhou, C.; Yang, P.; Ni, C.; Chen, S. Role of CDH13 promoter methylation in the carcinogenesis, progression and prognosis of colorectal cancer: A systematic meta-analysis under PRISMA guidelines. *Medicine* **2017**, *96*, e5956. [CrossRef] [PubMed]

66. Choe, E.K.; Yi, J.W.; Chai, Y.J.; Park, K.J. Upregulation of the adipokine genes ADIPOR1 and SPP1 is related to poor survival outcomes in colorectal cancer. *J. Surg. Oncol.* **2018**, *117*, 1833–1840. [CrossRef] [PubMed]

67. Sugiyama, M.; Takahashi, H.; Hosono, K.; Endo, H.; Kato, S.; Yoneda, K.; Nozaki, Y.; Fujita, K.; Yoneda, M.; Wada, K.; et al. Adiponectin inhibits colorectal cancer cell growth through the AMPK/mTOR pathway. *Int. J. Oncol.* **2009**, *34*, 339–344.

68. Kim, A.Y.; Lee, Y.S.; Kim, K.H.; Lee, J.H.; Lee, H.K.; Jang, S.-H.; Kim, S.-E.; Lee, G.Y.; Lee, J.-W.; Jung, S.-A.; et al. Adiponectin represses colon cancer cell proliferation via AdipoR1- and -R2-mediated AMPK activation. *Mol. Endocrinol.* **2010**, *24*, 1441–1452. [CrossRef]

69. Davis, J.E.; Gabler, N.K.; Walker-Daniels, J.; Spurlock, M.E. Tlr-4 deficiency selectively protects against obesity induced by diets high in saturated fat. *Obesity* **2008**, *16*, 1248–1255. [CrossRef]

70. Priego, T.; Sánchez, J.; Picó, C.; Palou, A. Sex-differential expression of metabolism-related genes in response to a high-fat diet. *Obesity* **2008**, *16*, 819–826. [CrossRef]

71. Boddicker, R.L.; Whitley, E.M.; Davis, J.E.; Birt, D.F.; Spurlock, M.E. Low-dose dietary resveratrol has differential effects on colorectal tumorigenesis in adiponectin knockout and wild-type mice. *Nutr. Cancer* **2011**, *63*, 1328–1338. [CrossRef] [PubMed]

72. Hwang, J.-T.; Ha, J.; Park, I.-J.; Lee, S.-K.; Baik, H.W.; Kim, Y.M.; Park, O.J. Apoptotic effect of EGCG in HT-29 colon cancer cells via AMPK signal pathway. *Cancer Lett.* **2007**, *247*, 115–121. [CrossRef] [PubMed]

73. Nigro, E.; Schettino, P.; Polito, R.; Scudiero, O.; Monaco, M.L.; De Palma, G.D.; Daniele, A. Adiponectin and colon cancer: Evidence for inhibitory effects on viability and migration of human colorectal cell lines. *Mol. Cell. Biochem.* **2018**, *448*, 125–135. [CrossRef]

74. Saxena, A.; Chumanevich, A.; Fletcher, E.; Larsen, B.; Lattwein, K.; Kaur, K.; Fayad, R. Adiponectin deficiency: Role in chronic inflammation induced colon cancer. *Biochim. Biophys. Acta Mol. Basis Dis.* **2012**, *1822*, 527–536. [CrossRef] [PubMed]

75. Ogunwobi, O.O.; Beales, I.L.P. Adiponectin stimulates proliferation and cytokine secretion in colonic epithelial cells. *Regul. Pept.* **2006**, *134*, 105–113. [CrossRef] [PubMed]

76. Habeeb, B.S.; Kitayama, J.; Nagawa, H. Adiponectin supports cell survival in glucose deprivation through enhancement of autophagic response in colorectal cancer cells. *Cancer Sci.* **2011**, *102*, 999–1006. [CrossRef] [PubMed]

77. Dossus, L.; Franceschi, S.; Biessy, C.; Navionis, A.-S.; Travis, R.C.; Weiderpass, E.; Scalbert, A.; Romieu, I.; Tjønneland, A.; Olsen, A.; et al. Adipokines and inflammation markers and risk of differentiated thyroid carcinoma: The EPIC study. *Int. J. Cancer* **2018**, *142*, 1332–1342. [CrossRef]

78. Abooshahab, R.; Yaghmaei, P.; Ghadaksaz, H.G.; Hedayati, M. Lack of Association between Serum Adiponectin/Leptin Levels and Medullary Thyroid Cancer. *Asian Pac. J. Cancer Prev.* **2016**, *17*, 3861–3864.

79. Cheng, S.-P.; Liu, C.-L.; Hsu, Y.-C.; Chang, Y.-C.; Huang, S.-Y.; Lee, J.-J. Expression and biologic significance of adiponectin receptors in papillary thyroid carcinoma. *Cell Biochem. Biophys.* **2013**, *65*, 203–210. [CrossRef]

80. Lin, S.-Y.; Huang, S.-C.; Sheu, W.H.-H. Circulating adiponectin concentrations were related to free thyroxine levels in thyroid cancer patients after thyroid hormone withdrawal. *Metabolism* **2010**, *59*, 195–199. [CrossRef]

81. Mitsiades, N.; Pazaitou-Panayiotou, K.; Aronis, K.N.; Moon, H.-S.; Chamberland, J.P.; Liu, X.; Diakopoulos, K.N.; Kyttaris, V.; Panagiotou, V.; Mylvaganam, G.; et al. Circulating adiponectin is inversely associated with risk of thyroid cancer: In vivo and in vitro studies. *J. Clin. Endocrinol. Metab.* **2011**, *96*, E2023–E2028. [CrossRef] [PubMed]

82. Kim, W.G.; Park, J.W.; Willingham, M.C.; Cheng, S. Diet-induced obesity increases tumor growth and promotes anaplastic change in thyroid cancer in a mouse model. *Endocrinology* **2013**, *154*, 2936–2947. [CrossRef] [PubMed]

83. Liao, W.; Yu, C.; Wen, J.; Jia, W.; Li, G.; Ke, Y.; Zhao, S.; Campell, W. Adiponectin induces interleukin-6 production and activates STAT3 in adult mouse cardiac fibroblasts. *Biol. Cell* **2009**, *101*, 263–272. [CrossRef] [PubMed]

84. Disney-Hogg, L.; Sud, A.; Law, P.J.; Cornish, A.J.; Kinnersley, B.; Ostrom, Q.T.; Labreche, K.; Eckel-Passow, J.E.; Armstrong, G.N.; Claus, E.B.; et al. Influence of obesity-related risk factors in the aetiology of glioma. *Br. J. Cancer* **2018**, *118*, 1020–1027. [CrossRef] [PubMed]

85. Porcile, C.; Di Zazzo, E.; Monaco, M.L.; D'Angelo, G.; Passarella, D.; Russo, C.; Di Costanzo, A.; Pattarozzi, A.; Gatti, M.; Bajetto, A.; et al. Adiponectin as novel regulator of cell proliferation in human glioblastoma. *J. Cell. Physiol.* **2014**, *229*, 1444–1454. [CrossRef] [PubMed]

86. Di Zazzo, E.; Galasso, G.; Giovannelli, P.; Di Donato, M.; Castoria, G. Estrogens and Their Receptors in Prostate Cancer: Therapeutic Implications. *Front. Oncol.* **2018**, *8*, 2. [CrossRef] [PubMed]

© 2019 by the authors. Licensee MDPI, Basel, Switzerland. This article is an open access article distributed under the terms and conditions of the Creative Commons Attribution (CC BY) license (http://creativecommons.org/licenses/by/4.0/).

MDPI

St. Alban-Anlage 66

4052 Basel

Switzerland

Tel. +41 61 683 77 34

Fax +41 61 302 89 18

www.mdpi.com

International Journal of Molecular Sciences Editorial Office

E-mail: ijms@mdpi.com

www.mdpi.com/journal/ijms

CPSIA information can be obtained
at www.ICGtesting.com
Printed in the USA
LVHW010404261119
638513LV00001B/5/P